INTRODUCING EINSTEIN'S RELAT

T0091997

Albert Einstein (1879–1955)

Introducing Einstein's Relativity

A deeper understanding

Ray d'Inverno and James Vickers

School of Mathematical Sciences, University of Southampton

OXFORD

UNIVERSITY PRESS

OXFORD

UNIVERSITY PRESS

Great Clarendon Street, Oxford, OX2 6DP,
United Kingdom

Oxford University Press is a department of the University of Oxford.
It furthers the University's objective of excellence in research, scholarship,
and education by publishing worldwide. Oxford is a registered trade mark of
Oxford University Press in the UK and in certain other countries

© Ray d'Inverno and James Vickers 2022

The moral rights of the authors have been asserted

First Edition published in 1992

Published in the United States of America by Oxford University Press
198 Madison Avenue, New York, NY 10016, United States of America

British Library Cataloguing in Publication Data
Data available

Library of Congress Control Number: 2021936302

ISBN 978–0–19–886202–4 (hbk)
ISBN 978–0–19–886203–1 (pbk)

DOI: 10.1093/oso/9780198862024.001.0001

Printed and bound by
CPI Group (UK) Ltd, Croydon, CR0 4YY

Ray: This book is dedicated to my beloved wife Pauline

*James: This second edition is dedicated to my children, Tom and Eleanor,
and in memory of my wife Gill*

Foreword

This book provides an excellent introduction to the ideas and basic mathematical techniques needed for a study of Einstein's superb – and now widely observationally confirmed – general theory of relativity. The underlying concepts and basic mathematics are presented with utmost clarity and by numerous greatly illuminating diagrams. The reader is taken on a gentle but comprehensive route up to cosmology, as currently understood, and to the strange features of rotating black holes and to gravitational waves. I am sure that it will inspire many students and other readers to enter into the beauties and of the power of this subject, which deeply underlies much of the physics of our world, and perhaps it will inspire others to carry this understanding further into what is currently unknown.

Roger Penrose
July 2021

Contents

Part B   The Formalism of Tensors

The organization of the book

1

1.1 The evolution of the book

There is little doubt that relativity theory captures the imagination. Nor is it surprising: the counter-intuitive properties of special relativity, the bizarre characteristics of black holes, the new era of gravitational wave detection and with it the advent of gravitational wave astronomy, and the sheer scope and nature of cosmology and its posing of ultimate questions; these and other issues combine to excite the minds of the inquisitive. Yet, if we are to look at these issues meaningfully, then we really require both physical insight and a sound mathematical foundation. The aim of this book is to help provide these.

This book is a substantial extension of the book *Introducing Einstein's Relativity*. The original book grew out of some notes written in the mid-1970s to accompany a UK course on general relativity. Originally, the course was a third-year undergraduate option aimed at mathematicians and physicists. It subsequently grew to include MSc students and some first-year PhD students. The notes were originally pitched principally at the undergraduate level, but the book contained sufficient depth and coverage to interest many students at the first-year graduate level. This book has been extended to include more advanced material which would be more appropriate for graduate-level students. To help fulfil this dual purpose, the more advanced sections (Level 2 material) are indicated by a hatched bar alongside the appropriate section. We emphasise that Level 1 material is essential to the understanding of the book. To help put a bit more light and shade into the book, the more important equations and results are given in tinted panels.

Part A on special relativity is designed to provide an introduction to special relativity sufficient for the needs of the rest of the book. The book is then designed to give students insight and confidence in handling the basic equations of the theory. From the mathematical viewpoint, this requires good manipulative ability with tensors. Part B is devoted to developing the necessary expertise in tensors for the rest of the book. It is essentially written as a self-study unit. Students are urged to attempt all the exercises which accompany the various sections. Experience has shown that this is the only real way to be in a position to deal confidently with the ensuing material. Part C then starts by using tensors to reformulate special relativity. From the physical viewpoint, in our view, the best route to understanding relativity theory is to follow the one taken by Einstein.

Introducing Einstein's Relativity. Second Edition. Ray d'Inverno and James Vickers, Oxford University Press.
© Ray d'Inverno and James Vickers (2022). DOI: 10.1093/oso/9780198862024.003.0001

Thus, the second chapter of Part C is devoted to discussing the principles which guided Einstein in his search for a relativistic theory of gravitation. The field equations are approached first from a largely physical viewpoint using these principles and subsequently from a purely mathematical viewpoint using the variational principle approach. After a chapter devoted to investigating the quantity which goes on the 'right-hand side' of the equations, the structure of the equations is discussed as a prelude to solving them in the simplest case. This part ends by considering solar system tests of the experimental status of general relativity. The main purpose of the book is to develop the theory in such a way that it is possible to reach three major topics of current interest, namely, black holes, gravitational waves, and cosmology. These topics form the subject matter of Parts D, E, and F, respectively.

Each of the chapters is supported by exercises, numbering over 350 in total. The bulk of these are straightforward calculations used to fill in parts omitted in the text. The numbers in parentheses indicate the sections to which the exercises refer. Although the exercises in general are important in aiding understanding, their status is different from those in Part B. Those exercises are absolutely essential for understanding the rest of the book and they should not be omitted. The remaining exercises are desirable. The book is neither exhaustive nor complete, since there are topics in the theory which are not covered or only met briefly. However, it is hoped that it provides the reader with a sound understanding of the basics of the theory.

1.2 Acknowledgements

Very little of this book is wholly original in character. Thus, to take an example right from the beginning of the book, the k-calculus provides the best introduction to special relativity because it offers insight from the outset through the simple diagrams that can be drawn. Indeed, one of the themes of this book is the provision of a large number of illustrative diagrams (over 250, in fact). The visual sense is the most immediate we possess and helps lead directly to a better comprehension. A good subtitle for the book would be *An approach to relativity via space-time pictures*. The k-calculus is an approach developed by Herman Bondi from the earlier ideas of A. Milne. So the fact that this and many of the approaches in the book have been borrowed from one author or another has been to organize the material in such a way that it is more readily accessible to the majority of students.

General relativity has the reputation of being intellectually very demanding. There is the apocryphal story attributed to Sir Arthur Eddington, who, when asked whether he believed it true that only three people in the world understood general relativity, replied, "Who is the third?" Indeed, the intellectual leap required by Einstein to move from the special theory to the general theory is, there can be little doubt, one of the greatest in the history of human thought. So it is not surprising that the theory has the reputation it does. However, general relativity has been with us for over a century, and our understanding is such that we can now build it up in a series of simple logical steps. This brings the theory within the grasp of most undergraduates equipped with the right background. So the book has been written in the spirit that any explanation that aids understanding should ultimately reside in the pool of human knowledge and thence in the public domain and therefore not belong to any one author.

1.3   The status of scientific research

Einstein's theory of relativity is arguably the greatest scientific achieve-ment of the human mind. It comprises the 'special theory' developed around 1905, concerned with physics in the absence of gravitation, and the 'general theory', developed some ten years later, which incorporates gravitation. Most surprisingly, it was the product of the work of just one theoretical physicist – Albert Einstein. The development of special rel-ativity was remarkable enough since it was achieved when Einstein was working in a patent office, and not in a scientific community or a uni-versity. However, the move to the general theory, which took Einstein ten years of endeavour, was a colossal achievement not just involving a deeper insight into the underlying physical principles but requiring a whole new mathematical machinery to make these ideas explicit. This book attempts to retrace the ideas of Einstein in leading up to the special and general theories. It is our belief that this route leads to a deeper understanding of the theory.

However, the question arises: Would we have arrived at these theories without Einstein? It was already clear at the turn of the twentieth century that something was wrong with the current understanding at the time of basic physical ideas, especially as it related to motion involving high veloc-ities and the propagation of light. The new physics required was encoded in the Lorentz transformations, which had been produced on an ad hoc basis to reconcile underlying inconsistencies. Einstein's key contribution was to derive them from two physical principles and demonstrate that they rested on a deeper understanding of the concept of simultaneity. Most historians of science would agree that, sooner or later, the new physics of the special theory would likely have been arrived at. However, whether the move to the general theory, and with it the accompanying revolution in our understanding of basic physical ideas, would have been achieved without Einstein is less clear. This raises the question: How does science develop and will it necessarily refine our ideas and thereby lead to an 'ul-timate' understanding of the world we live in? Einstein's work led to the development of the field of cosmology – modelling the universe – which is the science of the very large. He also made significant contributions to the other great theory of the twentieth-century quantum theory – the sci-ence of the very small. Yet these two theories remain in basic conflict and considerable research effort has gone into trying to find a theory of quan-tum gravity which reconciles the two. We end this section by exploring the question: Where are we currently in the search?

Table 1.1

GR0	1955	Bern, Switzerland
GR1	1957	Chapel Hill, USA
GR2	1959	Royaumont, France
GR3	1962	Jablonna, Poland
GR4	1965	London, UK
GR5	1968	Tbilisi, USSR
GR6	1971	Copenhagen, Denmark
GR7	1974	Tel-Aviv, Israel
GR8	1977	Waterloo, Canada
GR9	1980	Jena, DDR
GR10	1983	Padova, Italy
GR11	1986	Stockholm, Sweden
GR12	1989	Boulder, USA
GR13	1992	Cordoba, Argentina
GR14	1995	Florence, Italy
GR15	1997	Pune, India
GR16	2001	Durban, South Africa
GR17	2004	Dublin, Ireland
GR18	2007	Sydney, Australia
GR19	2010	Mexico City, Mexico
GR20	2013	Warsaw, Poland
GR21	2016	New York, USA
GR22	2020	Valencia, Spain

In 1955 a conference on general relativity and gravitation was held in Bern, Switzerland, now referred to as GR0. Two other conferences were held in 1957 and 1959, named GR1 and GR2, respectively, and after that they have been held generally every 3 years, with subsequent conferences being numbered accordingly (see Table 1.1). GR0 was held some forty years after the discovery of general relativity and, at the time, involved a relatively small community of scholars. Such has the world of scientific research grown in the interim that the conferences now include more than a thousand attendees. Even so, the field of classical general relativity research is a relatively small one in physics, although there is growth in the field of detection of gravitational waves, and cosmology has essentially become a discipline in its own right. So there are many thousands of people involved in fundamental research and, not surprisingly, there is a spread of opinion as to the progress that has been made. The biggest field of research in this area currently is in 'string theory', and its adherents would likely consider this to be the right way forward, but the jury is out on its efficacy, especially as regards to any experimental verification. In contradistinction, general relativity now has a considerable weight of experimental support. There have been a score or more attempts to provide an alternative classical theory of gravitation, but the consensus is that Einstein's theory is both consistent with current observations and is, in some sense, the simplest theory. But the issue of a theory of quantum gravity is more complex. First of all, there is an explosion in the research literature, and keeping track of it is a tall order. How would one know if someone had made the equivalent 'Einsteinian' breakthrough to a theory of quantum gravity? Many would agree that two of the most important theoreticians since Einstein are the UK mathematical physicists Stephen Hawking and Roger Penrose. Indeed, in 2020 Penrose was awarded a Nobel Prize for his pioneering work in showing 'that black hole formation is a robust prediction of the general theory of relativity', although he did not receive the prize until he was at the advanced age of 89. Perhaps Hawking did not live long enough to receive his recognition. In fact, Einstein was only awarded a Nobel Prize for his work on the photoelectric effect and not for his more important, although possibly controversial at the time, work on relativity. Both Hawking and Penrose have independently suggested a route to a theory of quantum gravity but neither is, what one might say, currently in the main stream of scientific endeavour in that relatively few researchers are continuing work using their suggested approaches. Indeed, the authors of this book have suggested a potential canonical quantization programme but it has received scant attention. We are not saying that our approach, or that of Hawking or Penrose, or string theory is 'right'. What we are saying is that the world of research is much more complex than in the time of Einstein and we are into other areas such as 'reputation' and 'fashion' (see Penrose 2017). There is also the question as to whether there will ever be a 'Theory of Everything'. In the very large – the world of cosmology – there appears to be a need for a theory of 'dark matter' and 'dark energy' and, at the time of writing, no such compelling theory exists. In the other direction – the world of the small – we have, in turn, theories of atoms, fundamental particles, quarks, and so on ... but does it necessarily lead to an ultimate theory? Looking back and doing so with our current understanding, you could argue that general relativity is almost forced on you. The hope is that, if and when a successful theory of quantum gravity is produced, it will also force itself on you, but detecting that is likely to be a more challenging task. Fundamental science does not unfold at a constant pace, but rather it does so in fits and starts. It is all the more remarkable that the work of just one man led to such a giant leap forward in our understating of the physical world, and this is the focus of this book.

1.4 A note for students on studying from a book

A few words of advice if you find studying from a book hard going. Remember that understanding is not an all-or-nothing process. One understands things at deeper and deeper levels, as various connections are made or ideas are seen in different contexts or from a different perspective. So do not simply attempt to study a section by going through it line by line and expect it all to make sense at the first go. It is better to begin by reading through a few sections quickly – skimming – thereby trying to get a general feel for the scope, level, and coverage of the subject matter. A second reading should be more thorough, but should not stop if ideas are met which are not clear straightaway. In a final pass, the sections should be studied in depth with the exercises attempted at the end of each section. However, if you get stuck, do not stop there; press on. You will often find that the penny will drop later, sometimes on its own, or that subsequent work will produce the necessary understanding. Many exercises (and exam questions) are hierarchical in nature. They require you to establish a result at one stage which is then used at a subsequent stage. If you cannot establish the result, then do not give up. Try and use it in the subsequent section. You will often find that this will give you the necessary insight to allow you to go back and establish the earlier result. For most students, frequent study sessions of not too long a duration are more productive than occasional long, drawn-out sessions. The best study environment varies greatly from one individual to another. Try experimenting with different environments to find out what is the most effective for you.

As far as initial conditions are concerned, that is, assumptions about your background, it is difficult to be precise, because you can probably get by with much less than the book might seem to indicate (see §1.5). Added to which, there is a big difference between understanding a topic fully and only having some vague acquaintance with it. On the mathematical side, you certainly need to know calculus, up to and including partial differentiation, and solution of simple ordinary differential equations. Basic algebra is assumed and some matrix theory, although you can probably take eigenvalues and diagonalization on trust. Familiarity with vectors and some exposure to vector fields is assumed. It would also be good to have met the ideas of a vector space and bases. We use Taylor's theorem a lot, but probably knowledge of Maclaurin's theorem will be sufficient. On the physics side, you obviously need to know Newton's laws and Newtonian

gravitation. It would be helpful also to know a little about the potential formulation of gravitation (though, again, just the basics will do). The book
assumes some familiarity with electromagnetism (Maxwell's equations,
in particular) and fluid dynamics (the Navier-Stokes equation, in particular), but neither of these are absolutely essential. It would be very helpful
to have met some ideas about waves (such as the fundamental relationship
$c = \lambda\nu$) and the wave equation in particular. In cosmology, it is assumed
that you know something about basic astronomy but, to gain an understanding of modern cosmology in the final chapter, you will need much
more of a background in contemporary physics.

Having listed all these topics, then, if you are still unsure about your
background, try the book and see how you get on. If it gets beyond you
(and it is not a Level 2 section) press on for a bit and, if things do not get
any better, then cut out. Hopefully, you may still have learnt a lot, and you
can always come back to it when your background is stronger. In fact, it
should not require much background to get started, for Part A on special
relativity assumes very little. After that, you hit Part B, and this is where
your motivation will be seriously tested. If you make it through the first
half of the book, then the pickings on the other side are very rich indeed.

1.5   A final note for the less able student from Ray

I was far from being a child prodigy, and yet I learnt relativity at the age of
15! Let me elaborate. As testimony to my intellectual ordinariness, I had
left my junior school at the age of 11, having achieved the unremarkable
feat of coming 22nd in the class in my final set of examinations. Yet I really
did know some relativity four years on – and I don't just mean the special
theory, but the general theory (up to and including the Schwarzschild
solution and the classical tests). I remember detecting a hint of disbelief
when I recounted this to Alan Tayler, who was later to become my tutor,
in an Oxford entrance interview. He followed up by asking me to define a
tensor and, when I rattled off a definition, he seemed somewhat surprised.
As it turned out, Alan was instrumental in enabling *Introducing Einstein's
Relativity* to be published by Oxford University Press thirty years later.
In fact, we did not cover very much more than I first knew in the Oxford
third-year specialist course on general relativity. So how was this possible?

I, too, had heard the story about how only a few people in the world
really understood relativity, and it had aroused my curiosity. I went to
the local library and, as luck would have it, I pulled out a book entitled
Einstein's Theory of Relativity by Lillian Lieber (2008; originally published
in 1945). This is a very bizarre book in appearance. The book is not set
out in the usual way but rather as though it were concrete poetry. Moreover, it is interspersed by surrealist drawings by Hugh Lieber involving
the symbols from the text (Fig. 1.1). I must confess that at first sight the
book looks rather cranky; but it is not. Indeed, it has been reprinted in
recent years (see Further reading). I worked through the book, filling in

Fig. 1.1 'The product of two tensors
is equal to another', according to Hugh
Lieber.

all the details missing from the calculations as I went. What was amazing was that the book did not make too many assumptions about what mathematics the reader needed to know. For example, I had not then met partial differentiation in my school mathematics, and yet there was sufficient coverage in the book for me to cope. It felt almost as if the book had been written just for me. The combination of the intrinsic interest of the material and the success I had in doing the intervening calculations provided sufficient motivation for me to see the enterprise through to the end.

Perhaps, if you consider yourself a less able student, you are a bit daunted by the intellectual challenge that lies ahead. I will not deny that the book includes some very demanding ideas (indeed, I do not understand every facet of all of these ideas myself). But I hope the two facts that the arguments are broken down into small steps and that the calculations are doable will help you on your way. Even if you decide to cut out after Part C, you will have come a long way. Take heart from my little story – I am certain that, if you persevere, you will consider it worth the effort in the end.

1.6 A final note for the more able student from James

In revising and extending Ray's book *Introducing Einstein's Relativity*, I wanted to keep to the style of the original version, which attempted as far as possible to give a self-contained account of the key areas of general relativity and which provided all the details of the calculations either in the text or in the exercises. However, since Ray wrote the original version of the book nearly thirty years ago, the range of topics that deserve attention has expanded considerably and this has made it harder to keep the material quite as self-contained so you may need to do more background reading. There are two reasons why the scope has expanded.

The first reason is that, from a mathematical point of view, the idea of general relativity as an entirely geometric theory of gravity described in terms of the curvature of space-time has been supplemented by an increasing emphasis on an approach where one thinks of Einstein's equations as a system of partial differential equations. This is needed in attacking outstanding theoretical problems such as 'cosmic censorship' as well as constructing stable numerical relativity codes to simulate events such as colliding black holes. This has resulted in extending the chapter on 'The structure of the field equations' and adding a new chapter, 'The 3+1 and 2+2 formalisms', which goes into more detail about the passage from the geometrical formulation of Einstein's equations to a description in terms of evolution equations.

The second reason comes from the increasingly detailed experimental information that we now have about the structure of the universe. For example, we now know considerably more about gravitational waves from the direct measurements, made by the LIGO gravitational wave observatories, of radiation from colliding black holes, as well as the indirect measurements of gravitational radiation coming from the orbits of binary pulsars (see Chapter 21 for details). We also have much more accurate information about the expansion of the universe and the cosmic microwave background (CMB). These developments have resulted in enlarging the chapter on gravitational waves considerably to provide details of both the generation and the detection of gravitational waves, as well as adding a new chapter describing the modern approach to cosmology, in which one uses experimental evidence to determine the cosmological parameters of our universe. Understanding the sources of gravitational waves and cosmological models requires input from a broader range of physics than the rest of the book. Rather than get bogged down in the details, we have tried to keep to a summary of the essential points, but much of the current research in these areas involves an understanding of other areas of physics in extreme relativistic conditions.

Much of the additional content is Level 2 material, which is more suitable for graduate students. As a result, we decided to extend the further reading sections for all the chapters in the book and include more references to ongoing research. In order to ensure that these references are easily accessible and remain up to date, our first port of call has been to the online journal *Living Reviews in Relativity* (see Further reading). This journal contains full, open-access, online articles that provide critical reviews of the current state of research, and available sources in all areas of relativity. Furthermore, as the name implies, authors are encouraged to update these reviews to take account of any recent developments. My advice to all readers of this book is to keep a broad outlook and to try and maintain an interest in both the mathematics and the underlying physics. Although Einstein's equations have remained unchallenged as the cornerstone of relativity, our approach to analysing them has changed over time. Once it was sufficient to know about differential geometry and tensor analysis. Although these still provide the key mathematical tools, current research in relativity now covers a wide spectrum and involves a variety of different formalisms. One also requires a knowledge of areas such as algebraic and differential topology and mathematical analysis, as well as more applied areas such as signal processing and relativistic astrophysics. As alluded to earlier, one of the big challenges is to provide a quantum theory of gravity and this brings in still other areas of mathematics and physics. Despite some progress, it would seem that we are still quite a long way from having an an accepted theory of quantum gravity and it might well be that this requires both new mathematics and new physics.

Like Ray, I would like to end by saying something about the books that have influenced me. Before going to university, I read a short book entitled *Space-Time Algebra* by David Hestenes. While there was much in the book I did not understand, it introduced me to the concept of 'spinors' as a way of describing both classical and quantum physics. Several years later, this drew me to the work of Roger Penrose, who saw conformal geometry and spinors as playing a key role in understanding gravitational physics, which influenced my own work on using spinors to investigate gravitational energy. The two-volume book *Spinors and Space-Time* by Penrose and Rindler provides a comprehensive treatment of these topics and more. A second source which was profoundly influential to both Ray and myself was *The Large Scale Structure of Space-Time* by Stephen Hawking and George Ellis. Indeed, for researchers of our generation, this was regarded as something akin to the status of the Bible in the field and, like the work of Penrose, it continues to inspire the current generation. However, it is written at a level which is perhaps too sophisticated for most undergraduates (in parts too sophisticated for many specialists!). Part D of the book owes much to the approach of Hawking and Ellis and we hope that this part of the book will provide a small stepping stone to *The Large Scale Structure of Space-Time*. To that end, and because we cannot improve on it, we have in places included extracts from that source virtually verbatim. We felt that, if students were to consult this text, then the familiarity of some of the material might instil confidence and encourage them to delve deeper. We are hugely indebted to the authors for allowing us to borrow from their superb book.

1.7 Research interests of the authors

To provide some background about the authors to our readers and scientific colleagues, here is a summary of our fields of research interests.

Ray
Computer algebra in general relativity
Exact solutions and their invariant classification
The 2+2 formalism
Numerical relativity and the CCM (Cauchy-Characteristic Matching) approach
A 2+2 canonical quantization programme

James
Quasi-local mass in general relativity
Low regularity solutions of Einstein's equations
Gravitational singularities
Numerical relativity and the CCM (Cauchy-Characteristic Matching) approach
A 2+2 canonical quantization programme

Exercises

1.1 Go online and look at the latest two volumes of *Living Reviews in Relativity* at https://www.springer.com/journal/41114/.

1.2 Go online and look at the titles of the new submissions in the general relativity and quantum cosmology section of arXiv at https://arxiv.org/list/gr-qc/new and see if you can find articles that relate to the topics you found when looking at *Living Reviews in Relativity*.

1.3 Read a biography of Einstein (see Part A of the Bibliography at the end of this book).

Further reading

The first four references relate to the discussion in §1.3 on quantum gravity. For a non-technical review of various approaches to quantum gravity, see also the article by Kiefer (2005). The remaining references are formative influences on the authors of this book.

Hawking, S. W. (1979). 'Euclidean quantum gravity', in Lévy, M., and Deser, S., eds, *Recent Developments in Gravitation*. NATO Advanced Study Institutes Series (Series B: Physics), vol 44. Springer, Boston, MA, 145–73.

Penrose, R. (1968). Twistor quantisation and curved space-time. *International Journal of Theoretical Physics*, *1*(1), 61–99.

d'Inverno, R. A., and Vickers, J. A. (1995). 2+2 decomposition of Ashtekar variables. *Classical and Quantum Gravity*, *12*(3), 753.

Penrose, R. (2016). *Fashion, Faith and Fantasy in the New Physics of the Universe*. Princeton University Press, Princeton, NJ.

Kiefer, C. (2006). Quantum gravity: General introduction and recent developments. Annalen der Physik, 15(1–2), 129–48.

Lieber, L. R. (2008). *The Einstein Theory of Relativity* (reprint of 1945 edition). Paul Dry Books, Philadelphia, PA.

Iya, B. (ed.) *Living Reviews in Relativity*, https://www.springer.com/journal/41114/, accessed 16 April 2021.

Hestenes, D. (2015) *Space-Time Algebra* (2nd edn). Birkhäuser, Basel.

Penrose, R., and Rindler, W. (1986). *Spinors and Space-Time*. Vols 1 and 2, Cambridge University Press, Cambridge.

Hawking, S. W., and Ellis, G. F. R. (1973). *The Large Scale Structure of Space-Time*. Cambridge University Press, Cambridge.

Part A

Special Relativity

The *k*-calculus

2

2.1 Model building

Before we start, we should be clear what we are about. The essential activity of mathematical physics, or theoretical physics, is that of **modelling** or **model building**. The activity consists of constructing a mathematical model which we hope in some way captures the essentials of the phenomena we are investigating. I think we should never fail to be surprised that this turns out to be such a productive activity. After all, the first thing you notice about the world we inhabit is that it is an extremely complex place. The fact that so much of this rich structure can be captured by what are, in essence, a set of simple formulae is quite astonishing. Just think how simple Newton's universal law of gravitation is and yet it encompasses a whole spectrum of phenomena, from a falling apple to the shape of a globular cluster of stars. As Einstein said, 'The most incomprehensible thing about the world is that it is comprehensible' (Einstein, 1954).

The very success of the activity of modelling has, throughout the history of science, turned out to be counterproductive. Time and again, the successful model has been confused with the ultimate reality, and this, in turn, has stultified progress. Newtonian theory provides an outstanding example of this. So successful had it been in explaining a wide range of phenomena that, after more than two centuries of success, the laws had taken on an absolute character. Thus it was that, when at the end of the nineteenth century it was becoming increasingly clear that something was fundamentally wrong with the current theories, there was considerable reluctance to make any fundamental changes to them. Instead, a number of artificial assumptions were made in an attempt to explain the unexpected phenomena. It eventually required the genius of Einstein to overthrow the prejudices of centuries and demonstrate in a number of simple thought experiments that some of the most cherished assumptions of Newtonian theory were untenable. This he did in a number of brilliant papers written in 1905, proposing a theory which has become known today as the **special theory of relativity**. Of course, the special theory of relativity was not the end of the story, and Einstein went on to develop general relativity – a relativistic theory of gravitation.

We should perhaps be discouraged from using words like 'right' or 'wrong' when discussing a physical theory. If we remember that the essential activity is model building, a model should then rather be described as 'good' or 'bad' depending on how well it describes the phenomena

Introducing Einstein's Relativity. Second Edition. Ray d'Inverno and James Vickers, Oxford University Press.
© Ray d'Inverno and James Vickers (2022). DOI: 10.1093/oso/9780198862024.003.0002

it encompasses. Thus, Newtonian theory is an excellent theory for describing a whole range of phenomena. For example, if one is concerned with describing the motion of a car, then the Newtonian framework is likely to be the appropriate one. However, it fails to be appropriate if we are interested in very high speeds (comparable with the speed of light) or very intense gravitational fields (such as in a neutron star). To put it another way, together with every theory, there should go its **range of validity**. Thus, to be more precise, we should say that Newtonian theory is an excellent theory within its range of validity. From this point of view, developing our models of the physical world does not involve us in constantly throwing theories out, since they are perceived to be wrong, or in unlearning them, but rather it consists more of a process of refinement in order to increase their range of validity. So the moral of this section is that, for all their remarkable success, one must not confuse theoretical models with the ultimate reality they seek to describe.

2.2 Historical background

In 1865, James Clerk Maxwell put forward the theory of electromagnetism. One of the triumphs of the theory was the discovery that light waves are electromagnetic in character. Since all other known wave phenomena required a material medium in which the oscillations were carried, it was postulated that there existed an all-pervading medium, called the 'luminiferous ether', which carried the oscillations of electromagnetism. It was then anticipated that experiments with light would allow the absolute motion of a body through the ether to be detected. Such hopes were upset by the null result of the famous (and technically difficult) Michelson–Morley experiment in 1881, which attempted to measure the velocity of the Earth relative to the ether and found it to be undetectably small. In order to explain this null result, two ad hoc hypotheses were put forward by Lorentz, Fitzgerald, and Poincaré in 1895, namely, the contraction of rigid bodies and the slowing down of clocks when moving through the ether. These effects were contained in some simple formulae called the 'Lorentz transformations'. This would affect every apparatus designed to measure the motion relative to the ether so as to neutralize exactly all expected results. Although this theory was consistent with the observations, it had the philosophical defect that its fundamental assumptions were unverifiable.

In fact, the essence of the special theory of relativity is contained in the Lorentz transformations. However, Einstein was able to derive them from two postulates, the first being called the 'principle of special relativity' – a principle which Poincaré had also suggested independently in 1904 – and the second concerning the constancy of the velocity of light. In so doing, he was forced to re-evaluate our ideas of space and time and he demonstrated through a number of simple thought experiments that the source of the limitations of the classical theory lay in the concept of **simultaneity**. Thus, although in a sense Einstein found nothing new, in

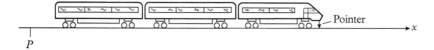

Fig. 2.1 Train travels in straight line.

that he rederived the Lorentz transformations, his derivation was physically meaningful and in the process revealed the inadequacy of some of the fundamental assumptions of classical thought. Herein lies his chief contribution.

2.3 Newtonian framework

We start by outlining the Newtonian framework. An **event** intuitively means something happening in a fairly limited region of space and for a short duration in time. Mathematically, we idealize this concept to become a point in space and an instant in time. Everything that happens in the universe is an event or collection of events. Consider a train travelling from one station P to another R, leaving at 10 a.m. and arriving at 11 a.m. We can illustrate this in the following way: for simplicity, let us assume that the motion takes place in a straight line (say, along the x-axis (Fig. 2.1)); then we can represent the motion by a **space-time diagram** (Fig. 2.2) in which we plot the position of some fixed point on the train, which we represent by a pointer, against time. The curve in the diagram is called the **history** or **world-line** of the pointer. Notice that at Q the train was stationary for a period.

We shall call individuals equipped with a method of measuring time (an ideal clock) and a method of measuring distance (an ideal ruler) **observers**. Had we looked out of the train window on our journey at a clock in a passing station, we would have expected it to agree with our watch. One of the central assumptions of the Newtonian framework is that two observers will, once they have synchronized their clocks, always agree about the time of an event, irrespective of their relative motion. This implies that, for all observers, time is an **absolute** concept. In particular, all observers can agree to synchronize their clocks so that they all agree on the time of an event. In order to fix an event in space, an observer may choose a convenient origin in space together with a set of three Cartesian coordinate axes. We shall refer to an observer's clock and coordinate axes as a **frame of reference** (Fig. 2.3). Then an observer is able to **coordinatize** events, i.e. determine the time t an event occurs and its relative position (x, y, z).

We have set the stage with space and time; they provide the backcloth, but what is the story about? The stuff which provides the events of the universe is **matter**. For the moment, we shall idealize lumps of matter into objects called **bodies**. If the body has no physical extent, we refer to it as a **point particle** or **point mass**. Thus, the role of observers in Newtonian theory is to chart the history of bodies.

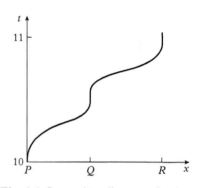

Fig. 2.2 Space-time diagram of pointer.

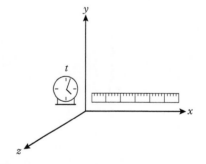

Fig. 2.3 Observer's frame of reference.

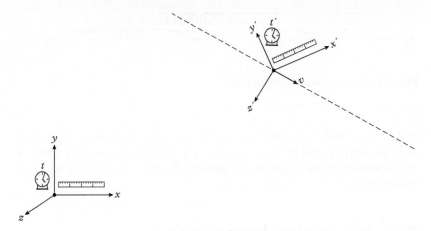

Fig. 2.4 Two observed bodies and their inertial frames.

2.4 Galilean transformations

Now, relativity theory is concerned with the way different observers see the same phenomena. One can ask: are the laws of physics the same for all observers or are there preferred states of motion, preferred reference systems, and so on? Newtonian theory postulates the existence of preferred frames of reference. The existence of these is essentially implied by the first law, which we shall call N1 and state in the following form:

N1: Every body continues in its state of rest or of uniform motion in a straight line unless it is compelled to change that state by forces acting on it.

Thus, there exists a privileged set of bodies, namely, those not acted on by forces. The frame of reference of a co-moving observer is called an **inertial frame** (Fig. 2.4). It follows that, once we have found one inertial frame, then all others are at rest or travel with constant velocity relative to it (since otherwise, Newton's first law would no longer be true). The transformation which connects one inertial frame with another is called a **Galilean transformation**. To fix ideas, let us consider two inertial frames called S and S' in **standard configuration**, i.e. with axes parallel and S' moving along S's positive x-axis with constant velocity (Fig. 2.5). We also assume that the observers synchronize their clocks so that the origins of time are set when the origins of the frames coincide. It follows from Fig. 2.5 that $x = x' + vt$ so the Galilean transformation connecting the two frames is given by

$$x' = x - vt, \quad y' = y, \quad z' = z, \quad t' = t. \tag{2.1}$$

The last equation provides a manifestation of the assumption of absolute time in Newtonian theory. Now, Newton's laws hold only in inertial

Fig. 2.5 Two frames in standard configuration at time t.

frames. From a mathematical viewpoint, this means that Newton's laws must be **invariant** under a Galilean transformation.

2.5   The principle of special relativity

We begin by stating the relativity principle which underpins Newtonian theory.

Restricted principle of special relativity:
All inertial observers are equivalent as far as dynamical experiments are concerned.

This means that, if one inertial observer carries out some dynamical experiments and discovers a physical law, then any other inertial observer performing the same experiments must discover the same law. Put another way, these laws must be invariant under a Galilean transformation. That is to say, if the law involves the coordinates x, y, z, t of an inertial observer S, then the law relative to another inertial observer S' will be the same, with x, y, z, t replaced by x', y', z', t', respectively. Many fundamental principles of physics are statements of impossibility, and the above statement of the relativity principle is equivalent to the statement of the **impossibility** of deciding, by performing dynamical experiments, whether a body is absolutely at rest or in uniform motion. In Newtonian theory, we cannot determine the **absolute** position in space of an event, but only its position **relative** to some other event. In exactly the same way, uniform velocity has only a relative significance; we can only talk about the velocity of a body relative to some other. Thus, both position and velocity are **relative** concepts.

Einstein realized that the principle as stated above is empty because there is no such thing as a purely **dynamical** experiment. Even on a very elementary level, any dynamical experiment we think of performing involves observation, i.e. **looking**, and looking is a part of optics, not dynamics. In fact, the more one analyses any one experiment, the more it becomes apparent that practically all the branches of physics are involved in the experiment. Thus, Einstein took the logical step of removing the restriction of dynamics in the principle and took the following as his first postulate.

> **Postulate I. Principle of special relativity:**
> All inertial observers are equivalent.

Hence we see that this principle is in no way a contradiction of Newtonian thought, but rather constitutes its logical completion.

2.6   The constancy of the velocity of light

We previously defined an observer in Newtonian theory as someone equipped with a clock and a way of measuring distance with which to map the events of the universe. In many textbooks, the concept of a 'rigid ruler' is introduced to do this. However, as pointed out by Bondi (1966), although quantum theory gives us a practical mechanism for producing an ideal clock (such as an atomic clock) the concept of a 'rigid ruler' is fraught with difficulty. What is rigidity anyway? If a moving frame appears non-rigid in another frame, which, if either, is the rigid one? The approach of the *k*-calculus is to dispense with the rigid ruler and use radar methods for measuring distances. In the **radar method**, an observer measures the distance of an object by sending out a light signal which is reflected off the object and received back by the observer. The **distance** is then simply defined as **half the time difference between emission and reception**. Note that, by this method, the speed of light is automatically one and distances are measured in intervals of time, like the light year or the light second ($\sim 3 \times 10^8$ m).

Why use light? The reason is that we know that the velocity of light is independent of many things. Observations from double stars tell us that the velocity of light **in vacuo** is independent of the motion of the sources as well as independent of colour, intensity, etc. For, if we suppose that the velocity of light were dependent on the motion of the source relative to an observer (so that if the source were coming towards us, the light would be travelling faster, and vice versa), then we would no longer see double stars moving in Keplerian orbits (circles, ellipses) about each other: their orbits would appear distorted; yet, no such distortion is observed. There are many experiments which confirm this assumption. However, these were not known to Einstein in 1905, who adopted the second postulate mainly on philosophical grounds. We state the second postulate in the following form.

> **Postulate II. Constancy of velocity of light:**
> The velocity of light is the same in all inertial systems.

The speed of light is conventionally denoted by c (from the Latin *celeritas* meaning 'speed') and, in SI units, it has the exact numerical value 2.997924580×10^8 ms^{-1} (so that the metre is defined in the SI system as the distance travelled by light in a vacuum in 1/299792458 of a second). In this book, we shall mostly work in **relativistic units**, in which c is taken to be unity (i.e. $c = 1$). Note, in passing, that another reason for using radar methods is that other methods are totally impracticable for large distances. In fact, these days, distances from the Earth to the Moon and Venus can be measured very accurately by bouncing radar signals off them.

2.7 The *k*-factor

For simplicity, we shall begin by working in two dimensions, one spatial dimension and one time dimension. Thus, we consider a system of observers distributed along a straight line, each equipped with a clock and a flashlight. We plot the events they map in a two-dimensional space-time diagram. Let us assume we have two observers, A at rest and B moving away from A with uniform (constant) speed. Then, in a space-time diagram, the world-line of A will be represented by a vertical straight line, and the world-line of B by a straight line at an angle to A's, as shown in Fig. 2.6.

A light signal in the diagram will be denoted by a straight line making an angle of 45° ($\pi/4$ radians) with the axes, because we are taking the speed of light to be 1. Now, suppose A sends out a series of flashes of light to B, where the interval between the flashes is denoted by T according to A's clock. Then it is plausible to assume that the intervals of reception by B's clock are proportional to T, say, kT. Moreover, the quantity k, which we call the **k-factor**, is clearly a characteristic of the motion of B relative to A. We now assume that if A and B are **inertial** observers, then the k-factor is a **constant** in time and **independent** of T. Indeed, we will go further and assume that it is independent of the point in space-time where the measurement is made and only depends on the relative speed of the two inertial observers. From a mathematical point of view, this is the assumption that space-time is **homogeneous**, i.e. the same at every point. From B's point of view, A is moving away from B with the same relative speed, so the principle of special relativity requires that the relationship between A and B must be reciprocal. So that, if B emits two signals with a time lapse of T according to B's clock, then A receives them after a time lapse of kT according to A's clock (Fig. 2.7). Note that, in interchanging the roles of A and B, we are assuming that there are no directional effects, which amounts to the assumption that space-time is **isotropic**, i.e. it is the same in any direction.

In the radar method, an observer A assigns coordinates to an event P by bouncing a light signal off it. Suppose a light signal is sent out at a time $t = t_1$, and received back at a time $t = t_2$ (Fig. 2.8); then, since the velocity of light in both directions is the same, the time (as measured by A) at the point P is halfway between t_1 and t_2. Furthermore, since by assumption the speed of light is 1, the distance to P is half the time for the round

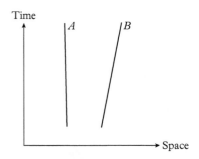

Fig. 2.6 The world-lines of observers A and B.

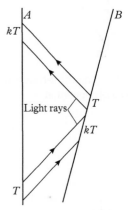

Fig. 2.7 The reciprocal nature of the k-factor.

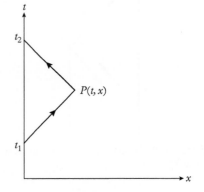

Fig. 2.8 Coordinatizing events.

trip. Hence, according to our radar definition of distances, the space-time coordinates of P are given by

$$(t, x) = (\tfrac{1}{2}(t_1 + t_2), \tfrac{1}{2}(t_2 - t_1)). \tag{2.2}$$

We now use the k-factor to develop the k-calculus.

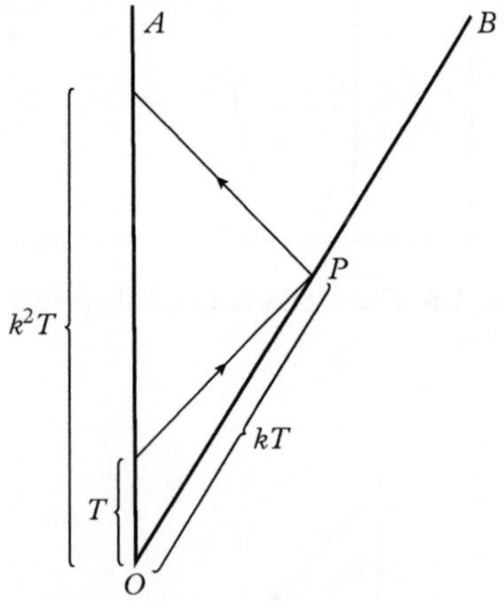

2.8   Relative speed of two inertial observers

Consider the configuration shown in Fig. 2.9 and assume that A and B synchronize their clocks to zero when they cross at event O. After a time T, A sends a signal to B, which is reflected back at event P. From B's point of view, a light signal is sent to A after a time lapse of kT by B's clock. It follows from the definition of the k-factor that A receives this signal after a time lapse of $k(kT)$. Then, using (2.2) with $t_1 = T$ and $t_2 = k^2 T$, we find the coordinates of P according to A's clock are given by

$$(t, x) = (\tfrac{1}{2}(k^2 + 1)T, \tfrac{1}{2}(k^2 - 1)T). \tag{2.3}$$

Thus, as T varies, this gives the coordinates of the events which constitute B's world-line. Hence, if v is the velocity of B relative to A, we find

$$v = \frac{x}{t} = \frac{k^2 - 1}{k^2 + 1}.$$

Solving for k^2 in terms of v, and taking the positive square root in order to have the same direction of time for A and B, we find

$$k = \left(\frac{1 + v}{1 - v}\right)^{\frac{1}{2}}. \tag{2.4}$$

We shall see in the next chapter that this is the usual relativistic formula for the radial Doppler shift. If B is moving away from A, then $k > 1$, which represents a 'red' shift, whereas, if B is approaching A, then $k < 1$, which represents a 'blue' shift. Note that the transformation $v \rightarrow -v$ corresponds to interchanging the roles of A and B and results in $k \rightarrow 1/k$. Moreover,

$$v = 0 \quad \Longleftrightarrow \quad k = 1,$$

as we should expect for observers relatively at rest: once they have synchronized their clocks, the synchronization remains (Fig. 2.10).

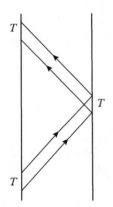

Fig. 2.9 Relating the k-factor to the relative speed of separation.

Fig. 2.10 Observers relatively at rest ($k = 1$).

2.9 Composition law for velocities

Consider the situation in Fig. 2.11, where k_{AB} denotes the k-factor between A and B, with k_{BC} and k_{AC} defined similarly. It follows immediately that

$$k_{AC} = k_{AB}k_{BC}. \qquad (2.5)$$

Using (2.4), we find the corresponding **composition law** for velocities:

$$v_{AC} = \frac{v_{AB} + v_{BC}}{1 + v_{AB}v_{BC}}. \qquad (2.6)$$

This formula has been verified experimentally to very high precision. Indeed, formula (2.6) was first proposed empirically (prior to the theory of special relativity) by Fizeau in 1851 in order to explain the results of an experiment measuring the speed of light in a rapidly moving fluid. Note that, if v_{AB} and v_{BC} are small compared with the speed of light, i.e.

$$v_{AB} \ll 1, \quad v_{BC} \ll 1,$$

then we obtain the classical Newtonian formula

$$v_{AC} = v_{AB} + v_{BC},$$

to lowest order. Although the composition law for velocities is not simple, the one for k-factors is and, in special relativity, it is the k-factors which are the directly measurable quantities. Note also that, formally, if we substitute $v_{BC} = 1$, representing the speed of a light signal **relative to B**, in (2.6), then the resulting speed of the light signal relative to A is

$$v_{AC} = \frac{v_{AB} + 1}{1 + v_{AB}} = 1,$$

in agreement with the constancy of the velocity of light postulate.

From the composition law, we can show that, if we add two speeds less than the speed of light, then we again obtain a speed less than the speed of light (exercise). This does not mean, as is sometimes stated, that nothing can move faster than the speed of light in special relativity, but rather that the speed of light is a border which can not be crossed or even reached. More precisely, special relativity allows for the existence of three classes of particles:

1. Particles that move slower than the speed of light are called **subluminal** particles. They include material particles and elementary particles such as electrons and neutrons.

2. Particles that move with the speed of light are called **luminal** particles. They include the carrier of the electromagnetic field interaction, the photon, other zero rest-mass particles (see §4.5) and, theoretically, the carrier of the gravitational field interaction, called the graviton.

3. Particles that move faster than the speed of light are called **superluminal** particles or **tachyons**. There was some excitement in the 1970s surrounding the possible existence of tachyons, but all attempts to detect

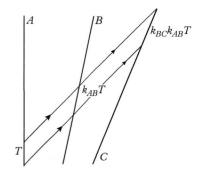

Fig. 2.11 Composition of k-factors.

them to date have failed. This suggests two likely possibilities: either tachyons do not exist or, if they do, they do not interact with ordinary matter. This would seem to be just as well, for otherwise they could be used to signal back into the past and so would appear to violate causality. For example, it would be possible theoretically to construct a device which sent out a tachyon at a given time and which would trigger a mechanism in the device to blow it up **before** the tachyon was sent out! We will therefore assume for the rest of this book that tachyons do not exist and that **nothing can travel faster than the speed of light**.

2.10   The relativity of simultaneity

For Einstein, the relativity of simultaneity was at the very heart of special relativity and resolves many of the paradoxes that the classical theory gives rise to. Consider two events P and Q which take place at the same time, according to A, at points which are equal but opposite distances away. A could establish this by sending out and receiving the light rays as shown in Fig. 2.12 (continuous lines). Suppose now that another inertial observer B meets A at the time these events occur **according to** A. B also sends out light rays RQU and SPV to illuminate the events, as shown (dashed lines). By symmetry, $RU = SV$ and so these events are equidistant, according to B. However, the signal RQ was sent before the signal SP and since the events are equidistant B concludes that the event Q took place **before** P. Hence, although A judges P and Q to be simultaneous, B considers Q to have occurred first. Indeed, it is not hard to see that, by making a very small change in the time of P (according to A), one can have P occurring before Q for observer A but after Q to occur before P for observer B. This is an example of the **relativity of simultaneity**.

Einstein realized the crucial role that simultaneity plays in the theory and, in his popular work *Relativity: The Special and General Theory*, gave the following simple thought experiment (which we slightly update) to illustrate its dependence on the observer. Imagine a train travelling along a straight track with velocity v relative to a stationary observer A on the bank of the track. In the train, B is an observer situated at the centre of one of the carriages. We assume that there are two electrical devices on the track that are the length of the carriage apart and equidistant from A. When the carriage containing B goes over these devices, they fire and activate two light sources which are situated at the end of the carriage (Fig. 2.13) and which each emit a photon. From the configuration, it is clear that, according to Observer A, the two photons will be emitted **simultaneously**. However, from A's point of view, B is travelling towards the light emanating from light source 2 and away from the light emanating from light source 1. Since the speed of light is a constant, A will observe B meeting the light from source 2 before the light from source 1. Hence, B will observe the photon from light source 1 strike the front of the train **before** the other photon strikes the back. This is in accordance with the space-time diagram given above where P is the photon hitting the back of the train and Q is the photon hitting the front. These are simultaneous for observer A on the bank but Q occurs before P for observer B on the train.

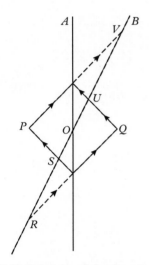

Fig. 2.12 Relativity of simultaneity.

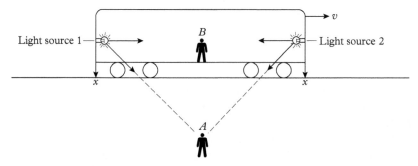

Fig. 2.13 Photons emanating from the two sources.

2.11 Causality

In Newtonian theory, the notion of absolute time, which all observers agree on, enables one to unambiguously say that a 'cause' precedes its 'effect'. Given the example of the previous section in which different inertial observers can disagree about the order of events, one might worry about how the notion of causality survives in special relativity. However, the assumption that nothing can travel faster than the speed of light comes to our rescue. Given an event O, we say that an event E (say) is on the **future light cone** of O if it lies on a light ray starting from O that then reaches E (see Fig. 2.14). The fact that it is a cone will become clearer later when we take all the spatial dimensions into account. Since **all** inertial observers agree on the speed of light, the future light cone does not depend on the particular choice of inertial observer but is **invariantly defined**. Similarly, we say an event H (say) is on the **past light cone** of O if there is a light ray starting from H that reaches O. The observer-independent concept of light cone thus divides space-time into three regions. The **future** of O consists of points E, F, and so on. that can be reached by travelling at speeds less than or equal to the speed of light. Since nothing can travel faster than the speed of light, these are the points in space-time that can be influenced by what happens at O. For this reason, we often call these points the **causal future** of O. Similarly, points in the **past** of O are points such as H, J, and so on, where it is possible to reach O by travelling at less than or equal to the speed of light. Again, since nothing can travel faster than the speed of light, these are the only points in space-time that can have an effect on O, so justifying the name **causal past**. Note that the world-line of any inertial observer or material particle passing through O must lie within the light cone at O. Finally, points in the region labelled 'elsewhere' in Fig. 2.14 consist of points that cannot affect or cannot be affected by what happens at O. This is because to go from O to a point G in the 'elsewhere' region (or vice versa) would require travelling faster than the speed of light. The temporal relationship to O of events in the 'elsewhere' region will not be something all observers will agree upon. For example, one class of observers will say that G took place after O, another class will say that G took place before O, and yet another will say they took place simultaneously.

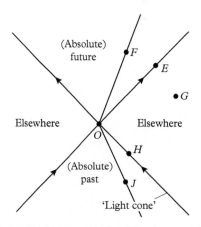

Fig. 2.14 Event relationships in special relativity.

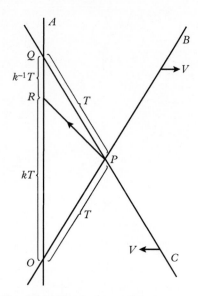

Fig. 2.15 The clock paradox.

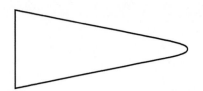

Fig. 2.16 Spatial analogue of clock paradox.

2.12 The clock paradox

Consider three inertial observers as shown in Fig. 2.15, with the relative velocity $v_{AC} = -v_{AB}$. Assume that A and B synchronize their clocks at O and that C's clock is synchronized with B's at P. Let B and C meet after a time T according to B, whereupon they emit a light signal to A. According to the k-calculus, A receives the signal at R after a time kT since meeting B. Remembering that C is moving with the opposite velocity to B (so that $k \to k^{-1}$), then A will meet C at Q after a subsequent time lapse of $k^{-1}T$. The total time that A records between events O and Q is therefore $(k + k^{-1})T$. For $k \neq 1$, this is **greater** than the combined time intervals $2T$ recorded between events OP and PQ by B and C. But should not the time lapse between the two events agree? This is one form of the so-called **clock paradox**.

However, it is not really a paradox; rather, what it shows is that, in relativity, time, like distance, is a route-dependent quantity. The point is that the $2T$ measurement is made by **two** inertial observers, not one. Some people have tried to reverse the argument by setting B and C to rest, but this is not possible since they are in relative motion to each other. Another argument says that, when B and C meet, C should take B's clock and use it. But, in this case, the clock would have to be **accelerated** when being transferred to C and so it is no longer inertial. Some opponents of special relativity have argued that the short period of acceleration should not make such a difference, but this is analogous to saying that a journey between two points which is straight nearly all the time is about the same length as one which is wholly straight (as shown), which is clearly not true (Fig. 2.16). The moral is that, in special relativity, time is a more difficult concept to work with than the absolute time of Newton.

A more subtle point revolves around the implicit assumption that the clocks of A and B are 'good' clocks, i.e. that the seconds of A's clock are the same as those of B's clock. One suggestion is that A has two clocks, adjusts the tick rate until they are the same, and then sends one of them to B at a very slow rate of acceleration. The assumption here is that the very slow rate of acceleration will not affect the tick rate of the clock. However, what is there to say that a clock may not be able to somehow add up the small bits of acceleration and so affect its performance? A more satisfactory approach would be for A and B to use identically constructed atomic clocks (which is, after all, what physicists use today to measure time). The objection then arises that their construction is based on ideas in quantum physics which is, a priori, outside the scope of special relativity. However, this is a manifestation of a point raised earlier, that virtually any real experiment which one can imagine carrying out involves more than one branch of physics. The whole structure is intertwined in a way which cannot easily be separated.

2.13   The Lorentz transformations

We have derived a number of important results in special relativity, which only involve one spatial dimension, by use of the k-calculus. Other results follow essentially from the transformations connecting inertial observers, the famous Lorentz transformations. We shall finally use the k-calculus to derive these transformations.

Let event P have coordinates (t, x) measured by A, and (t', x') measured by B (Fig. 2.17). Assume that A and B both set their clocks to zero when they meet. Let A send out a light signal at time t_1 to illuminate P which is reflected back and received by A at time t_2. Since, according to the radar method, $t = \frac{1}{2}(t_1 + t_2)$ and $x = \frac{1}{2}(t_2 - t_1)$, we can solve these to obtain $t_1 = t - x$ and $t_2 = t + x$. An identical calculation for observer B, using the primed coordinates, gives $t_1' = t' - x'$ and $t_2' = t' + x'$. On the other hand, according to the k-calculus, $t_1' = kt_1$ while $t_2 = kt_2'$ This gives

$$t' - x' = k(t - x), \quad t + x = k(t' + x'). \tag{2.7}$$

After some rearrangement, and using equation (2.4), we obtain (exercise) the so-called **special Lorentz transformation**

$$t' = \frac{t - vx}{(1 - v^2)^{\frac{1}{2}}}, \quad x' = \frac{x - vt}{(1 - v^2)^{\frac{1}{2}}}. \tag{2.8}$$

This is also referred to as a **boost in the x-direction with speed v**, since it takes one from A's coordinates to B's coordinates, and B is moving away from A, with speed v. Some simple algebra reveals the result (exercise)

$$t'^2 - x'^2 = t^2 - x^2,$$

showing that the quantity $t^2 - x^2$ is an invariant under a special Lorentz transformation or boost.

To obtain the corresponding formulae in the case of three spatial dimensions, we consider Fig. 2.5 with two inertial frames in standard configuration. Now, since, by assumption, the xz-plane ($y = 0$) of A must coincide with the $x'z'$-plane ($y' = 0$) of B, then the y and y' coordinates must be connected by a transformation of the form

$$y = ny', \tag{2.9}$$

because

$$y = 0 \iff y' = 0.$$

We now use the assumption that space is isotropic. We then reverse the direction of the x- and y-axes of A and B and consider the motion from B's point of view (see Figs. 2.18 and 2.19). Clearly, from B's point of view, the roles of A and B have interchanged. Hence, by symmetry, we must have

$$y' = ny. \tag{2.10}$$

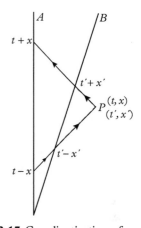

Fig. 2.17 Coordinatization of events by inertial observers.

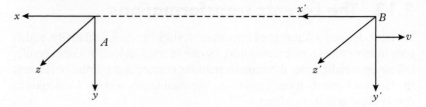

Fig. 2.18 The *x*- and *y*-axes from Figure 2.5, reversed.

Fig. 2.19 Figure 2.18 from *B*'s point of view.

Combining (2.9) and (2.10), we find

$$n^2 = 1 \quad \Rightarrow \quad n = \pm 1.$$

The negative sign can be dismissed since, as $v \to 0$, we must have $y' \to y$, in which case $n = 1$. Hence, we find $y' = y$, and a similar argument for z produces $z' = z$.

2.14 The four-dimensional world view

We now compare the special Lorentz transformation of the last section (using relativistic units in which the speed of light is one) with the Galilean transformation connecting inertial observers in standard configuration (see Table 2.1). In a Galilean transformation, the absolute time coordinate remains invariant. However, in a Lorentz transformation, the time and space coordinates get mixed up (note the symmetry in x and t). In the words of Minkowski (1952), 'Henceforth space by itself, and time by itself are doomed to fade away into mere shadows, and only a kind of union of the two will preserve an independent reality.'

In the old Newtonian picture, time is split off from three-dimensional Euclidean space. Moreover, since we have an absolute concept of simultaneity, we can consider two simultaneous events with coordinates (t, x_1, y_1, z_1), and (t, x_2, y_2, z_2); then, the square of the Euclidean **distance** between them,

$$\sigma^2 = (x_1 - x_2)^2 + (y_1 - y_2)^2 + (z_1 - z_2)^2, \tag{2.11}$$

Table 2.1

Galilean transformation	Lorentz transformation
$t' = t$	$t' = \dfrac{t - vx}{(1 - v^2)^{1/2}}$
$x' = x - vt$	$x' = \dfrac{x - vt}{(1 - v^2)^{1/2}}$
$y' = y$	$y' = y$
$z' = z$	$z' = z$

is invariant under a Galilean transformation. In the new special relativity picture, time and space merge together into a four-dimensional continuum called **space-time**. In this picture, the square of the **interval** between any two events (t_1, x_1, y_1, z_1) and (t_2, x_2, y_2, z_2) is defined by

$$s^2 = (t_1 - t_2)^2 - (x_1 - x_2)^2 - (y_1 - y_2)^2 - (z_1 - z_2)^2, \qquad (2.12)$$

and it is this quantity which is invariant under a Lorentz transformation. Note that, formally, we always denote the 'square' of the interval by s^2, but the quantity s is only defined if the right-hand side of (2.12) is non-negative. If we consider two events separated infinitesimally, (t, x, y, z) and $(t + dt, x + dx, y + dy, z + dz)$, then this equation becomes

$$ds^2 = dt^2 - dx^2 - dy^2 - dz^2, \qquad (2.13)$$

where all the infinitesimals are squared in (2.13). A four-dimensional space-time continuum in which the above form is invariant is called **Minkowski space-time** and provides the background geometry for special relativity. We will discuss this in more detail in the next chapter.

So far, we have only met a special Lorentz transformation which connects two inertial frames in standard configuration. A **full Lorentz transformation** connects two frames in general position (Fig. 2.20). It can be shown that a full Lorentz transformation can be decomposed into an ordinary spatial rotation, followed by a boost, followed by a further ordinary rotation. Physically, the first rotation lines up the x-axis of S with the velocity v of S'. Then a boost in this direction with speed v transforms S to a frame which is at rest relative to S'. A final rotation lines up the coordinate frame with that of S'. The spatial rotations introduce no new physics. The only new physical information arises from the boost and that is why we can, without loss of generality, restrict our attention to a special Lorentz transformation.

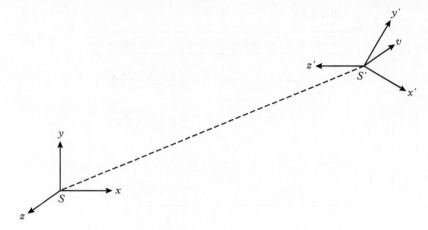

Fig. 2.20 Two frames in general position.

Exercises

2.1 (§2.4) Write down the Galilean transformation from observer S to observer S', where S' has velocity v_1 relative to S. Find the transformation from S' to S and state in simple terms how the transformations are related. Write down the Galilean transformation from S' to S'' where S'' has velocity v_2 relative to S'. Find the transformation from S to S''. Prove that the Galilean transformations form an Abelian (commutative) group.

2.2 (§2.7) Draw the four fundamental k-factor diagrams (see Fig. 2.7) for the cases of two inertial Observers A and B approaching and receding with uniform velocity v:
(i) as seen by A;
(ii) as seen by B.

2.3 (§2.8) Show that $v \to -v$ corresponds to $k \to k^{-1}$. If $k > 1$ corresponds physically to a red shift of recession, what does $k < 1$ correspond to?

2.4 (§2.9) Show that (2.6) follows from (2.5). Use the composition law for velocities to prove that, if $0 < v_{AB} < 1$ and $0 < v_{BC} < 1$, then $0 < v_{AC} < 1$.

2.5 (§2.9) Establish the fact that, if v_{AB} and v_{BC} are small compared with the velocity of light, then the composition law for velocities reduces to the standard additive law of Newtonian theory.

2.6 (§2.10) In the event diagram of Fig. 2.14, find a geometrical construction for the world-line of an inertial observer passing through O who considers event G as occurring simultaneously with O. Hence, describe the world-lines of inertial observers passing through O who consider G as occurring before or after O.

2.7 (§2.12) Draw Fig. 2.15 from B's point of view. Coordinatize the events O, R, and Q with respect to B and find the times between O and R, and R and Q, and compare them with A's timings.

2.8 (§2.13) Deduce (2.8) from (2.7). Use (2.7) to deduce directly that

$$t'^2 - x'^2 = t^2 - x^2.$$

Confirm the equality under the transformation formula (2.8).

2.9 (§2.13) In S, two events occur at the origin and a distance X along the x-axis simultaneously at $t = 0$. The time interval between the events in S' is T. Show that the spatial distance between the events in S' is $(X^2 + T^2)^{1/2}$ and determine the relative velocity v of the frames in terms of X and T.

2.10 (§2.14) Show that the interval between two events (t_1, x_1, y_1, z_1) and (t_2, x_2, y_2, z_2) defined by

$$s^2 = (t_1 - t_2)^2 - (x_1 - x_2)^2 - (y_1 - y_2)^2 - (z_1 - z_2)^2,$$

is invariant under a special Lorentz transformation. Deduce the Minkowski line element (2.13) for infinitesimally separated events. What does s^2 become if $t_1 = t_2$ and how is it related to the Euclidean distance σ between the two events?

Further reading

This chapter is based on Bondi's article in the Brandeis lectures (Trautmann et al. 1964). A slightly more popular version is in the book of his 1965 Tarner lectures (Bondi 1967). We consider a more conventional introduction to special relativity in the next chapter. The classical text by Einstein (100th anniversary edition of the 1915 original) gives an insight into his views on various aspects of relativity theory and, in particular, the importance of the relativity of simultaneity. The 100th anniversary edition has a useful commentary by Guttfreund and Renn.

Trautmann A., Pirani F. A. E., and Bondi, H. (1964). *Lectures on General Relativity*. Brandeis Summer Institute on Theoretical Physics, 1964, vol. 1. Prentice-Hall, Englewood Cliffs, NJ.

Bondi, H. (1967). *Assumption and Myth in Physical Theory*. Cambridge University Press, Cambridge.

Einstein, A. (2015). *Relativity: The Special and General Theory* (100th anniversary edn). Princeton University Press, Princeton, NJ.

The key attributes of special relativity

3

3.1 Standard derivation of the Lorentz transformations

We start this chapter by deriving again the Lorentz transformations, but this time by using a more standard approach. We shall work in non-relativistic units in which the speed of light is denoted by c. We restrict attention to two inertial observers S and S' in standard configuration. As before, we shall show that the Lorentz transformations follow from the two postulates, namely, the principle of special relativity and the constancy of the velocity of light.

Now, by the first postulate, if the observer S sees a **free** particle, i.e. a particle with no forces acting on it, travelling in a straight line with constant velocity, then so will S'. Thus, using vector notation, it follows that, under a transformation connecting the two frames,

$$\boldsymbol{r} = \boldsymbol{r}_0 + \boldsymbol{u}t \iff \boldsymbol{r}' = \boldsymbol{r}'_0 + \boldsymbol{u}'t'.$$

Since straight lines get mapped into straight lines, it suggests that the transformation between the frames is **linear** and so we shall assume that the transformation from S to S' can be written in matrix form

$$\begin{bmatrix} t' \\ x' \\ y' \\ z' \end{bmatrix} = L \begin{bmatrix} t \\ x \\ y \\ z \end{bmatrix}, \tag{3.1}$$

where L is a 4×4 matrix of quantities which can only depend on the speed of separation v. Using exactly the same argument as we used at the end of §2.13, the assumption that space is isotropic leads to the transformations of y and z being

$$y' = y \quad \text{and} \quad z' = z. \tag{3.2}$$

We next use the second postulate. Let us assume that, when the origins of S and S' are coincident, they zero their clocks, i.e. $t = t' = 0$, and emit a flash of light. Then, according to S, the light flash moves out radially from the origin with speed c. The wave front of light will constitute a sphere. If we define the quantity I by

$$I(t, \ x, \ y, \ z) = x^2 + y^2 + z^2 - c^2 \, t^2,$$

Introducing Einstein's Relativity. Second Edition. Ray d'Inverno and James Vickers, Oxford University Press.
© Ray d'Inverno and James Vickers (2022). DOI: 10.1093/oso/9780198862024.003.0003

then the events comprising this sphere must satisfy $I = 0$. By the second postulate, S' must also see the light move out in a spherical wave front with speed c and satisfy

$$I' = x'^2 + y'^2 + z'^2 - c^2 t'^2 = 0.$$

Thus it follows that, under a transformation connecting S and S',

$$I = 0 \quad \Longleftrightarrow \quad I' = 0, \tag{3.3}$$

and, since the transformation is linear by (3.1), we may conclude

$$I = nI', \tag{3.4}$$

where n is a quantity which can only depend on v. Using the same argument as we did in §2.13, we can reverse the role of S and S' and so, by the relativity principle, we must also have

$$I' = nI. \tag{3.5}$$

Combining the last two equations, we find

$$n^2 = 1 \quad \Rightarrow \quad n = \pm 1.$$

In the limit as $v \to 0$, the two frames coincide and $I' \to I$, from which we conclude that we must take $n = 1$.

Substituting $n = 1$ in (3.4), this becomes

$$x^2 + y^2 + z^2 - c^2 t^2 = x'^2 + y'^2 + z'^2 - c^2 t'^2,$$

and, using (3.2), this reduces to

$$x^2 - c^2 t^2 = x'^2 - c^2 t'^2,$$

or, in relativistic units with $c = 1$,

$$x^2 - t^2 = x'^2 - t'^2. \tag{3.6}$$

In the same way that two points (x, y) and (x', y') on a circle are related by rotations (so that $x' = x \cos \theta + y \sin \theta$, and $y' = -x \sin \theta + y \cos \theta$), two points on the hyperbola $x^2 - t^2 = $ constant are related by **hyperbolic rotations** (Fig 3.1) so that

$$x' = x \cosh \alpha - t \sinh \alpha, \tag{3.7}$$

$$t' = -x \sinh \alpha + t \cosh \alpha. \tag{3.8}$$

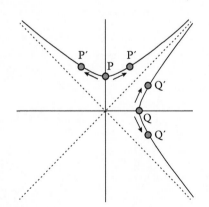

Fig. 3.1 A hyperbolic rotation for points on $t^2 - x^2 = 1$.

Indeed, using (3.7) and (3.8), one can verify that (3.6) is satisfied (exercise). Now, the origin of S' ($x' = 0$), as seen by S, moves along the positive x-axis of S with speed v and so $x' = 0$ must correspond to $x = vt$. Substituting into (3.7), we see

$$0 = vt \cosh \alpha - t \sinh \alpha,$$

so that

$$\tanh \alpha = v. \qquad (3.9)$$

Using the identity $\cosh^2 \alpha - \sinh^2 \alpha = 1$ and (3.9), we obtain

$$\cosh \alpha = \frac{1}{(1 - \tanh^2 \alpha)^{1/2}} = \frac{1}{(1 - v^2)^{1/2}},$$

$$\sinh \alpha = \tanh \alpha \cosh \alpha = \frac{v}{(1 - v^2)^{1/2}}.$$

Substituting in (3.7) and (3.8) gives

$$x' = \frac{(x - vt)}{(1 - v^2)^{1/2}},$$

$$t' = \frac{(t - vx)}{(1 - v^2)^{1/2}},$$

as we found in §2.13.

Writing the above equation in non-relativistic units by inserting factors of c to give the variables the correct dimensions, we obtain the formula for a special Lorentz transformation or boost:

$$t' = \beta(t - vx/c^2), \quad x' = \beta(x - vt), \quad y' = y, \quad z' = z. \qquad (3.10)$$

where we have introduced the standard quantity β given by

$$\beta(v) := \frac{1}{(1 - v^2/c^2)^{\frac{1}{2}}},$$

and the symbol := here means 'is defined to be'.

3.2 Mathematical properties of Lorentz transformations

From the results of the last section, we find the following properties of a special Lorentz transformation or boost.

1. A boost along the x-axis of speed v is equivalent to a hyperbolic rotation in (x, t)-space through an amount α (called the **rapidity** in some textbooks) given by $\tanh \alpha = v/c$.

2. If we consider v to be very small compared with c, for which we use the notation $v \ll c$, and neglect terms of order v^2/c^2, then we regain a Galilean transformation

$$t' = t, \quad x' = x - vt, \quad y' = y, \quad z' = z.$$

We can obtain this result formally by taking the limit $c \to \infty$ in (3.10).

3. If we solve (3.10) for the unprimed coordinates, we get

$$t = \beta(t' + vx'/c^2), \quad x = \beta(x' + vt'), \quad y = y', \quad z = z'.$$

This can also be obtained formally from (3.10) by interchanging primed and unprimed coordinates and replacing v by $-v$. This is what we should expect from physical reasons, since, if S' moves along the positive x-axis of S with speed v, then S moves along the negative x'-axis of S' with speed v or, equivalently, S moves along the positive x'-axis of S' with speed $-v$.

4. Special Lorentz transformations form a **group**:

(a) The identity element is given by $v = 0$.

(b) The inverse element is given by $-v$ (as in 3 above).

(c) The product of two boosts with velocities v and v' is another boost with velocity v''. Since boosts with velocities v and v' correspond to hyperbolic rotations in (x, t)-space with rapidities α and α', where

$$\tanh \alpha = v/c \quad \text{and} \quad \tanh \alpha' = v'/c,$$

then their resultant is a hyperbolic rotation of $\alpha'' = \alpha + \alpha'$, where

$$v''/c = \tanh \alpha'' = \tanh(\alpha + \alpha') = \frac{\tanh \alpha + \tanh \alpha'}{1 + \tanh \alpha \tanh \alpha'},$$

from which we immediately obtain

$$v'' = \frac{v + v'}{1 + vv'/c^2}. \tag{3.11}$$

Compare this with equation (2.6) in relativistic units.

(d) Associativity is left as an exercise.

5. The square of the infinitesimal interval between infinitesimally separated events (see (2.13)),

$$ds^2 = c^2 dt^2 - dx^2 - dy^2 - dz^2, \tag{3.12}$$

is invariant under a Lorentz transformation.

We now turn to the key physical attributes of Lorentz transformations. Throughout the remaining sections, we shall assume that S and S' are in standard configuration with non-zero relative velocity v.

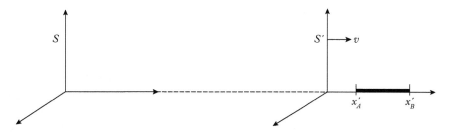

Fig. 3.2 A rod moving with velocity v relative to S.

3.3 Length contraction

Consider a rod fixed in S' with endpoints x'_A and x'_B, as shown in Fig. 3.2. In S, the ends have coordinates x_A and x_B (which, of course, vary in time) given by the Lorentz transformations

$$x'_A = \beta(x_A - vt_A), \quad x'_B = \beta(x_B - vt_B). \tag{3.13}$$

In order to measure the lengths of the rod according to S, we have to find the x-coordinates of the end points at the same time according to S. If we denote the **rest length**, namely the length in S', by

$$\ell_0 = x'_B - x'_A,$$

and the length in S at time $t = t_A = t_B$ by

$$\ell = x_B - x_A.$$

Then, subtracting the formulae in (3.13), we find the result

$$\ell = \beta^{-1} \ell_0. \tag{3.14}$$

Since

$$|v| < c \iff \beta > 1 \iff \ell < \ell_0,$$

the result shows that the length of a body in the direction of its motion with uniform velocity v is **reduced by a factor** $(1 - v^2/c^2)^{1/2}$. This phenomenon is called **length contraction**. Clearly, the body will have greatest length in its rest frame, in which case it is called the rest length, or **proper length**. Note also that the length approaches zero as the velocity approaches the velocity of light.

In an attempt to explain the null result of the Michelson–Morley experiment, Fitzgerald had suggested the shortening of a body in motion relative to the ether. He speculated that the intermolecular forces are possibly of electrical origin so that material bodies would contract in the line of motion. These ideas were subsequently developed by Lorentz and Poincaré using various modifications to the electromagnetic forces. Einstein was the first to completely remove the ad hoc character from the contraction hypothesis, by demonstrating that this contraction did not require motion through a supposed ether, but could be explained using special relativity, which changed our notions of space, time, and simultaneity. Unlike the Fitzgerald contraction, the special relativistic effect is **relative**, i.e. a rod fixed in S appears contracted in S' and, since the space-time interval $c^2t^2 - x^2$ remains unchanged, is better regarded as a change of perspective in Minkowski space-time. Note also that there are no contraction effects in directions transverse to the direction of motion.

3.4   Time dilation

Let a clock fixed at $x' = x'_A$ in S' record two successive events separated by an interval of time T_0 (Fig. 3.3). The successive events in S' are (x'_A, t'_1) and $(x'_A, t'_1 + T_0)$, say. Using the Lorentz transformation, we have in S

$$t_1 = \beta(t'_1 + vx'_A/c^2), \quad t_2 = \beta(t'_1 + T_0 + vx'_A/c^2).$$

On subtracting, we find the time interval in S defined by

$$T = t_2 - t_1,$$

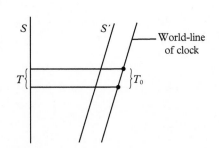

Fig. 3.3 Successive events recorded by a clock fixed in S'.

is given by

$$T = \beta T_0. \tag{3.15}$$

Thus, **moving clocks go slow by a factor** $(1 - v^2/c^2)^{-1/2}$. This phenomenon is called **time dilation**. The fastest rate of a clock is in its rest frame and is called its **proper rate**. Again, the effect has a reciprocal nature.

Let us now consider an accelerated clock. We define an **ideal clock** to be one unaffected by its acceleration; in other words, its instantaneous rate depends only on its instantaneous speed v, in accordance with the above phenomenon of time dilation. This is often referred to as the **clock hypothesis**. The time recorded by an ideal clock is called the **proper time** τ (Fig. 3.4). If at time t the clock is moving with speed $v(t)$, then the infinitesimal version of (3.15) is

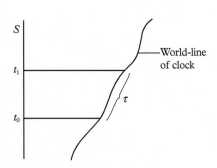

Fig. 3.4 Proper time recorded by an accelerated clock.

$$d\tau = \left(1 - \frac{v(t)^2}{c^2}\right)^{1/2} dt.$$

So that just as in vector calculus, by approximating the world-line of the clock by a number of short straight lines and taking the limit, the proper time of an ideal clock between t_0 and t_1 is given by

$$\tau = \int_{t_0}^{t_1} \left(1 - \frac{v(t)^2}{c^2}\right)^{1/2} dt. \tag{3.16}$$

The general question of what constitutes a clock or an ideal clock is a non-trivial one. However, an experiment has been performed where an atomic clock was flown round the world and then compared with an identical clock left back on the ground. The travelling clock was found on return to be running slow by precisely the amount predicted by relativity. Another instance occurs in the study of cosmic rays. Certain mesons reaching us from the top of the Earth's atmosphere are so short-lived that, even had they been travelling at the speed of light, their travel time in the absence of time dilation would exceed their known proper lifetimes by factors of the order of 10. However, these particles are, in fact, detected at the Earth's surface because their very high velocities, relative to observers on the Earth, keep them young, as it were. Of course, whether or not time dilation affects the human clock, that is, biological ageing, is still an open question. But the fact that we are ultimately made up of atoms, which do appear to suffer time dilation, would suggest that there is no reason by which we should be an exception.

3.5 Transformation of velocities

Consider a particle in motion (Fig. 3.5) with its Cartesian components of velocity being

$$(u_1, u_2, u_3) = \left(\frac{\mathrm{d}x}{\mathrm{d}t}, \frac{\mathrm{d}y}{\mathrm{d}t}, \frac{\mathrm{d}z}{\mathrm{d}t} \right) \text{ in } S,$$

and

$$(u_1', u_2', u_3') = \left(\frac{\mathrm{d}x'}{\mathrm{d}t'}, \frac{\mathrm{d}y'}{\mathrm{d}t'}, \frac{\mathrm{d}z'}{\mathrm{d}t'} \right) \text{ in } S'.$$

Taking differentials of a Lorentz transformation

$$t' = \beta(t - vx/c^2), \quad x' = \beta(x - vt), \quad y' = y, \quad z' = z,$$

we get

$$\mathrm{d}t' = \beta(\mathrm{d}t - v\mathrm{d}x/c^2), \quad \mathrm{d}x' = \beta(\mathrm{d}x - v\mathrm{d}t), \quad \mathrm{d}y' = \mathrm{d}y, \quad \mathrm{d}z' = \mathrm{d}z,$$

and hence

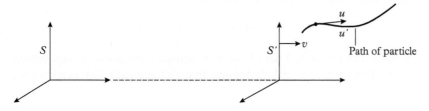

Fig. 3.5 Particle in motion relative to S and S'.

$$u_1' = \frac{\mathrm{d}x'}{\mathrm{d}t'} = \frac{\beta(\mathrm{d}x - v\,\mathrm{d}t)}{\beta(\mathrm{d}t - v\,\mathrm{d}x/c^2)} = \frac{\mathrm{d}x/\mathrm{d}t - v}{1 - v/c^2\,(\mathrm{d}x/\mathrm{d}t)} = \frac{u_1 - v}{1 - u_1 v/c^2}, \quad (3.17)$$

$$u_2' = \frac{\mathrm{d}y'}{\mathrm{d}t'} = \frac{\mathrm{d}y}{\beta(\mathrm{d}t - v\,\mathrm{d}x/c^2)} = \frac{\mathrm{d}y/\mathrm{d}t}{\beta\left[1 - v/c^2\,(\mathrm{d}x/\mathrm{d}t)\right]} = \frac{u_2}{\beta\left(1 - u_1 v/c^2\right)},$$
$$(3.18)$$

$$u_3' = \frac{\mathrm{d}z'}{\mathrm{d}t'} = \frac{\mathrm{d}z}{\beta(\mathrm{d}t - v\,\mathrm{d}x/c^2)} = \frac{\mathrm{d}z/\mathrm{d}t}{\beta\left[1 - v/c^2\,(\mathrm{d}x/\mathrm{d}t)\right]} = \frac{u_3}{\beta\left(1 - u_1 v/c^2\right)}.$$
$$(3.19)$$

Notice that the velocity components u_2 and u_3 transverse to the direction of motion of the frame S' are affected by the transformation. This is due to the time difference in the two frames. To obtain the inverse transformations, simply interchange primes and unprimes and replace v by $-v$.

3.6   Relationship between space-time diagrams of inertial observers

We now show how to relate the space-time diagrams of S and S' (see Fig. 3.6). We start by taking ct and x as the coordinate axes of S, so that a light ray has slope $45°$ (in relativistic units). Then, to draw the ct'- and x'-axes of S', we note from the Lorentz transformation equations (3.10)

$$ct' = 0 \quad \Longleftrightarrow \quad ct = (v/c)x,$$

that is, the x'-axis, given by $ct' = 0$, is the straight line $ct = (v/c)x$ with slope $v/c < 1$. Similarly,

$$x' = 0 \quad \Leftrightarrow \quad ct = (c/v)x,$$

that is, the ct'-axis, given by $x' = 0$, is the straight line $ct = (c/v)x$ with slope $c/v > 1$. The lines parallel to Oct' are the world-lines of fixed points in S. The lines parallel to Ox' are the lines connecting points at a fixed time according to S' and are called **lines of simultaneity in S'**. The coordinates of a general event P are $(ct, x) = (OR, OQ)$ relative to S and $(ct', x') = (OV, OU)$ relative to S'. However, the diagram is somewhat misleading because the length scales along the axes are not the same. To relate them, we draw in the hyperbolae

$$x^2 - c^2 t^2 = x'^2 - c^2 t'^2 = \pm 1$$

as shown in Fig. 3.7. Then, if we first consider the positive sign, setting $ct' = 0$, we get $x' = \pm 1$. It follows that OA is a unit distance on Ox'.

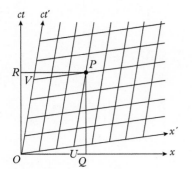

Fig. 3.6 The world-lines in S of the fixed points and simultaneity lines of S'.

Similarly, taking the negative sign and setting $x' = 0$ we get $ct' = \pm 1$ and so OB is the unit measure on Oct'. Then the coordinates of P in the frame S' are given by

$$(ct', x') = \left(\frac{OU}{OA}, \frac{OV}{OB}\right).$$

Note the following properties from Fig. 3.7.

1. A boost can be thought of as a hyperbolic rotation given by (3.7) and (3.8) in the (x, ct) plane through an amount given by the rapidity α. Thus, a boost is equivalent to a skewing of both the coordinate axes inwards through the angle $\tan(v/c)$. (This was not appreciated by some past opponents of special relativity, who gave some erroneous counter arguments based on the mistaken idea that a boost could be represented by a real rotation in the (x, ct)-plane.)

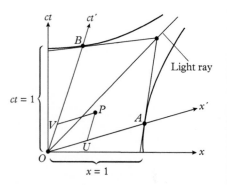

Fig. 3.7 Length scales in S and S'.

2. The hyperbolae are the same for all frames and so we can draw in any number of frames in the same diagram and use the hyperbolae to calibrate them.

3. The length contraction and time dilation effects can be read off directly from the diagram. For example, the world-lines of the end points of a unit rod OA in S', namely $x' = 0$ and $x' = 1$, cut Ox in less than unit distance. Similarly, world-lines $x = 0$ and $x = 1$ in S cut Ox' inside OE, from which the reciprocal nature of length contraction is evident.

4. Event A has coordinates $(ct', x') = (0, 1)$ relative to S' and hence, by a Lorentz transformation, coordinates $(ct, x) = (\beta v/c, \beta)$ relative to S. The quantity OA defined by

$$OA = (c^2 t^2 + x^2)^{1/2} = \beta(1 + v^2/c^2)^{1/2},$$

is a measure of the calibration factor

$$\left(\frac{1 + v^2/c^2}{1 - v^2/c^2}\right)^{1/2},$$

which can be used to compare distance measures in S' with those in S.

3.7   Acceleration in special relativity

We start with the inverse transformation of (3.17), namely

$$u_1 = \frac{u_1' + v}{1 + u_1' v/c^2},$$

from which we find the differential

$$du_1 = \frac{du_1'}{1 + u_1' v/c^2} - \left(\frac{u_1' + v}{(1 + u_1' v/c^2)^2} \right) \frac{v}{c^2} du_1'$$

$$= \frac{1}{\beta^2} \frac{du_1'}{(1 + u_1' v/c^2)^2}.$$

Similarly, from the inverse Lorentz transformation

$$t = \beta(t' + x'v/c^2),$$

we find the differential

$$dt = \beta(dt' + (dx')v/c^2) = \beta(1 + u_1' v/c^2)dt'.$$

Combining these results, we find that the x-component of the acceleration transforms according to

$$\frac{du_1}{dt} = \frac{1}{\beta^3 (1 + u_1' v/c^2)^3} \frac{du_1'}{dt'}. \tag{3.20}$$

Similarly, we find

$$\frac{du_2}{dt} = \frac{1}{\beta^2 (1 + u'_1 v/c^2)^2} \frac{du_2'}{dt'} - \frac{vu'_2}{c^2 \beta^2 (1 + u_1' v/c^2)^3} \frac{du_1'}{dt'}, \tag{3.21}$$

$$\frac{du_3}{dt} = \frac{1}{\beta^2 (1 + u_1' v/c^2)^2} \frac{du_3'}{dt'} - \frac{vu_3'}{c^2 \beta^2 (1 + u'_1 v/c^2)^3} \frac{du_1'}{dt'}. \tag{3.22}$$

The inverse transformations can be found in the usual way.

It follows from the above formulae that acceleration does not transform in the expected way under a Lorentz transformation, so does not correspond to a vector in Minkowski space. However, it is clear from the formulae that the existence or not of acceleration is an **absolute** quantity, that is, all inertial observers agree whether a body is accelerating or not. Put another way, if the acceleration is zero in one frame, then it is necessarily zero in any other frame. We shall see that this is no longer the case in general relativity. We summarize the situation in Table 3.1, which indicates why the subject matter of the book is 'relativity' theory.

3.8   Uniform acceleration

The Newtonian definition of a particle moving under uniform acceleration is

$$\frac{du}{dt} = \text{constant}.$$

Table 3.1

Theory	Position	Velocity	Time	Acceleration
Newtonian	Relative	Relative	Absolute	Absolute
Special relativity	Relative	Relative	Relative	Absolute
General relativity	Relative	Relative	Relative	Relative

This turns out to be inappropriate in special relativity since it would imply that $u \to \infty$ as $t \to \infty$, which we know is impossible. We therefore adopt a different definition. Acceleration is said to be **uniform** in special relativity if it has the same value in any instantaneously **co-moving frame**, that is, at each instant, the acceleration in an inertial frame travelling with the same velocity as the particle has the same value. This is analogous to the idea in Newtonian theory of motion under a constant force. For example, a spaceship whose motor is set at a constant emission rate would be uniformly accelerated in this sense. Taking the velocity of the particle to be $u = u(t)$ relative to an inertial frame S, then at any instant in a co-moving frame S' it follows that the velocity relative to S' is 0, that is, $u' = 0$, $v = u$ and the acceleration is a constant, a, say, i.e. $\mathrm{d}u'/\mathrm{d}t' = a$. Using (3.20), we find

$$\frac{\mathrm{d}u}{\mathrm{d}t} = \frac{1}{\beta^3}a = \left(1 - \frac{u^2}{c^2}\right)^{3/2}a.$$

We can solve this differential equation by separating the variables

$$\frac{\mathrm{d}u}{(1 - u^2/c^2)^{3/2}} = a\mathrm{d}t$$

and integrating both sides. Assuming that the particle starts from rest at $t = t_0$, we find

$$\frac{u}{(1 - u^2/c^2)^{1/2}} = a(t - t_0).$$

Solving for u, we get

$$u = \frac{\mathrm{d}x}{\mathrm{d}t} = \frac{a(t - t_0)}{[1 + a^2(t - t_0)^2/c^2]^{1/2}}.$$

Next, integrating with respect to t, and setting $x = x_0$ at $t = t_0$, produces

$$(x - x_0) = \frac{c}{a}[c^2 + a^2(t - t_0)^2]^{1/2} - \frac{c^2}{a}.$$

This can be rewritten in the form

$$\frac{\left(x - x_0 + c^2/a\right)^2}{\left(c^2/a\right)^2} - \frac{(ct - ct_0)^2}{\left(c^2/a\right)^2} = 1, \tag{3.23}$$

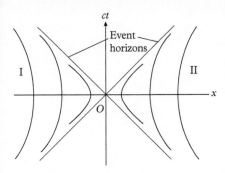

Fig. 3.8 Hyperbolic motions.

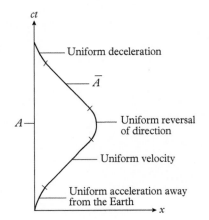

Fig. 3.9 The twin paradox.

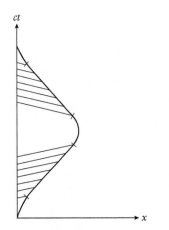

Fig. 3.10 Simultaneity lines of $\bar{A}$ on the outward and return journeys.

which is a hyperbola in (x, ct)-space. If, in particular, we take $x_0 - c^2/a = t_0 = 0$, then we obtain a family of hyperbolae for different values of a (Fig. 3.8). These world-lines are known as **hyperbolic motions** and, as we shall see in Chapter 25, they have significance in cosmology. It can be shown that the radar distance between the world-lines is a constant. Moreover, consider the regions I and II bounded by the light rays passing through O, and a system of particles undergoing hyperbolic motions as shown in Fig. 3.8 (in some cosmological models, the particles would be galaxies). Then, remembering that light rays emanating from any point in the diagram do so at 45°, no particle in region I can communicate with another particle in region II, and vice versa. The light rays are called **event horizons** and act as barriers beyond which no knowledge can ever be gained. We shall see that event horizons will play an important role later in this book.

3.9 The twin paradox

The twin paradox is a form of the clock paradox which has caused the most controversy – a controversy which raged on and off for over fifty years. The paradox concerns two twins whom we shall call A and $\bar{A}$. The twin $\bar{A}$ takes off in a spaceship for a return trip to some distant star. The assumption is that $\bar{A}$ is uniformly accelerated to some given velocity which is retained until the star is reached, whereupon the motion is uniformly reversed, as shown in Fig. 3.9. According to A, $\bar{A}$'s clock records slowly on the outward and return journeys and so, on return, $\bar{A}$ will be younger than A. If the periods of acceleration are negligible compared with the periods of uniform velocity, then could not $\bar{A}$ reverse the argument and conclude that it is A who should appear to be the younger? This is the basis of the paradox.

The resolution rests on the fact that the accelerations, however brief, have immediate and finite effects on $\bar{A}$ but not on A, who remains inertial throughout. One striking way of seeing this effect is to draw in the simultaneity lines of $\bar{A}$ for the periods of uniform velocity, as in Fig. 3.10. Clearly, the period of uniform reversal has a marked effect on the simultaneity lines. Another way of looking at it is to see the effect that the periods of acceleration have on shortening the length of the journey as viewed by $\bar{A}$. Let us be specific: we assume that the periods of acceleration are T_1, T_2, and T_3, and that, after the period T_1, $\bar{A}$ has attained a speed $v = \sqrt{3}c/2$. Then, from A's viewpoint, during the period T_1, $\bar{A}$ finds that more than half the outward journey has been accomplished, in that $\bar{A}$ has transferred to a frame in which the distance between the Earth and the star is more than halved by length contraction. Thus, $\bar{A}$ accomplishes the outward trip in about half the time which A ascribes to it, and the same applies to the return trip. In fact, we could use the machinery of previous sections to calculate the time elapsed in both the periods of uniform acceleration and uniform velocity, and we would again reach the conclusion that on return $\bar{A}$ will be younger than A. As we have said before, this points out the fact that, in special relativity, time is a route-dependent quantity.

The fact that in Fig. 3.9 $\bar{A}$'s world-line is longer than A's, and yet takes **less** time to travel, can appear at first counterintuitive. However, this can be shown to be a consequence of the fact that the usual three-dimensional Euclidean metric appears with **negative** signs in the Minkowski metric

$$ds^2 = c^2 dt^2 - dx^2 - dy^2 - dz^2,$$

which means that moving in space **reduces** the space-time length s.

3.10 The Doppler effect

All kinds of waves appear lengthened when the source recedes from the observer: sounds are deepened, light is reddened. Exactly the opposite occurs when the source, instead, approaches the observer. We first of all calculate the **classical** (non-relativistic) Doppler effect.

Consider a source of light emitting radiation whose wavelength in its rest frame is λ_0. Consider an observer S relative to whose frame the source is in motion with radial velocity u_r. Then, if two successive pulses are emitted at times differing by dt' as measured by S', the distance these pulses have to travel will differ by an amount $u_r dt'$ (see Fig. 3.11). Since the pulses travel with speed c, it follows that they arrive at S with a time difference

$$\Delta t = dt' + u_r dt'/c,$$

giving

$$\Delta t/dt' = 1 + u_r/c.$$

Now, using the fundamental relationship between wavelength and velocity, set

$$\lambda = c\Delta t \quad \text{and} \quad \lambda_0 = cdt'.$$

We then obtain the **classical Doppler formula**

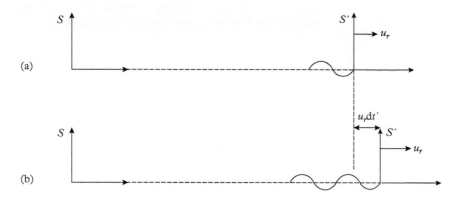

Fig. 3.11 The Doppler effect: (a) first pulse; (b) second pulse.

$$\lambda/\lambda_0 = 1 + u_{\mathrm{r}}/c. \tag{3.24}$$

Let us now consider the special relativistic formula. Because of time dilation (see Fig. 3.3), the time interval between successive pulses according to S is $\beta \mathrm{d}t'$ (Fig. 3.12). Hence, by the same argument, the pulses arrive at S with a time difference

$$\Delta t = \beta \mathrm{d}t' + u_r \beta \mathrm{d}t'/c,$$

and so this time we find that the **special relativistic Doppler formula** is

$$\frac{\lambda}{\lambda_0} = \frac{1 + u_{\mathrm{r}}/c}{\left(1 - v^2/c^2\right)^{1/2}}. \tag{3.25}$$

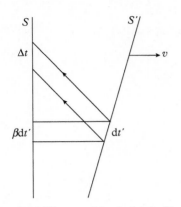

Fig. 3.12 The special relativistic Doppler shift.

If the velocity of the source is purely radial, then $u_r = v$ and (3.25) reduces to

$$\frac{\lambda}{\lambda_0} = \left(\frac{1 + v/c}{1 - v/c}\right)^{1/2}. \tag{3.26}$$

This is the **radial Doppler shift**, and, if we set $c = 1$, we obtain (2.4), which is the formula for the **k-factor**. Combining Figs. 2.7 and 3.12, the radial Doppler shift is illustrated in Fig. 3.13, where $\mathrm{d}t'$ is replaced by T. From (3.25), we see that there is also a change in wavelength, even when the radial velocity of the source is zero. For example, if the source is moving in a circle about the origin of S with speed v (as measured by an instantaneous co-moving frame), then the **transverse Doppler shift** is given by

$$\frac{\lambda}{\lambda_0} = \frac{1}{\left(1 - v^2/c^2\right)^{1/2}}. \tag{3.27}$$

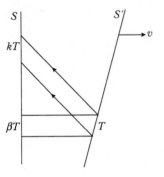

Fig. 3.13 The radial Doppler shift k.

This is a purely relativistic effect due to the time dilation of the moving source. Experiments with revolving apparatus using the so-called Mössbauer effect have directly confirmed the transverse Doppler shift in full agreement with the relativistic formula, thus providing another striking verification of the phenomenon of time dilation.

Exercises

3.1 (§3.1) Verify that, if $x' = x \cosh \alpha - t \sinh \alpha$, and $t' = -x \sinh \alpha + t \cosh \alpha$, then

$$x^2 - t^2 = x'^2 - t'^2.$$

3.2 (§3.1) S and S' are in standard configuration with $v = \alpha c \, (0 < \alpha < 1)$. If a rod at rest in S' makes an angle of $45°$ with Ox in S, and $30°$ with $O'x$ in S', then find α.

3.3 (§3.1) Note from the previous question that perpendicular lines in one frame need not be perpendicular in another frame. This shows that there is no obvious meaning to the phrase 'two inertial frames are parallel', unless their relative velocity is along a common axis, because the axes of either frame need not appear rectangular in the other. Verify that the Lorentz transformation between frames in standard configuration with relative velocity $\boldsymbol{v} = (v, 0, 0)$ may be written in vector form as

$$\boldsymbol{r}' = \boldsymbol{r} + \left(\frac{\boldsymbol{v \cdot r}}{v^2}(\beta - 1) - \beta t \right) \boldsymbol{v}, \qquad t' = \beta \left(t - \frac{\boldsymbol{v \cdot r}}{c^2} \right),$$

where $\boldsymbol{r} = (x, y, z)$. The formulae are said to comprise the 'Lorentz transformation without relative rotation'. Justify this name by showing that the formulae remain valid when the frames are not in standard configuration, but are parallel in the sense that the same rotation must be applied to each frame to bring the two into standard configuration (in which case, v is the velocity of S' relative to S, but $\boldsymbol{v} = (v, 0, 0)$ no longer applies).

3.4 (§3.1) Aberration refers to the fact that the direction of travel of a light ray depends on the motion of the observer. Hence, if a telescope observes a star at an inclination θ' to the horizontal, then show that, **classically**, the 'true' inclination θ of the star is related to θ' by

$$\tan \theta' = \frac{\sin \theta}{\cos \theta + v/c},$$

where v is the velocity of the telescope relative to the star. Show that the corresponding relativistic formula is

$$\tan \theta' = \frac{\sin \theta}{\beta(\cos \theta + v/c)}.$$

3.5 (§3.2) Show that special Lorentz transformations are associative, that is, if $O(v_1)$ represents the transformation from observer S to S', then

$$(O(v_1)O(v_2))\, O(v_3) = O(v_1)\, (O(v_2)O(v_3))\,.$$

3.6 (§3.3) An athlete carrying a horizontal 8 m-long pole runs at a speed v such that $(1 - v^2/c^2)^{-1/2} = 2$ into a 4 m-long room and closes the door. Explain, in the athlete's frame, in which the room is only 2 m long, how this is possible. [Hint: no effect travels faster than light.] Show that the minimum length of the room for the performance of this trick is $8/(\sqrt{3} + 2)$m. Draw a space-time diagram to indicate what is going on in the rest frame of the athlete. [Hint: You may find it helpful to look up the "pole in the barn paradox" on the web for a detailed discussion of this problem.]

3.7 (§3.5) A particle has velocity $\boldsymbol{u} = (u_1, u_2, u_3)$ in S and $\boldsymbol{u}' = (u_1', u_2', u_3')$ in S'. Prove from the velocity transformation formulae that

$$c^2 - u^2 = \frac{c^2(c^2 - u'^2)(c^2 - v^2)}{(c^2 + u_1'v)^2}\,.$$

Deduce that, if the speed of a particle is less than c in any one inertial frame, then it is less than c in every inertial frame.

3.8 (§3.7) Check the transformation formulae for the components of acceleration (3.20)–(3.22).
Deduce that acceleration is an absolute quantity in special relativity.

3.9 (§3.8) A particle moves from rest at the origin of a frame S along the x-axis, with constant acceleration α (as measured in an instantaneous rest frame). Show that the equation of motion is

$$\alpha x^2 + 2c^2 x - \alpha c^2 t^2 = 0,$$

and prove that the light signals emitted after time $t = c/\alpha$ at the origin will never reach the receding particle. A standard clock carried along with the particle is set to read 0 at the beginning of the motion and reads τ at time t in S. Using the clock hypothesis, prove the following relationships:

$$\frac{u}{c} = \tanh\frac{\alpha\tau}{c}, \qquad\qquad \left(1 - \frac{u^2}{c^2}\right)^{-1/2} = \cosh\frac{\alpha\tau}{c},$$

$$\frac{\alpha t}{c} = \sinh\frac{\alpha\tau}{c}, \qquad\qquad x = \frac{c^2}{\alpha}\left(\cosh\frac{\alpha\tau}{c} - 1\right),$$

where u is the speed of the particle. Show that, if $T^2 \ll c^2/\alpha^2$, then, during an elapsed time T in the inertial system, the particle clock will record approximately the time $T(1 - \alpha^2 T^2/6c^2)$.

If $\alpha = 3g$, find the difference in recorded times by the spaceship clock and those of the inertial system

(a) after 1 hour;
(b) after 10 days.

3.10 (§3.9) A space traveller $\bar{A}$ travels through space with uniform acceleration g (to ensure maximum comfort). Find the distance covered in twenty-two years of $\bar{A}$'s time. [Hint: using years and light years are used as time and distance units, respectively, then $g = 1.03$]. If on the other hand, $\bar{A}$ describes a straight double path $XYZYX$, with acceleration g on XY and ZY, and deceleration g on YZ and YX, for six years each, then draw a space-time diagram as seen from the Earth and find by how much the Earth would have aged in twenty-four years of $\bar{A}$'s time.

3.11 (§3.10) Let the relative velocity between a source of light and an observer be u, and establish the **classical** Doppler formulae for the frequency shift:

$$\text{source moving, observer at rest:} v = \frac{\nu_0}{1 + u/c},$$

$$\text{observer moving, source at rest:} v = (1 - u/c)\nu_0,$$

where ν_0 is the frequency in the rest frame of the source. What are the corresponding relativistic results?

3.12 (§3.10) How fast would you need to drive towards a red traffic light for the light to appear green? [Hint: $\lambda_{\text{red}} \simeq 7 \times 10^{-7}$m, $\lambda_{\text{green}} \simeq 5 \times 10^{-7}$m.]

Further reading

There are many fine texts around on special relativity. One is a book by Rindler (1982). Another excellent book is written by a Southampton ex-colleague, Les Marder (1968).

Marder, L. (1968). *An Introduction to Relativity*. Longman, London.

Rindler, W. (1982). *Introduction to Special Relativity*. Oxford University Press, Oxford.

The elements of relativistic mechanics

4

4.1 Newtonian theory

Before discussing relativistic mechanics, we shall review some basic ideas of Newtonian theory. We have met Newton's first law in §2.4, and it states that a body not acted upon by a force moves in a straight line with uniform velocity. The second law describes what happens if an object changes its velocity. In this case, something is causing it to change its velocity and this something is called a **force**. For the moment, let us think of a force as something tangible like a push or a pull. Now, we know from experience that it is more difficult to push a more massive body and get it moving than it is to push a less massive body. This resistance of a body to motion, or rather change in motion, is called its **inertia**. To every body, we can ascribe, at least at one particular time, a number measuring its inertia, which (again, for the moment) we shall call its **mass** m. If a body is moving with velocity $\boldsymbol{v}$, we define its **linear momentum** $\boldsymbol{p}$ to be the product of its mass and velocity. Then Newton's second law (N2) states that the force acting on a body is equal to the rate of change of linear momentum. The third law (N3) is less general and talks about a restricted class of forces called **internal** forces, namely, forces acting on a body due to the influence of other bodies in a system. The third law states that the force acting on a body due to the influence of the other bodies, the so-called **action**, is equal and opposite to the force acting on these other bodies due to the influence of the first body, the so-called **reaction**. We state the two laws below.

N2: The rate of change of momentum of a body is equal to the force acting on it, and is in the direction of the force.

N3: To every action there is an equal and opposite reaction.

Then, for a body of mass m with a force $\boldsymbol{F}$ acting on it, Newton's second law states

$$\boldsymbol{F} = \frac{\mathrm{d}\boldsymbol{p}}{\mathrm{d}t} = \frac{\mathrm{d}(m\boldsymbol{v})}{\mathrm{d}t}. \tag{4.1}$$

Introducing Einstein's Relativity. Second Edition. Ray d'Inverno and James Vickers, Oxford University Press.
© Ray d'Inverno and James Vickers (2022). DOI: 10.1093/oso/9780198862024.003.0004

If, in particular, the mass is a constant, then we obtain the well-known formula

$$\boldsymbol{F} = m\frac{d\boldsymbol{v}}{dt} = m\boldsymbol{a}, \tag{4.2}$$

where $\boldsymbol{a}$ is the acceleration.

Now, strictly speaking, in Newtonian theory, all observable quantities should be defined in terms of their measurement. We have seen how an observer equipped with a frame of reference, a ruler, and a clock can map the events of the universe, and hence measure such quantities as position, velocity, and acceleration. However, Newton's laws introduce the new concepts of force and mass, and so we should give a prescription for their measurement. Unfortunately, any experiment designed to measure these quantities involves Newton's laws themselves in its interpretation. Thus, Newtonian mechanics has the rather unexpected property that the operational definitions of force and mass which are required to make the laws physically significant are actually contained in the laws themselves.

To make this more precise, let us discuss how we might use the laws to measure the mass of a body. We consider two bodies isolated from all other influences other, than the force acting on one due to the influence of the other, and vice versa (Fig. 4.1). Since the masses are assumed to be constant, we have, by Newton's second law in the form (4.2),

Fig. 4.1 Measuring mass by mutually induced accelerations.

$$\boldsymbol{F}_1 = m_1\boldsymbol{a}_1 \quad \text{and} \quad \boldsymbol{F}_2 = m_2\boldsymbol{a}_2.$$

In addition, by Newton's third law, $\boldsymbol{F}_1 = -\boldsymbol{F}_2$. Hence, we have

$$m_1\boldsymbol{a}_1 = -m_2\boldsymbol{a}_2. \tag{4.3}$$

Therefore, if we take one standard body and define it to have **unit** mass, then we can find the mass of the other body, by using (4.3). We can keep doing this with any other body and in this way we can calibrate masses. In fact, this method is commonly used for comparing the masses of elementary particles. Of course, in practice, we cannot remove all other influences, but it may be possible to keep them almost constant and so neglect them.

We have described how to use Newton's laws to measure mass. How do we measure force? One approach is simply to use Newton's second law, work out $m\boldsymbol{a}$ for a body, and then read off from the law the force acting on m. This is consistent, although rather circular, especially since a force has independent properties of its own. For example, Newton has provided us with a way for working out the force in the case of gravitation in his **universal law of gravitation** (UG).

UG: Two particles attract each other with a force directly proportional to their masses and inversely proportional to the distance between them.

If we denote the constant of proportionality by G (with value approximately $6.67 \times 10^{-11}\,\text{m}^3\text{Kg}^{-1}\text{s}^{-2}$ in SI units), the so-called Newtonian constant, then the law is (see Fig. 4.2)

$$F = -G\frac{m_1 m_2}{r^2}\hat{r}, \qquad (4.4)$$

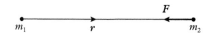

Fig. 4.2 Newton's universal law of gravitation.

where a hat denotes a unit vector. There are other force laws which can be stated separately. Again, another independent property which holds for certain forces is contained in Newton's third law. The standard approach to defining force is to consider it as being fundamental, in which case force laws can be stated separately or they can be worked out from other considerations. We postpone a more detailed critique of Newton's laws until Part C of the book.

Special relativity is concerned with the behaviour of material bodies and light rays **in the absence of gravitation**. So we shall also postpone a detailed consideration of gravitation until we discuss general relativity in Part C of the book. However, since we have stated Newton's universal laws of gravitation in (4.4), we should, for completeness, include a statement of Newtonian gravitation for a distribution of matter. A distribution of matter of mass density $\rho = \rho(x, y, z, t)$ gives rise to a gravitational potential ϕ which satisfies **Poisson's equation**

$$\nabla^2 \phi = 4\pi G\rho, \qquad (4.5)$$

at points inside the distribution, where the Laplacian operator ∇^2 is given in Cartesian coordinates by

$$\nabla^2 = \frac{\partial^2}{\partial x^2} + \frac{\partial^2}{\partial y^2} + \frac{\partial^2}{\partial z^2}.$$

At points external to the distribution, this reduces to **Laplace's equation**

$$\nabla^2 \phi = 0. \qquad (4.6)$$

We assume that the reader is familiar with this background to Newtonian theory.

4.2 Isolated systems of particles in Newtonian mechanics

In this section, we shall, for completeness, derive the conservation of linear momentum in Newtonian mechanics for a system of n particles. Let the ith particle have constant mass m_i and position vector r_i relative to some arbitrary origin. Then the ith particle possesses linear momentum p_i defined by $p_i = m_i\dot{r}_i$, where the dot denotes differentiation with respect to time t. If F_i is the total force on m_i, then, by Newton's second law, we have

$$F_i = \dot{p}_i = m_i \ddot{r}_i. \tag{4.7}$$

The total force F_i on the ith particle can be divided into an external force F_i^{ext} due to any external fields present and to the resultant of the internal forces. We write

$$F_i = F_i^{\text{ext}} + \sum_{j=1}^{n} F_{ij},$$

where F_{ij} is the force on the ith particle due to the jth particle and where, for convenience, we define $F_{ii} = 0$. If we sum over i in (4.7), we find

$$\frac{\mathrm{d}}{\mathrm{d}t} \sum_{i=1}^{n} p_i = \sum_{i=1}^{n} \frac{\mathrm{d}p_i}{\mathrm{d}t} = \sum_{i=1}^{n} F_i^{\text{ext}} + \sum_{i,j=1}^{n} F_{ij}.$$

Using Newton's third law, namely, $F_{ij} = -F_{ji}$, then the last term is zero and we obtain $\dot{P} = F^{\text{ext}}$, where $P = \sum_{i=1}^{n} P_i$ is termed the **total linear momentum** of the system, and $F^{\text{ext}} = \sum_{i=1}^{n} F_i^{\text{ext}}$ is the **total external force** on the system. If, in particular, the system of particles is **isolated**, then

$$F^{\text{ext}} = 0 \quad \Rightarrow \quad P = C,$$

where C is a constant vector. This leads to the law of the **conservation of linear momentum** of the system, namely,

$$P_{\text{initial}} = P_{\text{final}}. \tag{4.8}$$

4.3 Relativistic mass

The transition from Newtonian to relativistic mechanics is not, in fact, completely straightforward, because it involves at some point or another the introduction of ad hoc assumptions about the behaviour of particles in relativistic situations. We shall adopt the approach of trying to keep as close to the non-relativistic definition of energy and momentum as we can. This leads to results which in the end must be confronted with experiment. The ultimate justification of the formulae we shall derive resides in the fact that they have been repeatedly confirmed in numerous experiments, for example in particle physics. We shall only derive them in a simple case and state that the arguments can be extended to a more general situation.

It would seem plausible that, since length and time measurements are dependent on the observer, then mass should also be an observer-dependent quantity. We thus assume that a particle which is moving with

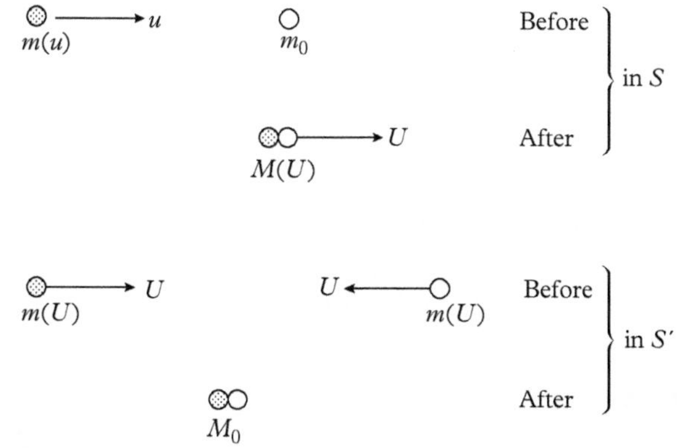

Fig. 4.3 The inelastic collision in the frames S and S'.

a velocity $\boldsymbol{u}$ relative to an inertial observer has a mass, which we shall term its **relativistic mass**, which is some function of $\boldsymbol{u}$, that is,

$$m = m(\boldsymbol{u}), \tag{4.9}$$

where the problem is to find the explicit dependence of m on $\boldsymbol{u}$. We restrict attention to motion along a straight line and consider the special case of two equal particles colliding **inelastically** (in which case they stick together), and look at the collision from the point of view of two inertial observers S and S' (see Fig. 4.3). Let one of the particles be at rest in the frame S and the other possess a velocity u before they collide. We then assume that they coalesce and that the combined object moves with velocity U. The masses of the two particles are respectively $m(0)$ and $m(u)$ by (4.9). We denote $m(0)$ by m_0 and term it the **rest mass** of the particle. In addition, we denote the mass of the combined object by $M(U)$. If we take S' to be the **centre-of-mass frame**, then it should be clear that, relative to S', the two equal particles collide with equal and opposite speeds, leaving the combined object with mass M_0 at rest. It follows that S' must have velocity U relative to S.

We shall assume both conservation of relativistic mass and conservation of linear momentum and see what this leads to. In the frame S, we obtain

$$m(u) + m_0 = M(U), \qquad m(u)u + 0 = M(U)U,$$

from which we get, eliminating $M(U)$,

$$m(u) = m_0 \left(\frac{U}{u - U} \right). \tag{4.10}$$

The left-hand particle has a velocity U relative to S', which in turn has a velocity U relative to S. Hence, using the composition of velocities law, we can compose these two velocities, and the resultant velocity must be identical with the velocity u of the left-hand particle in S. Thus, by (2.6) in non-relativistic units,

$$u = \frac{2U}{(1 + U^2/c^2)}.$$

Solving for U in terms of u, we obtain the quadratic

$$U^2 - \left(\frac{2c^2}{u}\right) U + c^2 = 0,$$

which has roots

$$U = \frac{c^2}{u} \pm \left[\left(\frac{c^2}{u}\right)^2 - c^2\right]^{1/2} = \frac{c^2}{u}\left[1 \pm \left(1 - \frac{u^2}{c^2}\right)^{1/2}\right].$$

In the limit $u \to 0$, this must produce a finite result, so we must take the negative sign (check), and, substituting in (4.10), we find finally

$$m(u) = \gamma m_0, \tag{4.11}$$

where

$$\gamma(u) := (1 - u^2/c^2)^{-1/2}. \tag{4.12}$$

This is the basic result which relates the relativistic mass of a moving particle to its rest mass. Note that this is the same in structure as the time dilation formula (3.15), i.e. $T = \beta T_0$, where $\beta(v) = (1 - v^2/c^2)^{-1/2}$, except that time dilation involves the factor $\beta(v)$, which depends on the velocity v of the frame S' relative to S, whereas $\gamma(u)$ depends on the velocity u of the particle relative to S. If we plot m against u, we see that relativistic mass increases without bound as u approaches c (Fig. 4.4).

It is possible to extend the above argument to establish (4.11) in more general situations. However, we emphasize that it is not possible to derive the result a priori, but only with the help of extra assumptions. However it is produced, the only real test of the validity of the result is in the experimental arena and here it has been extensively confirmed.

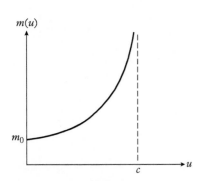

Fig. 4.4 Relativistic mass as a function of velocity.

4.4 Relativistic energy

Let us expand the expression for the relativistic mass, namely,

$$m(u) = \gamma m_0 = m_0(1 - u^2/c^2)^{-1/2},$$

in the case when the velocity u is small compared with the speed of light c. Then we get

$$m(u) = m_0 + \frac{1}{c^2}(\tfrac{1}{2}m_0 u^2) + O\left(\frac{u^4}{c^4}\right), \tag{4.13}$$

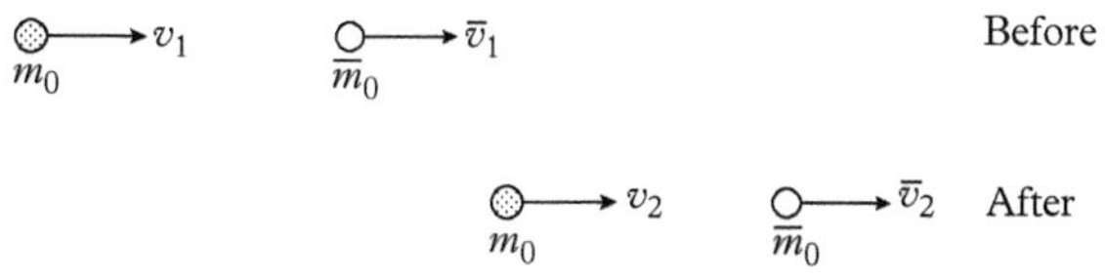

Fig. 4.5 Two colliding particles.

where the final term stands for all terms of order $(u/c)^4$ and higher. If we multiply both sides by c^2, then, apart from the constant $m_0 c^2$, the right-hand side is to first approximation the classical kinetic energy (k.e.), that is,

$$mc^2 = m_0 c^2 + \tfrac{1}{2} m_0 u^2 + \cdots \simeq \text{constant} + \text{k.e.} \tag{4.14}$$

We have seen that relativistic mass contains within it the expression for classical kinetic energy. In fact, it can be shown that the conservation of relativistic mass leads to the conservation of kinetic energy in the Newtonian approximation. As a simple example, consider the collision of two particles with rest mass m_0 and $\bar{m}_0$, initial velocities v_1 and $\bar{v}_1$, and final velocities v_2 and $\bar{v}_2$, respectively (Fig. 4.5).

Conservation of relativistic mass gives

$$m_0\left(1 - v_1^2/c^2\right)^{-1/2} + \bar{m}_0\left(1 - \bar{v}_1^2/c^2\right)^{-1/2} = m_0\left(1 - v_2^2/c^2\right)^{-1/2}$$
$$+ \bar{m}_0\left(1 - \bar{v}_2^2/c^2\right)^{-1/2}. \tag{4.15}$$

If we now assume that v_1, $\bar{v}_1$, v_2, and $\bar{v}_2$ are all small compared with c, then we find (exercise) that the leading terms in the expansion of (4.15) give

$$\tfrac{1}{2} m_0 v_1^2 + \tfrac{1}{2} \bar{m}_0 \bar{v}_1^2 = \tfrac{1}{2} m_0 v_2^2 + \tfrac{1}{2} \bar{m}_0 \bar{v}_2^2, \tag{4.16}$$

which is the usual conservation of energy equation. Thus, in this sense, conservation of relativistic mass includes within it conservation of energy. Now, since energy is only defined up to the addition of a constant, the result (4.14) suggest that we regard the **energy** E of a particle as given by

$$E = mc^2. \tag{4.17}$$

This is one of the most famous equations in physics. However, it is not just a mathematical relationship between two different quantities, namely energy and mass, but rather states that energy and mass are **equivalent** concepts. Because of the arbitrariness in the actual value of E, a better way of stating the relationship is to say that a change in energy is equal to a change in relativistic mass, namely,

$$\Delta E = \Delta(mc^2).$$

Using conventional units, c^2 is a large number and indicates that a small change in mass is equivalent to an enormous change in energy. As is well known, this relationship and the deep implications it carries with it for peace and war, have been amply verified. For obvious reasons, the term m_0c^2 is termed the **rest energy** of the particle. Finally, we point out that conservation of linear momentum, using relativistic mass, leads to the usual conservation law in the Newtonian approximation. For example (exercise), the collision problem considered above leads to the usual conservation of linear momentum equation for slow-moving particles:

$$m_0 v_1 + \bar{m}_0 \bar{v}_1 = m_0 v_2 + \bar{m}_0 \bar{v}_2. \tag{4.18}$$

Extending these ideas to three spatial dimensions, then a particle moving with velocity $\boldsymbol{u}$ relative to an inertial frame S has relativistic mass m, energy E, and linear momentum $\boldsymbol{p}$ given by

$$m = \gamma m_0, \quad E = mc^2, \quad \boldsymbol{p} = m\boldsymbol{u}. \tag{4.19}$$

Some straightforward algebra (exercise) reveals that

$$(E/c)^2 - p_x^2 - p_y^2 - p_z^2 = (m_0 c)^2, \tag{4.20}$$

where $m_0 c^2$ is an invariant, since it is the same for all inertial observers. If we compare this with the invariant (3.12), i.e.

$$(ct)^2 - x^2 - y^2 - z^2 = s^2,$$

then it suggests that the quantities $(E/c, p_x, p_y, p_z)$ transform under a Lorentz transformation in the same way as the quantities (ct, x, y, z). We shall see in Part C that the language of tensors provides a better framework for discussing transformation laws. For the moment, we shall assume that energy and momentum transform in an identical manner, and quote the results. Thus, in a frame S' moving in standard configuration with velocity v relative to S, the transformation equations are (see (3.10))

$$E' = \beta(E - vp_x), \quad p_x' = \beta(p_x - vE/c^2), \quad p_y' = p_y, \quad p_z' = p_z. \tag{4.21}$$

The inverse transformations are obtained in the usual way, namely, by interchanging primes and unprimes and replacing v by $-v$, which gives

$$E = \beta(E' + vp_x'), \quad p_x = \beta(p_x' + vE'/c^2), \quad p_y = p_y', \quad p_z = p_z'. \tag{4.22}$$

If, in particular, we take S' to be the instantaneous rest frame of the particle, then $\boldsymbol{p}' = 0$ and $E' = E_0 = m_0c^2$. Substituting in (4.22), we find

$$E = \beta E' = \frac{m_0 c^2}{(1 - v^2/c^2)^{1/2}} = mc^2,$$

where $m = m_0(1 - v^2/c^2)^{-1/2}$ and $\boldsymbol{p} = \left(\beta v E'/c^2, 0, 0\right) = (mv, 0, 0) = m\boldsymbol{v}$, which are precisely the values of the energy, mass, and momentum arrived at in (4.19) with $\boldsymbol{u}$ replaced by $\boldsymbol{v}$.

4.5   Photons

At the end of the 19th century, there was considerable conflict between theory and experiment in the investigation of radiation in enclosed volumes. In an attempt to resolve the difficulties, Max Planck proposed that light and other electromagnetic radiation consisted of individual 'packets' of energy, which he called **quanta**. He suggested that the energy E of each quantum was to depend on its frequency ν, and proposed the simple law, called **Planck's hypothesis**,

$$E = h\nu, \tag{4.23}$$

where h is a universal constant known now as **Planck's constant**. The idea of the quantum was developed further by Einstein, especially in attempting to explain the photoelectric effect. The effect is to do with the ejection of electrons from a metal surface by incident light (especially ultraviolet) and is strongly in support of Planck's quantum hypothesis. Nowadays, the quantum theory is well established and applications of it to explain properties of molecules, atoms, and fundamental particles are at the heart of modern physics. Theories of light now give it a dual wave–particle nature. Some properties, such as diffraction and interference, are wavelike in nature, while the photoelectric effect and other cases of the interaction of light and atoms are best described on a particle basis.

The particle description of light consists in treating it as a stream of quanta called **photons**. Using equation (4.19) and substituting in the speed of light, $u = c$, we find

$$m_0 = \gamma^{-1}m = \left(1 - u^2/c^2\right)^{1/2}m = 0, \tag{4.24}$$

that is, the rest mass of a photon must be zero! This is not so bizarre as it first seems, since no inertial observer ever sees a photon at rest – its speed is always c – and so the rest mass of a photon is merely a notional quantity. If we let $\hat{\boldsymbol{n}}$ be a unit vector denoting the direction of travel of the photon, then

$$\boldsymbol{p} = (p_x, p_y, p_z) = p\hat{\boldsymbol{n}},$$

and (4.20) becomes

$$(E/c)^2 - p^2 = 0.$$

Taking square roots (and remembering c and p are positive), we find that the energy E of a photon is related to the magnitude p of its momentum by

$$E = pc. \tag{4.25}$$

Combining these results with Planck's hypothesis $E = h\nu$, we obtain the following formulae for the energy E, and linear momentum $\boldsymbol{p}$ of the photon:

$$E = h\nu, \quad \boldsymbol{p} = (h\nu/c)\hat{\boldsymbol{n}}. \tag{4.26}$$

It is gratifying to discover that special relativity, which was born to reconcile conflicts in the kinematical properties of light and matter, also includes their mechanical properties in a single all-inclusive system.

We finish this section with an argument which shows that Planck's hypothesis can be derived directly within the framework of special relativity. We have already seen in the last chapter that the radial Doppler effect for a moving source is given by (3.26), namely

$$\frac{\lambda}{\lambda_0} = \left(\frac{1 + v/c}{1 - v/c}\right)^{1/2},$$

where λ_0 is the wavelength in the frame of the source and λ is the wavelength in the frame of the observer. We write this result, instead, in terms of frequency, using the fundamental relationships $c = \lambda\nu$ and $c = \lambda_0\nu_0$, to obtain

$$\frac{\nu_0}{\nu} == \left(\frac{1 + v/c}{1 - v/c}\right)^{1/2}. \tag{4.27}$$

Now, suppose that the source emits a light flash of total energy E_0. Let us use (4.22) to find the energy received in the frame of the observer S. Since, recalling Fig. 3.11, the light flash is travelling along the negative x-direction of both frames, (4.25) leads to the result $p'_x = -E_0/c$, with the other primed components of momentum zero. Substituting in the first equation of (4.22), namely,

$$E = \beta(E' + vp'_x),$$

we get

$$E = \beta(E_0 - vE_0/c) = \frac{E_0(1 - v/c)}{(1 - v^2/c^2)^{1/2}} = E_0\left(\frac{1 - v/c}{1 + v/c}\right)^{1/2},$$

or

$$\frac{E_0}{E} = \left(\frac{1 + v/c}{1 - v/c}\right)^{1/2}. \tag{4.28}$$

Combining this with (4.27), we obtain

$$\frac{E_0}{\nu_0} = \frac{E}{\nu}.$$

Since this relationship holds for **any** pair of inertial observers, it follows that the ratio must be a universal constant, which we call h. Thus, we have derived Planck's hypothesis, $E = h\nu$.

We leave our considerations of special relativity at this point and turn our attention to the formalism of tensors. This will enable us to reformulate special relativity in a way which will aid our transition to general relativity, that is, to a theory of gravitation consistent with special relativity.

Exercises

4.1 (§4.1) Discuss the possibility of using force rather than mass as the basic quantity, taking, for example, a standard weight at a given latitude as the unit of force. How should one then define and measure the mass of a body?

4.2 (§4.3) Show that, in the inelastic collision considered in §4.3, the rest mass of the combined object is greater than the sum of the original rest masses. Where does this increase derive from?

4.3 (§4.3) A particle of rest mass $\bar{m}_0$ and speed u strikes a stationary particle of rest mass m_0. If the collision is perfectly inelastic, then find the rest mass of the composite particle.

4.4 (§4.4)
(i) Establish the transition from (4.15) to (4.16).
(ii) Establish the Newtonian approximation equation (4.18).

4.5 (§4.4) Show that (4.19) leads to (4.20). Deduce (4.21).

4.6 (§4.4) Newton's second law for a particle of relativistic mass m is

$$\boldsymbol{F} = \frac{\mathrm{d}}{\mathrm{d}t}(m\boldsymbol{u}).$$

Define the work done $\mathrm{d}E$ in moving the particle from $\boldsymbol{r}$ to $\boldsymbol{r} + \mathrm{d}\boldsymbol{r}$. Show that the rate of doing work is given by

$$\frac{\mathrm{d}E}{\mathrm{d}t} = \frac{\mathrm{d}(m\boldsymbol{u})}{\mathrm{d}t}\cdot\boldsymbol{u}.$$

Use the definition of relativistic mass to obtain the result

$$\frac{\mathrm{d}E}{\mathrm{d}t} = \frac{m_0}{(1 - u^2/c^2)^{3/2}}u\frac{\mathrm{d}u}{\mathrm{d}t}. \qquad \left[\text{Hint: } \boldsymbol{u}\cdot\frac{\mathrm{d}\boldsymbol{u}}{\mathrm{d}t} = u\frac{\mathrm{d}u}{\mathrm{d}t}.\right]$$

Express this last result in terms of dm/dt and integrate to obtain

$$E = mc^2 + \text{constant}.$$

4.7 (§4.4) Two particles whose rest masses are m_1 and m_2, move along a straight line with velocities u_1 and u_2, respectively, measured in the same direction. They collide inelastically to form a new particle. Show that the rest mass and velocity of the new particle are m_3 and u_3, respectively, where

$$m_3^2 = m_1^2 + m_2^2 + 2m_1 m_2 \gamma_1 \gamma_2 (1 - u_1 u_2/c^2),$$

$$u_3 = \frac{m_1 \gamma_1 u_1 + m_2 \gamma_2 u_2}{m_1 \gamma_1 + m_2 \gamma_2},$$

with

$$\gamma_1 = (1 - u_1^2/c^2)^{-1/2}, \qquad \gamma_2 = (1 - u_2^2/c^2)^{-1/2}.$$

4.8 (§4.4) A particle of rest mass m_0, energy e_0, and momentum p_0 suffers a head-on elastic collision (i.e. masses of particles unaltered) with a stationary mass M. In the collision, M is knocked straight forward, with energy E and momentum P, leaving the first particle with energy e and momentum p.

(i) Show that

$$p + P = p_0,$$

$$e + E = e_0 + Mc^2.$$

(ii) Squaring the above equations and using (4.20) show that

$$\frac{eE}{c^2} = pP + e_0 M,$$

(iii) Squaring the above equation and again using (4.20) show that

$$m^2 c^2 P^2 + M^2 c^2 p^2 = 2pP e_o M + M^2 p_0^2 c^2.$$

(iv) Letting $P = p_0 - p$ in the above show that

$$p = \frac{p_0(m^2 c^2 - M^2 c^2)}{2Me_0 + M^2 c^2 + m_0^2 c^2},$$

and

$$P = \frac{2p_0 M(e_0 + Mc^2)}{2Me_0 + M^2 c^2 + m_0^2 c^2}.$$

What do these formulae become in the classical limit?

4.9 (§4.4) Assume that the formulae (4.19) holds for a tachyon, which travels with speed $v > c$. Taking the energy to be a measurable quantity, then deduce that the rest mass of a tachyon is imaginary and define the real quantity μ_0 by $m_0 = i\mu_0$.

If the tachyon moves along the x-axis and if we assume that the x-component of the momentum is a real positive quantity, then deduce

$$m = \frac{v}{|v|}\alpha\mu_0, \quad p = \mu_0|v|\alpha, \quad E = mc^2,$$

where $\alpha = (v^2/c^2 - 1)^{-1/2}$.

Plot E/m_0c^2 against v/c for both tachyons and subluminal particles.

4.10 (§4.5) Two light rays in the xy-plane of an inertial observer, making angles θ and $-\theta$, respectively, with the positive x-axis, collide at the origin. What is the velocity v of the inertial observer (travelling in standard configuration) who sees the light rays collide head on?

4.11 (§4.5) An atom of rest mass m_0 is at rest in a laboratory and absorbs a photon of frequency ν. Find the velocity and mass of the recoiling particle.

4.12 (§4.5) An atom at rest in a laboratory emits a photon and recoils. If its initial mass is m_0 and it loses the rest energy e in the emission, prove that the frequency of the emitted photon is given by

$$\nu = \frac{e}{h}(1 - e/2m_0c^2).$$

Further reading

Again, the main reference is Rindler (1982), but the book by Dixon (1978) and also the one by Taylor and Wheeler (1966) give alternative approaches.

Dixon, W. G. (1978). *Special Relativity, the Foundation of Modern Physics.* Cambridge University Press, Cambridge.

Rindler, W. (1982). *Introduction to Special Relativity.* Oxford University Press, Oxford.

Taylor, E. F., and Wheeler, J. A. (1966). *Spacetime Physics.* Freeman, San Francisco, CA.

Part B

The Formalism
of Tensors

Tensor algebra

5

5.1 Introduction

To work effectively in Newtonian theory, one really needs the language of vectors. This language, first of all, is more succinct, since it summarizes a set of three equations in one. Moreover, the formalism of vectors helps to solve certain problems more readily, and, most important of all, the language reveals structure and thereby offers insight. In exactly the same way, in relativity theory, one needs the language of tensors. Again, the language helps to summarize sets of equations succinctly and to solve problems more readily, and it reveals structure in the equations. This part of the book is devoted to learning the formalism of tensors, which is a pre condition for the rest of the book.

The approach we adopt is to concentrate on the technique of tensors without fully taking into account the deeper geometrical significance behind the theory. We shall be concerned more with what you do with tensors rather than what tensors actually are. There are two distinct approaches to the teaching of tensors: the abstract or index-free (coordinate-free) approach and the more common approach in relativity text books, which uses indices. The main advantage of the more abstract approach is that it is based on the existence of an underlying geometrical object defined on the whole manifold and thus offers deeper geometrical insight, particularly when it comes to looking at global structure. However, it has a number of disadvantages. First of all, it requires much more of a mathematical background, which in turn takes time to develop. The other disadvantage is that the tensorial objects used in relativity have objects with large numbers of indices and complicated contractions which are hard to write down in an index-free fashion. Finally, for all its elegance, when one wants to do a real calculation with tensors, as one frequently needs to, then recourse often has to be made to using a particular coordinate system adapted to the problem in hand. We shall adopt the more conventional index approach based on how tensors transform under a change of coordinate system, because it will prove faster and more practical. In some ways, it also accords more with Einstein's ideas that the laws of physics should not depend on how one constructs the local coordinate system. Furthermore, it also provides a quick route to the geometrical and global 'abstract index' approach of Penrose (1968) in which indices are used simply to indicate the type of a tensor and are not related to the use

Introducing Einstein's Relativity. Second Edition. Ray d'Inverno and James Vickers, Oxford University Press.
© Ray d'Inverno and James Vickers (2022). DOI: 10.1093/oso/9780198862024.003.0005

of a particular coordinate system. In any case, we advise those who wish to take their study of the subject further to look at a more geometrical approach at the first opportunity.

We repeat that **the exercises are seen as integral to this part of the book and should not be omitted**.

5.2   Manifolds and coordinates

We shall start by working with tensors defined in n dimensions since, and it is part of the power of the formalism, there is little extra effort involved. A tensor is an object defined on a geometric entity called a (differential) **manifold**. We shall not define a manifold precisely because it would involve us in too much of a digression. But, in simple terms, a manifold is something which 'locally' looks like a bit of n-dimensional Euclidean space $\mathbb{R}^n$. For example, compare a 2-sphere S^2 with the Euclidean plane $\mathbb{R}^2$. They are clearly different. But a small bit of S^2 looks very much like a small bit of $\mathbb{R}^2$ (if we neglect metrical properties). The fact that S^2 is 'compact', i.e. in some sense finite, whereas $\mathbb{R}^2$ 'goes off to infinity' is a **global** property rather than a local property. We shall not say anything precise about global properties – the **topology** of the manifold – although the issue will surface when we start to look carefully at solutions of Einstein's equations in general relativity.

We shall simply take an n-dimensional manifold M to be a set of points such that each point possesses a set of n **coordinates** $(x^1, x^2, \ldots, x^n)$ where each coordinate ranges over a subset of the reals, which may, in particular, range from $-\infty$ to $+\infty$. The reason why the coordinates are written as superscripts rather than subscripts will become clear later. Now the key thing about a manifold is that it may not be possible to cover the whole manifold by one **non-degenerate** coordinate system, namely, one which ascribes a **unique** set of n coordinate numbers to each point. Sometimes it is simply convenient to use coordinate systems with **degenerate** points. For example, plane polar coordinates (R, ϕ) in the plane have a degeneracy at the origin because ϕ is indeterminate there (Fig. 5.1). However, here we could avoid the degeneracy at the origin by using Cartesian coordinates. But in other circumstances we have no choice in the matter. For example, it can be shown that there is no single coordinate system which covers the whole of a 2-sphere S^2 without degeneracy. The smallest number needed is two, which is shown schematically in Fig. 5.2. We therefore work with coordinate systems which cover only a portion of the manifold and which are called **coordinate patches**. Figure 5.3 indicates this schematically. A set of coordinate patches which covers the whole manifold is called an **atlas**. The theory of manifolds tells us how to get from one coordinate patch to another by a differentiable coordinate transformation in the overlap region. The behaviour of geometric quantities under coordinate transformations lies at the heart of tensor calculus.

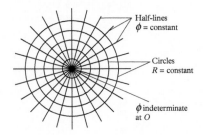

Half-lines
ϕ = constant

Circles
R = constant

ϕ indeterminate
at O

Fig. 5.1 Plane polar coordinate curves.

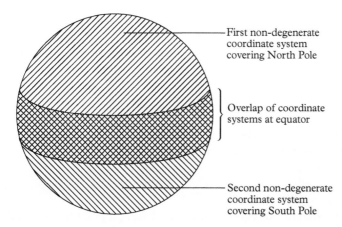

Fig. 5.2 Two non-degenerate coordinate systems covering an S^2.

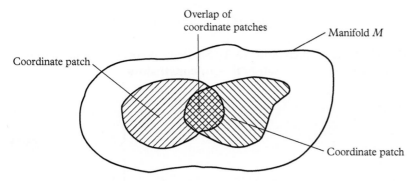

Fig. 5.3 Overlapping coordinate patches in a manifold.

5.3   Curves and surfaces

Given a manifold, we shall be concerned with points in it and subsets of points which define curves and surfaces of different dimensions. We shall frequently define these curves and surfaces **parametrically**. Thus (in exactly the same way as is done in Euclidean 2- and 3-space), since a curve has one degree of freedom, it depends on one parameter and so we define a **curve** by the parametric equations

$$x^a = x^a(u), \quad (a = 1, 2, \ldots, n), \tag{5.1}$$

where u is a parameter and $x^1(u), x^2(u), \ldots, x^n(u)$ denote n functions of u. Similarly, since a **subspace** or **surface** of m dimensions ($m < n$) has m degrees of freedom,, it depends on m parameters and it is given by the parametric equations

$$x^a = x^a(u^1, u^2, \ldots, u^m), \quad (a = 1, 2, \ldots, n). \tag{5.2}$$

If, in particular, $m = n - 1$, the subspace is called a **hypersurface**. In this case,

$$x^a = x^a(u^1, u^2, \ldots, u^{n-1}), \quad (a = 1, 2, \ldots, n)$$

and the $n - 1$ parameters can be eliminated from these n equations to give one equation connecting the coordinates, i.e.

$$f(x^1, x^2, \ldots, x^n) = 0. \tag{5.3}$$

From a different but equivalent point of view, a point in a general position in a manifold has n degrees of freedom. If it is restricted to lie in a hypersurface, an $(n - 1)$-subspace, then its coordinates must satisfy **one constraint**, namely,

$$f(x^1, x^2, \ldots, x^n) = 0,$$

which is the same as equation (5.3). Similarly, points in an m-dimensional subspace $(m < n)$ must satisfy $\boldsymbol{n - m}$ **constraints**

$$\left.\begin{array}{l} f^1(x^1, x^2, \ldots, x^n) = 0, \\ f^2(x^1, x^2, \ldots, x^n) = 0, \\ \qquad \vdots \\ f^{n-m}(x^1, x^2, \ldots, x^n) = 0, \end{array}\right\} \tag{5.4}$$

which is an alternative to the parametric representation (5.2).

5.4 Transformation of coordinates

As we have seen, a point in a manifold can be covered by many different coordinate patches. The essential point about tensor calculus is that, when we make a statement about tensors, we do not wish it simply to hold just for one coordinate system but rather for **all** coordinate systems. Consequently, we need to find out how quantities behave when we go from one coordinate system to another one. We therefore consider the **change of coordinates** $x^a \rightarrow x'^a$ given by the n equations

$$x'^a = f^a(x^1, x^2, \ldots, x^n) \; (a = 1, 2, \ldots, n), \tag{5.5}$$

where the f's are single-valued continuous differentiable functions, at least for certain ranges of their arguments. Hence, at this stage, we view a coordinate transformation **passively** as assigning to a point of the manifold whose old coordinates are $(x^1, x^2, \ldots, x^n)$ the **new** primed coordinates $(x'^1, x'^2, \ldots, x'^n)$. We can write (5.5) more succinctly as $x'^a = f^a(x)$, where, from now on, lower-case Latin indices are assumed to run from 1 to n, the dimension of the manifold, and the f^a are all functions of the old un-primed coordinates. Furthermore, we can write the equation more simply still as

$$x'^a = x'^a(x),\tag{5.6}$$

where $x'^a(x)$ denote the n functions $f^a(x)$. Notation plays an important role in tensor calculus, and equation (5.6) is clearly easier to write than equation (5.5).

We next contemplate differentiating (5.6) with respect to each of the coordinates x^b. This produces the $n \times n$ **transformation matrix** of coefficients:

$$\left[\frac{\partial x'^a}{\partial x^b}\right] = \begin{bmatrix} \dfrac{\partial x'^1}{\partial x^1} & \dfrac{\partial x'^1}{\partial x^2} & \cdots & \dfrac{\partial x'^1}{\partial x^n} \\[2mm] \dfrac{\partial x'^2}{\partial x^1} & \dfrac{\partial x'^2}{\partial x^2} & \cdots & \dfrac{\partial x'^2}{\partial x^n} \\[2mm] \vdots & & & \\[2mm] \dfrac{\partial x'^n}{\partial x^1} & \dfrac{\partial x'^n}{\partial x^2} & \cdots & \dfrac{\partial x'^n}{\partial x^n} \end{bmatrix}.\tag{5.7}$$

The determinant J' of this matrix is called the **Jacobian** of the transformation:

$$J' = \left|\frac{\partial x'^a}{\partial x^b}\right|.\tag{5.8}$$

We shall assume that this is non-zero for some range of the coordinates x^b. Then it follows from the inverse function theorem that we can (in principle) solve equation (5.6) uniquely for the old coordinates x^a and obtain the **inverse** transformation equations

$$x^a = x^a(x').\tag{5.9}$$

It follows from the product rule for determinants that, if we define the Jacobian of the inverse transformation by

$$J = \left|\frac{\partial x^a}{\partial x'^b}\right|,$$

then $J = 1/J'$.

It is convenient to assume that the functions $x'^a = x'^a(x)$ and $x^a = x^a(x')$ are not just differentiable but **smooth**, which means that we can differentiate the functions as often as we wish. A manifold for which the transformation functions (5.6) and (5.9) are smooth is called a **smooth manifold** and, from now onwards, unless we say otherwise, we will assume that we are working on a smooth manifold.

In three dimensions, the equation of a surface is given by $z = f(x, y)$; then its total differential is defined to be

$$dz = \frac{\partial f}{\partial x} dx + \frac{\partial f}{\partial y} dy.$$

Then, in an exactly analogous manner, starting from (5.6), we define the total differential

$$dx'^a = \frac{\partial x'^a}{\partial x^1} dx^1 + \frac{\partial x'^a}{\partial x^2} dx^2 + \cdots + \frac{\partial x'^a}{\partial x^n} dx^n,$$

for each a running from 1 to n. We can write this more economically by introducing an explicit summation sign:

$$dx'^a = \sum_{b=1}^{n} \frac{\partial x'^a}{\partial x^b} dx^b. \tag{5.10}$$

This can be written more economically still by introducing the **Einstein summation convention:** whenever a literal index is repeated, it is understood to imply a summation over the index from 1 to n, the dimension of the manifold. Hence, we can write (5.10) simply as

$$dx'^a = \frac{\partial x'^a}{\partial x^b} dx^b. \tag{5.11}$$

The index a occurring on each side of this equation is said to be **free** and may take on separately any value from 1 to n. The index b on the right-hand side is **repeated** and hence there is an implied summation from 1 to n. A repeated index is called **bound** or **dummy** because it can be replaced by any other index not already in use. For example,

$$\frac{\partial x'^a}{\partial x^b} dx^b = \frac{\partial x'^a}{\partial x^c} dx^c,$$

because c was not already in use in the expression. We define the **Kronecker delta** δ_b^a to be a quantity which is either 0 or 1 according to

$$\delta_b^a = \begin{cases} 1 & \text{if } a = b, \\ 0 & \text{if } a \neq b. \end{cases} \tag{5.12}$$

It therefore follows directly from the definition of partial differentiation (check) that

$$\frac{\partial x'^a}{\partial x'^b} = \frac{\partial x^a}{\partial x^b} = \delta_b^a. \tag{5.13}$$

5.5 Contravariant tensors

The approach we are going to adopt is to define a geometrical quantity in terms of its transformation properties under a coordinate transformation (5.6). We shall start with a prototype and then give the general definition.

Let P be a point on the manifold and let $\gamma(u)$ be a differentiable curve parameterized by u such that $P = \gamma(0)$. Now let $(x^1, \ldots, x^n)$ be a local coordinate system in a neighbourhood of P; then we can write the curve γ in these coordinates as being given by

$$x^a = x^a(u).$$

Then the derivative at $u = 0$ defines a **tangent vector** to the curve at P (Fig. 5.4) which in these coordinates is given by

$$T^a = \frac{\mathrm{d}x^a}{\mathrm{d}u}(0). \tag{5.14}$$

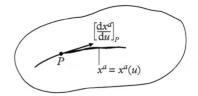

Fig. 5.4 The tangent vector to the curve $x^a = x^a(u)$ at P.

Now suppose we introduce a new coordinate system $(x'^1, \ldots, x'^n)$ in a neighbourhood of P and look at the **same curve** γ in the x'^a coordinate system. Then using (5.6) we may write the curve $\gamma(u)$ in this coordinate system as

$$x'^a(u) = x'^a(x(u)). \tag{5.15}$$

In terms of these coordinates, the tangent vector at P is given by

$$T'^a = \frac{\mathrm{d}x'^a}{\mathrm{d}u}(0). \tag{5.16}$$

Then, by the function of a function rule for derivatives, we have

$$T'^a = \left[\frac{\partial x'^a}{\partial x^b}\right]_P \left[\frac{\mathrm{d}x^b}{\mathrm{d}u}(0)\right] = \left[\frac{\partial x'^a}{\partial x^b}\right]_P T^b. \tag{5.17}$$

Remember that the **repeated** index b is summed over. Thus, the components of the tangent vector T'^a in the new coordinate system are nothing but the components in the old coordinates multiplied by the $n \times n$ transformation matrix $[\partial x'^a/\partial x^b]_P$. We will take (5.17) as our prototype for how a vector transforms at a point P and use it in future as our definition.

We now define a **contravariant vector** or **contravariant tensor of rank (order) 1** at P as a set of quantities, written X^a in the x^a-coordinate system, which transforms under a change of coordinates according to

$$X'^a = \frac{\partial x'^a}{\partial x^b} X^b, \tag{5.18}$$

where the transformation matrix is evaluated at P. The tangent vector to a curve is just a special case of (5.18). It is important to distinguish between the actual invariant geometric object like the tangent vector and its representation in a particular coordinate system. This is given by the n numbers $[\mathrm{d}x^a/\mathrm{d}u]_P$ in the x^a-coordinate system and the (in general) different numbers $[\mathrm{d}x'^a/\mathrm{d}u]_P$ in the x'^a-coordinate system. When we want to talk about the tangent vector to the curve without referring to a specific coordinate system, we will write it as $\dot\gamma(0)$ and we depict it with an arrow in Fig. 5.5.

We now generalize the definition (5.18) to obtain contravariant tensors of higher rank or order. Thus, a **contravariant tensor of rank 2** is a set of n^2 quantities associated with a point P, denoted by X^{ab} in the x^a-coordinate system, which transform according to

$$X'^{ab} = \frac{\partial x'^a}{\partial x^c}\frac{\partial x'^b}{\partial x^d}X^{cd}. \tag{5.19}$$

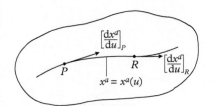

Fig. 5.5 The tangent vector at two points of a curve $x^a = x^a(u)$.

The quantities X'^{ab} are the components in the x'^a-coordinate system, the transformation matrices are evaluated at P, and the law involves two dummy indices c and d. An example of such a quantity is provided by the so-called tensor product $Y^a Z^b$, of two contravariant vectors Y^a and Z^b (exercise). The definition of third- and higher-order contravariant tensors proceeds in an analogous manner. An important case is a tensor of zero rank, called a **scalar** or **scalar invariant** ϕ, which transforms according to

$$\phi' = \phi. \tag{5.20}$$

at P.

5.6   Covariant tensors

In ordinary vector calculus in $\mathbb{R}^3$, there are two obvious geometric constructions which give rise to a vector. The first is taking the tangent to a curve and the other is taking the gradient of a scalar function defining a surface. The first provided our prototype for a contravariant vector while the second will provide us with the prototype for a covariant vector. Let ϕ be a differentiable scalar (real-valued) function on the manifold; then, at a general point P in the manifold, $\phi(P) = k$, where k is a real number. Now, in $\mathbb{R}^3$ in standard Euclidean coordinates (x^1, x^2, x^3), the equation $\phi(x^a) = k$ defines a surface S and

$$\mathrm{grad}\phi = \left(\frac{\partial\phi}{\partial x^1}, \frac{\partial\phi}{\partial x^2}, \frac{\partial\phi}{\partial x^3}\right),$$

gives a vector normal to the surface. We will show that this remains true in the general case, but things are a bit more complicated using a general coordinate system. Let ϕ be a scalar field and let $(x^1, \ldots, x^n)$ be a coordinate system in the neighbourhood of the point P. Then in these coordinates we may write

$$\phi = \phi(x^a), \tag{5.21}$$

and by (5.2) the equation $\phi(x^a) = k$ defines a hypersurface S through P. The derivative of ϕ defines a **covariant vector** or **co-vector** at P normal to S given in these coordinates by

$$N_a = \left[\frac{\partial \phi}{\partial x^a}\right]_P. \tag{5.22}$$

Note that the index on the covariant vector N_a is below while that on the contravariant vector T^a was above. As we will see below, the difference in the position of the index is important and indicates that it transforms in a different way.

Now let us introduce a different coordinate system $(x'^1, \ldots, x'^n)$ in the neighbourhood of P. We now look at the **same** hypersurface but this time described in terms of the x'^a coordinate system so the hypersurface is now given by $\phi(x'^a) = k$. Then in this coordinate system the components of the normal co-vector are given by

$$N'_a = \left[\frac{\partial \phi}{\partial x'^a}\right]_P. \tag{5.23}$$

Remembering from equation (5.9) that x^a can be thought of as a function of x'^b, equation (5.21) can be written equivalently as

$$\phi = \phi(x^a(x')).$$

Differentiating this with respect to x'^b, using the function of a function rule, we obtain

$$\frac{\partial \phi}{\partial x'^b} = \frac{\partial \phi}{\partial x^a}\frac{\partial x^a}{\partial x'^b},$$

where $\partial x^a / \partial x'^b$ is evaluated at P. Then changing the order of the terms, the dummy index, and the free index (from b to a) gives

$$\frac{\partial \phi}{\partial x'^a} = \frac{\partial x^b}{\partial x'^a}\frac{\partial \phi}{\partial x^b}, \tag{5.24}$$

so that by (5.22) and (5.23)

$$N'_a = \left[\frac{\partial x^b}{\partial x'^a}\right]_P N_b. \tag{5.25}$$

This is the prototype equation we are looking for. Notice that compared to (5.17) it involves the inverse transformation matrix $\partial x^b / \partial x'^a$.

We therefore define a **covariant vector** or **covariant tensor of rank (order) 1** to be a set of quantities, written X_a in the x^a-coordinate system, associated with a point P, which transforms according to

$$X'_a = \frac{\partial x^b}{\partial x'^a} X_b, \tag{5.26}$$

where the transformation matrix occurring is assumed to be evaluated at P. The normal co-vector to a hypersurface is just a special case of (5.26). Again, it is important to distinguish between the actual invariant geometric object like the normal co-vector (which we will write as $d\phi$) and its representation in a particular coordinate system, given by the n numbers $[\partial\phi/\partial x^a]_P$ in the x^a-coordinate system and the (in general) different numbers $[\partial\phi/\partial x'^a]_P$ in the x'^a-coordinate system.

Similarly, we define a covariant tensor of rank 2 by the transformation law

$$X'_{ab} = \frac{\partial x^c}{\partial x'^a}\frac{\partial x^d}{\partial x'^b}X_{cd},\qquad(5.27)$$

and so on for higher-rank tensors. Note the convention that contravariant tensors have raised indices whereas covariant tensors have lowered indices. (The way to remember this is that **co** goes bel**ow**.) The fact that according to (5.10) the differentials dx^a formally transform as the components of a contravariant vector explains the convention that the coordinates themselves are written as x^a rather than x_a, although note that it is only the differentials and not the coordinates which have tensorial character.

5.7 Mixed tensors

Following the pattern of (5.19) and (5.27), we go on to define **mixed** tensors in the obvious way. For example, a mixed tensor of rank 3 – one contravariant rank and two covariant rank – satisfies

$$X'^a{}_{bc} = \frac{\partial x'^a}{\partial x^d}\frac{\partial x^e}{\partial x'^b}\frac{\partial x^f}{\partial x'^c}X^d{}_{ef}.\qquad(5.28)$$

If a mixed tensor has contravariant rank p and covariant rank q, then it is said to have **type** or **valence** (p, q).

We now come to the reason why tensors are important in mathematical physics. Let us illustrate the reason by way of an example. Suppose we find in one coordinate system that two tensors, X_{ab} and Y_{ab}, say, are equal, i.e.

$$X_{ab} = Y_{ab}.\qquad(5.29)$$

Let us multiply both sides by the matrices $\partial x^a/\partial x'^c$ and $\partial x^b/\partial x'^d$ and take the implied summations to get

$$\frac{\partial x^a}{\partial x'^c}\frac{\partial x^b}{\partial x'^d}X_{ab} = \frac{\partial x^a}{\partial x'^c}\frac{\partial x^b}{\partial x'^d}Y_{ab}.$$

Since X_{ab} and Y_{ab} are both covariant tensors of rank 2, it follows that $X'_{ab} = Y'_{ab}$. In other words, the equation (5.29) holds in **any** other coordinate system. In short, a tensor equation which holds in one coordinate system necessarily holds in **all** coordinate systems. Thus, although we introduce coordinate systems for convenience in tackling particular problems, if we work with tensorial equations, then they hold in all coordinate systems. Put another way, tensorial equations are coordinate independent. This is something that the index-free or coordinate-free approach makes clear from the outset.

5.8 Tensor fields

In vector analysis, a fixed vector is a vector associated with a point, whereas a vector **field** defined over a region is an association of a vector to every point in that region. In exactly the same way, a tensor is a set of quantities defined at one point in the manifold. A **tensor field** defined over some region of the manifold is an association of a tensor of the same valence to every point of the region, i.e.

$$P \to T^{a\cdots}_{b\cdots}(P),$$

where $T^{a\cdots}_{b\cdots}(P)$ is the value of the tensor at P. The tensor field is called continuous or differentiable if its components in all coordinate systems are continuous or differentiable functions of the coordinates. The tensor field is called **smooth** if its components are differentiable to all orders, which is denoted mathematically by saying that all the components are C^∞. Thus, for example, a contravariant vector field defined over a region is a set of n **functions** defined over that region, and the vector field is smooth if the functions are all C^∞. The transformation law for a contravariant vector field then becomes

$$X'^a(x') = \left[\frac{\partial x'^a}{\partial x^b}\right]_P X^b(x), \tag{5.30}$$

at each point P in the region, since the old components X^a are functions of the old x^a-coordinates and the new components X'^a are functions of the new x'^a-coordinates.

As in the case of vectors and vector fields in vector analysis, the distinction between a tensor and a tensor field is not always made completely clear. We shall for the most part be dealing with tensor fields from now on, but to conform with general usage we shall often refer to tensor fields simply as tensors. We will again shorten the transformation law such as (5.30) to the form (5.26) with everything else being implied. If we wish to emphasize that a tensor is a tensor field, we shall write it in functional form, namely, as $T^{a\cdots}_{b\cdots}(x)$.

5.9 Elementary operations with tensors

Tensor calculus is concerned with **tensorial operations**, that is, operations on tensors which result in quantities which are still tensors. In this section, we will look at algebraic operations on tensors, i.e. operations that can be performed at a point. A simple way of establishing whether or not a quantity is a tensor is to see how it transforms under a coordinate transformation. For example, we can deduce directly from the transformation law that two tensors of the same type can be added together to give a tensor of the same type, e.g.

$$X^a{}_{bc} = Y^a{}_{bc} + Z^a{}_{bc}. \tag{5.31}$$

The same holds true for subtraction and multiplication by a real number.

A covariant tensor of rank 2 is said to be **symmetric** if $X_{ab} = X_{ba}$, in which case it has only $\frac{1}{2}n(n+1)$ independent components (check this by establishing how many independent components there are of a symmetric matrix of order n). Symmetry is a tensorial property. A similar definition holds for a contravariant tensor X^{ab}. The tensor X_{ab} is said to be **anti-symmetric** or **skew symmetric** if $X_{ab} = -X_{ba}$, which has only $\frac{1}{2}n(n-1)$ independent components; this is again a tensorial property. A notation frequently used to denote the symmetric part of a tensor is

$$X_{(ab)} = \tfrac{1}{2}(X_{ab} + X_{ba}) \tag{5.32}$$

and the anti-symmetric part is

$$X_{[ab]} = \tfrac{1}{2}(X_{ab} - X_{ba}). \tag{5.33}$$

In general,

$$X_{(a_1 a_2 \cdots a_r)} = \frac{1}{r!} \quad \text{(sum over all permutations of the indices } a_1 \text{ to } a_r\text{)}$$

and

$$X_{[a_1 a_2 \cdots a_r]} = \frac{1}{r!} \quad \text{(alternating sum over all permutations of the indices}$$
$$a_1 \text{ to } a_r\text{).}$$

For example, we shall need to make use of the result

$$X_{[abc]} = \tfrac{1}{6}(X_{abc} - X_{acb} + X_{cab} - X_{cba} + X_{bca} - X_{bac}). \tag{5.34}$$

(A way to remember the above expression is to note that the positive terms are obtained by cycling the indices to the right and the corresponding negative terms by flipping the last two indices.) **A totally symmetric tensor** is defined to be one equal to its symmetric part, and a **totally anti-symmetric tensor** is one equal to its anti-symmetric part.

Given a tensor field of type (p, q), we may multiply it by a scalar field ϕ (i.e. a tensor field of type $(0, 0)$) to obtain a tensor field also of type (p, q). More generally, we can multiply the components of two tensors of type (p_1, q_1) and (p_2, q_2) together and obtain the components of a tensor of type $(p_1 + p_2, q_1 + q_2)$, for example

$$X^a{}_{bcd} = Y^a{}_b Z_{cd}. \tag{5.35}$$

This is an example of the **tensor product** of two tensors. One can readily show (exercise) that the above definition does not depend on the coordinate system used to undertake the multiplication.

Another particularly important example of a tensorial operation is **contraction**. We start with an example. Let X^a be a contravariant vector field and let Y_a be a covariant vector field. Then at each point they define a real number

$$\phi = X^a Y_a, \tag{5.36}$$

(remember that the repeated index a is summed over), called the contraction of Y_a with X^a. What does this look like in another set of coordinates? Using the transformation laws (5.18) and (5.26) for X'^a and Y'_a, we find

$$
\begin{aligned}
X'^a Y'_a &= \left(\frac{\partial x'^a}{\partial x^b} X^b \right) \left(\frac{\partial x^c}{\partial x'^a} Y_c \right) \\
&= \left(\frac{\partial x'^a}{\partial x^b} \frac{\partial x^c}{\partial x'^a} \right) X^b Y_c \\
&= \delta^c_b X^b Y_c \\
&= X^c Y_c \\
&= X^a Y_a.
\end{aligned}
$$

Thus

$$(X'^a Y'_a)_P = (X^a Y_a)_P = \phi(P). \tag{5.37}$$

This is a very important result as it shows that the contraction of a covariant vector with a contravariant vector gives a scalar field ϕ, which does not depend on the coordinates. This is important physically as it shows how to obtain coordinate independent results using tensors. Although we will not make use of it in this book, this result is also important mathematically for the coordinate-free approach to differential geometry as it shows that covariant vectors are, in the language of linear algebra, **dual** to contravariant vectors and explains the alternative name of **co-vectors**.

We now consider the contraction of two general tensors. Given a tensor of mixed type (p, q), we can form an object of type $(p - 1, q - 1)$ by the process of **contraction**, which simply involves setting a raised and lowered index equal. For example,

$$X^a{}_{bcd} \longrightarrow X^a{}_{acd} = Y_{cd}, \quad \text{contraction on } a \text{ and } b.$$

One can show that by doing this one obtains a **tensor** of type $(p-1, q-1)$, i.e. in the above example, a tensor of type $(1, 3)$ has become a tensor of type $(0, 2)$ and that the tensor one obtains does not depend upon the coordinate system in which one does the contraction. Notice that we can also contract a tensor by multiplying by the Kronecker tensor δ^a_b, e.g.

$$X^a{}_{acd} = \delta^b_a X^a{}_{bcd}. \tag{5.38}$$

In effect, multiplying by δ_b^a turns the index b into a (or, equivalently, the index a into b). If one starts with some tensorial object of type (p, p) and contracts all the indices to obtain an object of type $(0, 0)$ this gives a scalar field or **tensor invariant** whose value does not depend upon the coordinate system. It was the fact that contracting tensorial objects results in scalar quantities, which can in principle be measured and do not depend on the coordinates used, that led Einstein to use tensors as a way of formulating the laws of physics.

5.10   Index-free interpretation of contravariant vector fields

As we pointed out in §5.5, we must distinguish between the actual geometric object itself and its components in a particular coordinate system. The important point about tensors is that we want to make statements which are independent of any particular coordinate system being used. This is abundantly clear in the index-free approach to tensors. We shall get a feel for this approach in this section by considering the special case of a contravariant vector field, although similar index-free interpretations can be given for any tensor field. The key idea is to interpret the vector field as an **operator** which maps real-valued functions into real-valued functions. Thus, if X represents a contravariant vector field, then X operates on any real-valued function f to produce another function g, i.e. $Xf = g$. We shall show how actually to compute Xf by introducing a coordinate system. However, as we shall see, we could equally well introduce any other coordinate system, and the computation would lead to the identical result.

In the x^a-coordinate system, we introduce the notation

$$\partial_a := \frac{\partial}{\partial x^a},$$

and then X is defined as the operator

$$X = X^a \partial_a, \tag{5.39}$$

so that, for any real function f,

$$Xf = (X^a \partial_a) f = X^a (\partial_a f), \tag{5.40}$$

and, in the x^a-coordinate system, X gives the **directional derivative** in the X^a direction. Let us compute X in some other x'^a-coordinate system. We need to use the result (5.13) expressed in the following form: we may take x^a to be a function of x'^b by (5.9) and x'^b to be a function of x^c by (5.6), and so, using the function of a function rule, we find

$$\delta_b^a = \frac{\partial x^a}{\partial x^b} = \frac{\partial}{\partial x^b} x^a (x'^c (x^d)) = \frac{\partial x^a}{\partial x'^c} \frac{\partial x'^c}{\partial x^b}. \tag{5.41}$$

Then, using the transformation law (5.18) and (5.24) together with the above trick, we get

$$
\begin{aligned}
X'^a \partial'_a &= X'^a \frac{\partial}{\partial x'^a} \\
&= \frac{\partial x'^a}{\partial x^b} X^b \frac{\partial x^c}{\partial x'^a} \frac{\partial}{\partial x^c} \\
&= \frac{\partial x^c}{\partial x'^a} \frac{\partial x'^a}{\partial x^b} X^b \frac{\partial}{\partial x^c} \\
&= \delta^c_b X^b \frac{\partial}{\partial x^c} \\
&= X^b \frac{\partial}{\partial x^b} \\
&= X^a \partial_a.
\end{aligned}
$$

Thus, the result of operating on f by X will be the same **irrespective** of the coordinate system employed in (5.39), and this provides the key idea in the coordinate-free approach to differential geometry.

In any coordinate system, we may think of the quantities $[\partial / \partial x_a]_P$ as forming a basis for all the vectors at P, since any vector at P is, by (5.39), given by

$$
X_P = [X^a]_P \left[\frac{\partial}{\partial x^a} \right]_P,
$$

that is, a linear combination of the $[\partial / \partial x^a]_P$. The vector space of all the contravariant vectors at P is known as the **tangent space** at P and is written $T_P M$ (Fig. 5.6). In general, the tangent space at any point in a manifold is different from the underlying manifold. For this reason, we need to be careful in representing a finite contravariant vector by an arrow in our figures since, strictly speaking, the arrow lies in the tangent space, not the manifold. Two exceptions to this are Euclidean space and Minkowski space-time, where the tangent space at each point coincides with the manifold.

As we remarked at the end of §5.9, equation (5.37) shows that covariant vectors at P are elements of the so-called **dual vector space** to $T_P M$, which is called the **cotangent space** and is denoted $T^\star_p M$. General tensors in the coordinate-free approach are then constructed by taking tensor products of elements of $T_P M$ and $T^\star_P M$ or, equivalently, by considering multi-linear maps from copies of $T_P M$ and $T^\star_P M$ to $\mathbb{R}$. We will not pursue the coordinate-free approach here but see, for example, Wald (1984) for further details.

Given two vector fields, X and Y, we can define a new vector field called the **commutator** or **Lie bracket** of X and Y by

$$
[X, Y] = XY - YX. \tag{5.42}
$$

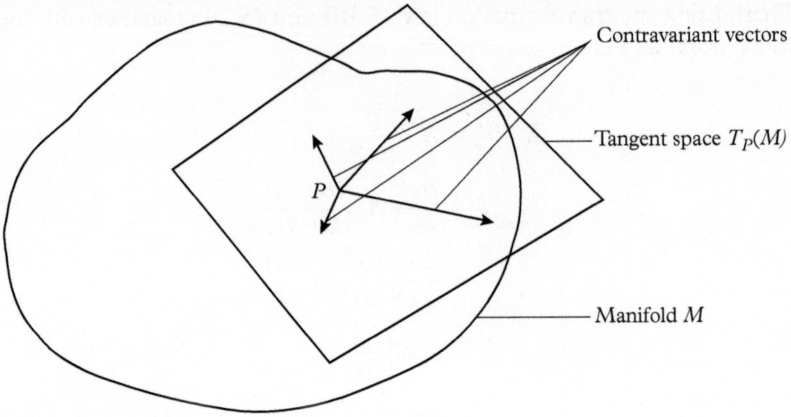

Contravariant vectors

Tangent space $T_P(M)$

Manifold M

P

Fig. 5.6 The tangent space at P.

Letting $[X, Y] = Z$ and operating with it on some arbitrary function f

$$
\begin{aligned}
Zf &= [X, Y]f \\
&= (XY - YX)f \\
&= X(Yf) - Y(Xf) \\
&= X(Y^a\partial_a f) - Y(X^a\partial_a f) \\
&= X^b\partial_b(Y^a\partial_a f) - Y^b\partial_b(X^a\partial_a f) \\
&= (X^b\partial_b Y^a - Y^b\partial_b X^a)\partial_a f - X^a Y^b(\partial_b\partial_a f - \partial_a\partial_b f) \\
&= (X^b\partial_b Y^a - Y^b\partial_b X^a)\partial_a f,
\end{aligned}
$$

since the last term vanishes from the commutativity of second mixed partial derivatives, i.e.

$$
\partial_a\partial_b = \frac{\partial^2}{\partial x^a\partial x^b} = \frac{\partial^2}{\partial x^b\partial x^a} = \partial_b\partial_a.
$$

We therefore see that the Lie bracket of two vector fields also defines a directional derivative and is therefore itself a vector field with components Z^a given by

$$
Z^a = [X, Y]^a = X^b\partial_b Y^a - Y^b\partial_b X^a, \tag{5.43}
$$

since f is arbitrary. It also follows, directly from the definition (5.42) (exercise), that

$$
[X, X] \equiv 0, \tag{5.44}
$$

$$
[X, Y] \equiv -[Y, X], \tag{5.45}
$$

$$
[X, [Y, Z]] + [Z, [X, Y]] + [Y, [Z, X]] \equiv 0. \tag{5.46}
$$

Equation (5.45) shows that the Lie bracket is anti-commutative. The result (5.46) is known as **Jacobi's identity**. Notice it states that the left-hand side is not just equal to zero but is **identically** zero. What does this mean? The equation $x^2 - 4 = 0$ is only satisfied by particular values of x, namely, $+2$ and -2. The identity $x^2 - x^2 \equiv 0$ is satisfied for all values of x. But, you may argue, the x^2 terms cancel out, and this is precisely the point. An expression is identically zero if, when all the terms are written out fully, they all cancel in pairs.

Exercises

5.1 (§5.3) In Euclidean 3-space $\mathbb{R}^3$:
(i) Write down the equation of a circle of radius a lying in the xy-plane centred at the origin in (a) parametric form and (b) constraint form.
(ii) Write down the equation of a hypersurface consisting of a sphere of radius a centred at the origin in (a) parametric form and (b) constraint form. Eliminate the parameters in form (a) to obtain form (b).

5.2 (§5.4) Write down the change of coordinates from Cartesian coordinates $(x^a) = (x, y, z)$ to spherical polar coordinates $(x'^a) = (r, \theta, \phi)$ in $\mathbb{R}^3$. Obtain the transformation matrices and express them both in terms of the primed coordinates. Obtain the Jacobians J and J'. Where is J' zero or infinite?

5.3 (§5.4) Show by manipulating the dummy indices that

$$(Z_{abc} + Z_{cab} + Z_{bca})X^a X^b X^c = 3Z_{abc}X^a X^b X^c.$$

5.4 (§5.4) Show that
(i) $\delta_a^b X^a = X^b$,
(ii) $\delta_a^b X_b = X_a$,
(iii) $\delta_a^b \delta_b^c \delta_c^d = \delta_a^d$.

5.5 (§5.5) If Y^a and Z^a are contravariant vectors, then show that $Y^a Z^b$ is a contravariant tensor of rank 2.

5.6 (§5.5) Write down the change of coordinates from Cartesian coordinates $(x^a) = (x, y)$ to plane polar coordinates $(x'^a) = (R, \phi)$ in $\mathbb{R}^2$ and obtain the transformation matrix $[\partial x'^a / \partial x^b]$ expressed as a function of the primed coordinates. Find the components of the tangent vector to the curve consisting of a circle of radius a centred at the origin with the standard parametrization (see Exercise 5.1 (i)) and use (5.18) to find its components in the primed coordinate system.

5.7 (§5.6) Write down the definition of a tensor of type (2,1).

5.8 (§5.6) Show that, if one assumes that δ_a^b is defined by (5.12) in the x^a coordinates and has the tensor character indicated, then (5.12) is true in any coordinate system. Thus, δ_a^b is a **constant** or **numerical** tensor, that is, it has the same components in all coordinate systems.

5.9 (§5.6) Show, by differentiating (5.24) with respect to x'^c, that $\partial^2 \phi / \partial x^a \partial x^b$ is not a tensor.

5.10 (§5.9) Show that, if $Y^a{}_{bc}$ and $Z^a{}_{bc}$ are tensors of the type indicated, then so is their sum and difference.

5.11 (§5.9)
(i) Show that the fact that a covariant second-rank tensor is symmetric in one coordinate system is a tensorial property.
(ii) If X^{ab} is anti-symmetric and Y_{ab} is symmetric, then prove that $X^{ab}Y_{ab} = 0$.

5.12 (§5.9) Prove that any covariant (or contravariant) tensor of rank 2 can be written as the sum of a symmetric and an anti-symmetric tensor. [Hint: consider the identity $X_{ab} = \frac{1}{2}(X_{ab} + X_{ba}) + \frac{1}{2}(X_{ab} - X_{ba})$.]

5.13 (§5.9) Verify that the definition of the tensor product given by (5.35) does not depend on the coordinate system used.

5.14 (§5.9) If $X^a{}_{bc}$ is a tensor of the type indicated, then prove that the contracted quantity $Y_c = X^a{}_{ac}$ is a covariant vector.

5.15 (§5.9) Evaluate δ_a^a and $\delta_b^a \delta_a^b$ in n dimensions.

5.16 (§5.10) Check that the definition of the Lie bracket leads to the results (5.44), (5.45), and (5.46).

5.17 (§5.10) In $\mathbb{R}^2$, let $(x^a) = (x, y)$ denote Cartesian and $(x'^a) = (R, \phi)$ plane polar coordinates (see Exercise 5.6).
(i) If the vector field X has components $X^a = (1, 0)$, then find X'^a.
(ii) The operator grad can be written in each coordinate system as

$$\text{grad} f = \frac{\partial f}{\partial x}\boldsymbol{i} + \frac{\partial f}{\partial y}\boldsymbol{j} = \frac{\partial f}{\partial R}\hat{\boldsymbol{R}} + \frac{1}{R}\frac{\partial f}{\partial \phi}\hat{\boldsymbol{\phi}}$$

where f is an arbitrary function and

$$\hat{\boldsymbol{R}} = \cos \phi \boldsymbol{i} + \sin \phi \boldsymbol{j}, \qquad \hat{\boldsymbol{\phi}} = -\sin \phi \boldsymbol{i} + \cos \phi \boldsymbol{j}.$$

Take the scalar product of grad f with $\boldsymbol{i}, \boldsymbol{j}, \hat{\boldsymbol{R}}$, and $\hat{\boldsymbol{\phi}}$ in turn to find relationships between the operators $\partial/\partial x, \partial/\partial y, \partial/\partial R$, and $\partial/\partial \phi$.
(iii) Express the vector field $\boldsymbol{X}$ as an operator in each coordinate system. Use Part (ii) to show that these expressions are the same.
(iv) If $Y^a = (0, 1)$ and $Z^a = (-y, x)$, then find Y'^a, Z'^a, Y and Z.
(v) Evaluate all the Lie brackets of X, Y, and Z.

Further reading

As discussed in the book, we consider tensors via the index approach, as we consider it the quickest route to being proficient in using tensors in practice. The older texts adopt the same approach, and one example of a classic text on differential geometry, which was a major source for this book, is the one by Synge and Schild (1949). Many of the modern books which introduce tensors using the index-free approach are, in our opinion, quite sophisticated for a first course in general relativity. One exception, however, is the excellent book of Schutz (1985). This is written at about the same level as this book and contains material not covered in this book, so may be considered as a companion text to this book. The earlier book of Schutz (1980) provides a more solid grounding in differential geometry. The book by Wald (1984) is also excellent and contains some more advanced material. Adopting a completely index-free approach is notationally difficult in many calculations, so that the abstract index notation of Penrose (1968) provides an excellent and practical coordinate-free method of doing tensorial calculations. The most advanced and complete treatment of this and other geometrical methods can be found in the two volumes of Penrose and Rindler (1986). Our treatment has one important omission, and that is the topic of differential forms (which is omitted because we do not use it). The book by Hughston and Tod (1990) on general relativity includes both a treatment and a subsequent application in discussing anisotropic cosmologies. The various sign conventions can be found on the inside cover of Misner et al. (1973). We use the time-like convention of Landau and Lifshitz (1971). We also list some other more mathematical texts on Lorentzian geometry that readers who want to go into more detail may find useful. These include the books by O'Neil (1983) and by Choquet-Bruhat, De Witt-Morette, and Dillard-Bleick, (1977).

Choquet-Bruhat, Y., De Witt-Morette, C., and Dillard-Bleick, M. (1977). *Analysis, Manifolds and Physics*. North-Holland, Amsterdam.

Hughston, L. P., and Tod, K. P. (1990). *An Introduction to General Relativity*. Cambridge University Press, Cambridge.

Landau, L. D., and Lifshitz, E. M. (1971). *The Classical Theory of Fields*. Pergamon, Oxford.

Misner, C. W., Thorne, K. S., and Wheeler, J. A. (1973). *Gravitation*. Freeman, San Francisco, CA.

O'Neil, B. (1983). Semi-Riemannian Geometry: *With Application to Relativity*. Pure and Applied Mathematics Series. Academic Press, New York, NY.

Penrose, R. (1968). 'Structure of space-time', in DeWitt, C. M., and Wheeler, J. A., eds, *Battelle Rencontres* 1967 *Lectures in Mathematics and Physics*. W. A. Benjamin, New York, NY, 121–235.

Penrose, R., and Rindler, W. (1986). *Spinors and Space-Time*. Vols 1 and 2, Cambridge University Press, Cambridge.

Schutz, B. F. (1980). *Geometrical Methods in Mathematical Physics*. Cambridge University Press, Cambridge.

Schutz, B. F. (1985). *A First Course in General Relativity*. Cambridge University Press, Cambridge.

Synge, J. L., and Schild, A. (1949). *Tensor Calculus*. University of Toronto Press, Toronto.

Tensor calculus

6

6.1 Partial derivative of a tensor

In the last chapter, we met algebraic operations which are tensorial, that is, which convert tensors into tensors. The operations are addition, subtraction, contraction, and tensor products. The next question which arises is, What differential operations are there that are tensorial? The answer to this turns out to be very much more involved. The first thing we shall see is that partial differentiation of tensors is **not** tensorial. Different authors denote the partial derivative of a contravariant vector X^a by

$$\partial_b X^a \quad \text{or} \quad \frac{\partial X^a}{\partial x^b} \quad \text{or} \quad X^a{}_{,b} \quad \text{or} \quad X^a{}_{|b}$$

and similarly for higher-rank tensors. We shall use a mixture of all the first three notations. (Note that, in the literature, the partial derivative of a tensor is often referred to as the **ordinary** derivative of a tensor, to distinguish it from the tensorial differentiation we shall shortly meet). Now differentiating (5.18) with respect to x'^c, we find

$$
\begin{aligned}
\partial'_c X'^a &= \frac{\partial}{\partial x'^c}\left(\frac{\partial x'^a}{\partial x^b} X^b\right) \\
&= \frac{\partial x^d}{\partial x'^c}\frac{\partial}{\partial x^d}\left(\frac{\partial x'^a}{\partial x^b} X^b\right) \\
&= \frac{\partial x'^a}{\partial x^b}\frac{\partial x^d}{\partial x'^c}\partial_d X^b + \frac{\partial^2 x'^a}{\partial x^b \partial x^d}\frac{\partial x^d}{\partial x'^c} X^b.
\end{aligned}
\tag{6.1}
$$

If the first term on the right-hand side alone were present, then this would be the usual tensor transformation law for a tensor of type $(1, 1)$. However, the presence of the second term prevents $\partial_b X^a$ from behaving like a tensor.

There is a fundamental reason why this is the case. By definition, the process of differentiation involves comparing a quantity evaluated at two neighbouring points, P and Q, say, dividing by some parameter representing the separation of P and Q, and then taking the limit as this parameter goes to zero. In the case of a contravariant vector field X^a, this would involve computing

$$\lim_{\delta u \to 0} \frac{[X^a]_P - [X^a]_Q}{\delta u},$$

Introducing Einstein's Relativity. Second Edition. Ray d'Inverno and James Vickers, Oxford University Press.
© Ray d'Inverno and James Vickers (2022). DOI: 10.1093/oso/9780198862024.003.0006

for some appropriate parameter δu. However, from the transformation law in the form (5.30), we see that

$$X_P'^a = \left[\frac{\partial x'^a}{\partial x^b}\right]_P X_P^b \quad \text{and} \quad X_Q'^a = \left[\frac{\partial x'^a}{\partial x^b}\right]_Q X_Q^b.$$

This involves the transformation matrix evaluated at **different** points, from which it should be clear that $X_P^a - X_Q^a$ is not a tensor. Similar remarks hold for differentiating tensors in general.

It turns out that, if we wish to differentiate a tensor in a tensorial manner, then we need to introduce some auxiliary structure onto the manifold. We shall meet three different types of differentiation. First of all, in the next section, we shall introduce a contravariant vector field onto the manifold and use it to define the **Lie derivative**. Then we shall introduce a quantity called an **affine connection** and use it to define **covariant differentiation**. Finally, we shall introduce a tensor called a **metric** and from it build a special affine connection, called the **metric connection**, and again define covariant differentiation but relative to this specific connection.

6.2 The Lie derivative

The argument we present in this section is rather intricate. It rests on the idea of interpreting a coordinate transformation **actively** as a point transformation, rather than **passively**, as we have done up to now. The important results occur at the end of the section and consist of the formula for the Lie derivative of a general tensor field and the basic properties of Lie differentiation.

We start by considering a **congruence of curves** defined such that only one curve goes through each point in the manifold. Then, given any one curve of the congruence,

$$x^a = x^a(u),$$

we can use it to define the tangent vector field $\mathrm{d}x^a/\mathrm{d}u$ along the curve. If we do this for every curve in the congruence, then we end up with a vector field X^a (given by $\mathrm{d}x^a/\mathrm{d}u$ at each point) defined over the whole manifold (Fig. 6.1).

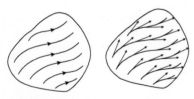

Fig. 6.1 The tangent vector field resulting from a congruence of curves.

Conversely, given a non-zero vector field $X^a(x)$ defined over the manifold, then this can be used to define a congruence of curves in the manifold called the **integral curves** or **trajectories** of X^a. The procedure is exactly the same as the way in which a vector field gives rise to field lines or streamlines in vector analysis. These curves are obtained by solving the ordinary differential equations

$$\frac{\mathrm{d}x^a}{\mathrm{d}u} = X^a\left(x\left(u\right)\right). \tag{6.2}$$

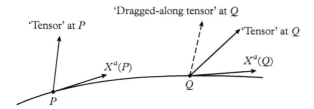

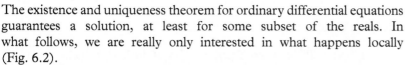

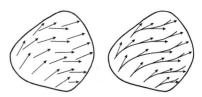

Fig. 6.3 Using the congruence to compare tensors at neighbouring points.

Fig. 6.2 The local congruence of curves resulting from a vector field.

The existence and uniqueness theorem for ordinary differential equations guarantees a solution, at least for some subset of the reals. In what follows, we are really only interested in what happens locally (Fig. 6.2).

We therefore assume that X^a has been given and we have used it to construct the local congruence of integral curves. Suppose we have some tensor field $T^{a\cdots}_{b\cdots}(x)$, which we wish to differentiate using X^a. Then the essential idea is to use the congruence of curves to **drag** the tensor at some point P (i.e. $T^{a\cdots}_{b\cdots}(P)$) along the curve passing through P to some neighbouring point Q, and then compare this 'dragged-along tensor' with the tensor already there (i.e. $T^{a\cdots}_{b\cdots}(Q)$) (Fig. 6.3). Since the dragged-along tensor will be of the same type as the tensor already at Q, we can **subtract the two tensors at Q** and so define a derivative by some limiting process as Q tends to P. The technique for dragging involves viewing the coordinate transformation from P to Q **actively**, and applying it to the usual transformation law for tensors. We shall consider the detailed calculation in the case of a contravariant tensor field of rank 2, $T^{ab}(x)$, say.

Consider the transformation

$$\tilde{x}^a = x^a + \delta u\, X^a(x), \tag{6.3}$$

where δu is small. This is called a **point transformation** and is to be regarded actively as sending the point P, with coordinates x^a, to the point Q, with coordinates $x^a + \delta u X^a(x)$, where the coordinates of each point are given in the **same** x^a-coordinate system, i.e.

$$P \to Q,$$
$$x^a \to x^a + \delta u X^a(x).$$

The point Q clearly lies on the curve of the congruence through P which X^a generates (Fig. 6.4). Differentiating (6.3), we get

$$\frac{\partial \tilde{x}^a}{\partial x^b} = \delta^a_b \; + \; \delta u\, \partial_b X^a. \tag{6.4}$$

Next, consider the tensor field T^{ab} at the point P. Then its components at P are $T^{ab}(x)$ and, under the point transformation (6.3), we have the mapping

$$T^{ab}(x) \to \tilde{T}^{ab}(\tilde{x}),$$

i.e. the transformation 'drags' the tensor T^{ab} along from P to Q. The components of the dragged-along tensor are given by the usual transformation law for tensors (see (5.30)), and so, using (6.4)

$$
\begin{aligned}
\tilde{T}^{ab}(\tilde{x}) &= \frac{\partial \tilde{x}^a}{\partial x^c}\frac{\partial \tilde{x}^b}{\partial x^d}T^{cd}(x) \\
&= (\delta_c^a + \delta u \partial_c X^a)(\delta_d^b + \delta u \partial_d X^b)T^{cd}(x) \\
&= T^{ab}(x) + \left[\partial_c X^a\, T^{cb}(x) + \partial_d X^b\, T^{ad}(x)\right]\delta u + O(\delta u^2).
\end{aligned}\tag{6.5}
$$

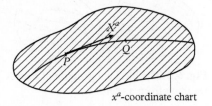

Fig. 6.4 The point P transported to Q in the same x^a coordinate system.

Applying Taylor's theorem to first order, we get

$$
T^{ab}(\tilde{x}) = T^{ab}(x^c + \delta u\, X^c(x)) = T^{ab}(x) + \delta u\, X^c\, \partial_c\, T^{ab}(x).\tag{6.6}
$$

We are now in a position to define the Lie derivative of T^{ab} with respect to X^a, which is denoted by $\mathrm{L}_X T^{ab}$, as

$$
\mathrm{L}_X T^{ab} = \lim_{\delta u \to 0}\frac{T^{ab}(\tilde{x}) - \tilde{T}^{ab}(\tilde{x})}{\delta u}.\tag{6.7}
$$

This involves comparing the tensor $T^{ab}(\tilde{x})$ already at Q with $\tilde{T}^{ab}(\tilde{x})$, the dragged-along tensor at Q. Using (6.5) and (6.6), we find

$$
\mathrm{L}_X T^{ab} = X^c\, \partial_c\, T^{ab} - T^{ac}\, \partial_c\, X^b - T^{cb}\, \partial_c\, X^a.\tag{6.8}
$$

It can be shown that it is always possible to introduce a coordinate system such that the curve passing through P is given by x^1 varying, with $x^2, x^3, \ldots, x^n$ all constant along the curve, and such that

$$
X^a \overset{*}{=} \delta_1^a = (1,\ 0,\ 0,\ldots,0)\tag{6.9}
$$

along this curve. The notation $\overset{*}{=}$ used in (6.9) means that the equation holds only in a particular coordinate system. Then it follows that

$$
X = X^a \partial_a \overset{*}{=} \partial_1,
$$

and equation (6.8) reduces to

$$
\mathrm{L}_X T^{ab} \overset{*}{=} \partial_1\, T^{ab}.\tag{6.10}
$$

Thus, in this special coordinate system, Lie differentiation reduces to ordinary differentiation. In fact, one can define Lie differentiation starting from this viewpoint.

We end the section by collecting together some important properties of Lie differentiation with respect to X which follow from its definition.

1. It is **linear**; for example

$$L_X(\lambda Y^a + \mu Z^a) = \lambda L_X Y^a + \mu L_X Z^a, \qquad (6.11)$$

where λ and μ are **constants**. Thus, in particular, the Lie derivative of the sum and difference of two tensors is the sum and difference, respectively, of the Lie derivatives of the two tensors.

2. It is **Leibniz**; that is, it satisfies the usual product rule for differentiation, for example

$$L_X(Y^a Z_{bc}) = Y^a(L_X Z_{bc}) + (L_X Y^a)Z_{bc}. \qquad (6.12)$$

3. It is **type preserving**; that is, the Lie derivative of a tensor of type (p, q) is again a tensor of type (p, q).

4. It commutes with contraction; for example

$$\delta_b^a L_X T^a{}_b = L_X T^a{}_a. \qquad (6.13)$$

5. The Lie derivative of a **scalar field** ϕ is given by

$$L_X\phi = X\phi = X^a\,\partial_a\,\phi. \qquad (6.14)$$

6. The Lie derivative of a **contravariant vector field** Y^a is given by the Lie bracket of X and Y, that is,

$$L_X Y^a = [X, Y]^a = X^b\,\partial_b\,Y^a - Y^b\,\partial_b X^a. \qquad (6.15)$$

7. The Lie derivative of a **covariant vector field** Y_a is given by

$$L_X Y_a = X^b\partial_b\,Y_a + Y_b\partial_a X^b. \qquad (6.16)$$

8. The Lie derivative of a **general tensor field** $T^{a\cdots}_{b\cdots}$ is obtained as follows: we first partially differentiate the tensor and contract it with X. We then get an additional term for each index of the form of the last two terms in (6.15) and (6.16), where the corresponding sign is negative for a contravariant index and positive for a covariant index, that is,

$$L_X T^{a\cdots}_{b\cdots} = X^c\,\partial_c\,T^{a\cdots}_{b\cdots} - T^{c\cdots}_{b\cdots}\,\partial_c X^a - \cdots + T^{a\cdots}_{c\cdots}\,\partial_b X^c + \cdots. \qquad (6.17)$$

6.3 The affine connection and covariant differentiation

Consider a contravariant vector field $X^a(x)$ evaluated at a point Q, with coordinates $x^a + \delta x^a$, near to a point P, with coordinates x^a. Then, by Taylor's theorem,

$$X^a(x + \delta x) = X^a(x) + \delta x^b \partial_b X^a \tag{6.18}$$

to first order. If we denote the second term by $\delta X^a(x)$, i.e.

$$\delta X^a(x) = \delta x^b \, \partial_b X^a = X^a(x + \delta x) - X^a(x), \tag{6.19}$$

then it is not tensorial, since it involves subtracting tensors evaluated at two different points. We are going to define a tensorial derivative by introducing a vector at Q which in some general sense is 'parallel' to X^a at P. Since $x^a + \delta x^a$ is close to x^a, we can assume that the parallel vector only differs from $X^a(x)$ by a small amount, which we denote $\bar{\delta} X^a(x)$ (Fig. 6.5). By the same argument as in §6.1 above, $\bar{\delta} X^a(x)$ is not tensorial, but we shall construct it in such a way as to make the **difference** vector

$$[X^a(x) + \delta X^a(x)] - [X^a(x) + \bar{\delta} X^a(x)] = \delta X^a(x) - \bar{\delta} X^a(x) \tag{6.20}$$

tensorial. It is natural to require that $\bar{\delta} X^a(x)$ should vanish whenever $X^a(x)$ or δx^a does. Then the simplest definition is to assume that $\bar{\delta} X^a$ is linear in both X^a and δx^a, which means that there exist multiplicative factors Γ^a_{bc} where

$$\bar{\delta} X^a(x) = -\Gamma^a_{bc}(x) X^b(x) \delta x^c \tag{6.21}$$

and the minus sign is introduced to agree with convention.

We have therefore introduced a set of n^3 functions $\Gamma^a_{bc}(x)$ on the manifold, whose transformation properties have yet to be determined. We now define the **covariant derivative** of X^a, written in one of the notations (where we shall use a mixture of the first two)

$$\nabla_c X^a \quad \text{or} \quad X^a{}_{;c} \quad \text{or} \quad X^a{}_{||c}$$

by the limiting process

$$\nabla_c X^a = \lim_{\delta x^c \to 0} \frac{1}{\delta x^c} \left\{ X^a(x + \delta x) - \left[X^a(x) + \bar{\delta} X^a(x) \right] \right\}.$$

In other words, it is the difference between the vector $X^a(Q)$ and the vector at Q parallel to $X^a(P)$, divided by the coordinate differences, in the limit as these differences tend to zero. Using (6.18) and (6.21), we find

$$\nabla_c X^a = \partial_c X^a + \Gamma^a_{bc} X^b. \tag{6.22}$$

Note that in the formula the differentiation index c comes second in the downstairs indices of Γ. If we now demand that $\nabla_c X^a$ is a **tensor** of type $(1, 1)$, then a straightforward calculation (exercise) reveals that Γ^a_{bc} must transform according to

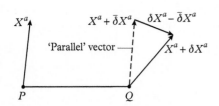

X^a $X^a + \bar{\delta} X^a$ $\delta X^a - \bar{\delta} X^a$

'Parallel' vector ——— $X^a + \delta X^a$

P Q

Fig. 6.5 The parallel vector $X^a + \delta X^a$ at Q.

$$\Gamma'^{a}_{bc} = \frac{\partial x'^{a}}{\partial x^{d}} \frac{\partial x^{e}}{\partial x'^{b}} \frac{\partial x^{f}}{\partial x'^{c}} \Gamma^{d}_{ef} - \frac{\partial x^{d}}{\partial x'^{b}} \frac{\partial x^{e}}{\partial x'^{c}} \frac{\partial^{2} x'^{a}}{\partial x^{d} \partial x^{e}}, \qquad (6.23)$$

or, equivalently (exercise),

$$\Gamma'^{a}_{bc} = \frac{\partial x'^{a}}{\partial x^{d}} \frac{\partial x^{e}}{\partial x'^{b}} \frac{\partial x^{f}}{\partial x'^{c}} \Gamma^{d}_{ef} + \frac{\partial x'^{a}}{\partial x^{d}} \frac{\partial^{2} x^{d}}{\partial x'^{b} \partial x'^{c}}. \qquad (6.24)$$

If the second term on the right-hand side were absent, then this would be the usual transformation law for a tensor of type $(1, 2)$. However, the presence of the second term reveals that the transformation law is linear **inhomogeneous,** and so Γ^{a}_{bc} is not a tensor. Note, however, the inhomogeneous term in (6.23) is exactly what is needed to cancel the inhomogeneous term in (6.1) and so guarantees that (6.22) defines a tensor. Any quantity Γ^{a}_{bc} which transforms according to (6.23) or (6.24) is called an **affine connection** or sometimes simply a **connection** or **affinity**. A manifold with a continuous connection prescribed on it is called an **affine manifold**.

We next define the covariant derivative of a scalar field to be the same as its ordinary derivative, i.e.

$$\nabla_{a} \phi = \partial_{a} \phi. \qquad (6.25)$$

If we now demand that covariant differentiation satisfies the Leibniz rule, then we find (exercise)

$$\nabla_{c} X_{a} = \partial_{c} X_{a} - \Gamma^{b}_{ac} X_{b}. \qquad (6.26)$$

Notice again that the differentiation index comes last in the Γ-term and that this term enters with a minus sign. The name covariant derivative stems from the fact that the derivative of a tensor of type (p, q) is of type $(p, q + 1)$, that is, it has one extra covariant rank. The expression in the case of a general tensor is (compare and contrast with (6.17))

$$\nabla_{c} T^{a\cdots}_{b\cdots} = \partial_{c} T^{a\cdots}_{b\cdots} + \Gamma^{a}_{dc} T^{d\cdots}_{b\cdots} + \cdots - \Gamma^{d}_{bc} T^{a\cdots}_{d\cdots} - \cdots. \qquad (6.27)$$

It follows directly from the transformation laws that the sum of two connections is not a connection or a tensor. However, the **difference** of two connections is a tensor of valence $(1, 2)$, because the inhomogeneous term cancels out in the transformation. For the same reason, the antisymmetric part of a Γ^{a}_{bc}, namely,

$$T^{a}{}_{bc} = \Gamma^{a}_{bc} - \Gamma^{a}_{cb}$$

is a tensor called the **torsion tensor**. If the torsion tensor vanishes, then the connection is **symmetric**, i.e.

$$\Gamma^{a}_{bc} = \Gamma^{a}_{cb}. \qquad (6.28)$$

From now on, unless we state otherwise, we shall **restrict ourselves to symmetric connections**, in which case the torsion vanishes. The assumption that the connection is symmetric leads to the following useful result. In the expression for a Lie derivative of a tensor, **all** occurrences of the partial derivatives may be replaced by covariant derivatives. For example, in the case of a vector (exercise),

$$L_X Y^a = X^b \partial_b Y^a - Y^b \partial_b X^a = X^b \nabla_b Y^a - Y^b \nabla_b X^a. \qquad (6.29)$$

6.4 Affine geodesics

If $T^{a\cdots}_{b\cdots}$ is any tensor, then we introduce the notation

$$\nabla_X T^{a\cdots}_{b\cdots} = X^c \nabla_c T^{a\cdots}_{b\cdots}, \qquad (6.30)$$

that is, ∇_X of a tensor is its covariant derivative contracted with X. Now in §6.2 we saw that a contravariant vector field X determines a local congruence of curves,

$$x^a = x^a(u),$$

where the tangent vector field to the congruence is

$$\frac{dx^a}{du} = X^a.$$

We next define the **absolute derivative** of a tensor $T^{a\cdots}_{b\cdots}$ along a curve C of the congruence, written $DT^{a\cdots}_{b\cdots}/Du$, by

$$\frac{D}{Du}\left(T^{a\cdots}_{b\cdots}\right) = \nabla_X T^{a\cdots}_{b\cdots}. \qquad (6.31)$$

The tensor $T^{a\cdots}_{b\cdots}$ is said to be **parallely propagated** or **transported** along the curve C if

$$\frac{D}{Du}(T^{a\cdots}_{b\cdots}) = 0. \qquad (6.32)$$

This is a first-order ordinary differential equation for $T^{a\cdots}_{b\cdots}$, and so given an initial value for $T^{a\cdots}_{b\cdots}$, say, $T^{a\cdots}_{b\cdots}(P)$, equation (6.32) determines a tensor along C which is everywhere parallel to $T^{a\cdots}_{b\cdots}(P)$.

Using this notation, an **affine geodesic** is defined as a privileged curve along which the direction of the tangent vector is propagated parallel to itself. In other words, the parallely propagated vector at any point of the curve is parallel, that is, proportional, to the tangent vector at that point:

$$\frac{D}{Du}\left(\frac{dx^a}{du}\right) = \lambda(u)\frac{dx^a}{du}.$$

Using (6.31), the equation for an affine geodesic can be written in the form

$$\nabla_X X^a = \lambda X^a, \qquad (6.33)$$

or, equivalently (exercise)

$$\frac{\mathrm{d}^2 x^a}{\mathrm{d}u^2} + \Gamma^a_{bc}\frac{\mathrm{d}x^b}{\mathrm{d}u}\frac{\mathrm{d}x^c}{\mathrm{d}u} = \lambda\frac{\mathrm{d}x^a}{\mathrm{d}u}. \tag{6.34}$$

Note that Γ^a_{bc} appears in the equation multiplied by the symmetric quantity $(\mathrm{d}x^b/\mathrm{d}u)(\mathrm{d}x^c/\mathrm{d}u)$, and so, even if we had not assumed that Γ^a_{bc} was symmetric, the equation picks out just its symmetric part.

The property of being a geodesic as defined by (6.34) does not depend on the choice of parameter. If we introduce a new parameter $\tilde{u}$ along the curve by an invertible transformation $\tilde{u} = \tilde{u}(u)$, then

$$\frac{\mathrm{d}x^a}{\mathrm{d}\tilde{u}} = \frac{\mathrm{d}u}{\mathrm{d}\tilde{u}}\frac{\mathrm{d}x^a}{\mathrm{d}u}, \quad \text{and} \quad \frac{\mathrm{d}^2 x^a}{\mathrm{d}\tilde{u}^2} = \frac{\mathrm{d}^2 u}{\mathrm{d}\tilde{u}^2}\frac{\mathrm{d}x^a}{\mathrm{d}u} + \left(\frac{\mathrm{d}u}{\mathrm{d}\tilde{u}}\right)^2 \frac{\mathrm{d}^2 x^a}{\mathrm{d}u^2}.$$

Hence

$$\frac{\mathrm{d}^2 x^a}{\mathrm{d}\tilde{u}^2} + \Gamma^a_{bc}\frac{\mathrm{d}x^b}{\mathrm{d}\tilde{u}}\frac{\mathrm{d}x^c}{\mathrm{d}\tilde{u}} = \left(\frac{\mathrm{d}u}{\mathrm{d}\tilde{u}}\right)^2 \left(\frac{\mathrm{d}^2 x^a}{\mathrm{d}u^2} + \Gamma^a_{bc}\frac{\mathrm{d}x^b}{\mathrm{d}u}\frac{\mathrm{d}x^c}{\mathrm{d}u}\right) + \frac{\mathrm{d}^2 u}{\mathrm{d}\tilde{u}^2}\frac{\mathrm{d}x^a}{\mathrm{d}u}$$

$$= \left[\left(\frac{\mathrm{d}u}{\mathrm{d}\tilde{u}}\right)^2 \lambda + \frac{\mathrm{d}^2 u}{\mathrm{d}\tilde{u}^2}\right]\frac{\mathrm{d}x^a}{\mathrm{d}u},$$

using (6.34). So defining a new parameter

$$\tilde{\lambda} = \frac{\mathrm{d}u}{\mathrm{d}\tilde{u}}\lambda + \frac{\mathrm{d}\tilde{u}}{\mathrm{d}u}\frac{\mathrm{d}^2 u}{\mathrm{d}\tilde{u}^2}, \tag{6.35}$$

we obtain (check)

$$\frac{\mathrm{d}^2 x^a}{\mathrm{d}\tilde{u}^2} + \Gamma^a_{bc}\frac{\mathrm{d}x^b}{\mathrm{d}\tilde{u}}\frac{\mathrm{d}x^c}{\mathrm{d}\tilde{u}} = \tilde{\lambda}\frac{\mathrm{d}x^a}{\mathrm{d}\tilde{u}}, \tag{6.36}$$

which has the same form as (6.34). From (6.35) we see that by choosing $\tilde{u}$ suitably it is possible to parameterize the curve in such a way that λ vanishes and hence the tangent vector is **covariantly constant** along the curve. Such a parameter is a privileged parameter called an **affine parameter**, often conventionally denoted by s, and the affine geodesic equation reduces to

$$\frac{\mathrm{d}^2 x^a}{\mathrm{d}s^2} + \Gamma^a_{bc}\frac{\mathrm{d}x^b}{\mathrm{d}s}\frac{\mathrm{d}x^c}{\mathrm{d}s} = 0 \tag{6.37}$$

or, equivalently,

$$\nabla_X X^a = 0. \tag{6.38}$$

We can also see from (6.35) that an affine parameter s is only defined up to an **affine transformation** (exercise)

$$s \to \alpha s + \beta,$$

where α and β are constants. We can use the affine parameter s to define the **affine length** of the geodesic between two points P_1 and P_2 by $\int_{P_1}^{P_2} \mathrm{d}s$, and so we can compare lengths on the **same geodesic**. However, we cannot compare lengths on different geodesics (without a metric) because of the arbitrariness in the parameter s. From the existence and uniqueness theorem for ordinary differential equations, it follows that, corresponding to every direction at a point, there is a unique geodesic passing through the point (Fig. 6.6). Similarly, any point can be joined to any other point, as long as the points are sufficiently 'close', by a unique geodesic. However, in the large, geodesics may **focus**, that is, meet again (Fig. 6.7).

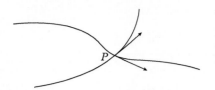

Fig. 6.6 Two affine geodesics passing through P, with given directions.

Fig. 6.7 Two affine geodesics from P, refocusing at Q.

6.5   The Riemann tensor

Covariant differentiation, unlike partial differentiation, is not in general commutative. For any tensor $T^{a\cdots}_{b\cdots}$, we define its **commutator** to be

$$\nabla_c \nabla_d T^{a\cdots}_{b\cdots} - \nabla_d \nabla_c T^{a\cdots}_{b\cdots} \ .$$

Let us work out the commutator in the case of a vector X^a. From (6.22), we see that

$$\nabla_c X^a = \partial_c X^a + \Gamma^a_{bc} X^b.$$

Remembering that this is a tensor of type $(1, 1)$ and using (6.27), we find

$$\nabla_d \nabla_c X^a = \partial_d \left(\partial_c X^a + \Gamma^a_{bc} X^b \right) + \Gamma^a_{ed} \left(\partial_c X^e + \Gamma^e_{bc} X^b \right) - \Gamma^e_{cd} \left(\partial_e X^a + \Gamma^a_{be} X^b \right),$$

with a similar expression for $\nabla_c \nabla_d X^a$, namely,

$$\nabla_c \nabla_d X^a = \partial_c \left(\partial_d X^a + \Gamma^a_{bd} X^b \right) + \Gamma^a_{ec} \left(\partial_d X^e + \Gamma^e_{bd} X^b \right) - \Gamma^e_{dc} \left(\partial_e X^a + \Gamma^a_{be} X^b \right).$$

Subtracting these last two equations and assuming that

$$\partial_d \partial_c X^a = \partial_c \partial_d X^a,$$

we obtain the result

$$\nabla_c \nabla_d X^a - \nabla_d \nabla_c X^a = R^a{}_{bcd} X^b + \left(\Gamma^e_{cd} - \Gamma^e_{dc} \right) \nabla_e X^a, \tag{6.39}$$

where $R^a{}_{bcd}$ is defined by

$$R^a{}_{bcd} = \partial_c \Gamma^a_{bd} - \partial_d \Gamma^a_{bc} + \Gamma^e_{bd} \Gamma^a_{ec} - \Gamma^e_{bc} \Gamma^a_{ed}. \tag{6.40}$$

Moreover, since we are only interested in torsion-free connections, the last term in (6.39) vanishes, so using (5.33) we have

$$\nabla_{[c}\nabla_{d]}X^a = \tfrac{1}{2}R^a{}_{bcd}X^b. \qquad (6.41)$$

Since the left-hand side of (6.41) is a tensor, and X^a is an arbitrary vector, it follows that $R^a{}_{bcd}$ is a tensor of type $(1,3)$. It is called the **Riemann tensor**. It can be shown that, for a symmetric connection, the commutator of any tensor can be expressed in terms of the tensor itself and the Riemann tensor. Thus, the vanishing of the Riemann tensor is a necessary and sufficient condition for the vanishing of the commutator of any tensor. In Section 6.7, we shall search for a geometrical characterization of the vanishing of the Riemann tensor.

6.6 Geodesic coordinates

We first prove a very useful result. At any point P in a manifold, we can introduce a special coordinate system, called a **geodesic coordinate system**, in which

$$[\Gamma^a_{bc}]_P \overset{*}{=} 0.$$

To see this result we can, without loss of generality, choose P to be at the origin of coordinates $x^a \overset{*}{=} 0$ and consider a transformation to a new coordinate system

$$x^a \to x'^a = x^a + \tfrac{1}{2}Q^a_{bc}\,x^b x^c, \qquad (6.42)$$

where $Q^a_{bc} = Q^a_{cb}$ are constants to be determined. Differentiating (6.42), we get

$$\frac{\partial x'^a}{\partial x^d} = \delta^a_d + Q^a_{bd}x^d \quad \text{and} \quad \frac{\partial^2 x'^a}{\partial x^d \partial x^e} = Q^a_{de}.$$

Then, since x^a vanishes at P, we have

$$\left[\frac{\partial x'^a}{\partial x^b}\right]_P = \delta^a_b,$$

from which it follows immediately that the inverse matrix

$$\left[\frac{\partial x^a}{\partial x'^b}\right]_P = \delta^a_b.$$

Substituting these results in (6.23), we find

$$[\Gamma'^a_{bc}]_P = [\Gamma^a_{bc}]_P - Q^a_{bc}.$$

Since the connection is symmetric, we can choose the constants so that

$$Q^a_{bc} = [\Gamma^a_{bc}]_P,$$

and hence we obtain the promised result

$$[\Gamma'^a_{bc}]_P \overset{*}{=} 0. \tag{6.43}$$

Many tensorial equations can be established most easily in geodesic coordinates. Note that, although the connection vanishes at P,

$$\left[\Gamma'^a_{bc,d}\right]_P \overset{*}{\neq} 0$$

in general. It can be shown that the result can be extended to obtain a coordinate system in which the connection vanishes along a curve, but not in general to a neighbourhood of P. If, however, there exists a special coordinate system in which the connection vanishes everywhere, then the manifold is called **affine flat** or simply **flat**. We shall next see that this is intimately connected with the vanishing of the Riemann tensor.

6.7 Affine flatness

In a general affine manifold, the intuitive concept of parallelism breaks down. For, if we parallely transport a vector from one point to another along two different curves, we will obtain two different vectors (Fig. 6.8). If, however, we can transport a vector from one point to any other and the resulting vector is independent of the path taken, then the connection is called **integrable**. Thus, for the usual concept of parallelism to hold, the manifold must possess an integrable connection. We now prove two lemmas which connect together the concepts of affine flatness, integrability, and vanishing Riemann tensor.

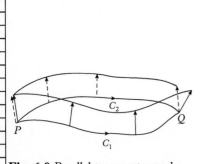

Fig. 6.8 Parallel transport round two curves in a general affine manifold.

Lemma: A necessary and sufficient condition for a connection to be integrable is that the Riemann tensor vanishes.

We consider, first, necessity. Since Γ^a_{bc} is integrable, we can start with a vector X^a at any point and from it construct a unique vector field $X^a(x)$ over the manifold by parallely propagating X^a. The equation for parallely propagating X^a is

$$\frac{\mathrm{D}X^a}{\mathrm{D}u} = \frac{\mathrm{d}x^c}{\mathrm{d}u}\nabla_c X^a = 0,$$

and, since dx^a/du is arbitrary, it follows that the covariant derivative of X^a must vanish, i.e.

$$\nabla_c X^a = \partial_c X^a + \Gamma^a_{bc} X^b = 0. \tag{6.44}$$

Hence, this equation must possess solutions. A necessary condition for a solution of this first-order partial differential equation is the so-called integrability condition

$$\partial_d \partial_c X^a = \partial_c \partial_d X^a, \tag{6.45}$$

namely, the second mixed partial derivatives should commute. In the previous section, we met the identity for the commutator of a vector field (6.39), which for a torsion-free connection gives

$$\nabla_c \nabla_d X^a - \nabla_d \nabla_c X^a = \partial_c \partial_d X^a - \partial_d \partial_c X^a + R^a{}_{bcd} X^b.$$

The left-hand side of this equation vanishes by construction, that is, by (6.44); hence, it follows that (6.45) will hold if and only if

$$R^a{}_{bcd} X^b = 0.$$

Finally, since X^b is arbitrary at every point, a necessary condition for integrability is $R^a{}_{bcd} = 0$ everywhere.

We next prove sufficiency. We start by considering the difference in parallelly propagating a vector X^a around an infinitesimal loop connecting x^a to $x^a + \delta x^a + \Delta x^a$, first via $x^a + \delta x^a$ and then via $x^a + \Delta x^a$ (Fig. 6.9). From §6.3, if we parallelly transport X^a from x^a to $x^a + \delta x^a$, we obtain the vector

$$X^a(x + \delta x) = X^a + \bar{\delta} X^a(x),$$

where, by (6.21),

$$\bar{\delta} X^a(x) = -\Gamma^a_{bc}(x) X^b(x) \delta x^c.$$

Similarly, if we transport this vector subsequently to $x^a + \delta x^a + \Delta x^a$, we obtain the vector

$$X^a(x + \delta x + \Delta x) = X^a(x + \delta x) + \bar{\delta} X^a(x + \delta x),$$

where, in this case,

$$\bar{\delta} X^a(x + \delta x) = -\Gamma^a_{bc}(x + \delta x) X^b(x + \delta x) \Delta x^c.$$

Expanding by Taylor's theorem and using the previous results, we obtain (where everything is assumed evaluated at x^a)

$$\bar{\delta} X^a(x + \delta x) = -(\Gamma^a_{bc} + \partial_d \Gamma^a_{bc} \delta x^d)(X^b - \Gamma^b_{ef} X^e \delta x^f) \Delta x^c$$

$$= -\Gamma^a_{bc} X^b \Delta x^c - \partial_d \Gamma^a_{bc} X^b \delta x^d \Delta x^c$$

$$+ \Gamma^a_{bc} \Gamma^b_{ef} X^e \delta x^f \Delta x^c + \partial_d \Gamma^a_{bc} \Gamma^b_{ef} X^e \delta x^d \delta x^f \Delta x^c.$$

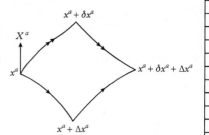

Fig. 6.9 Transporting X^a around an infinitesimal loop.

Neglecting the last term, which is third order, we have

$$X^a(x + \delta x + \Delta x) = X^a - \Gamma^a_{bc}X^b\delta x^c - \Gamma^a_{bc}X^b\Delta x^c - \partial_d\Gamma^a_{bc}X^b\delta x^d\Delta x^c$$
$$+ \Gamma^a_{bc}\Gamma^b_{ef}X^e\delta x^f\Delta x^c.$$

To obtain the equivalent result for the path connecting x^a to $x^a + \delta x^a + \Delta x^a$ via $x^a + \delta x^a$, we simply interchange δx^a and Δx^a to give

$$X^a(x + \Delta x + \delta x) = X^a - \Gamma^a_{bc}X^b\Delta x^c - \Gamma^a_{bc}X^b\delta x^c - \partial_d\Gamma^a_{bc}X^b\Delta x^d\delta x^c$$
$$+ \Gamma^a_{bc}\Gamma^b_{ef}X^e\Delta x^f\delta x^c.$$

Hence, the difference between these two vectors is

$$\begin{aligned}\Delta X^a &= X^a(x + \delta x + \Delta x) - X^a(x + \Delta x + \delta x)\\ &= (\partial_d\Gamma^a_{bc} - \partial_c\Gamma^a_{bd} + \Gamma^a_{ed}\Gamma^e_{bc} - \Gamma^a_{ec}\Gamma^e_{bd})\,X^b\delta x^c\Delta x^d\\ &= R^a{}_{bdc}X^b\delta x^c\Delta x^d\\ &= -R^a{}_{bcd}X^b\delta x^c\Delta x^d,\end{aligned}$$

by (6.40) and the fact that the Riemann tensor is antisymmetric on its last pair of indices (see (6.78)). Thus, the vector X^a will be the same at $x^a + \delta x^a + \Delta x^a$, irrespective of which path is taken, if and only if $R^a{}_{bcd} = 0$. It follows that, if the Riemann tensor vanishes, then the vector X^a will not change if parallely transported around **any** infinitesimal closed loop. Using this result and assuming the manifold has no holes (i.e. the manifold is **simply connected**), then we can continuously deform one curve into another by deforming the curves infinitesimally at each stage (Fig. 6.10), which establishes that the connection is integrable (check).

The second lemma is as follows.

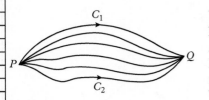

Fig. 6.10 Deforming C_1 into C_2 (infinitesimally at each stage).

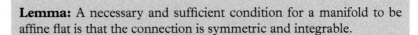

Lemma: A necessary and sufficient condition for a manifold to be affine flat is that the connection is symmetric and integrable.

Sufficiency is established by first choosing n linearly independent vectors

$$X_i{}^a \quad (\boldsymbol{i} = 1, 2, \dots, n)$$

at P, where the bold index $\boldsymbol{i}$ runs from 1 to n and labels the vectors. Using the integrability assumption, we can construct the parallel vector fields $X_i{}^a(x)$ and these will also be linearly independent everywhere. Therefore, at each point P, $X_i{}^a(P)$ is a non-singular matrix of numbers and so we can construct its inverse, denoted by $X^i{}_b$, which must satisfy

$$X^i{}_b X_i{}^a = \delta^a_b, \tag{6.46}$$

where there is a summation over $\boldsymbol{i}$. Multiplying the propagation equation

$$\partial_b X_i{}^a + \Gamma^a_{eb}X_i{}^e = 0,$$

by $X^i_{\ c}$ produces

$$\Gamma^a_{cb} = -X^i_{\ c}\partial_b X_i^{\ a}. \tag{6.47}$$

Differentiating (6.46), we obtain

$$X_i^{\ a}\partial_c X^i_{\ b} = -X^i_{\ b}\partial_c X_i^{\ a} = \Gamma^a_{bc}, \tag{6.48}$$

by (6.47). Using (6.48), we find that

$$X_i^{\ a}\left(\partial_c X^i_{\ b} - \partial_b X^i_{\ c}\right) = \Gamma^a_{bc} - \Gamma^a_{cb} = 0,$$

because the connection is symmetric by assumption. Since the determinant of $X_i^{\ a}$ is non-zero, it follows that the quantity in brackets must vanish, from which we get

$$\partial_c X^i_{\ b} = \partial_b X^i_{\ c}.$$

This in turn implies that $X^i_{\ b}$ must (locally) be the gradient of n scalar fields, $f^i(x)$, say, that is,

$$X^i_{\ b} = \partial_b f^i(x).$$

If we consider the transformation

$$x^a \rightarrow x'^a = f^a(x),$$

then

$$\frac{\partial x'^a}{\partial x^b} = \partial_b f^a(x) = X^a_{\ b}, \tag{6.49}$$

and so, taking inverses,

$$\frac{\partial x^a}{\partial x'^b} = X_b^{\ a}. \tag{6.50}$$

Multiplying (6.23) by $X_a^{\ h}$ and using (6.49) and (6.50) and then (6.46) and (6.48), we find

$$X_a^{\ h}\Gamma'^a_{bc} = X_a^{\ h}(X^a_{\ d}X_b^{\ e}X_c^{\ f}\Gamma^d_{ef} - X_b^{\ e}X_c^{\ f}\partial_e X^a_{\ f})$$
$$= \delta^h_d X_b^{\ e}X_c^{\ f}\Gamma^d_{ef} - X_b^{\ e}X_c^{\ f}\Gamma^h_{ef} \equiv 0.$$

Again, since the determinant of $X_a^{\ h}$ is non-zero, Γ'^a_{bc} vanishes everywhere in this coordinate system and hence the manifold is affine flat. The necessity is straightforward and is left as an exercise.

If we put these two lemmas together, we get the result we have been looking for.

Theorem: A necessary and sufficient condition for a manifold to be affine flat is that the Riemann tensor vanishes.

6.8 The metric

Consider a symmetric covariant tensor field of type $(0, 2)$, say $g_{ab}(x)$. The **determinant** of g_{ab} is denoted by

$$g = \det(g_{ab}), \tag{6.51}$$

and, provided the determinant $\det g_{ab} \neq 0$, then g_{ab} is said to define a (non-singular) **metric**. Since g_{ab} is a symmetric matrix, at every point P we may calculate the eigenvalues at P. We define the **signature** of the metric (at P) to be the number of positive eigenvalues minus the number of negative eigenvalues (there are no zero eigenvalues – why?). Although the eigenvalues themselves depend on the choice of coordinates the signature does not (why?). For a **Riemannian metric**, all the eigenvalues are positive. A manifold endowed with such a metric is called a **Riemannian manifold**. A Riemannian metric can be used to define distances and lengths of vectors. The infinitesimal **distance** (or **interval** in relativity), which we call ds, between two neighbouring points x^a and $x^a + \mathrm{d}x^a$ is defined by

$$\mathrm{d}s^2 = g_{ab}(x)\, \mathrm{d}x^a\, \mathrm{d}x^b. \tag{6.52}$$

Note that this gives the square of the infinitesimal distance, $(\mathrm{d}s)^2$, which is conventionally written as ds^2. The equation (6.52) is also known as the **line element** and g_{ab} is also called the **metric form** or **first fundamental form**. The square of the **length** or **norm** of a contravariant vector X^a is defined by

$$X^2 := g_{ab}(x) X^a X^b. \tag{6.53}$$

The metric is said to be **positive definite** if, for all non-zero vectors X, $X^2 > 0$. It then follows from the definition that a Riemannian metric is nothing but a positive definite metric. For relativity theory, as we will see, one has one positive and three negative eigenvalues, so the signature is -2. We call such a metric **Lorentzian**. (Note that some authors adopt a different convention in which a Lorentzian metric has three positive and one negative eigenvalues so the signature is $+2$). Because a Lorentzian metric has eigenvalues of different signs, one can find non-zero vectors such that

$$g_{ab} X^a X^b = 0. \tag{6.54}$$

We call such vectors **null vectors**. Just as in special relativity, the set of null vectors at a point P form a null cone (in the tangent space $T_P M$) which divides the vectors at P into 'timelike', 'spacelike', or 'null'.

We may also use the metric to define the **angle** between two vectors X^a and Y^a with $X^2 \neq 0$ and $Y^2 \neq 0$. This is given by

$$\cos(X, Y) = \frac{g_{ab}X^a Y^b}{\left(|g_{cd}X^c X^d|\right)^{\frac{1}{2}}\left(|g_{ef}Y^e Y^f|\right)^{\frac{1}{2}}}. \tag{6.55}$$

In particular, the vectors X^a and Y^a are said to be **orthogonal** if

$$g_{ab} X^a Y^b = 0. \tag{6.56}$$

So that a null vector is orthogonal to itself.

Since a metric satisfies $g \neq 0$, at every point we may define the **inverse metric** g^{ab} in the x^a-coordinates, by

$$g_{ab} g^{bc} = \delta_a^c. \tag{6.57}$$

Although it is not completely obvious, it follows from this definition that g^{ab} is a contravariant tensor of rank 2 and it is called the **contravariant metric** (exercise). We may now use g_{ab} and g^{ab} to lower and raise tensorial indices by defining

$$T_{\dots\ a\ \dots}^{\dots\ \dots} = g_{ab} T_{\dots\ \dots}^{\dots\ b\ \dots}, \tag{6.58}$$

and

$$T_{\dots\ \dots}^{\dots\ a\ \dots} = g^{ab} T_{\dots\ b\ \dots}^{\dots\ \dots}, \tag{6.59}$$

where we use the same kernel letter for the tensor irrespective of the position of the indices. Since from now on we shall be working with a manifold endowed with a metric, we shall regard such associated contravariant and covariant tensors as representations of essentially the **same** geometric object. Thus, in particular, T_{ab}, $T_a{}^b$, $T^a{}_b$, and T^{ab} may all be thought of as different representations of the same geometric object. Since we can raise and lower indices freely with the metric, we must be careful about the order in which we write contravariant and covariant indices. For example, in general, $X_a{}^b$ will be different from $X^b{}_a$.

6.9 Metric geodesics

Consider a timelike curve γ (i.e. a curve with timelike tangent vector) with parametric equation $x^a = x^a(u)$. Dividing equation (6.52) by the square of du, we get

$$\left(\frac{ds}{du}\right)^2 = g_{ab}\frac{dx^a}{du}\frac{dx^b}{du}. \tag{6.60}$$

Then the interval s between two points P_1 and P_2 on γ is given by

$$s = \int_{P_1}^{P_2} \mathrm{d}s = \int_{P_1}^{P_2} \frac{\mathrm{d}s}{\mathrm{d}u} \mathrm{d}u = \int_{P_1}^{P_2} \left(g_{ab} \frac{\mathrm{d}x^a}{\mathrm{d}u} \frac{\mathrm{d}x^b}{\mathrm{d}u} \right)^{1/2} \mathrm{d}u. \tag{6.61}$$

We define a **timelike metric geodesic** between any two points P_1 and P_2 as the privileged curve joining them whose interval is **stationary** under small variations that vanish at the end points. Hence, the interval may be a maximum, a minimum, or a saddle point. Deriving the geodesic equations involves the calculus of variations and we postpone this to the next chapter. In that chapter, we shall see that the Euler-Lagrange equations result in the second-order differential equations

$$g_{ab} \frac{\mathrm{d}^2 x^b}{\mathrm{d}u^2} + \{bc, a\} \frac{\mathrm{d}x^b}{\mathrm{d}u} \frac{\mathrm{d}x^c}{\mathrm{d}u} = \left(\frac{\mathrm{d}^2 s}{\mathrm{d}u^2} \Big/ \frac{\mathrm{d}s}{\mathrm{d}u} \right) g_{ab} \frac{\mathrm{d}x^b}{\mathrm{d}u}, \tag{6.62}$$

where the quantities in curly brackets are called the **Christoffel symbols of the first kind** and are defined in terms of derivatives of the metric by

$$\{ab, c\} = \tfrac{1}{2} \left(\partial_b g_{ac} + \partial_a g_{bc} - \partial_c g_{ab} \right). \tag{6.63}$$

Multiplying through by g^{ad} and using (6.57), we get the equations

$$\frac{\mathrm{d}^2 x^a}{\mathrm{d}u^2} + \left\{ \begin{matrix} a \\ bc \end{matrix} \right\} \frac{\mathrm{d}x^b}{\mathrm{d}u} \frac{\mathrm{d}x^c}{\mathrm{d}u} = \left(\frac{\mathrm{d}^2 s}{\mathrm{d}u^2} \Big/ \frac{\mathrm{d}s}{\mathrm{d}u} \right) \frac{\mathrm{d}x^a}{\mathrm{d}u}, \tag{6.64}$$

where $\left\{ \begin{matrix} a \\ bc \end{matrix} \right\}$ are the **Christoffel symbols of the second kind** defined by

$$\left\{ \begin{matrix} a \\ bc \end{matrix} \right\} = g^{ad} \{bc, d\}. \tag{6.65}$$

In addition, the norm of the tangent vector $\mathrm{d}x^a/\mathrm{d}u$ is given by (6.60). If, in particular, we choose a parameter u which is linearly related to the interval s, that is,

$$u = \alpha s + \beta, \tag{6.66}$$

where α and β are constants, then the right hand side of (6.64) vanishes. In the special case when $u = s$, the **equations for a metric geodesic** become

$$\frac{\mathrm{d}^2 x^a}{\mathrm{d}s^2} + \left\{ \begin{matrix} a \\ bc \end{matrix} \right\} \frac{\mathrm{d}x^b}{\mathrm{d}s} \frac{\mathrm{d}x^c}{\mathrm{d}s} = 0 \tag{6.67}$$

and

$$g_{ab}\frac{\mathrm{d}x^a}{\mathrm{d}s}\frac{\mathrm{d}x^b}{\mathrm{d}s} = 1. \tag{6.68}$$

Apart from trivial sign changes, similar results apply for spacelike geodesics, except that we replace s by σ, say, where

$$\mathrm{d}\sigma^2 = -g_{ab}\mathrm{d}x^a\mathrm{d}x^b.$$

However, in the case of an indefinite metric, there exist geodesics for which the distance between any two points is zero called **null geodesics**. It can also be shown that these curves can be parametrized by a special parameter u, called an **affine parameter**, such that their equation does not possess a right hand side, and again takes the form

$$\frac{\mathrm{d}^2 x^a}{\mathrm{d}u^2} + \left\{ \begin{matrix} a \\ bc \end{matrix} \right\} \frac{\mathrm{d}x^b}{\mathrm{d}u}\frac{\mathrm{d}x^c}{\mathrm{d}u} = 0, \tag{6.69}$$

where

$$g_{ab}\frac{\mathrm{d}x^a}{\mathrm{d}u}\frac{\mathrm{d}x^b}{\mathrm{d}u} = 0. \tag{6.70}$$

The last equation follows since the distance between any two points is zero, or, equivalently, the tangent vector is **null**. Again, any other affine parameter is related to u by the transformation

$$u \to \alpha u + \beta,$$

where α and β are constants.

6.10   The metric connection

In general, if we have a manifold endowed with both an affine connection and metric, then it possesses two classes of curves, affine geodesics and metric geodesics, which can be different (Fig. 6.11). In standard Euclidean space, both classes are given by straight lines. Affine geodesics generalize the notion of a straight line as one which does not change direction, while metric geodesics generalise the notion of a straight line as the shortest distance between two points. However, comparing (6.37) with (6.67), the two classes will coincide if we take

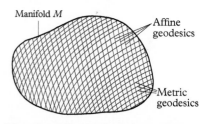

Fig. 6.11 Affine and metric geodesics on a manifold.

$$\Gamma^a_{bc} = \left\{ \begin{matrix} a \\ bc \end{matrix} \right\}, \tag{6.71}$$

or, using (6.65) and (6.63), if

$$\Gamma^a_{bc} = \tfrac{1}{2}g^{ad}\left(\partial_b g_{dc} + \partial_c g_{db} - \partial_d g_{bc}\right). \tag{6.72}$$

It follows from the last equation that the connection is necessarily symmetric, i.e.

$$\Gamma^a_{bc} = \Gamma^a_{cb}. \tag{6.73}$$

In fact, if one checks the transformation properties of $\left\{ {a \atop bc} \right\}$ from first principles, it does indeed transform like a connection (exercise). This special connection built out of the metric and its derivatives is called the **metric connection** and ensures that the two notions of geodesic coincide. From now on, we shall **always** work with the metric connection and we shall denote it by Γ^a_{bc} rather than $\left\{ {a \atop bc} \right\}$ where Γ^a_{bc} is defined by (6.72). This definition leads immediately to the identity (exercise)

$$\nabla_c g_{ab} \equiv 0, \tag{6.74}$$

so that the metric is 'covariantly constant'. Conversely, if we require that (6.74) holds for an arbitrary symmetric connection, then it can be deduced (exercise) that the connection is necessarily the metric connection. Thus, we have the following important result.

Theorem: If ∇_a denotes the covariant derivative with respect to the symmetric affine connection Γ^a_{bc}, then the necessary and sufficient condition for the covariant derivative of the metric to vanish is that the connection is the metric connection.

In addition, we can show that

$$\nabla_c \delta^a_b \equiv 0, \tag{6.75}$$

and

$$\nabla_c g^{ab} \equiv 0. \tag{6.76}$$

6.11   Metric flatness

Now, at any point P of a manifold, g_{ab} is a symmetric matrix of real numbers. Therefore, by standard matrix theory, there exists a transformation which reduces the matrix at P to diagonal form with every diagonal term either $+1$ or -1. The excess of plus signs over minus signs in this form is just the signature that we defined earlier. Assuming that the metric is continuous over the manifold, then, since the determinant is non-zero, it follows that the signature is an invariant. In general, it will not be possible to find a coordinate system in which the metric reduces to this diagonal form everywhere. If, however, there does exist a coordinate system in which the metric reduces to diagonal form with ± 1 diagonal elements everywhere, then the metric is called **flat**.

How does metric flatness relate to affine flatness in the case we are interested in, that is, when the connection is the metric connection? The answer is contained in the following result.

Theorem: A necessary and sufficient condition for a metric to be flat is that its Riemann tensor vanishes.

Necessity follows from the fact that there exists a coordinate system in which the metric is diagonal with ± 1 diagonal elements. Since the metric is constant everywhere, its partial derivatives vanish and therefore the metric connection Γ^a_{bc} vanishes as a consequence of the definition (6.72). Since Γ^a_{bc} vanishes **everywhere** then so must its derivatives. (One way to see this is to recall the definition of partial differentiation which involves subtracting quantities at neighbouring points. If the quantities are always zero, then their difference vanishes, and so does the resulting limit.) The Riemann tensor therefore vanishes by the definition (6.40).

Conversely, if the Riemann tensor vanishes, then by the theorem of §6.7, there exists a special coordinate system in which the connection vanishes everywhere. Since this is the metric connection, by (6.74),

$$\nabla_c g_{ab} = \partial_c g_{ab} - \Gamma^d_{ac} g_{db} - \Gamma^d_{bc} g_{ad} = 0,$$

from which we get

$$\partial_c g_{ab} = \Gamma^d_{ac} g_{db} + \Gamma^d_{bc} g_{ad}, \tag{6.77}$$

and it follows that $\partial_c g_{ab} = 0$. The metric is therefore constant everywhere and hence can be transformed into diagonal form with diagonal elements ± 1. Note the result (6.77), which expresses the ordinary derivative of the metric in terms of the connection. This equation will prove useful later.

Combining this theorem with the theorem of §6.7, we see that, if we use the metric connection, then metric flatness coincides with affine flatness.

6.12 The curvature tensor

The **curvature tensor** or **Riemann–Christoffel tensor** (Riemann tensor for short) is defined by (6.40), namely,

$$R^a{}_{bcd} = \partial_c \Gamma^a_{bd} - \partial_d \Gamma^a_{bc} + \Gamma^e_{bd} \Gamma^a_{ec} - \Gamma^e_{bc} \Gamma^a_{ed},$$

where Γ^a_{bc} is the metric connection, which by (6.72) is given as

$$\Gamma^a_{bc} = \tfrac{1}{2} g^{ad}(\partial_b g_{dc} + \partial_c g_{db} - \partial_d g_{bc}).$$

Thus, $R^a{}_{bcd}$ depends on the metric and its first and second derivatives. It follows immediately from the definition that it is anti-symmetric on its last pair of indices

$$R^a{}_{bcd} = -R^a{}_{bdc}. \tag{6.78}$$

The fact that the connection is symmetric leads to the identity (exercise)

$$R^a{}_{bcd} + R^a{}_{dbc} + R^a{}_{cdb} \equiv 0. \tag{6.79}$$

Lowering the first index with the metric, then it is easy to establish, for example by using geodesic coordinates, that the lowered tensor is symmetric under interchange of the first and last pair of indices, that is (exercise),

$$R_{abcd} = R_{cdab}. \tag{6.80}$$

Combining this with equation (6.78), we see that the lowered tensor is anti-symmetric on its first pair of indices as well:

$$R_{abcd} = -R_{bacd}. \tag{6.81}$$

Collecting these symmetries together, we see that the lowered curvature tensor satisfies

$$R_{abcd} = -R_{abdc} = -R_{bacd} = R_{cdab},$$
$$R_{abcd} + R_{adbc} + R_{acdb} \equiv 0. \tag{6.82}$$

These symmetries considerably reduce the number of independent components; in fact, in n dimensions, the number is reduced from n^4 to $\frac{1}{12}n^2(n^2 - 1)$. In addition to the algebraic identities, it can be shown, again most easily by using geodesic coordinates, that the curvature tensor satisfies a set of **differential** identities called the **Bianchi identities**:

$$\nabla_a R_{debc} + \nabla_c R_{deab} + \nabla_b R_{deca} \equiv 0. \tag{6.83}$$

We can use the curvature tensor to define several other important tensors. The **Ricci tensor** is defined by the contraction

$$R_{ab} = R^c{}_{acb} = g^{cd} R_{dacb}, \tag{6.84}$$

which by (6.80) is symmetric. A final contraction defines the **curvature scalar** or **Ricci scalar** R by

$$R = g^{ab} R_{ab}. \tag{6.85}$$

These two tensors can be used to define the **Einstein tensor**

$$G_{ab} = R_{ab} - \tfrac{1}{2} g_{ab} R, \tag{6.86}$$

which is also symmetric and, by (6.83), the Einstein tensor can be shown to satisfy the **contracted Bianchi identities**

$$\nabla_b G_a{}^b \equiv 0. \tag{6.87}$$

Note that some authors adopt a different sign convention, which leads to the Riemann tensor or the Ricci tensor having the opposite sign to ours.

6.13 The Weyl tensor

We shall mostly be concerned with tensors in four dimensions or less. The algebraic identities (6.82) lead to the following special cases for the curvature tensor:

(1) if $n = 1$, $R_{abcd} = 0$;

(2) if $n = 2$, R_{abcd} has one independent component – essentially R;

(3) if $n = 3$, R_{abcd} has six independent components – essentially R_{ab};

(4) if $n = 4$, R_{abcd} has twenty independent components – ten of which are given by R_{ab} and the remaining ten by the Weyl tensor.

The **Weyl tensor** or **conformal tensor** C_{abcd} is defined in n dimensions, $(n \geqslant 4)$ by

$$C_{abcd} = R_{abcd} + \frac{1}{n-2}(g_{ad}R_{cb} + g_{bc}R_{da} - g_{ac}R_{db} - g_{bd}R_{ca})$$
$$+ \frac{1}{(n-1)(n-2)}(g_{ac}g_{db} - g_{ad}g_{cb})R.$$

Thus, in four dimensions, this becomes

$$C_{abcd} = R_{abcd} + \tfrac{1}{2}(g_{ad}R_{cb} + g_{bc}R_{da} - g_{ac}R_{db} - g_{bd}R_{ca})$$
$$+ \tfrac{1}{6}(g_{ac}g_{db} - g_{ad}g_{cb})R. \tag{6.88}$$

It is straightforward to show that the Weyl tensor possesses the same symmetries as the Riemann tensor, namely,

$$C_{abcd} = -C_{abdc} = -C_{bacd} = C_{cdab},$$
$$C_{abcd} + C_{adbc} + C_{acdb} \equiv 0. \tag{6.89}$$

However, it possesses an additional symmetry

$$C^{a}{}_{bad} \equiv 0. \tag{6.90}$$

Combining this result with the previous symmetries, it then follows that the Weyl tensor is **trace-free**; in other words, it vanishes if one contracts **any** pair of indices. One can think of the Weyl tensor as that part of the curvature tensor for which all contractions vanish.

Two metrics g_{ab} and $\bar{g}_{ab}$ are said to be **conformally related** or **conformal** to each other if

$$\bar{g}_{ab} = \Omega^2 g_{ab}, \tag{6.91}$$

where $\Omega(x)$ is a non-zero differentiable function. Given a manifold with two metrics defined on it which are conformal, then it is straightforward from (6.53) and (6.55) to show that angles between vectors and ratios of magnitudes of vectors, but not lengths, are the same for each metric. Moreover, the null geodesics of one metric **coincide** with the null geodesics of the other (exercise). The metrics also possess the same Weyl tensor, i.e.

$$\bar{C}^a{}_{bcd} = C^a{}_{bcd}. \tag{6.92}$$

Any quantity which satisfies a relationship like (6.92) is called **conformally invariant**. Note the position of the indices on the Weyl tensor is important (with one index up and three indices down). The Weyl tensor with, for example, two indices up and two down is **not** conformally invariant. Other examples of quantities which are not conformally invariant are g_{ab}, Γ^a_{bc}, and $R^a{}_{bcd}$. A metric is said to be **conformally flat** if it can be reduced to the form

$$g_{ab} = \Omega^2 \eta_{ab}, \tag{6.93}$$

where η_{ab} is a flat metric. We end this section by quoting two results concerning conformally flat metrics.

Theorem: A necessary and sufficient condition for a metric to be conformally flat is that its Weyl tensor vanishes everywhere.

Theorem: Any two-dimensional Riemannian manifold is conformally flat.

Exercises

6.1 (§6.2) Prove (6.13) by showing that $L_X \delta^a_b = 0$ in two ways: (i) using (6.17), and (ii) from first principles (remembering Exercise 5.8).

6.2 (§6.2) Use (6.17) to find expressions for $L_X Z_{bc}$ and $L_X(Y^a Z_{bc})$. Use these expressions and (6.15) to check the Leibniz property in the form (6.12).

6.3 (§6.3) Establish (6.23) by assuming that the quantity defined by (6.22) has the tensor character indicated. Take the partial derivative of

$$\delta'^a_c = \frac{\partial x'^a}{\partial x'^c} = \frac{\partial x'^a}{\partial x^d}\frac{\partial x^d}{\partial x'^c},$$

with respect to x'^b to establish the alternative form, (6.24).

6.4 (§6.3) Show that covariant differentiation commutes with contraction by checking that $\nabla_c \delta^a_b = 0$.

6.5 (§6.3) Assuming (6.22) and (6.25), apply the Leibniz rule to the covariant derivative of $X_a Y^a$, where Y^a is arbitrary, to verify (6.26).

6.6 (§6.3) Check (6.29).

6.7 (§6.4) If X, Y and Z are vector fields, f and g smooth functions, and λ and μ constants, then show that
(i) $\nabla_X(\lambda Y + \mu Z) = \lambda\nabla_X Y + \nu\nabla_X Z$,
(ii) $\nabla_{fX+gY}Z = f\nabla_X Z + g\nabla_Y Z$,
(iii) $\nabla_X(fY) = (Xf)Y + f\nabla_X Y$.

6.8 (§6.4) Show that (6.33) leads to (6.34).

6.9 (§6.4) If s is an affine parameter, then show that, under the transformation

$$s \to \bar{s} = \bar{s}(s),$$

the parameter will be affine only if $\bar{s} = \alpha s + \beta$ where α and β are constants.

6.10 (§6.5) Show that

$$\nabla_c\nabla_d X^a{}_b - \nabla_d\nabla_c X^a{}_b = R^a{}_{ecd}X^e{}_b - R^e{}_{bcd}X^a{}_e.$$

[Hint: write out all the terms on the LHS and many should cancel in pairs leaving the terms on the RHS.]

6.11 (§6.5) Show that

$$\nabla_X(\nabla_Y Z^a) - \nabla_Y(\nabla_X Z^a) - \nabla_{[X,Y]}Z^a = R^a{}_{bcd}Z^b X^c Y^d.$$

[Hint: write out all the terms on the LHS and many should cancel in pairs leaving the terms on the RHS.]

6.12 (§6.7) Prove that, if a manifold is affine flat, then the connection is necessarily integrable and symmetric.

6.13 (§6.8) Show that if g^{ab} is defined by (6.57) then it is a rank 2 contravariant tensor. [Hint: one method is to start from the primed version of (6.57).]

6.14 (§6.8) Show that if g_a is **diagonal**, i.e. $g_{ab} = 0$ if $a \neq b$, then g^{ab} is also diagonal with corresponding reciprocal diagonal elements.

6.15 (§6.8) The line elements of $\mathbb{R}^3$ in Cartesian, cylindrical polar, and spherical polar coordinates are given respectively by
(i) $ds^2 = dx^2 + dy^2 + dz^2$,
(ii) $ds^2 = dR^2 + R^2 d\phi^2 + dz^2$,
(iii) $ds^2 = dr^2 + r^2 d\theta^2 + r^2 \sin^2\theta d\phi^2$.
Find g_{ab}, g^{ab}, and g in each case.

6.16 (§6.8) Express T_{ab} in terms of T^{cd}.

6.17 (§6.9) Write down the tensor transformation law of g_{ab}. Show directly that

$$\left\{ \begin{matrix} a \\ bc \end{matrix} \right\} = \tfrac{1}{2}g^{ad}(\partial_b g_{dc} + \partial_c g_{db} - \partial_d g_{bc}),$$

transforms like a connection.

6.18 (§6.9) Find the geodesic equation for $\mathbb{R}^3$ in cylindrical polars. [Hint: use the results of Exercise 6.15(ii) to compute the metric connection and substitute in (6.69).]

6.19 (§6.9) Consider a 3-space with coordinates $(x^a) = (x, y, z)$ and line element

$$ds^2 = dx^2 + dy^2 + dz^2.$$

Prove that the null geodesics are given by

$$x = \ell u + \ell', \quad y = mu + m', \quad z = nu + n',$$

where u is a parameter and ℓ, ℓ', m, m', n, and n' are arbitrary constants satisfying $\ell^2 + m^2 - n^2 = 0$.

6.20 (§6.10) Prove that $\nabla_c g_{ab} \equiv 0$. Deduce that $\nabla_b X_a = g_{ac} \nabla_b X^c$.

6.21 (§6.10) Suppose we have an arbitrary symmetric connection Γ^a_{bc} satisfying $\nabla_c g_{ab} = 0$. Deduce that Γ^a_{bc} must be the metric connection. [Hint: use the equation to find expressions for $\partial_b g_{dc}$, $\partial_c g_{db}$, and $-\partial_d g_{bc}$, as in (6.77), add the equations together, and multiply by $\frac{1}{2} g^{ad}$.]

6.22 (§6.11) The Minkowski line element in Minkowski coordinates

$$(x^a) = (x^0, x^1, x^2, x^3) = (t, x, y, z),$$

is given by

$$ds^2 = dt^2 - dx^2 - dy^2 - dz^2.$$

(i) What is the signature?
(ii) Is the metric non-singular?
(iii) Is the metric flat?

6.23 (§6.11) The line element of $\mathbb{R}^3$ in a particular coordinate system is

$$ds^2 = (dx^1)^2 + (x^1)^2 (dx^2)^2 + (x^1 \sin x^2)^2 (dx^3)^2.$$

(i) Identify the coordinates.
(ii) Is the metric flat?

6.24 (§6.12) Establish the identities (6.79) and (6.80). [Hint: choose an arbitrary point P and introduce geodesic coordinates at P.] Show that (6.79) is equivalent to $R^a{}_{[bcd]} \equiv 0$.

6.25 (§6.12) Establish the identity (6.83). [Hint: use geodesic coordinates.] Show that (6.83) is equivalent to $R_{de[ab;c]} \equiv 0$. Deduce (6.87).

6.26 (§6.12) Show that $G_{ab} = 0$ if and only if $R_{ab} = 0$.

6.27 (§6.13) Establish the identity (6.90). Deduce that the Weyl tensor is trace-free on **all** pairs of indices.

6.28 (§6.13) Show that angles between vectors and ratios of lengths of vectors, but not lengths, are the same for conformally related metrics.

6.29 (§6.13) Prove that the null geodesics of two conformally related metrics coincide. [Hint: the two classes of geodesics need not both be affinely parametrized.]

6.30 (§6.13) Given two metrics with are conformally related i.e.

$$\bar{g}_{ab} = \Omega^2 g_{ab},$$

then defining

$$W = \ln \Omega, \quad W_c = \nabla_c(\ln \Omega), \quad W_{cd} = \nabla_c(\nabla_d(\ln \Omega)),$$

show that
(i)

$$W_{cd} = W_{dc}.$$

(ii)

$$\bar{\Gamma}^a_{bc} = \Gamma^a_{bc} + \delta^a_b W_c + \delta^a_c W_b - g_{bc} W^a.$$

(iii)

$$\bar{R}^a{}_{bcd} = R^a{}_{bcd} + \delta^a_d W_{cb} - \delta^a_c W_{bd} - g_{bd} W^a{}_c - g_{bc} W^a{}_d + \delta^a_d W_b W_d - \delta^a_c g_{bd} W_e W^e$$
$$- g_{bc} W_d W^a - \delta^a_d W_b W^c - \delta^a_d g_{bc} W_e W^e - g_{bd} W_c W^a.$$

(iv) Use the definition (6.88) to deduce (6.92). [Hint: parts (iii) and (iv) involve quite long but straightforward calculations (that is if you are careful about symmetries and dummy indices) and eventually some of the terms cancel in pairs to give simpler expressions.]

6.31 (§6.13) Establish the theorem that any two-dimensional Lorentzian manifold (i.e. at any point the metric can be reduced to the diagonal form $(+1, -1)$) is conformally flat. [Hint: use null curves as coordinate curves, that is, change to new coordinates

$$\lambda = \lambda(x^0, x^1), \quad \nu = \nu(x^0, x^1),$$

satisfying

$$g^{ab}\lambda_{,a}\lambda_{,b} = g^{ab}\nu_{,a}\nu_{,b} = 0,$$

and show that the line element reduces to the form

$$ds^2 = e^{2\mu}d\lambda d\nu,$$

and, finally, introduce new coordinates $\frac{1}{2}(\lambda + \nu)$ and $\frac{1}{2}(\lambda - \nu)$.]

6.32 This final exercise consists of a long calculation which will be needed later in the book. If we take coordinates

$$(x^a) = (x^0, x^1, x^2, x^3) = (t, r, \theta, \phi),$$

then the four-dimensional spherically symmetric line element can be shown to have the form (see Chapter 15 equation (15.37))

$$ds^2 = e^{\nu}dt^2 - e^{\lambda}dr^2 - r^2d\theta^2 - r^2\sin^2\theta d\phi^2,$$

where $\nu = \nu(t, r)$ and $\lambda = \lambda(t, r)$ are arbitrary functions of t and r.
(i) Find g_{ab}, g and g^{ab} (see Exercise 6.14).
(ii) Use the expressions in (i) to find Γ^a_{bc}. [Hint: remember $\Gamma^a_{bc} = \Gamma^a_{cb}$.]
(iii) Calculate R_{abcd} [Hint: use the symmetry relations (6.82).]
(iv) Calculate R_{ab}, R, and G_{ab}.
(v) Calculate $G^a{}_b (= g^{ac}G_{cb} = G_b{}^a)$.

Further reading

Here again we recommend the same set of books as suggested for Chapter 5.

Choquet-Bruhat, Y., De Witt-Morette, C., and Dillard-Bleick, M. (1977). *Analysis, Manifolds and Physics*. North-Holland, Amsterdam.

Hughston, L. P., and Tod, K. P. (1990). *An Introduction to General Relativity*. Cambridge University Press, Cambridge.

Landau, L. D., and Lifshitz, E. M. (1971). *The Classical Theory of Fields*. Pergamon, Oxford.

Misner, C. W., Thorne, K. S., and Wheeler, J. A. (1973). *Gravitation*. Freeman, San Francisco.

O'Neil, B. (1983). *Semi-Riemannian Geometry: With Application to Relativity*. Pure and Applied Mathematics Series. Academic Press, New York, NY.

Penrose, R. (1968). 'Structure of space-time' in DeWitt, C. M., and Wheeler, J. A., eds, *Battelle Rencontres 1967 Lectures in Mathematics and Physics*. W. A. Benjamin, New York, NY, 121–235.

Penrose, R., and Rindler, W. (1986). *Spinors and Space-Time*. Vols 1 and 2, Cambridge University Press, Cambridge.

Schutz, B. F. (1980). *Geometrical Methods in Mathematical Physics.* Cambridge University Press, Cambridge.

Schutz, B. F. (1985). *A First Course in General Relativity.* Cambridge University Press, Cambridge.

Synge, J. L., and Schild, A. (1949). *Tensor Calculus.* University of Toronto Press, Toronto.

Integration, variation, and symmetry

7

7.1 Tensor densities

A tensor density of weight W, denoted conventionally by a gothic letter, $\mathfrak{T}^{a\cdots}_{b\cdots}$, transforms like an ordinary tensor, except that, in addition, the Wth power of the Jacobian

$$J = \left| \frac{\partial x^a}{\partial x'^b} \right| ,$$

appears as a factor, i.e.

$$\mathfrak{T}'^{a\cdots}_{b\cdots} = J^W \frac{\partial x'^a}{\partial x^c} \cdots \frac{\partial x^d}{\partial x'^b} \cdots \mathfrak{T}^{c\cdots}_{d\cdots}. \tag{7.1}$$

Then, with certain modifications, we can combine tensor densities in much the same way as we do tensors. One exception, which follows from (7.1), is that the product of two tensor densities of weight W_1 and W_2 is a tensor density of weight $W_1 + W_2$. There is some arbitrariness in defining the covariant derivative of a tensor density, but we shall adhere to the definition that, if $\mathfrak{T}^{a\cdots}_{b\cdots}$ is a tensor density of weight W, then

$$\nabla_c \mathfrak{T}^{a\cdots}_{b\cdots} = \text{usual terms if } \mathfrak{T}^{a\cdots}_{b\cdots} \text{ were a tensor} - W\Gamma^d_{dc}\mathfrak{T}^{a\cdots}_{b\cdots}. \tag{7.2}$$

For example, the covariant derivative of a vector density of weight W is

$$\nabla_c \mathfrak{T}^a = \partial_c \mathfrak{T}^a + \Gamma^a_{bc}\mathfrak{T}^b - W\Gamma^b_{bc}\mathfrak{T}^a.$$

In the special case when $W = +1$ and $c = a$, we get the important result (check)

$$\nabla_a \mathfrak{T}^a = \partial_a \mathfrak{T}^a, \tag{7.3}$$

that is, the **covariant divergence** of a vector density of weight $+1$ is identical to its **ordinary divergence**. It can be shown that both these quantities are scalar densities of weight $+1$ (exercise).

Introducing Einstein's Relativity. Second Edition. Ray d'Inverno and James Vickers, Oxford University Press.
© Ray d'Inverno and James Vickers (2022). DOI: 10.1093/oso/9780198862024.003.0007

7.2 The Levi-Civita alternating symbol

We introduce a quantity which is a generalization of the Kronecker delta δ^a_b, but which turns out to be a tensor density. The **Levi-Civita alternating symbol** ε^{abcd} is a completely anti-symmetric tensor density of weight $+1$ and contravariant rank 4, whose values in any coordinate system is $+1$ or -1 if $abcd$ is an even or odd permutation of 0123, respectively, and zero otherwise. Thus, for example, in four dimensions, if we let the coordinates range from 0 to 3 (as we shall), i.e.

$$(x^a) = (x^0, x^1, x^2, x^3),$$

then some of its values are

$$\varepsilon^{0123} = \varepsilon^{2301} = -\varepsilon^{0132} = -\varepsilon^{0321} = +1,$$

and

$$\varepsilon^{0120} = \varepsilon^{0331} = \varepsilon^{0101} = 0.$$

We can use ε^{abcd} to define the determinant of a second-rank covariant tensor T_{ab} by

$$\det(T_{ab}) = \frac{1}{4!}\varepsilon^{abcd}\varepsilon^{efgh} T_{ae}T_{bf}T_{cg}T_{dh}, \tag{7.4}$$

which can be shown to be equal to the standard definition where one expands in rows and etc. Since ε^{abcd} is a tensor density of weight $+1$, we see from (7.4) that $\det(T_{ab})$ is a scalar density of weight $+2$. This is in agreement with the fact that

$$T'_{ab}(x') = \frac{\partial x^c}{\partial x'^a}\frac{\partial x^d}{\partial x'^b} T_{cd}(x) \quad\Longrightarrow\quad \det T' = J^2 \det T.$$

Assuming the determinant is non-zero, we can construct the inverse of a second-rank tensor. Similarly, we can define the covariant version ε_{abcd}, which has weight -1. It can be used, in particular, to form the determinant of a second-rank contravariant tensor T^{ab},

$$\det T^{ab} = \frac{1}{4!}\varepsilon_{abcd}\varepsilon_{efgh} T^{ae}T^{bf}T^{cg}T^{dh},$$

which is a scalar density of weight -2. The covariant derivatives of both ε^{abcd} and ε_{abcd} vanish identically (exercise), which from one point of view motivates the definition (7.2).

We define the **generalized Kronecker delta** by

$$\delta^{ab}_{cd} = \begin{cases} +1 & \text{for } a \neq b,\, a = c,\, b = d \\ -1 & \text{for } a \neq b,\, a = d,\, b = c \\ 0 & \text{otherwise} \end{cases}$$

and similarly for higher-order tensors. They are constant tensors of the type indicated, and can be defined in terms of the Kronecker delta by the determinant relationships

$$\delta^{ab}_{cd} = \begin{vmatrix} \delta^a_c & \delta^b_c \\ \delta^a_d & \delta^b_d \end{vmatrix},$$

and

$$\delta^{abc}_{def} = \begin{vmatrix} \delta^a_d & \delta^b_d & \delta^c_d \\ \delta^a_e & \delta^b_e & \delta^c_e \\ \delta^a_f & \delta^b_f & \delta^c_f \end{vmatrix},$$

and so forth. In four dimensions they are related to products of the alternating symbols according to

$$\varepsilon^{abcd}\varepsilon_{efgh} = \delta^{abcd}_{efgh},$$

$$\varepsilon^{abcd}\varepsilon_{efgd} = \delta^{abc}_{efg},$$

$$\varepsilon^{abcd}\varepsilon_{efcd} = 2\delta^{ab}_{ef},$$

$$\varepsilon^{abcd}\varepsilon_{ebcd} = 3!\delta^a_e,$$

$$\varepsilon^{abcd}\varepsilon_{abcd} = 4!.$$

7.3 The metric determinant

If we have a Riemannian manifold with metric g_{ab}, then it transforms according to

$$g'_{ab}(x') = \frac{\partial x^c}{\partial x'^a}\frac{\partial x^d}{\partial x'^b}g_{cd}(x), \qquad (7.5)$$

and so, taking determinants, we have

$$g' = J^2 g.$$

Hence the metric determinant g is a scalar density of weight $+2$. In later chapters, we shall be working with metrics of **negative signature**, in which case g will be negative, and so we write the last equation in the equivalent form

$$(-g') = J^2(-g).$$

Since all these terms are now positive, we can take square roots, to get

$$\sqrt{-g'} = J\sqrt{-g}$$

and hence $\sqrt{-g}$ is a **scalar density of weight +1**. The quantity $\sqrt{-g}$ plays an important role in integration. Given any tensor $T^{a\cdots}_{b\cdots}$, we can form the product $\sqrt{-g}T^{a\cdots}_{b\cdots}$ which is then a **tensor density of weight + 1**. In particular, we can deduce an important result from equation (7.3), namely, for any **vector** T^a,

$$\nabla_a[\sqrt{-g}T^a] = \partial_a[\sqrt{-g}T^a]. \tag{7.6}$$

Now, at any point, the covariant and contravariant metrics are symmetric matrices which are inverse to each other by

$$g_{ab}g^{bc} = \delta^c_a.$$

Let us digress for a moment and consider the general case of finding the derivative of a determinant of a matrix whose elements are functions of the coordinates. Consider any square matrix $A = (a_{ij})$. Then its inverse, (b^{ij}), say, is defined by

$$(b^{ij}) = \frac{1}{a}(A^{ij})^{\mathrm{T}} = \frac{1}{a}(A^{ji}), \tag{7.7}$$

where a is the determinant of A, A^{ij} is the cofactor of a_{ij}, and T denotes the transpose. Let us **fix** i, and expand the determinant a by the ith row. Then

$$a = \sum_{j=1}^n a_{ij}A^{ij},$$

where the index i is not summed and we have explicitly used the summation sign for summing over j for clarity. If we partially differentiate both sides with respect to a_{ij}, then we get

$$\frac{\partial a}{\partial a_{ij}} = A^{ij}, \tag{7.8}$$

since a_{ij} does not occur in any of the cofactors A^{ij} (i fixed, j runs from 1 to n). Repeating the argument for every i, as i runs from 1 to n, we see that the formula (7.8) is quite general. Let us suppose that the a_{ij} are all functions of the coordinates x^k. Then the determinant is a functional of the a_{ij}, which in turn are functions of the x^k, that is,

$$a = a(a_{ij}(x^k)).$$

Differentiating this partially with respect to x^k, using the function of a function rule and equation (7.8), we obtain

$$\frac{\partial a}{\partial x^k} = \frac{\partial a}{\partial a_{ij}} \frac{\partial a_{ij}}{\partial x^k}$$

$$= A^{ij} \frac{\partial a_{ij}}{\partial x^k}$$

$$= ab^{ji} \frac{\partial a_{ij}}{\partial x^k},$$

by equation (7.7). Applying this result to the metric determinant g and remembering that g_{ab} is symmetric, we get the useful equation

$$\partial_c g = g g^{ab} \partial_c g_{ab}. \tag{7.9}$$

We now combine this result with (6.77) (which comes directly from the vanishing of the covariant derivative of the metric) and find

$$\partial_c g = g g^{ab} (\Gamma^d_{ac} g_{db} + \Gamma^d_{bc} g_{ad})$$

$$= g \delta^a_d \Gamma^d_{ac} + g \delta^b_d \Gamma^d_{bc}$$

$$= 2 g \Gamma^a_{ac}. \tag{7.10}$$

Let us compute the covariant derivative of g using (7.2). Then, since g is a scalar density of weight $+2$, we have

$$\nabla_c g = \partial_c g - 2 g \Gamma^a_{ac},$$

and so by equation (7.10) it follows that

$$\nabla_c g \equiv 0. \tag{7.11}$$

This is again intimately connected with the choice of the definition (7.2). Similarly, we find from equation (7.10) that

$$\partial_c \sqrt{-g} - \sqrt{-g} \, \Gamma^a_{ac} = 0,$$

that is, by (7.2),

$$\nabla_c \sqrt{-g} \equiv 0. \tag{7.12}$$

In particular, for any tensor $T^{a\cdots}_{b\cdots}$, this leads to the identity

$$\nabla_c [\sqrt{-g} \, T^{a\cdots}_{b\cdots}] = \sqrt{-g} (\nabla_c T^{a\cdots}_{b\cdots}), \tag{7.13}$$

that is, we can pull factors of $\sqrt{-g}$ and g through covariant derivatives in the same way as we can with factors involving the covariant or contravariant metric.

7.4 Integrals and Stokes' theorem

Unlike tensors in general, we can add a scalar field ϕ evaluated at two different points, x_1 and x_2, say, and the resulting quantity is still a scalar since, under a coordinate transformation, the sum transforms like

$$\phi'(x'_1) + \phi'(x'_2) = \phi(x_1) + \phi(x_2), \tag{7.14}$$

by (5.20). Hence, we might imagine that it is possible to integrate a scalar field ϕ over some n-dimensional region Ω of a manifold M. However, it turns out that the volume element $d\Omega$ is not a scalar but, as we shall see, a scalar density of weight -1. It follows that we can integrate a scalar density Φ of weight $+1$ over a region Ω,

$$\int_\Omega \Phi d\Omega, \tag{7.15}$$

since at each point $\Phi d\Omega$ is a scalar and can be added together by (7.14). There are analogous statements which can be made about integration over curves, surfaces, and hypersurfaces.

Consider an m-dimensional subspace of M whose parametric equation by (5.2) is

$$x^a = x^a(u^i), \quad (i = 1, 2, \ldots, m).$$

The **'volume' element** of this subspace is defined to be

$$d\tau^{a_1 a_2 \cdots a_m} = \delta^{a_1 a_2 \cdots a_m}_{b_1 b_2 \cdots b_m} \frac{\partial x^{b_1}}{\partial u^1} \frac{\partial x^{b_2}}{\partial u^2} \cdots \frac{\partial x^{b_m}}{\partial u^m} du^1 du^2 \cdots du^m. \tag{7.16}$$

This element is an mth rank contravariant tensor under coordinate transformations and behaves like a scalar under arbitrary change of parameter. Hence, if $X_{a_1 a_2 \cdots a_m}$ is an mth rank covariant tensor, then $X_{a_1 a_2 \cdots a_m} d\tau^{a_1 a_2 \cdots a_m}$ is a scalar under both coordinate and parameter transformations, and we can form the integral

$$\int_{\Omega_m} X_{a_1 a_2 \cdots a_m} d\tau^{a_1 a_2 \cdots a_m} \tag{7.17}$$

over some region Ω_m of the subspace.

We now state **Stokes' theorem** for a simply connected m-dimensional subspace Ω_m bounded by the $(m-1)$-dimensional subspace $\partial \Omega_m = \Omega_{m-1}$:

$$\int_{\partial \Omega_m} X_{a_1 a_2 \cdots a_{m-1}} d\tau^{a_1 a_2 \cdots a_{m-1}} = \int_{\Omega_m} \partial_{a_m} X_{a_1 a_2 \cdots a_{m-1}} d\tau^{a_1 a_2 \cdots a_m}. \tag{7.18}$$

Writing this in terms of the parameters u^i, this is nothing but the standard version of Stokes' theorem for a region in $\mathbb{R}^n$. We will be particularly interested in the special case of a four-dimensional region Ω of a four-dimensional manifold M, where Ω is bounded by the hypersurface $\partial \Omega$ (Fig 7.1). Stokes' theorem then becomes the **divergence theorem** or

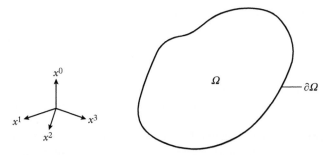

Fig. 7.1 A four-dimensional region Ω bounded by $\partial\Omega$.

Gauss's theorem for a contravariant vector density $\mathfrak{T}^a$ of weight $+1$, which we write in the form

$$\int_{\partial\Omega} \mathfrak{T}^a \mathrm{d}S_a = \int_\Omega \mathfrak{T}^a{}_{,a} \mathrm{d}\Omega, \qquad (7.19)$$

where

$$\mathrm{d}S_a = \frac{1}{3!}\varepsilon_{abcd}\mathrm{d}\tau^{bcd}, \qquad (7.20)$$

and

$$\mathrm{d}\Omega = \frac{1}{4!}\varepsilon_{abcd}\mathrm{d}\tau^{abcd}. \qquad (7.21)$$

If we use the **coordinates** x^a as parameters, then $\mathrm{d}\Omega$ is written as d^4x, where

$$\mathrm{d}^4x := \mathrm{d}x^0 \mathrm{d}x^1 \mathrm{d}x^2 \mathrm{d}x^3, \qquad (7.22)$$

and

$$\mathrm{d}S_a = (\mathrm{d}x^1 \mathrm{d}x^2 \mathrm{d}x^3, \mathrm{d}x^0 \mathrm{d}x^2 \mathrm{d}x^3, \mathrm{d}x^0 \mathrm{d}x^1 \mathrm{d}x^3, \mathrm{d}x^0 \mathrm{d}x^1 \mathrm{d}x^2). \qquad (7.23)$$

Note from the definition (7.21) that d^4x is a **scalar density of weight** -1.

A particularly important case of (7.19) is when we take $\mathfrak{T}^a = \sqrt{-g}T^a$ and we use (7.6) and (7.13) to write

$$\mathfrak{T}^a{}_{,a} = \partial_a(\sqrt{-g}T^a)$$
$$= \nabla_a(\sqrt{-g}T^a)$$
$$= \sqrt{-g}\nabla_a T^a,$$

in which case (7.19) becomes the **covariant divergence theorem**

$$\int_{\partial\Omega} T^a \sqrt{-g}\,\mathrm{d}S_a = \int_\Omega \nabla_a T^a \sqrt{-g}\,\mathrm{d}\Omega. \qquad (7.24)$$

7.5   The Euler-Lagrange equations

The variational principle and with it the Euler-Lagrange equations will play an important role in this book. So, although it is something of a digression, we shall, for completeness, include a brief discussion of their derivation. Then, as a first indication of their usefulness, we shall show in the next section how they provide an efficient method for obtaining geodesics.

A **functional** may be defined as a correspondence between a real number and a function belonging to some class. Thus, a functional is a kind of function where the independent variable is itself a function. One of the basic problems in the calculus of variations is that of finding the stationary values (maxima, minima, saddle points) of the action I defined by

$$I[y] = \int_{x_1}^{x_2} L(y, y', x) \, \mathrm{d}x, \tag{7.25}$$

where L is a functional of the **dynamical variable** $y(x)$, its derivative $y' = \mathrm{d}y/\mathrm{d}x$, and the coordinate x, and is called the **Lagrangian**. The problem is easily generalized. In order to solve the problem, we need to make use of the following result.

Lemma: If $\int_{x_1}^{x_2} \phi(x)\eta(x)\,\mathrm{d}x = 0$, where $\phi(x)$ is continuous and $\eta(x)$ is an **arbitrary** twice-differentiable function vanishing at the end points, i.e. $\eta(x_1) = \eta(x_2) = 0$, then $\phi(x) \equiv 0$.

To establish this, we suppose that $\phi(x) \neq 0$ for some $x = \xi$ in the interval (x_1, x_2). To fix ideas, let us assume $\phi(\xi) > 0$. Then, by continuity, there exists a neighbourhood of ξ, $(\xi_1 < \xi < \xi_2)$ for which $\phi(x) > 0$. Setting

$$\eta(x) = \begin{cases} (x - \xi_1)^4 (x - \xi_2)^4 & \text{for } x \in (\xi_1, \xi_2) \\ 0 & \text{otherwise,} \end{cases}$$

we find that $\eta(x)$ satisfies the conditions of the above lemma. Furthermore,

$$\int_{x_1}^{x_2} \phi(x)\eta(x)\,\mathrm{d}x = \int_{\xi_1}^{\xi_2} \phi(x)\eta(x)\,\mathrm{d}x > 0,$$

which produces a contradiction. Similarly, if we assume $\phi(\xi) < 0$, then again we get a contradiction, and so the result follows.

Returning to (7.25), we assume L is twice differentiable with respect to its three variables. Let us vary y by an **arbitrary small amount** and write

$$\bar{y} = y + \varepsilon\eta(x), \tag{7.26}$$

where ε is small and $\eta(x)$ satisfies the conditions of the lemma, that is, it has continuous second derivatives and vanishes at x_1 and x_2 but is otherwise arbitrary. We define a **variation** of y by

$$\delta y := \bar{y} - y = \varepsilon \eta(x). \tag{7.27}$$

Differentiating (7.26) with respect to x and using the prime notation, we get

$$\bar{y}' = y' + \varepsilon \eta',$$

so that

$$\delta(y') := \bar{y}' - y' = \varepsilon \eta' = (\delta y)',$$

from which we see that δ and $\mathrm{d}/\mathrm{d}x$ acting on y commute. Then, working to first order in ε,

$$
\begin{aligned}
I[\bar{y}] &= I[y + \delta y] \\
&= \int_{x_1}^{x_2} L(y + \varepsilon \eta, y' + \varepsilon \eta', x)\,\mathrm{d}x \\
&= \int_{x_1}^{x_2} \left(L(y, y', x) + \frac{\partial L}{\partial y}\varepsilon \eta + \frac{\partial L}{\partial y'}\varepsilon \eta' \right)\mathrm{d}x,
\end{aligned}
$$

by Taylor's theorem. Thus defining the quantity

$$\delta I := I[y + \delta y] - I[y],$$

we get

$$\delta I = \varepsilon \int_{x_1}^{x_2} \left(\frac{\partial L}{\partial y}\eta + \frac{\partial L}{\partial y'}\eta' \right)\mathrm{d}x.$$

The last term can be integrated by parts, to give

$$\int_{x_1}^{x_2} \frac{\partial L}{\partial y'}\eta'\,\mathrm{d}x = \left[\frac{\partial L}{\partial y'}\eta \right]_{x_1}^{x_2} - \int_{x_1}^{x_2} \frac{\mathrm{d}}{\mathrm{d}x}\left(\frac{\partial L}{\partial y'} \right)\eta\,\mathrm{d}x.$$

The term in square brackets vanishes since $\eta(x_1) = \eta(x_2) = 0$, and hence

$$\delta I = \varepsilon \int_{x_1}^{x_2} \left[\frac{\partial L}{\partial y} - \frac{\mathrm{d}}{\mathrm{d}x}\left(\frac{\partial L}{\partial y'} \right) \right]\eta\,\mathrm{d}x. \tag{7.28}$$

If $y = y(x)$ is a stationary curve, then δI must vanish to first order, and so, using the above lemma, we find that y must satisfy the **Euler–Lagrange equation** for L, that is,

$$\frac{\partial L}{\partial y} - \frac{\mathrm{d}}{\mathrm{d}x}\left(\frac{\partial L}{\partial y'} \right) = 0. \tag{7.29}$$

Introducing some further notation which serves as a useful abbreviation, we define the **variational derivative**, **functional derivative**, or **Euler-Lagrange derivative** of L by

$$\frac{\delta L}{\delta y} := \frac{\partial L}{\partial y} - \frac{\mathrm{d}}{\mathrm{d}x}\left(\frac{\partial L}{\partial y'}\right),$$

so that (7.28) can be written as

$$\delta I = \int_{x_1}^{x_2} \frac{\delta L}{\delta y}\delta y \mathrm{d}x. \tag{7.30}$$

Then, in this formalism, the **principle of stationary action** requires

$$\delta I = 0, \tag{7.31}$$

for arbitrary δy, which leads immediately by the lemma to the Euler-Lagrange equation

$$\frac{\delta L}{\delta y} = 0. \tag{7.32}$$

The argument can be generalized to n dynamical variables, each of which consists of functions of one variable $y_1(x), \ldots, y_n(x)$ in a straightforward manner. Then the action is defined in terms of the Lagrangian by

$$I[y_1, \ldots, y_n] = \int_{x_1}^{x_2} L(y_1, \ldots, y_n, y_1', \ldots, y_n', x)\mathrm{d}x \tag{7.33}$$

and the variations

$$y_i \to \bar{y}_i = y_i + \delta y_i \quad (i = 1, 2, \ldots, n),$$

where

$$\delta y_i = \varepsilon \eta_i(x), \qquad \eta_i(x_1) = \eta_i(x_2) = 0,$$

lead to

$$\delta I = \int_{x_1}^{x_2} \frac{\delta L}{\delta y_i}\delta y_i \mathrm{d}x \quad \text{(summed over } i\text{)},$$

with

$$\frac{\delta L}{\delta y_i} := \frac{\partial L}{\partial y_i} - \frac{\mathrm{d}}{\mathrm{d}x}\left(\frac{\partial L}{\partial y_i'}\right).$$

The principle of stationary action, $\delta I = 0$, for arbitrary independent variations δy_i, produces the Euler-Lagrange equations

$$\frac{\partial L}{\partial y_i} - \frac{\mathrm{d}}{\mathrm{d}x}\left(\frac{\partial L}{\partial y_i'}\right) = 0. \quad (i = 1, 2, \ldots, n). \qquad (7.34)$$

The further generalization to a system of m dynamical variables $y_A(x)$ ($A = 1, 2, \ldots m$), defined on an n-dimensional manifold M, starts from the action

$$I = \int_\Omega \mathcal{L}(y_A, y_{A,b}, x^a) \mathrm{d}\Omega, \qquad (7.35)$$

where a comma in the subscript denotes a partial derivative, i.e. $y_{A,b} = \partial_b y_A$, and the Lagrangian $\mathcal{L}$ is a scalar density of weight $+1$ and leads to the Euler-Lagrange equations

$$\frac{\delta \mathcal{L}}{\delta y_A} := \frac{\partial \mathcal{L}}{\partial y_A} - \left(\frac{\partial \mathcal{L}}{\partial y_{A,b}}\right)_{,b} = 0 \quad (A = 1, 2, \ldots m). \qquad (7.36)$$

The significance of the variational principle approach is that most, if not all, physical theories may be formulated by specifying a suitable Lagrangian. The Euler-Lagrange equations can then be computed in a straightforward manner and these constitute the **field equations** of the theory.

7.6 The variational method for geodesics

We now apply the technique of the last section to finding a convenient way for computing the geodesics of a given metric. We start from the Lagrangian functional (compare with (7.33))

$$L = L(x^a, \dot{x}^a, u),$$

where u is a parameter along a timelike curve and the dot denotes differentiation with respect to u, defined in terms of the metric by

$$L = [g_{ab}(x)\dot{x}^a \dot{x}^b]^{1/2}. \qquad (7.37)$$

It follows from (6.60) that the action is

$$\int_{P_1}^{P_2} L \mathrm{d}u = \int_{P_1}^{P_2} \mathrm{d}s = s, \qquad (7.38)$$

where s is the interval between any two points P_1 and P_2 on a curve connecting them. The **metric geodesic** between these points P_1 and P_2 is defined as that curve joining them whose interval is stationary under small variations which vanish at the end points. In other words, we need to solve the principle of stationary action problem $\delta s = 0$. The solution consists of the Euler-Lagrange equations (7.34) in the form

$$\frac{\partial L}{\partial x^a} - \frac{\mathrm{d}}{\mathrm{d}u}\left(\frac{\partial L}{\partial \dot{x}^a}\right) = 0. \tag{7.39}$$

In principle these equations solve the problem, but in practice there are a number of difficulties. First of all, it is much better to work where possible with L^2 rather than L to avoid square roots. Then there is the freedom in the choice of the parameter u. Finally, in the case of an indefinite metric, there is the distinction between null and non-null geodesics. Assuming $L \neq 0$ and multiplying (7.39) by $-2L$, we get

$$2L\left[\frac{\mathrm{d}}{\mathrm{d}u}\left(\frac{\partial L}{\partial \dot{x}^a}\right) - \frac{\partial L}{\partial x^a}\right] = 0 \tag{7.40}$$

which can be rewritten as

$$\frac{\mathrm{d}}{\mathrm{d}u}\left(\frac{\partial L^2}{\partial \dot{x}^a}\right) - \frac{\partial L^2}{\partial x^a} = 2\frac{\partial L}{\partial \dot{x}^a}\frac{\mathrm{d}L}{\mathrm{d}u}. \tag{7.41}$$

Substituting for L^2, the left-hand side of (7.41) produces

$$\frac{\mathrm{d}}{\mathrm{d}x}\left(\frac{\partial L^2}{\partial \dot{x}^a}\right) - \frac{\partial L^2}{\partial x^a} = \frac{\mathrm{d}}{\mathrm{d}u}\left[\frac{\partial}{\partial \dot{x}^a}(g_{bc}\dot{x}^b\dot{x}^c)\right] - \frac{\partial}{\partial x^a}(g_{bc}\dot{x}^b\dot{x}^c)$$

$$= \frac{\mathrm{d}}{\mathrm{d}u}(2g_{ab}\dot{x}^b) - (\partial_a g_{bc})\dot{x}^b\dot{x}^c$$

$$= 2g_{ab}\ddot{x}^b + 2\partial_c g_{ab}\dot{x}^b\dot{x}^c - \partial_a g_{bc}\dot{x}^b\dot{x}^c$$

$$= 2g_{ab}\ddot{x}^b + 2\dot{x}^b\dot{x}^c[\tfrac{1}{2}(\partial_c g_{ba} + \partial_b g_{ca} - \partial_a g_{bc})]$$

$$= 2g_{ab}\ddot{x}^b + 2\dot{x}^b\dot{x}^c\{bc,a\},$$

where we have used symmetry, interchange of dummy indices, and (6.63). If we again assume that $L \neq 0$, then the right-hand side of (7.41) produces

$$2\frac{\partial L}{\partial \dot{x}^a}\frac{\mathrm{d}L}{\mathrm{d}u} = 2\frac{\partial}{\partial \dot{x}^a}(g_{bc}\dot{x}^x\dot{x}^c)^{\frac{1}{2}}\frac{\mathrm{d}}{\mathrm{d}u}\left(\frac{\mathrm{d}s}{\mathrm{d}u}\right)$$

$$= 2(g_{bc}\dot{x}^b\dot{x}^c)^{-\frac{1}{2}}g_{ad}\dot{x}^d\frac{\mathrm{d}^2 s}{\mathrm{d}u^2}$$

$$= 2\left(\frac{\mathrm{d}^2 s}{\mathrm{d}u^2}\Big/\frac{\mathrm{d}s}{\mathrm{d}u}\right)g_{ab}\dot{x}^b.$$

Equating these two results and dividing by 2 gives the equation (6.62). Multiplying through by g^{ad} and using (6.65) leads to

$$\ddot{x}^a + \Gamma^a_{bc}\dot{x}^b\dot{x}^c = (\ddot{s}/\dot{s})\dot{x}^a. \tag{7.42}$$

If we choose the parameter $u = s$, then the right-hand side vanishes, giving

$$\ddot{x}^a + \Gamma^a_{bc}\dot{x}^b\dot{x}^c = 0, \tag{7.43}$$

and hence s is an affine parameter. It follows from (7.42) that any other affine parameter is related to s by

$$\bar{s} = \alpha s + \beta, \tag{7.44}$$

where α and β are constants. A similar argument applies to spacelike geodesics (exercise).

In the case of an indefinite metric, the interval ds between neighbouring points on a curve may be zero. A null geodesic is a geodesic whose interval between any of its two points is zero. It follows from (7.37) that L vanishes and so the argument given above breaks down. However, it is possible to modify the argument (we shall not do it) to show that the general equations of a null geodesic are

$$\ddot{x}^a + \Gamma^a_{bc}\dot{x}^b\dot{x}^c = \lambda(u)\dot{x}^a,$$

where $\lambda(u)$ is some function of the curve's parameter u and where the tangent vector $\dot{x}^a$ satisfies $g_{ab}\dot{x}^a\dot{x}^b = 0$. As before, if the geodesic equations do not possess a right-hand side, that is, $\lambda = 0$, then the parameter u is called affine. Any other parameter $\bar{u}$ will be affine if it is related to u by

$$\bar{u} = \alpha u + \beta, \tag{7.45}$$

where α and β are constants.

Summarizing, if we define the quantity K by

$$K := \tfrac{1}{2}g_{ab}\dot{x}^a\dot{x}^b, \tag{7.46}$$

and if we take u to be an **affine** parameter, then the most useful form of the geodesic equations is (exercise)

$$\frac{\partial K}{\partial x^a} - \frac{d}{du}\left(\frac{\partial K}{\partial \dot{x}^a}\right) = 0, \tag{7.47}$$

where along any geodesic the quantity K is a constant, with

$$2K = \begin{cases} 0, \\ +1, \\ -1, \end{cases} \tag{7.48}$$

depending on whether the tangent vector is null, or has positive or negative length, respectively, and where in the last two cases we take u to be the distance parameters s and σ. This is the approach we shall adopt in our ensuing work. It is possible, by (7.43), to read off directly from (7.47) the components of the connection Γ^a_{bc}, and this proves to be a very efficient way of calculating Γ^a_{bc}.

7.7 Isometries

Tensor calculus is largely concerned with how quantities change under co-ordinate transformations. It is of particular interest when a quantity does not change, i.e. remains invariant, under coordinate transformations. For example, coordinate transformations which leave a metric invariant are of importance since they contain information about the **symmetries** of a Riemannian manifold. Just as in an ordinary Euclidean space, there are two sorts of transformations: **discrete** ones, like reflections, and **continuous** ones, like translations and rotations. In most applications, these latter types are the more important ones and they can in principle be obtained systematically by obtaining the so-called Killing vectors of a metric, which we now discuss below.

Consider a map $\phi : M \rightarrow M$ that is **invertible** and **smooth** (i.e. can be differentiated as often as we want). Such a map is called a **diffeomorphism**. Suppose we introduce a coordinate system $(x^1, \ldots, x^n)$; then ϕ, treated as an active transformation, takes the point P with coordinates x^a to the point Q with coordinates $\tilde{x}^a$, say. Since $\tilde{x}^a$ depends on the coordinates of P, we may write

$$\tilde{x}^a = \tilde{x}^a(x^1, \ldots, x^n),$$

and, since ϕ is invertible,

$$x^a = x^a(\tilde{x}^1, \ldots, \tilde{x}^n).$$

We may use ϕ to take a tensor at the point P to a tensor at the point Q. For example, given the tensor T_{ab} at the point P, we define $\tilde{T}_{ab}$ at the point Q by

$$\tilde{T}_{ab} := \frac{\partial x^c}{\partial \tilde{x}^a} \frac{\partial x^d}{\partial \tilde{x}^b} T_{cd}. \tag{7.49}$$

This is called the **push-forward** map. Note, despite the similarity to the formula (5.27) for a change of coordinates, formula (7.49) takes the x^a-components of a tensor at the point P to the x^a-components of a tensor at the point Q, so is an **active** transformation describing a change of location rather than a passive one describing a change of coordinates (as in §6.2).

Conversely, we may take a tensor $\tilde{T}_{ab}$ back from Q to the point P using

$$T_{ab} := \frac{\partial \tilde{x}^c}{\partial x^a} \frac{\partial \tilde{x}^d}{\partial x^b} \tilde{T}_{cd}. \tag{7.50}$$

This is called the **pull-back** map.

Rather than consider a fixed tensor at P, we now consider a tensor field $T_{ab}(x)$. We say that ϕ is a **symmetry** of the tensor field if taking $\tilde{T}_{ab}$ in (7.50) to be the value of the tensor field at Q, i.e. $\tilde{T}_{ab} = T_{ab}(\tilde{x})$ and then pulling $\tilde{T}_{ab}$ back to the point P just gives the same as $T_{ab}(x)$ at the point P. Thus, ϕ is a symmetry if

$$T_{ab}(x) = \frac{\partial \tilde{x}^c}{\partial x^a}\frac{\partial \tilde{x}^d}{\partial x^b} T_{cd}(\tilde{x}(x)), \qquad (7.51)$$

where both sides are now a function of x.

Of special interest is the case where the tensor field is the metric g_{ab}. A symmetry of the metric is called an **isometry** and satisfies

$$g_{ab}(x) = \frac{\partial \tilde{x}^c}{\partial x^a}\frac{\partial \tilde{x}^d}{\partial x^b} g_{cd}(\tilde{x}(x)). \qquad (7.52)$$

In general, the condition (7.52) is very complicated, but it may be greatly simplified if we consider the special case of an **infinitesimal** coordinate transformation

$$x^a \rightarrow \tilde{x}^a = x^a + \varepsilon X^a(x), \qquad (7.53)$$

where ε is small and arbitrary and X^a is a vector field. Differentiating (7.53) gives

$$\frac{\partial \tilde{x}^a}{\partial x^b} = \delta_b^a + \varepsilon \partial_b X^a,$$

and so, substituting in (7.52) and using Taylor's theorem, we get

$$\begin{aligned}
g_{ab}(x) &= (\delta_a^c + \varepsilon \partial_a X^c)(\delta_b^d + \varepsilon \partial_b X^d) g_{cd}(x^e + \varepsilon X^e) \\
&= (\delta_a^c + \varepsilon \partial_a X^c)(\delta_b^d + \varepsilon \partial_b X^d)[g_{cd}(x) + \varepsilon X^e \partial_e g_{cd}(x) + \cdots] \\
&= g_{ab}(x) + \varepsilon[g_{ad}\partial_b X^d + g_{bd}\partial_a X^d + X^e \partial_e g_{ab}] + O(\varepsilon^2).
\end{aligned}$$

Working to first order in ε and subtracting $g_{ab}(x)$ from each side, it follows that the quantity in square brackets must vanish. This quantity is simply the Lie derivative of g_{ab} with respect to X by (6.17), namely,

$$\mathrm{L}_X g_{ab} = X^e \partial_e g_{ab} + g_{ad}\partial_b X^d + g_{bd}\partial_a X^d. \qquad (7.54)$$

Now we can replace ordinary derivatives by covariant derivatives in any expression for a Lie derivative and so, using (6.74) and (6.58), the condition for an infinitesimal isometry becomes

$$\mathrm{L}_X g_{ab} = \nabla_b X_a + \nabla_a X_b = 0. \qquad (7.55)$$

These are called **Killing's equations** and any solution of them is called a **Killing vector field** X^a. In the language of §6.2, equation (7.55) states that the metric is 'dragged into itself' by the vector field X^a. We have thus established the following important result.

> **Theorem**: An infinitesimal isometry is generated by a Killing vector $X^a(x)$ satisfying $L_X g_{ab} = 0$.

It proves sufficient to restrict attention to infinitesimal transformations because it can be shown that it is possible to build up any finite transformation with non-zero Jacobian (i.e. a continuous transformation) by an integration process involving an infinite sequence of infinitesimal transformations.

Exercises

7.1 (§7.1) Write down the expression for the covariant derivative of a scalar density Φ of weight $+1$.

7.2 (§7.2) Use the definition of the covariant derivative of a tensor density (7.2) to show that the covariant derivatives of both ε^{abcd} and ε_{abcd} vanish identically.

7.3 (§7.3) Denoting the transformation matrices by

$$J_{ab} = \left(\frac{\partial x^a}{\partial x'^b} \right), \quad J^{ab} = \left(\frac{\partial x'^a}{\partial x^b} \right),$$

use the argument of §7.3 to show that

$$\partial_c J = J J^{ba} \partial_c J_{ab},$$

where $J = \det(J_{ab})$ is the Jacobian. Hence show from first principles that, if $\mathfrak{T}^a$ is a vector density of weight $+1$, then $\partial_a \mathfrak{T}^a$ is a scalar density of weight $+1$.

7.4 (§7.3) Start from the assumption that, for an arbitrary vector field T^a,

$$\nabla_a \left[\sqrt{-g} T^a \right] = \partial_a \left[\sqrt{-g} T^a \right],$$

and show that this leads directly to the result

$$\nabla_a [\sqrt{-g}] = \partial_a \left[\sqrt{-g} \right] - \Gamma^b_{ba} \sqrt{-g}$$

(which is consistent with the definition in Exercise 7.1).

7.5 (§7.4) Show that, for any vector field T^a, the divergence theorem in four dimensions can be written in the form

$$\int_{\partial\Omega} T^a \sqrt{-g} dS_a = \int_{\Omega} \nabla_a T^a \sqrt{-g} d^4 x.$$

7.6 (§7.5) Find the Euler–Lagrange equations for the Lagrangians
(i) $L(y, y', x) = y^2 + y'^2$,
(ii) $L(y_1, y_2, y'_1, y'_2, x) = xy_1^3 + y_1 y_2 + y_1 (y'^2_1 + y'^2_2)$.

7.7 (§7.6) Trace the variational argument which leads to the equations for a spacelike geodesic. Defining K by (7.46) and (7.48), show that (7.41) can be written in the form (7.47). [Hint: if u is affine, then $\mathrm{d}L/\mathrm{d}u = 0$.]

7.8 (§7.7) Use (7.46), (7.47), and (7.48) to find the geodesic equations of the spherically symmetric line element given in Exercise 6.32. Use the equations to read off directly the components and check them with those obtained in Exercise 6.32(ii). [Hint: remember $\Gamma^a_{bc} = \Gamma^a_{cb}$.]

7.9 (§7.7) Find all Killing vector solutions of the metric

$$g_{ab} = \begin{pmatrix} x^2 & 0 \\ 0 & x \end{pmatrix},$$

where $(x^a) = (x^0, x^1) = (x, y)$.

7.10 (§7.7) Deduce (7.55) from (7.54).

7.11 (§7.7) Find all the Killing vectors X^a of the three-dimensional Euclidean line element

$$\mathrm{d}s^2 = \mathrm{d}x^2 + \mathrm{d}y^2 + \mathrm{d}z^2.$$

[Hint: deduce from Killing's equations that $\partial_b X_a + \partial_a X_b = 0$, differentiate with respect to x^c, permute the indices to show that $\partial_b \partial_c X_a = 0$, and integrate to get $X^a = \omega^a{}_b X^b + t^a$, where $\omega^a{}_b$ and t^a are constants of integration, usually termed **parameters**.]

Denoting the six independent constants of integration by $\lambda_1, \lambda_2, \lambda_3, \lambda_4, \lambda_5,$ and λ_6, respectively, write the general solution for X^a in the form

$$\lambda_1 \overset{1}{X^a} + \lambda_2 \overset{2}{X^a} + \lambda_3 \overset{3}{X^a} + \lambda_4 \overset{4}{X^a} + \lambda_5 \overset{5}{X^a} + \lambda_6 \overset{6}{X^a}.$$

Find expressions for the vector fields $\overset{i}{X}$, $(i = 1, 2, \ldots, 6)$ and hence, or otherwise, find all values of $[\overset{i}{X}, \overset{j}{X}]$. Interpret the six Killing vector fields in terms of geometrical transformations.

7.12 (§7.7) Show that, if X^a and Y^a are Killing vectors, then so is any linear combination $\lambda X^a + \mu Y^a$, where λ and μ are constants.

7.13 (§7.7) Consider the following operator identity:

$$L_X L_Y - L_Y L_X = L_{[X,Y]}.$$

(i) Check it holds when applied to an arbitrary scalar function f.
(ii) Check it holds when applied to an arbitrary contravariant vector field m^a. [Hint: use the Jacobi identity.]
(iii) Deduce that the identity holds when applied to a covariant vector field V_a. [Hint: let $f = W^a V_a$ where W^a is arbitrary.]

Use the identity to prove that, if X and Y are Killing vector fields, then so is their commutator $[X, Y]$.

Given that $\partial/\partial x$ and $-y\partial/\partial x + x\partial/\partial y$ are Killing vector fields, find another.

7.14 (§7.7) Express $(\nabla_c \nabla_b - \nabla_b \nabla_c) X_a$ in terms of the Riemann tensor. Use this result to prove that any Killing vector satisfies

$$g^{bc} \nabla_b \nabla_a X_c - R_{ab} X^b = 0.$$

7.15 (§7.7) By making use of the identity

$$R^a{}_{bcd} + R^a{}_{dbc} + R^a{}_{cdb} \equiv 0,$$

or otherwise, prove that a Killing vector satisfies

$$\nabla_c \nabla_b X_a = R_{abcd} X^d.$$

Further reading

Yet again we recommend the books as suggested for Chapter 5.

Choquet-Bruhat, Y., De Witt-Morette, C., and Dillard-Bleick, M. (1977). *Analysis, Manifolds and Physics*. North-Holland, Amsterdam.

Hughston, L. P., and Tod, K. P. (1990). *An Introduction to General Relativity*. Cambridge University Press, Cambridge.

Landau, L. D., and Lifshitz, E. M. (1971). *The Classical Theory of Fields*. Pergamon, Oxford.

Misner, C. W., Thorne, K. S., and Wheeler, J. A. (1973). *Gravitation*. Freeman, San Francisco, CA.

O'Neil, B. (1983). *Semi-Riemannian Geometry: With Application to Relativity*. Pure and Applied Mathematics Series. Academic Press, New York, NY.

Penrose, R. (1968). 'Structure of space-time', in DeWitt, C. M., and Wheeler, J. A., eds, *Battelle Rencontres 1967 Lectures in Mathematics and Physics*. W. A. Benjamin, New York, NY, 121–235.

Penrose, R., and Rindler, W. (1986). *Spinors and Space-Time*. Vols 1 and 2, Cambridge University Press, Cambridge.

Schutz, B. F. (1980). *Geometrical Methods in Mathematical Physics*. Cambridge University Press, Cambridge.

Schutz, B. F. (1985). *A First Course in General Relativity*. Cambridge University Press, Cambridge.

Synge, J. L., and Schild, A. (1949). *Tensor Calculus*. University of Toronto Press, Toronto.

Part C
General Relativity

Special relativity revisited

8

8.1 Minkowski space-time

As we saw in Chapter 2, special relativity discards the old Newtonian picture in which absolute time is split off from three-dimensional Euclidean space. Instead, we introduce a four-dimensional continuum called space-time in which an event has coordinates (t, x, y, z), and where the square of the infinitesimal interval ds between infinitesimally separated events satisfies the Minkowski line element (2.13). The essence of special relativity lies in the special Lorentz transformations, and the significance of the Minkowski line element is that it is invariant under such transformations. We now use the language of Part B to formulate this more precisely.

Minkowski space-time, or simply **flat space**, is defined as a four-dimensional manifold endowed with a **flat** metric of signature -2. Then, by definition, since the metric is flat, there exists a special coordinate system covering the whole manifold in which the metric is diagonal, with diagonal elements equal to ± 1. From now on, we shall use the convention that lower-case latin indices run from 0 to 3. The special coordinate system is called a **Minkowski coordinate** system and is written

$$(x^a) = (x^0,\ x^1,\ x^2,\ x^3) = (t,\ x,\ y,\ z). \tag{8.1}$$

We adopt the sign convention in which the **Minkowski line element** takes the form

$$ds^2 = dt^2 - dx^2 - dy^2 - dz^2. \tag{8.2}$$

We write this in tensorial form as

$$ds^2 = \eta_{ab}\, dx^a\, dx^b, \tag{8.3}$$

where from now on we will always take η_{ab} to denote the **Minkowski metric**

Introducing Einstein's Relativity. Second Edition. Ray d'Inverno and James Vickers, Oxford University Press.
© Ray d'Inverno and James Vickers (2022). DOI: 10.1093/oso/9780198862024.003.0007

$$\eta_{ab} := \begin{bmatrix} 1 & 0 & 0 & 0 \\ 0 & -1 & 0 & 0 \\ 0 & 0 & -1 & 0 \\ 0 & 0 & 0 & -1 \end{bmatrix} = \text{diag}\,(1,\ -1,\ -1,\ -1). \qquad (8.4)$$

If we use some other general coordinate system then we shall write the metric in the form

$$ds^2 = g_{ab}dx^a dx^b.$$

For example, in spherical polar coordinates,

$$(x^a) = (t, r, \theta, \phi),$$

where, as usual,

$$x = r \sin \theta \cos \phi, \quad x = r \sin \theta \sin \phi, \quad z = r \cos \theta,$$

the line element becomes

$$ds^2 = dt^2 - dr^2 - r^2 d\theta^2 - r^2 \sin^2 \theta d\phi^2,$$

and the metric is

$$g_{ab} = \text{diag}(1, -1, -r^2, -r^2 \sin^2 \theta).$$

One of the main results of Part B is the theorem of §6.11, which states that a necessary and sufficient condition for a metric to be flat is that its Riemann tensor vanishes. In Minkowski coordinates, the metric η_{ab} is constant and so the connection Γ^a_{bc} vanishes in this coordinate system, from which it is clear that the Riemann curvature tensor vanishes. However, in a general coordinate system, the connection components will not necessarily vanish. For example, in spherical polar coordinates, we find that Γ^a_{bc} has non-vanishing components

$$\begin{aligned} \Gamma^1_{22} &= -r, & \Gamma^1_{33} &= -r \sin^2 \theta, \\ \Gamma^2_{12} &= r^{-1}, & \Gamma^2_{33} &= -\sin \theta \cos \theta, \\ \Gamma^3_{13} &= r^{-1}, & \Gamma^3_{23} &= \cot \theta, \end{aligned} \qquad (8.5)$$

but, if we compute the Riemann tensor, we will again find

$$R^a_{\ bcd} = 0,$$

as required by the theorem.

8.2 The null cone

In Minkowski space-time, the 'square' of the length or norm of a vector is defined as usual by

$$X^2 = g_{ab}X^a X^b = X_a X^a. \qquad (8.6)$$

The vector is said to be

$$\textbf{timelike} \quad \text{if } g_{ab}X^a X^b > 0,$$
$$\textbf{spacelike} \quad \text{if } g_{ab}X^a X^b < 0, \qquad (8.7)$$
$$\textbf{null} \text{ or } \textbf{lightlike} \quad \text{if } g_{ab}X^a X^b = 0.$$

Two vectors X^a and Y^a are orthogonal if their **inner product** vanishes, that is,

$$g_{ab}X^a Y^b = 0,$$

from which it follows that a null vector is orthogonal to itself.

The set of all null vectors at a point P of a Minkowski manifold forms a double cone called the **null cone** or **light cone.** In Minkowski coordinates, the null vectors X^a at P satisfy

$$\eta_{ab}X^a X^b = 0,$$

that is,

$$(X^0)^2 - (X^1)^2 - (X^2)^2 - (X^3)^2 = 0, \qquad (8.8)$$

which is the equation of a double cone. This null cone lies in the tangent space T_p at P but, since it is easy to show that the tangent space is itself a Minkowski manifold (by (8.8)), we can identify the tangent space with the underlying manifold and regard the null cone as lying in the manifold. We will not be able to do this when we go on to consider non-flat manifolds. If we define the timelike vector T^a in Minkowski coordinates by $T^a = (1, 0, 0, 0)$, then a timelike or null vector X^a is said to be

$$\textbf{future-pointing} \quad \text{if } \eta_{ab}X^a T^b > 0,$$
$$\textbf{past-pointing} \quad \text{if } \eta_{ab}X^a T^b < 0.$$

The future-pointing vectors all lie inside or on one sheet of the cone called the **future sheet**, and past-pointing vectors lie inside or on the **past sheet** (Fig. 8.1).

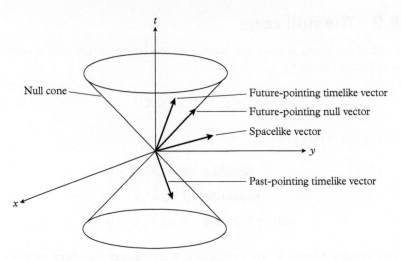

Fig. 8.1 The null cone with one dimension (the z-direction) suppressed.

8.3 The Lorentz group

The Lorentz transformations are defined as those linear homogeneous transformations

$$x^a \rightarrow x'^a = L^a{}_b x^b, \qquad (8.9)$$

of Minkowski coordinates which leave the Minkowski metric η_{ab} **invariant**. From (8.9),

$$\frac{\partial x'^a}{\partial x^b} = L^a{}_b,$$

and, substituting in the transformation formula for a metric (7.5) (with primes and unprimes interchanged), we get (exercise)

$$L^a{}_c L^b{}_d \eta_{ab} = \eta_{cd}, \qquad (8.10)$$

since the metric remains invariant. We see from (7.52) that Lorentz transformations are **isometries**. It follows immediately from (8.10) that Lorentz transformations preserve lengths and inner products of vectors. The Lorentz transformations form a group called the **Lorentz group L**. The identity element of the group is δ^a_b and the inverse element is given by the inverse matrix. The matrix $L^a{}_b$ is invertible because, if we take determinants of each side of (8.10), we get

$$(\det L^a{}_b)^2 = 1 \quad \Rightarrow \quad \det L^a{}_b = \pm 1,$$

and so the matrix is non-singular. If we set $c = d = 0$ in (8.10), we also find that

$$(L^0{}_0)^2 - \left[(L^1{}_0)^2 + (L^2{}_0)^2 + (L^3{}_0)^2 \right] = 1,$$

from which it follows that $(L^0{}_0)^2 \geqslant 1$ and so either $L^0{}_0 \geqslant 1$ or $L^0{}_0 \leqslant -1$. We divide Lorentz transformations into four separate classes depending on whether $\det L^a{}_b = \pm 1$ and $L^0{}_0 \geqslant 1$ or $L^0{}_0 \leqslant -1$. If $\det L^a{}_b = +1$, then $L^a{}_b$ is called **proper** or **orientation preserving**. An example of an improper Lorentz transformation is the discrete transformation

$$t' = t, \quad x' = -x, \quad y' = y, \quad z' = z,$$

which reverses the x-direction. If $L^0{}_0 \geqslant 1$, then $L^a{}_b$ is called **orthochronous** or **time-orientation preserving**. An example of a non-orthochronous Lorentz transformation is the discrete transformation

$$t' = -t, \quad x' = x, \quad y' = y, \quad z' = z,$$

which reverses the t-direction. The proper orthochronous transformations, denoted by $L^\uparrow_+$ (read 'L arrow plus'), form a **subgroup** of L. Clearly, $L^\uparrow_+$ contains the identity, whereas the other three subsets do not and hence are not subgroups.

In fact, $L^\uparrow_+$ is a six-parameter continuous group of transformations. We can interpret the parameters physically by considering the transformation actively as transforming one inertial frame S into another inertial frame S' in general position which is moving with velocity $\boldsymbol{v}$ relative to S (see Chapter 2, Fig. 2.20). Then two parameters correspond to the two Euler rotations required to line up the x-axis of S with the velocity of S', one parameter corresponds to a boost from S to a frame at rest relative to S' (and this parameter depends on the speed of S' relative to S), and the final three parameters correspond to the three Euler rotations required to rotate the frame into the same orientation that S' has. Another subgroup of L is the ordinary three-dimensional rotation group.

The **Poincaré group** P consists of those linear inhomogeneous transformations which leave η_{ab} invariant. A Poincaré transformation is made up of a Lorentz transformation together with an arbitrary translation (in space and time), i.e.

$$x^a \to x'^a = L^a{}_b x^b + t^a. \tag{8.11}$$

The Lorentz group L is a proper subgroup of P, and the translations form an invariant (normal) subgroup of P. The Poincaré group P is a ten-parameter group, consisting of six Lorentz parameters plus four translation parameters. The continuous Poincaré transformations constitute the full set of isometries of the Minkowski metric. Physically, a Poincaré transformation maps one inertial frame S into another inertial frame S' in general position.

8.4 Proper time

A timelike world-line or **timelike curve** is defined as a curve whose tangent vector is everywhere timelike. If, in particular, the curve is a geodesic, it is called a **timelike geodesic**. Timelike curves represent tracks on which material particles or observers can travel. From §8.2, we see that the velocity tangent vector to a timelike curve at any point P must lie within the null cone emanating from P (Fig. 8.2). This is a manifestation of the special relativity result that material particles travel with speeds always less than the speed of light. Spacelike and null curves and geodesics are defined in an analogous manner to timelike ones.

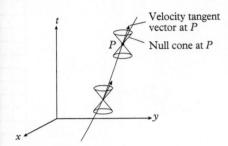

Fig. 8.2 World-line of a material particle.

At any point P, we define the **null cone** or **light cone**, which consists of all null geodesics passing through P. This coincides with the null cone of null vectors passing through P. Then the null cone divides space-time into three distinct regions, namely future, past, and elsewhere (Fig. 8.3). Any point in the **future** or **past** may be reached by a future-directed or past-directed timelike geodesic, respectively. Any point in the region exterior to the null cone, called **elsewhere**, can be reached by a geodesic which is everywhere spacelike. This is an invariant division of events which all observers agree upon. This follows because of the invariance of η_{ab} under a Lorentz transformation, which means that null cones get mapped onto null cones. Moreover, events to the future of P get mapped into events which are still to the future of P under an orthochronous Lorentz transformation. A similar result holds for past events. However, non-orthochronous Lorentz transformations reverse the past and future.

Since Γ^a_{bc} vanishes in Minkowski coordinates, the equations for a non-null geodesic (7.43), in these coordinates, reduce to

$$\frac{\mathrm{d}^2 x^a}{\mathrm{d}\mu^2} = 0, \tag{8.12}$$

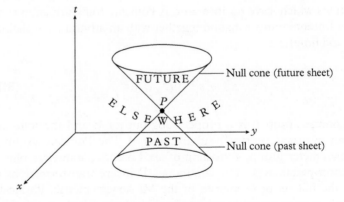

Fig. 8.3 Invariant classification of events relative to P.

for some affine parameter μ, where the tangent vector satisfies

$$\eta_{ab}\frac{\mathrm{d}x^a}{\mathrm{d}\mu}\frac{\mathrm{d}x^b}{\mathrm{d}\mu} = k. \qquad (8.13)$$

The geodesic is timelike or spacelike depending on whether $k > 0$ or $k < 0$, respectively. In the case when $k > 0$, we introduce a new parameter

$$\mu \to \bar{\mu} = \bar{\mu}(\mu),$$

satisfying

$$\left(\frac{\mathrm{d}\bar{\mu}}{\mathrm{d}\mu}\right)^2 = k.$$

It follows from (8.13) that the new tangent vector $\mathrm{d}x^a/\mathrm{d}\bar{\mu}$ has **unit** length. The parameter $\bar{\mu}$ is called the **proper time** and is denoted by τ. Thus, in relativistic units, from (8.3) and (8.13), the proper time satisfies

$$\mathrm{d}\tau^2 = \mathrm{d}s^2. \qquad (8.14)$$

This shows that proper time τ is an **affine parameter** along timelike geodesics.

In non-relativistic units, the equation for the proper time becomes

$$\mathrm{d}\tau^2 = \frac{1}{c^2}\mathrm{d}s^2, \qquad (8.15)$$

which checks dimensionally since s is a distance parameter. Let us see how proper time τ relates to coordinate time t for any observer whose 3-velocity at time t is $\boldsymbol{v}$, where

$$\boldsymbol{v} = \left(\frac{\mathrm{d}x}{\mathrm{d}t}, \frac{\mathrm{d}y}{\mathrm{d}t}, \frac{\mathrm{d}z}{\mathrm{d}t}\right).$$

From (8.15) and (3.12), we have

$$\begin{aligned}
\mathrm{d}\tau^2 &= \frac{1}{c^2}\mathrm{d}s^2 \\
&= \frac{1}{c^2}(c^2\mathrm{d}t^2 - \mathrm{d}x^2 - \mathrm{d}y^2 - \mathrm{d}z^2) \\
&= \mathrm{d}t^2\left\{1 - \frac{1}{c^2}\left[\left(\frac{\mathrm{d}x}{\mathrm{d}t}\right)^2 + \left(\frac{\mathrm{d}y}{\mathrm{d}t}\right)^2 + \left(\frac{\mathrm{d}z}{\mathrm{d}t}\right)^2\right]\right\} \\
&= \mathrm{d}t^2\left(1 - \frac{v^2}{c^2}\right).
\end{aligned}$$

So the proper time between t_0 and t_1, is given by

$$\tau = \int_{t_0}^{t_1} \left(1 - \frac{v^2}{c^2}\right)^{\frac{1}{2}} dt, \qquad (8.16)$$

in agreement with (3.16).

8.5 An axiomatic formulation of special relativity

We are now in a position to give a completely precise formulation of special relativity which will prove useful when we wish to generalize to the general theory. We do this by stating two sets of postulates or axioms.

> **Axiom I.** Space and time are represented by a four-dimensional manifold endowed with a symmetric affine connection Γ^a_{bc} and a metric tensor g_{ab} satisfying the following:
>
> (i) g_{ab} is non-singular with signature -2;
> (ii) $\nabla_c g_{ab} = 0$;
> (iii) $R^a{}_{bcd} = 0$.

> **Axiom II.** There exist privileged classes of curves in the manifold singled out as follows:
>
> (i)   ideal clocks travel along timelike curves and measure the parameter τ (called the 'proper time') defined by $d\tau^2 = g_{ab}dx^a dx^b$;
> (ii) free particles travel along timelike geodesics ;
> (iii) light rays travel along null geodesics.

The first axiom defines the **geometry** of the theory and the second axiom puts in the **physics**. Thus, the first axiom states that Γ^a_{bc} is the metric connection (by I(ii)) and that the metric is flat (by I(iii)) and defines a formal parameter whose physical significance is given in the second axiom. The first part of the second axiom makes physical the distinction between space and time in the manifold. In canonical (Minkowski) coordinates, it distinguishes the coordinate x^0 from the other three as the 'time' coordinate. More precisely, it states that it is the proper time τ which a clock measures in accordance with the clock hypothesis. The remainder of Axiom II singles out the privileged curves that free particles and light rays travel along.

Looking at this theory from a purely axiomatic viewpoint, one can ask, Is there any a priori reason for singling out timelike and null geodesics as trajectories for material particles and photons for light rays, or could one make some other choice (say, spacelike geodesics)? In Newtonian theory, free particles travel in straight lines, by Newton's first law. It would seem natural, therefore, to take geodesics as the analogue of straight lines. The significance of timelike geodesics is that their choice, unlike the case of spacelike geodesics, is consistent with **causality**. As we have seen, Minkowski space-time admits the Poincaré group as its invariance group. Hence, if two neighbouring events P and Q of the history of a free particle occur on a timelike geodesic at proper times τ and $\tau + d\tau$, respectively, then an orthochronous Poincaré transformation preserves the fact that Q occurs **after** P. This is consistent with causality, since we say that the arrival of the particle at Q is **caused** by its having previously been at P.

Null geodesics possess a special property which makes them natural candidates for light signals. The equation of a null geodesic in Minkowski coordinates is

$$\frac{d^2 x^a}{d\mu^2} = 0, \tag{8.17}$$

where

$$\eta_{ab}\frac{dx^a}{d\mu}\frac{dx^b}{d\mu} = 0, \tag{8.18}$$

for an affine parameter μ. Integrating (8.17), we get

$$\frac{dx^a}{d\mu} = k^a, \tag{8.19}$$

where the components of k^a are constants of integration. Substituting in (8.18), we obtain

$$\eta_{ab}\,k^a k^b = 0, \tag{8.20}$$

and so k^a is a null vector. Let us define the 3-velocity $\boldsymbol{u}$ along the null geodesic by

$$\boldsymbol{u} = (u^1,\ u^2,\ u^3) = \left(\frac{dx^1}{dx^0}, \frac{dx^2}{dx^0}, \frac{dx^3}{dx^0}\right) = \left(\frac{k^1}{k^0}, \frac{k^2}{k^0}, \frac{k^3}{k^0}\right), \tag{8.21}$$

using (8.19) and the fact that $k^0 \neq 0$ (why?). Writing (8.20) out fully, we find

$$(k^0)^2 - (k^1)^2 - (k^2)^2 - (k^3)^2 = 0,$$

and hence it follows from (8.21) that $\boldsymbol{u}^2 = 1$. Thus, null geodesics have associated with them a characteristic velocity of magnitude 1. Furthermore, this property is preserved under a Poincaré transformation, and so they seem natural candidates for encoding the constancy of the velocity of light.

8.6 A variational principle approach to classical mechanics

We met an introduction to relativistic mechanics in Chapter 4. We shall now look for a formulation which rests on a variational principle. The importance of the variational formulation of a physical theory is that it is often very simple and elegant and, moreover, it is one method which lends itself easily to generalization. Indeed, most current theories use the variational approach as their starting point. We start by summarizing the variational formulation of a classical system moving under a conservative force.

A mechanical system is described by n **generalized coordinates** x^a $(a = 1, 2, \ldots, n)$ which are functions of time t, n **generalized velocities** $\dot{x}^a$, the **kinetic energy** $T = \frac{1}{2} g_{ab} \dot{x}^a \dot{x}^b$, and the **potential energy** $V(x)$, which gives rise to n **generalized forces** $F_a = -\partial V / \partial x^a$. The **Lagrangian** L is defined to be

$$L := T - V.$$

Then the principle of stationary action is

$$\delta S = \delta \int_{t_1}^{t_2} L dt = 0,$$

and this leads to the Euler-Lagrange equations

$$\frac{\partial L}{\partial x^a} - \frac{\mathrm{d}}{\mathrm{d}t} \left(\frac{\partial L}{\partial \dot{x}^a} \right) = 0.$$

A straightforward calculation leads to the equations of motion

$$\ddot{x}^a + \Gamma^a_{bc} \dot{x}^b \dot{x}^c = F^a, \tag{8.22}$$

where Γ^a_{bc} is the metric connection of g_{ab}. If there are no external forces, then the above equations can be thought of as defining **geodesics** on an n-dimensional Riemannian manifold, with metric g_{ab} called **configuration space**. We define **generalized momenta**

$$p_a := \partial L / \partial \dot{x}^a \tag{8.23}$$

and the **Hamiltonian** H by

$$H := p_a \dot{x}^a - L. \tag{8.24}$$

If H is time-independent, then it can be shown to be equal to the total energy E of the system.

As an example of this formalism, let us consider the simple case of a **free particle** moving in three dimensions with velocity $\boldsymbol{u}$. Adopting Cartesian coordinates, we have

$$(x^a) = (x^1, x^2, x^3) = (x, y, z).$$

Then

$$T = \tfrac{1}{2}mu^2 = \tfrac{1}{2}m(\dot{x}^2 + \dot{y}^2 + \dot{z}^2),$$

from which we find

$$g_{ab} = \mathrm{diag}(m, m, m) = m\delta_{ab}.$$

By assumption, $V = 0$, and so

$$L = T = \tfrac{1}{2}mu^2, \qquad\qquad (8.25)$$

giving generalized momenta

$$p_x = \frac{\partial L}{\partial x} = m\dot{x}, \quad p_y = \frac{\partial L}{\partial y} = m\dot{y}, \quad p_z = \frac{\partial L}{\partial z} = m\dot{z}.$$

The Euler-Lagrange equations are

$$\frac{\mathrm{d}}{\mathrm{d}t}(m\dot{x}) = 0, \quad \frac{\mathrm{d}}{\mathrm{d}t}(m\dot{y}) = 0, \quad \frac{\mathrm{d}}{\mathrm{d}t}(m\dot{z}) = 0,$$

which are just the three components of Newton's second law. The Hamiltonian is

$$H = \boldsymbol{u} \cdot \boldsymbol{u} - L = m(\dot{x}^2 + \dot{y}^2 + \dot{z}^2) - T = \tfrac{1}{2}mu^2 = T = E.$$

In general, if we consider a system with no forces acting, then the Lagrangian reduces to

$$T = \tfrac{1}{2}g_{ab}\dot{x}^a\dot{x}^b.$$

This Lagrangian is identical to the quantity K defined in (7.46) of §7.6. In that section, we saw that (if we work with affine parameters) this gives the same Euler-Lagrange equations as the Lagrangian (7.38), namely, as

$$\frac{\mathrm{d}s}{\mathrm{d}t} = (g_{ab}\dot{x}^a\dot{x}^b)^{\frac{1}{2}},$$

does. Thus, for convenience, we may take the action S for a free particle to be

$$S = \int_{t_1}^{t_2} \frac{\mathrm{d}s}{\mathrm{d}t}\,\mathrm{d}t = \int_{t_1}^{t_2} \mathrm{d}s. \qquad\qquad (8.26)$$

8.7 A variational principle approach to relativistic mechanics

We now consider a free particle in relativistic mechanics moving on a curve

$$x^a = x^a(\tau),$$

where τ is the proper time. Since τ is an affine parameter, we assume from (8.26) of the last section that the action can be written as

$$S = -\alpha \int_{\tau_1}^{\tau_2} \mathrm{d}s, \tag{8.27}$$

where α is a constant to be determined. Working in Minkowski coordinates and introducing a new parameter μ, where $\mu = \mu(\tau)$, we can write the action as

$$S = -\alpha \int_{u_1}^{u_2} \left(\eta_{ab} \frac{\mathrm{d}x^a}{\mathrm{d}\mu} \frac{\mathrm{d}x^b}{\mathrm{d}\mu} \right)^{1/2} \mathrm{d}\mu.$$

The Euler-Lagrange equations

$$\frac{\partial L}{\partial x^a} - \frac{\mathrm{d}}{\mathrm{d}\mu} \left(\frac{\partial L}{\partial(\mathrm{d}x^a/\mathrm{d}\mu)} \right) = 0,$$

lead to

$$\frac{\mathrm{d}}{\mathrm{d}\mu} \left[\alpha \left(\eta_{cd} \frac{\mathrm{d}x^c}{\mathrm{d}\mu} \frac{\mathrm{d}x^d}{\mathrm{d}\mu} \right)^{-1/2} \eta_{ab} \frac{\mathrm{d}x^b}{\mathrm{d}\mu} \right] = 0. \tag{8.28}$$

Since

$$\eta_{cd} \frac{\mathrm{d}x^c}{\mathrm{d}\mu} \frac{\mathrm{d}x^d}{\mathrm{d}\mu} = \frac{\eta_{cd}\mathrm{d}x^c\mathrm{d}x^d}{\mathrm{d}\mu^2} = \frac{\mathrm{d}s^2}{\mathrm{d}\mu^2} = \frac{\mathrm{d}\tau^2}{\mathrm{d}\mu^2},$$

In relativistic units, the field equations give

$$0 = \frac{\mathrm{d}}{\mathrm{d}\mu} \left[\alpha\eta_{ab} \frac{\mathrm{d}\mu}{\mathrm{d}\tau} \frac{\mathrm{d}x^b}{\mathrm{d}\mu} \right] = \frac{\mathrm{d}}{\mathrm{d}\mu} \left[\alpha\eta_{ab} \frac{\mathrm{d}x^b}{\mathrm{d}\tau} \right] = \alpha\eta_{ab} \frac{\mathrm{d}^2x^b}{\mathrm{d}\tau^2} \frac{\mathrm{d}\tau}{\mathrm{d}\mu},$$

which leads to $\ddot{x}^b = 0$, where now we are using dot to denote differentiation with respect to τ, and which are the standard geodesic equations in Minkowski coordinates.

Instead of using the proper time τ as our time parameter, let us use instead the coordinate time t and see how various quantities are defined in terms of time and space coordinates. The equation of the world-line of the particle is now

$$x = x(t), \quad y = y(t), \quad z = z(t),$$

and it has a 3-velocity $\boldsymbol{u}$ defined by

$$\boldsymbol{u} = (u^1, u^2, u^3) = \left(\frac{\mathrm{d}x}{\mathrm{d}t}, \frac{\mathrm{d}y}{\mathrm{d}t}, \frac{\mathrm{d}z}{\mathrm{d}t} \right).$$

Using

$$
\begin{aligned}
\mathrm{d}s^2 &= \eta_{ab}\mathrm{d}x^a\mathrm{d}x^b \\
&= \mathrm{d}t^2 - \mathrm{d}x^2 - \mathrm{d}y^2 - \mathrm{d}z^2 \\
&= \mathrm{d}t^2(1 - u^2),
\end{aligned}
$$

we can write the action (8.27) as

$$S = -\alpha \int_{t_1}^{t_2} (1 - u^2)^{1/2} \mathrm{d}t,$$

where the new Lagrangian (which we shall also write as L) is

$$L = -\alpha(1 - u^2)^{1/2} = -\alpha + \tfrac{1}{2}\alpha u^2 + \cdots$$

for small velocities. Comparing this with the classical expression (8.25), namely $\tfrac{1}{2}mu^2$, we may identify α with the mass of the particle as $u \to 0$. Note that the additive constant $-\alpha$ in the Lagrangian is unimportant (see Exercise 8.9). Thus α is equal to the **rest mass** m_0 of the particle. Hence, we have

$$L = -m_0\left(1 - u^2\right)^{1/2}. \tag{8.29}$$

We define the 3-momentum $\boldsymbol{p}$ by (check)

$$\boldsymbol{p} = \left(\frac{\partial L}{\partial u^1}, \frac{\partial L}{\partial u^2}, \frac{\partial L}{\partial u^3} \right) = m_0\left(1 - u^2\right)^{-1/2}\boldsymbol{u}. \tag{8.30}$$

Comparing this with the classical relationship $\boldsymbol{p} = m\boldsymbol{u}$, we define the **relativistic mass** m by (see (4.11))

$$m = m_0\left(1 - u^2\right)^{1/2}.$$

Using the Hamiltonian to define the **energy** E (see (4.17)), we find

$$E = H = \boldsymbol{p} \cdot \boldsymbol{u} - L = m_0\left(1 - u^2\right)^{-1/2} = m, \tag{8.31}$$

after some simple algebra. We have thus regained the results of (4.19) in relativistic units.

8.8 Covariant formulation of relativistic mechanics

We finish this discussion of relativistic mechanics by giving a full **4-dimensional** or **covariant** formulation of the variational principle. The action S is a coordinate independent quantity, so in a general coordinate system (8.27) becomes

$$S = -m_0 \int_{\tau_1}^{\tau_2} (g_{ab}\dot{x}^a \dot{x}^b)^{1/2} \mathrm{d}\tau,$$

where g_{ab} is a flat metric and is used for raising and lowering indices. The **4-velocity** u^a is defined by

$$u^a := \frac{\mathrm{d}x^a}{\mathrm{d}\tau} = \dot{x}^a, \tag{8.32}$$

and the **4-acceleration** a^b by

$$a^b := \frac{\mathrm{d}u^b}{\mathrm{d}\tau} = \frac{\mathrm{d}^2 x^b}{\mathrm{d}\tau^2} = \ddot{x}^b. \tag{8.33}$$

The covariant **4-momentum** p_a is defined by

$$p_a := \frac{\partial L}{\partial \dot{x}^a},$$

from which we find that

$$p_a = -m_0 g_{ab} u^b (g_{cd} u^c u^d)^{1/2}.$$

So in Minkowski coordinates the spatial components are given by

$$(p_1, p_2, p_3) = m_0(u^1, u^2, u^3), \tag{8.34}$$

so that $\boldsymbol{p} = m_0 \boldsymbol{u}$ as expected.

If a particle is acted on by a force, then the four-dimensional version of Newton's second law becomes

$$f^a = \frac{\mathrm{d}p^a}{\mathrm{d}\tau}, \tag{8.35}$$

where f^a is called the **4-force**. If there is no external force acting, then

$$\frac{\mathrm{d}p^a}{\mathrm{d}\tau} = 0 \quad \Rightarrow \quad p^a = k^a, \tag{8.36}$$

where k^a is a constant 4-vector. This is the **conservation of 4-momentum** law and generalizes to an isolated system of n particles with 4-momenta $p_i{}^a$ $(i = 1, 2, \ldots, n)$

$$\sum_{=1}^{n} p^a = k^a,$$

where k^a is a constant 4-vector. Finally, we define the **angular momentum tensor** ℓ^{ab} of the particle in Minkowski coordinates by

$$\ell^{ab} = x^a p^b - x^b p^a. \tag{8.37}$$

If we now assume that m_0 is a **scalar**, then it follows that all the quantities have the tensor character indicated under a general coordinate transformation. If, in particular, we restrict attention to Minkowski coordinates, we can relate these four-dimensional quantities to the three-dimensional ones of the last section and Chapter 4. We can then consider how the four-dimensional quantities transform under a **Lorentz transformation** and so obtain the transformation law for the three-dimensional quantities (exercise). Thus, in particular, we can confirm the transformation equations (4.21) for the energy and momentum of a particle.

We have considered the main ingredients of special relativistic mechanics, but we shall not pursue the topic further. We shall, rather, concentrate on our main task – that of establishing the general theory.

Exercises

8.1 (§8.1) Check (8.5) and show that the Riemann tensor vanishes.

8.2 (§8.2) Show that a timelike vector cannot be orthogonal to a null vector or to another timelike vector. Show that two null vectors are orthogonal if and only if they are parallel.

8.3 (§8.2) The vectors T, X, Y, and Z have components

$$T^a = (1,0,0,0), \quad X^a = (0,1,0,0),$$
$$Y^a = (0,0,1,0), \quad Z^a = (0,0,0,1).$$

Show that the only non-vanishing inner products between the vectors are

$$T^2 = -X^2 = -Y^2 = -Z^2 = 1.$$

Define the following:

$$L^a = \frac{1}{\sqrt{2}}(T^a + Z^a), \quad N^a = \frac{1}{\sqrt{2}}(T^a - Z^a),$$
$$M^a = \frac{1}{\sqrt{2}}(X^a + iY^a), \quad \bar{M}^a = \frac{1}{\sqrt{2}}(X^a - iY^a),$$

where $i = \sqrt{-1}$. Treating M^a and $\bar{M}^a$ as vectors, show that all four vectors are null and the only non-vanishing inner products are

$$L^a N_a = -M^a \bar{M}_a = 1.$$

8.4 (§8.3)
(i) Check that (8.9) leads to (8.10), assuming invariance.
(ii) Show that the Lorentz transformations form a group.
(iii) Show that the Poincaré transformations form a group.

8.5 (§8.3) Show that a Killing vector X_a satisfies the equation $\partial_b \partial_c X_a = 0$ in flat space in Minkowski coordinates. [Hint: use Exercise 7.10 or Exercise 7.14.] Deduce that the Killing vectors are given by

$$X_a = \omega_{ab} x^b + t_a,$$

where $\omega_{ab} = -\omega_{ba}$ and t_a are arbitrary parameters (constants of integration). How many parameters are there in:
(a) an n-dimensional manifold?
(b) Minkowski space-time?

What do the parameters correspond to physically in Minkowski space-time?

8.6 (§8.4) Prove that the proper time τ is an affine parameter along a timelike geodesic for a general space-time. [Hint: Use (7.42).]

8.7 (§8.6) Establish the equation of motion (8.22).

8.8 (§8.6) Consider two masses m_1 and m_2 suspended on the ends of a rope passing over a frictionless pulley. Show that the Lagrangian can be written in the form

$$L = \tfrac{1}{2}(m_1 + m_2)\dot{x}^2 + m_1 g x + m_2 g(\ell - x),$$

where the mass m_1 is a distance x below the horizontal and ℓ is a constant. Find the Euler–Lagrange equation of motion. Define the generalized momentum for the system and hence obtain the Hamiltonian.

8.9 (§8.7) If L is a Lagrangian, then show that the Lagrangians L_1 and L_2, where (i) $L_1 = \lambda L$ and (ii) $L_2 = L + \mu$, with λ and μ constants, possess the same field equations as L. Show also that, if $L \neq 0$, then the Lagrangians (iii) $L_3 = L^2$ and (iv) $L_4 = L^{1/2}$ give rise to the same field equations.

8.10 (§8.8) Show that, in Minkowski space-time in Minkowski coordinates, $u^a = (u^0, u^1, u^2, u^3) = (\gamma, \gamma \boldsymbol{u})$, where $\gamma = (1 - u^2)^{-1/2}$. Show also that $p_a = (E, \boldsymbol{p})$.

By considering the invariant $p_a p^a$, deduce that (see (4.20))

$$E^2 - p^2 = m_0^2.$$

Use the four-dimensional version of Newton's second law to identify the 4-force in Minkowski coordinates as

$$f_a = (\gamma \boldsymbol{u} \cdot \boldsymbol{F}, \gamma \boldsymbol{F}),$$

where $\boldsymbol{F}$ is the force acting on the particle. Show also that

$$\frac{\mathrm{d}p_a}{\mathrm{d}\tau} = \left(\gamma \frac{\mathrm{d}E}{\mathrm{d}t}, \gamma \frac{\mathrm{d}\boldsymbol{p}}{\mathrm{d}t} \right)$$

and give a physical interpretation of the zero component of the four-dimensional Newton's law.

8.11 (§8.8)
(i) Use the tensor transformation law on the 4-velocity u^a to find the transformation properties of $\boldsymbol{u}$ under a special Lorentz transformation between two frames in standard configuration moving with velocity $\boldsymbol{v}$. Show in particular, that $\gamma'/\gamma = \beta(1 - u_x v)$, where $\beta = (1 - v^2)^{-1/2}$.
(ii) Find the transformation properties of E and $\boldsymbol{p}$ under a special Lorentz transformation.
(iii) Find the transformation properties of $\boldsymbol{F}$ under a special Lorentz transformation. Are forces still absolute quantities in special relativity?
(iv) A particle moves parallel to the x-axis under the influence of a force $\boldsymbol{F} = (F, 0, 0)$. What is the force in a frame co-moving with the particle?

Further reading

The axiomatic description of special relativity given here is that of Trautmann, Pirani, and Bondi (1964). For further consideration of the Lorentz group, see the book by Carmeli and Malin (1976).

Carmeli, M., and Malin, S. (1976). *Representation of the Rotation and Lorentz Groups: An Introduction*. Dekker, New York, NY.

Trautmann A., Pirani F. A. E., and Bondi, H. (1964). *Lectures on General Relativity*. Brandeis Summer Institute on Theoretical Physics, 1964, vol. 1. Prentice-Hall, Englewood Cliffs, NJ.

The principles of general relativity

9

9.1 The role of physical principles

We are at last ready to embark on our central task, namely, that of extending special relativity to a theory which incorporates gravitation. In this chapter, we shall undertake a detailed consideration of the physical principles which guided Einstein in his search for the general theory. There is a school of thought that considers this an unnecessary process, but rather argues that it is sufficient to state the theory and investigate its consequences. There seems little doubt, however, that consideration of these physical principles helps give insight into the theory and promotes understanding. The mere fact that they were important to Einstein would seem sufficient to justify their inclusion. If nothing else, it will help us to understand how one of the greatest achievements of the human mind came about. Many physical theories today start by specifying a Lagrangian from which everything else flows. Indeed, we could adopt the same attitude with general relativity, but in so doing we would miss out on gaining some understanding of how the framework of general relativity is different again from the framework of Newtonian theory or special relativity. Moreover, if we discover limitations in the theory, then there is more chance of rescuing it by investigating the physical basis of the theory rather than simply tinkering with the mathematics. It is perhaps significant that Einstein devoted much of his later life to an attempt to unify general relativity and electromagnetism by various mathematical devices, but without success.

There are five principles which, explicitly or implicitly, guided Einstein in his search. Their names are:

(1) Mach's principle,

(2) the principle of equivalence,

(3) the principle of covariance,

(4) the principle of minimal gravitational coupling,

(5) the correspondence principle.

The status of these principles has been the source of much controversy. For example, the principle of covariance is considered by some authors (e.g. Bondi, Fock) to be empty, whereas there are others (e.g. Anderson) who believe it possible to derive general relativity more or less solely from this principle. Similarly, although the ideas behind Mach's principle were important to Einstein in deriving the field equations, there are considerable doubts about whether relativity is a fully Machian theory (e.g. Bondi, Samuel). On the other hand, there is general agreement that the principle

Introducing Einstein's Relativity. Second Edition. Ray d'Inverno and James Vickers, Oxford University Press.
© Ray d'Inverno and James Vickers (2022). DOI: 10.1093/oso/9780198862024.003.0009

of equivalence is the key principle and we discuss in Chapter 16 how this leads to a metric theory of gravity. One source of confusion over the role of the various principles arises from the fact that their formulation differs quite markedly from author to author. Since some of the principles are more of a philosophical nature, this is perhaps not so surprising. We shall attempt to give some precise formulations of them in the hope that we can ultimately check the principles out against the theory. We now discuss the principles in turn.

9.2 Mach's principle

The essence of the first two principles comes from understanding the nature of Newton's laws more precisely. Do Newton's laws hold in all frames of reference? As we have seen before, they are stated only for a privileged class of frames called **inertial** frames. So the question arises as to how inertial frames are determined by the properties of the Universe and what form Newton's laws take in other, non-inertial, reference frames.

We shall investigate the status of Newton's second law for a non-inertial frame S' being uniformly accelerated relative to an inertial frame S with acceleration a. For simplicity, we shall assume the frames are in standard configuration with the acceleration along the common axis (Fig. 9.1). Assuming that the observers initialize their clocks when they meet, then the relationship between the frames is given by

$$x = x' + s, \quad y = y', \quad z = z', \quad t = t'. \tag{9.1}$$

Letting a dot denote differentiation with respect to t (or t', which is the same by the last equation), then we find from the first equation that

$$\dot{x} = \dot{x} + \dot{s},$$

and, differentiating again,

$$\ddot{x} = \ddot{x}' + \ddot{s} = \ddot{x}' + a, \tag{9.2}$$

by assumption. Consider a particle of mass m moving along the x-axis under the influence of a force $\boldsymbol{F} = (F, 0, 0)$. Then Newton's second law becomes $F = m\ddot{x}$, which by (9.2) gives

$$F = m\ddot{x} = m\ddot{x}' + ma.$$

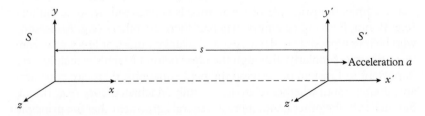

Fig. 9.1 Positions of S and S' at time t.

From the point of view of the observer S', this equation can be rewritten in a standard form with the term mass times **acceleration relative to S'** on the right-hand side, to give

$$F - ma = m\ddot{x}'. \tag{9.3}$$

Thus, compared to S, observer S' detects a **reduction** of the force on the particle by an amount ma. This additional force is called an **inertial force**. Other well-known inertial forces are **centrifugal** and **Coriolis** forces arising in a frame rotating relative to an inertial frame (exercise). Notice that all inertial forces have the **mass as a constant of proportionality** in them. The status of inertial forces is again a controversial one. One school of thought describes them as **apparent** or **fictitious** forces which arise in non-inertial frames of reference (and which can be eliminated mathematically by putting the terms back on the right-hand side). We shall adopt the attitude that, if you judge them by their effects, then they are very real forces. For, after all, inertial forces cause astronauts to blackout in rocket ships, and flywheels to break under centrifugal effects. Is it enough to describe these as being due to apparent forces or reference frame effects? There must be some interaction going on to cause such dramatic effects. The question arises, What is the physical origin of inertial forces? Newtonian theory makes no attempt to answer this question; the Machian viewpoint, as we shall see, does.

Let us ask another fundamental question. If Newton's laws only hold in inertial frames, then how do we detect inertial frames? Newton realized that this was a fundamental question and attempted to answer it by devising an ingenious thought experiment – the famous **bucket experiment**. He first of all postulated the existence of absolute space: 'Absolute space, in its own nature, without relation to anything external, remains always similar and immovable'. Thus, he saw absolute space as a fixed backcloth against which all motion is observed. An inertial observer then becomes an observer at rest or in uniform motion relative to absolute space. Inertial forces arise in the manner described above only when an observer is in **absolute acceleration** relative to absolute space. The bucket experiment is a device for detecting such motion. More precisely, the experiment determines whether or not a system is in **absolute rotation** relative to absolute space.

The experiment consists of suspending a bucket containing water by a rope in an inertial frame. The rope is twisted and the bucket is released. The motion divides into four phases:

B1: At first, the bucket rotates, but the water does not, its surface remaining **flat**.

B2: The frictional effects between the bucket and the water eventually communicate the rotation to the water. The centrifugal forces cause the water to pile up round the edges of the bucket and the surface becomes **concave** (Fig. 9.2). The faster the water rotates, the more concave the surface becomes.

B3: Eventually the bucket will slow down and stop, but the water will continue rotating for a while, its surface remaining concave.

Fig. 9.2 The bucket and water in absolute rotation.

B4:   Finally, the water will return to rest with a flat surface.

Newton's explanation of this experiment is that the curvature of the water surface in B2 and B3 arises from centrifugal effects due to the rotation of the **water** relative to absolute space. This curvature is not directly connected to local considerations such as the bucket's rotation since, in B1, the surface is flat when the bucket is rotating and, in B3, curved when the bucket is at rest. In this way, Newton gave a prescription for determining whether a system is in absolute rotation or not. Similar arguments apply to systems which are linearly accelerated relative to absolute space. Here, the surface becomes inclined at angle to the horizontal (Fig. 9.3) (see Exercise 9.1(ii)). In simple terms, all observers should be equipped with a bucket of water. Then an observer will be inertial if and only if the surface of the water is flat.

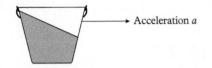

Fig. 9.3 Inclination of the surface of the water in absolute linear acceleration.

We now turn to the view which was proposed by Mach in 1893, although it grew out of similar ideas arrived at earlier by Bishop Berkeley. This is a semi-philosophical view, the starting point of which is that there is no meaning to the concept of motion, but only to that of **relative** motion. For example, a body in an otherwise empty universe cannot be said to be in motion according to Mach, since there is nothing to which the body's motion can be referred. Moreover, in a populated universe, it is the interaction between all the matter in the universe (over and above the usual gravitational interaction) which is the source of the inertial effects we have discussed above. In our universe, the bulk of the matter resides in what is historically called the 'fixed stars'. Then, from Mach's viewpoint, an inertial frame is a frame in some privileged state of motion relative to the average motion of the fixed stars. Thus, it is the fixed stars through their masses, distribution, and motion which **determine** a local inertial frame. This is Mach's principle in essence. Returning to the bucket experiment, Newton gives no reason why the surface curves up when it is in rotation relative to absolute space. Mach, however, says that the curvature stems from the fact that the water is in rotation **relative** to the fixed stars. One way of seeing the difference between the two viewpoints is to ask what would happen if the bucket was fixed and the **universe** (i.e. the fixed stars) rotated. Since all motion is relative, it follows from the Machian viewpoint that the surfaces of the water would be curved, whereas in Newtonian theory no such effect would be detected. Hence, Mach sees all matter coupled together in such a way that inertial forces have their physical origin in matter. The bucket has very little effect on the water's motion because its mass is so small. On the other hand, the fixed stars contain most of the matter in the universe and this counteracts the fact that they are a very long way away.

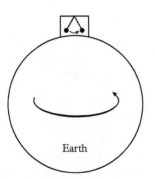

Fig. 9.4 Pendulum swinging in a non-rotating frame.

There is one very outstanding and simple fact that lends support to the Machian viewpoint. Consider a pendulum set swinging at the North Pole (Fig. 9.4). According to Newton, the pendulum swings in a frame which is not rotating relative to absolute space. In this frame, the Earth is rotating under the pendulum. An observer fixed on the Earth will see the pendulum rotating. The time taken for the pendulum to swing through 360° is therefore the time taken for the Earth to rotate through 360° with respect to **absolute space**. We can also measure how long the Earth takes to rotate through 360° relative to the **fixed stars**. The remarkable fact is that, within the limits of experimental accuracy, the two times are the **same**. In other words, the fixed stars are not rotating relative to absolute space, from which it follows that **inertial frames are those in which the fixed stars are not rotating**. In Newtonian theory, there is nothing a priori to predict this; it is simply a **coincidence**. Whenever we find coincidences in a physical theory, we should be highly suspicious of the theory – it is usually saying that something fundamental is going on. From the Machian viewpoint, it is the fixed stars which determine the inertial frames and the result is precisely what we would expect.

Can one say anything more precise about the interaction postulated by Mach? Since inertial forces involve the mass of the body experiencing them, it would seem likely for reasons of reciprocity that the effect of the stars should be due to their masses and proportional to them. On the other hand, inertial forces are unaffected (at least to the accuracy of experiment) by local masses such as the Earth or the Sun. Accordingly, the influence of the distant bodies preponderates. So we would not expect inertial effects to vary appreciably from place to place.

Consider the motion of a particle in an otherwise empty universe. Then, according to Mach, since there are no other masses in existence, the particle cannot experience any inertial effects. Now introduce another particle of tiny mass. It is inconceivable that the introduction of this very small mass would restore the inertial properties of the first particle to its customary magnitude – its effect can only be slight. This implies that the magnitude of an inertial force on a body is determined by the mass of the universe and its distribution. If, in particular, the universe were not isotropic, then inertial effects would not be isotropic. For example, if there were a preponderance of matter in a particular direction, then inertial effects would be direction dependent (as illustrated schematically in Fig. 9.5).

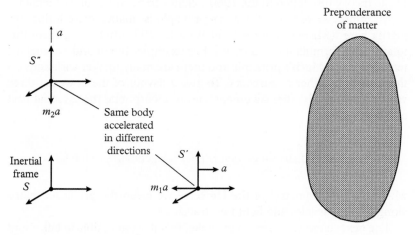

Fig. 9.5 Direction-dependent inertial effects in an anisotropic universe ($m_1 \neq m_2$).

Experiments were carried out separately by Hughes and Drever around 1960 which established that mass is isotropic to at least one part in 10^{18}. The Hughes–Drever experiment has been called the most precise null experiment ever performed. This null result can be interpreted in two ways. Either Mach's principle is untenable or the universe is highly isotropic. Indeed the uniform nature of the cosmic microwave background radiation (CMB) puts strong upper limits on the lack of isotropy in the Universe.

In Newtonian theory, the gravitational potential ϕ at a point a distance r from the origin due to a particle of mass m situated at the origin is $\phi = -Gm/r$, where G is Newton's gravitational constant. The potential at any point can only depend on the properties of the body itself. However, from the Machian point of view, the mass m of the body depends on the state of the universe. Hence, the ratio of these two effects, namely G, contains **information about the universe**. In particular, if the universe was in a different state at any earlier epoch, then the 'constant' G would have a different value. An evolutionary universe would require $G = G(t)$, i.e. a function of epoch. Again, if the universe did not present the same aspect from every point (except for local irregularities), G would vary from point to point. A fully Machian theory should essentially allow one to calculate G from a knowledge of the structure of the universe.

What is the status of Mach's principle? The biggest limitation of the principle is that it does not give a quantitative relation for the interaction of matter. Similarly, it can be argued that Mach's ideas do not really contribute to an understanding of why there appears to be such a fundamental distinction between unaccelerated and accelerated motion in nature, that is, it does not explain why the interaction should be velocity independent but acceleration dependent. Some critics claim that Mach only replaced Newton's absolute space by the distant stars and learnt nothing new thereby. However, the principle was considered to be of great importance to Einstein, who attempted to incorporate it into his general theory. This, as we shall see, he only partially succeeded in doing (although, an alternative theory to general relativity, called the Brans–Dicke theory developed in the 1960s, claims to be more fully Machian).

An imprecise version of Mach's principle is 'matter there influences inertia here' (Misner, Thorne, and Wheeler 1973) but going from this to a precise formulation is difficult. For example, Bondi and Samuel list ten versions of Mach's principle and there are many further variations on these given by other researchers. To give a flavour of these, we list three variants below. The first statement tries to incorporate the essential part of Mach's ideas.

M1. The matter distribution determines the geometry of the Universe,

where by the 'geometry' of the Universe we mean the privileged paths along which particles and light rays travel.

The next statement refers to the belief that it is impossible to talk about motion or geometry in an empty universe, so that there should be no solution corresponding to an empty universe.

M2. If there is no matter then there is no geometry.

The final statement refers to a universe containing just one body, then, since there is nothing for it to interact with, it should not possess any inertial properties.

M3. A body in an otherwise empty universe should possess no inertial properties.

9.3 Mass in Newtonian theory

Up to now, we have talked rather glibly about the mass m of a body. Even in Newtonian theory, we can ascribe three masses to any body which describe quite different properties. Their names, notation, and general description are:

(1) **inertial mass** m^I, which is a measure of the body's resistance to change in motion;
(2) **passive gravitational mass** m^P, which is a measure of its reaction to a gravitational field;
(3) **active gravitational mass** m^A, which is a measure of its source strength for producing a gravitational field.

We shall discuss each of these in turn.

Inertial mass m^I is the quantity occurring in Newton's second law, which we met in Chapter 4. It is at any one time a measure of a body's resistance to change in motion and is also called the body's **inertia**. Newton's second law, stated more precisely, is

$$F = \frac{\mathrm{d}(m^I\boldsymbol{v})}{\mathrm{d}t}, \tag{9.4}$$

or

$$\boldsymbol{F} = m^I\boldsymbol{a}, \tag{9.5}$$

for constant inertial mass m^I. Note that, a priori, m^I has **nothing** directly to do with gravitation. The next two masses, however, do.

Passive gravitational mass m^P measures a body's response to being placed in a gravitational field. Let the gravitational potential at some point be denoted by ϕ. Then, if m^P is placed at this point, it will experience a force on it given by

$$\boldsymbol{F} = -m^P \operatorname{grad} \phi. \tag{9.6}$$

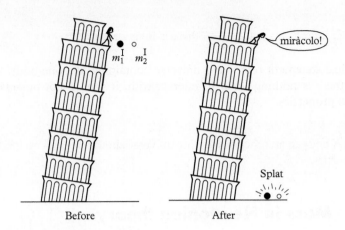

Fig. 9.6 Galileo's Pisa experiments.

Active gravitational mass m^A measures the strength of the gravitational field produced by the body itself. If m^A is placed at the origin, then the gravitational potential at any point distant r from the origin is given by

$$\phi = -\frac{Gm^A}{r}. \tag{9.7}$$

We shall now see how these three masses are related in the Newtonian framework. We start from the observational result that if we neglect non-fundamental forces, like air resistance, then two bodies dropped from the same height will reach the ground together. In other words they suffer the **same** acceleration irrespective of their internal composition. This empirical result is attributed to Galileo in his famous Pisa experiments (Fig. 9.6).

Of course, you would not get this result with a hammer and a feather, say, because the air resistance would slow down the fall of the feather. It would be possible on the Moon, however, since the Moon has no atmosphere. Indeed, readers may know of the incident on one of the Moon landings when an astronaut tried this 'experiment' and confirmed the anticipated result (Fig. 9.7).

Let us assume that two particles of inertial masses m_1^I and m_2^I, and passive gravitational masses m_1^P and m_2^P are dropped from the same height in a gravitational field. Then, from (9.5) and (9.6), we have

$$m_1^I \boldsymbol{a}_1 = \boldsymbol{F}_1 = -m_1^P \operatorname{grad} \phi,$$
$$m_2^I \boldsymbol{a}_2 = \boldsymbol{F}_2 = -m_2^P \operatorname{grad} \phi.$$

The observational result is $\boldsymbol{a}_1 = \boldsymbol{a}_2$, from which we get

$$m_1^I / m_1^P = m_2^I / m_2^P.$$

Repeating this experiment with other bodies, we see that the ratio m^I / m^P for any body is equal to a universal constant, α, say. By a suitable choice

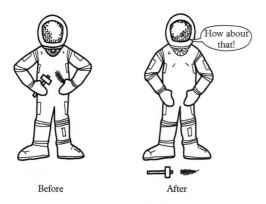

Before   After

Fig. 9.7 The Moon landing 'experiment'.

of units, we can, without loss of generality, take $\alpha = 1$, from which we obtain the result

> inertial mass = passive gravitational mass.   (9.8)

We discuss the experimental verification of (9.8) in Chapter 16 but here we simply note that this equality is one of the best attested results in physics and has been verified to one part in 10^{12} (see §16.3).

In order to relate passive gravitational mass to active gravitational mass, we make use of the observation that nothing can be shielded from a gravitational field. All matter is both acted upon by a gravitational field and is itself a source of a gravitational field. Consider two isolated bodies situated at points Q and R moving under their mutual gravitational interaction (Fig. 9.8). The gravitational potential due to each body is, by (9.7),

$$\phi_1 = -\frac{Gm_1^A}{r}, \quad \phi_2 = -\frac{Gm_2^A}{r}.$$

The force which each body experiences is, by (9.6),

$$\boldsymbol{F}_1 = -m_1^P \operatorname{grad}_Q \phi_2, \quad \boldsymbol{F}_2 = -m_2^P \operatorname{grad}_R \phi_1,$$

Q $\boldsymbol{F}_1$ $\boldsymbol{F}_2$ R

Body 1   Body 2

Fig. 9.8 The mutual gravitational interaction of two isolated bodies.

If we take Q to be the origin, then the gradient operators are

$$\text{grad}_R = \hat{\boldsymbol{r}}\frac{\partial}{\partial r} = -\text{grad}_Q,$$

so that

$$\boldsymbol{F}_1 = \frac{Gm_1^P m_2^A}{r^2}\hat{\boldsymbol{r}}, \quad \boldsymbol{F}_2 = -\frac{Gm_2^P m_1^A}{r^2}\hat{\boldsymbol{r}}.$$

But, by Newton's third law, $\boldsymbol{F}_1 = -\boldsymbol{F}_2$, and so we conclude

$$m_1^P/m_1^A = m_2^P/m_2^A.$$

Using the same argument as before, we see that

passive gravitational mass = active gravitational mass. (9.9)

This is why in Newtonian theory we can simply refer to the **mass** m of a body, where

$$m = m^I = m^P = m^A.$$

9.4   The principle of equivalence

We define a **gravitational test particle** to be a test particle which experiences a gravitational field but does **not** itself alter the field or contribute to the field. We wish to embody the empirical result of the Pisa experiments in a principle.

P1. The motion of a gravitational test particle in a gravitational field is independent of its mass and composition.

This is known as the **strong** form of the principle of equivalence, and we are going to build general relativity on this principle. Notice the difference in its status in the two theories. In Newtonian theory, it is an observational result – another **coincidence**. It could be possible, for example, that if we looked closer (with an accuracy greater than 1 in 10^{12}) then different bodies would possess different accelerations when placed in a gravitational field. This would not upset Newtonian theory, which could accommodate such a result. In general relativity, it forms an essential hypothesis of the theory and, if it falls, then so does the theory.

Next, we wish to make explicit the assumption that matter both responds to, and is a source of, a gravitational field. However, we have seen in special relativity that matter and energy are equivalent, so the statement about the gravitational field applies to energy as well. We incorporate this result into a statement which is known as the **weak** form of the principle of equivalence.

P2. The gravitational field is coupled to everything.

Thus, no body can be shielded from a gravitational field. However, it is possible to remove gravitational effects locally from our theory and so regain special relativity. This we do by considering a frame of reference which is in **free fall**, i.e. co-moving with a gravitational test particle. If, in particular, we choose a freely falling frame which is not rotating, then we regain the concept of an **inertial frame**, at least locally. We mean here by 'locally' that observations are confined to a region over which the **variation** of the gravitational field is unobservably small. In such inertial frames, test particles remain at rest or move in straight lines with uniform velocity. This leads to the following statement of the principle of equivalence.

P3. There are no local experiments which can distinguish non-rotating free fall in a gravitational field from uniform motion in space in the absence of a gravitational field.

In Einstein's words, 'for an observer falling freely from the roof of a house there exists no gravitational field'. He described this as the 'happiest thought of my life', and it played an important role in devising the general theory of relativity. Notice that once again in **P3** we have encoded our principle as a statement of impossibility.

Einstein noticed one other **coincidence** in Newtonian theory which proved to be of great importance in formulating a statement of the principle of equivalence. All inertial forces are proportional to the mass of the body experiencing them. There is one other force which behaves in the same way, that is, the force of **gravitation**. For, if we drop two bodies in the Earth's gravitational field, then they experience forces $m_1 g$ and $m_2 g$, respectively. This coincidence suggested to Einstein that the two effects should be considered as arising from the same origin. Thus he suggested that we treat gravitation as an **inertial effect** as well; in other words, it is an effect which arises from not using an inertial frame. Comparing the force mg of a falling body with the inertial force ma of (9.3) suggests the following version of the principle of equivalence.

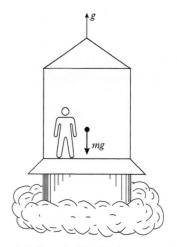

P4. A frame linearly accelerated relative to an inertial frame in special relativity is locally identical to a frame at rest in a gravitational field.

These last two versions of the principle of equivalence can be vividly clarified by considering the famous thought experiments (*Gedankenexperiment* in German) of Einstein, which are called the **lift experiments**.

We consider an observer confined to a lift or, more precisely, a room with no windows in it or other means of communication with the outside world. The observer is allowed equipment to carry out simple dynamical experiments. The object of the exercise is to try and determine the observer's state of motion. We consider four states of motion (Figs. 9.9–9.12).

Fig. 9.9 Case 1: The lift in an accelerated rocket ship.

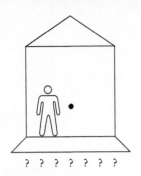

Fig. 9.10 Case 2: The lift in an unaccelerated rocket ship.

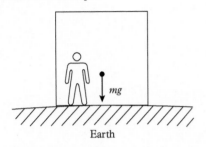

Fig. 9.11 Case 3: The lift placed on the Earth's surface.

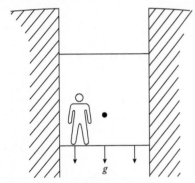

Fig. 9.12 Case 4: The lift dropped down an evacuated lift shaft.

Case 1. The lift is placed in a rocket ship in a part of the universe far removed from gravitating bodies. The rocket is accelerated forward with a constant acceleration g relative to an inertial observer. The observer in the lift releases a body from rest and (neglecting the influence of the lift, etc.) sees it fall to the floor with acceleration g.

Case 2. The rocket motor is switched off so that the lift undergoes uniform motion relative to the inertial observer. A released body is found to remain at rest relative to the observer.

Case 3. The lift is next placed on the surface of the Earth, whose rotational and orbital motions are ignored. A released body is found to fall to the floor with acceleration g.

Case 4. Finally, the lift is placed in an evacuated lift shaft and allowed to fall freely towards the centre of the Earth. A released body is found to remain at rest relative to the observer.

Clearly, from the point of view of the observer, Cases 1 and 3 are indistinguishable, as required by P4, and Cases 2 and 4 are indistinguishable, as required by P3. Let us trace the argument that shows that these requirements lead to the concept of a non-flat, i.e. a curved space-time.

In special relativity, in a coordinate system adapted to an inertial frame, namely, Minkowski coordinates, the equation for a test particle is

$$\frac{\mathrm{d}^2 x^a}{\mathrm{d}\tau^2} = 0.$$

If we use a non-inertial frame of reference, then this is equivalent to using a more general coordinate system. In this case, the equation becomes

$$\frac{\mathrm{d}^2 x^a}{\mathrm{d}\tau^2} + \Gamma^a_{bc} \frac{\mathrm{d}x^b}{\mathrm{d}\tau} \frac{\mathrm{d}x^c}{\mathrm{d}\tau} = 0,$$

where Γ^a_{bc} is the metric connection of g_{ab}, which is still a flat metric but not the Minkowski metric η_{ab}. The additional terms involving Γ^a_{bc} which appear are precisely the **inertial force** terms we have encountered before. Then the principle of equivalence requires that the gravitational forces, **as well as** the inertial forces, should be given by an appropriate Γ^a_{bc}. In this case, we can no longer take space-time to be flat, for otherwise there would be no distinction from the non-gravitational case. The **simplest** generalization is to keep Γ^a_{bc} as the metric connection, but now take it to be the metric connection of a **non-flat** metric. If we are to interpret the Γ^a_{bc} as force terms, then it follows that we should regard the g_{ab} as **potentials**. The field equations of Newtonian gravitation consist of second-order partial differential equations in the gravitational potential ϕ. In an analogous manner, we would expect general relativity also to involve second-order partial differential equations in the potentials g_{ab}. The remaining task which will allow us to build a relativistic theory of gravitation is to choose a likely set of second-order partial differential equations.

9.5 The principle of general covariance

Recall the principle of special relativity, namely, that all **inertial** observers are equivalent. As we have seen in the last section, general relativity attempts to include non-inertial observers into its area of concern in order to cope with gravitation. Einstein argued that all observers, whether inertial or not, should be capable of discovering the laws of physics. If this were not true, then we would have little chance in discovering them since we are bound to the Earth, whose motion is almost certainly not inertial. Thus, Einstein proposed the following as the logical completion of the principle of special relativity.

Principle of general relativity: All observers are equivalent.

Observers are intimately tied up with their reference systems or coordinate systems. So, if any observer can discover the laws of physics, then any old coordinate system should do. The situation is somewhat different in special relativity, where, because the metric is flat and the connection integrable, there exists a **canonical** or preferred coordinate system: namely, Minkowski coordinates. In a curved space-time, that is, a manifold with a non-flat metric, there is no canonical coordinate system. This is just another statement of the non-existence of a global inertial observer. However, the statement needs to be treated with caution, because in many applications, there will be preferred coordinate systems. For example, many problems possess symmetries and the simplest thing to do is to adapt the coordinate system to the underlying symmetry. It is not so much that any coordinate system will do, but rather that the theory should be invariant under a coordinate transformation. Thus, the full import of the principle of general relativity is contained in the following statement.

Principle of general covariance: The equations of physics should have tensorial form.

Some authors argue that this statement is empty, because it is possible to formulate any physical theory in tensorial form. (Of course, this realization only came **after** the advent of general relativity.) Whether or not this is the case, it was clearly of central importance to Einstein, as is evident from the name he gave it. We shall make use of it in the form of the principle of general covariance, which is why we undertook our major digression in Part B to learn the language of tensors.

9.6 The principle of minimal gravitational coupling

The principles we have discussed so far do not tell us how to obtain field equations of systems in general relativity when the corresponding equations are known in special relativity. The principle of minimal gravitational coupling is a **simplicity principle** or Occam's razor that essentially says we should not add unnecessary terms in making the transition from the special to the general theory. For example, we shall later meet the conservation law

$$\partial_b T^{ab} = 0, \tag{9.10}$$

in special relativity in Minkowski coordinates. The simplest generalization of this to the general theory is to take the tensor equation

$$\nabla_b T^{ab} = 0. \tag{9.11}$$

However, we could equally well take

$$\nabla_b T^{ab} + g^{be} R^a{}_{bcd} \nabla_e T^{cd} = 0, \tag{9.12}$$

since $R^a{}_{bcd} = 0$ in special relativity and (9.12) again reduces to (9.10) in Minkowski coordinates. We therefore adopt the following principle.

Principle of minimal gravitational coupling: No terms explicitly containing the curvature tensor should be added in making the transition from the special to the general theory.

The principle was not stated by Einstein but was used implicitly. Unfortunately, it is rather vague and ambiguous and needs to be used with care.

9.7   The correspondence principle

As we stated from the outset, we are engaged with modelling, and together with any model should go its range of validity. Then any new theory must be consistent with any acceptable earlier theories within their range of validity. General relativity must agree on the one hand with special relativity in the absence of gravitation and on the other hand with Newtonian gravitational theory in the limit of weak gravitational fields and low velocities (compared with the speed of light). This gives rise to a **correspondence principle**, as indicated in Fig. 9.13, where arrows indicate directions of increased specialization.

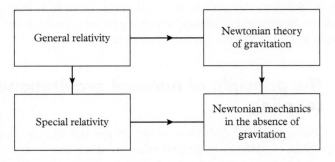

Fig. 9.13 The correspondence principle for general relativity.

Exercises

9.1 (§9.2)
(i) A pendulum is suspended from the roof of a car moving in a straight line with uniform acceleration a. Find the angle the pendulum makes with the vertical. Explain what is happening from the viewpoint of an inertial observer external to the car and a non-inertial observed fixed in the car.
(ii) A bucket of water is located in the car as well. Find the angle which the surface of the water makes with the horizontal.
(iii) A bucket of water slides freely under gravity down a slope of fixed angle α to the horizontal. What is the angle of inclination of the surface of the water relative to the base of the bucket?

9.2 (§9.2)
(i) Consider a body rotating relative to an inertial frame about a fixed point O with angular velocity $\boldsymbol{\omega}$ in Newtonian theory. The velocity $\boldsymbol{v}$ of any point P in the body with position vector $\vec{OP} = \boldsymbol{r}$ is given by

$$\boldsymbol{v} = \boldsymbol{\omega} \times \boldsymbol{r}.$$

Let $\boldsymbol{i}, \boldsymbol{j}, \boldsymbol{k}$ denote unit vectors in the inertial frame S, and $\boldsymbol{i}', \boldsymbol{j}', \boldsymbol{k}'$ denote unit vectors in a frame S' fixed in the body, where both origins are at O. If $\boldsymbol{u} = \boldsymbol{u}(t)$ is a general vector with components

$$\boldsymbol{u} = u_1' \boldsymbol{i}' + u_2' \boldsymbol{j}' + u_3' \boldsymbol{k}',$$

in S', show, by differentiating this equation, that

$$\left[\frac{d\boldsymbol{u}}{dt} \right]_S = \left[\frac{d\boldsymbol{u}}{dt} \right]_{S'} + \boldsymbol{\omega} \times \boldsymbol{u}.$$

(ii) Consider a non-inertial frame S' moving arbitrarily relative to an inertial frame S, where the position of the origin O' of S' relative to the origin O of S is $\boldsymbol{s}(t)$ and its angular velocity is $\boldsymbol{\omega}(t)$. A particle of constant mass m situated at a point with position vectors $\boldsymbol{r}$ and $\boldsymbol{r}'$ relative to S and S', respectively, is acted on by a force $\boldsymbol{F}$. Show that S' can write the equation of motion of the particle in the form

$$\boldsymbol{F} - \left[m\boldsymbol{a} + 2m\boldsymbol{\omega} \times \dot{\boldsymbol{r}}' + m\boldsymbol{\omega} \times (\boldsymbol{\omega} \times \boldsymbol{r}') + m\dot{\boldsymbol{\omega}} \times \boldsymbol{r}' \right] = m\ddot{\boldsymbol{r}}',$$

where $\boldsymbol{a}$ is the acceleration of O' relative to O and a dot denotes differentiation with respect to time in the frame of S'. What are the quantities in square brackets? Interpret these quantities physically.

9.3 (§9.3) Fill in the details that lead to the equalities (9.8) and (9.9).

9.4 (§9.3) Write down the equations of motion for an isolated system of three bodies of inertial masses m_1^I, m_2^I, and m_3^I. Eliminate the internal forces from these equations and demonstrate that, if two of the bodies are rigidly bound to form a composite system, then the inertial mass is additive.

9.5 (§9.4) In the lift experiments, explain the motion of the released body from the point of view of: case (1) an inertial observer, case (2) an inertial observer who initially sees the rocket moving away with constant velocity v, and case (4), an observer at rest on the surface of the Earth.

9.6 (§9.4) Consider a sphere of non-interacting particles falling towards the Earth's surface. Taking into account the different accelerations of particles in the sphere, what is the ensuing shape of the enclosing volume?

9.7 (§9.4) Find the geodesic equations for $\mathbb{R}^3$ in cylindrical polar coordinates (see Exercise 6.18). Interpret the terms occurring which involve Γ^a_{bc}.

9.8 (§9.4) What is the path of a free particle
(i) in an inertial frame?
(ii) in the presence of a uniform gravitational field?

Use the principle of equivalence and the particle theory of light to find the path of a light ray in the above two cases and hence deduce light bending in a gravitational field.

9.9 (§9.6) Write down a generalization of (9.10) to a curved space which involves a term quadratic in the Riemann tensor.

9.10 (§9.6) An anti-symmetric tensor F_{ab} satisfies the equation in special relativity in Minkowski coordinates

$$\partial_{[a}F_{bc]} = 0.$$

Write down the simplest generalization to a curved space-time and show that it is identical to the original equation.

9.11 (§9.7) Write down the correspondence principle for the transition from special relativity (in non-relativistic units) to Newtonian theory in the absence of gravitation. Express this transition as a limit involving the speed of light. Draw a sequence of diagrams to indicate what happens to the null cone in this limit. What happens to the three regions defined by the null cone in special relativity? What happens to the concept of simultaneity in the limit?

Further reading

There are many excellent textbooks suitable for a first course in general relativity. The book *A first course in general relativity* by Schutz (1985) is at a similar level to this book. The book by Carroll (2004) is also suitable for a first course and covers a slightly different range of topics. The book by Hartle adopts what is calls a 'physics first' approach, so those with a strong physics background will find it a useful alternative. For those wanting something at a more advanced level, the first recommendation would be the book by Wald (1984).

Carroll, S. M. (2004). *Spacetime and Geometry: An Introduction to General Relativity.* Addison Wesley, San Francisco, CA.

Hartle, J. B. (2003). *Gravity: An Introduction to Einstein's General Relativity.* Addison Wesley, San Francisco, CA.

Schutz, B. F. (1985). *A First Course in General Relativity.* Cambridge University Press, Cambridge.

Wald, R. M. (1984). *General Relativity.* University of Chicago Press, Chicago, IL.

The field equations of general relativity

10

10.1 Non-local lift experiments

The considerations of the last chapter led us to conclude that, locally, i.e. neglecting variations in the gravitational field, we can regain special relativity. However, in a non-local situation, we require a non-flat metric which may be thought of as the potentials of the gravitational field. Correspondence with Newtonian theory then suggests that we require second-order field equations in these potentials, and, moreover, from the principle of covariance, these equations must be tensorial in character. In this chapter, we shall pursue the Newtonian correspondence further and reformulate Newtonian theory in such a way that it leads naturally to the particular set of field equations of general relativity.

We return to the lift experiments and consider performing the following **non-local** experiments. We assume that the observer's equipment is sufficiently sensitive to detect variations in the gravitational field. The four experiments take the same form as before, but this time the observer releases two bodies, whose mutual interactions we ignore (Figs. 10.1–10.4).

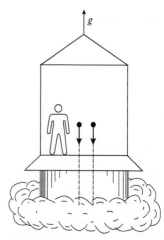

Fig. 10.1 Case 1: The lift in an accelerated rocket ship.

Case 1. From the point of view of the observer in the lift, the two bodies fall to the ground parallel to each other.

Case 2. The bodies remain at rest relative to the observer.

Case 3. The two bodies fall towards the centre of the Earth and hence fall on paths which **converge**.

Case 4. The bodies appear to the observer to move closer together, because they are falling on lines which converge towards the centre of the earth.

It follows that the observer can distinguish the **uniform** inertial field of Case 1 from the Earth's **non-uniform** gravitational field of Case 3 by considering the relative motion of test particles. Again, in free fall, bodies travel on geodesics in a gravitational field which **converge** (or diverge), as in Case 4. The point of these thought experiments is that the presence of a genuine gravitational field, as distinct from an inertial field, is verified by the observation of the **variation** of the field rather than by the observation of the field itself. We shall see that in general relativity this variation is described by the Riemann tensor through the equation of geodesic deviation.

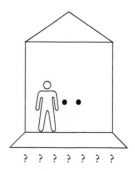

Fig. 10.2 Case 2: The lift in an unaccelerated rocket ship.

Introducing Einstein's Relativity. Second Edition. Ray d'Inverno and James Vickers, Oxford University Press.
© Ray d'Inverno and James Vickers (2022). DOI: 10.1093/oso/9780198862024.003.0010

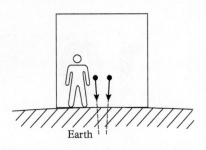

Fig. 10.3 Case 3: The lift placed on the Earth's surface.

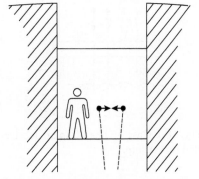

Fig. 10.4 Case 4: The lift dropped down an evacuated lift shaft.

10.2 The Newtonian equation of deviation

The non-local lift experiments reveal that we should focus our attention on two neighbouring test particles in free fall in a gravitational field. We look at this motion first of all in Newtonian theory using the tensor apparatus of Part B. We introduce Cartesian coordinates

$$(x^a) = (x^1, x^2, x^3) = (x, y, z),$$

where, for the rest of this chapter, Greek indices run from 1 to 3, and then the line element of Euclidean 3-space $\mathbb{R}^3$ is

$$d\sigma^2 = dx^2 + dy^2 + dz^2,$$

from which we obtain the Euclidean metric

$$g_{\alpha\beta} = \delta_{\alpha\beta} = \text{diag}(1, 1, 1). \tag{10.1}$$

We therefore raise and lower indices with the three-dimensional Kronecker delta. This means that in Newtonian theory there is really no distinction between raised and lowered indices, but we will retain the notation in order to help us compare results later with the general theory. We consider the paths of two neighbouring gravitational test particles of unit mass travelling **in vacuo** in a gravitational field whose potential is ϕ.

Let the particles travel on curves C_1 and C_2 so that they reach the points P and Q at time t (Fig. 10.5). If we use the time t as the parameter along the curves, then the parametric equations of C_1 are

$$x^\alpha = x^\alpha(t), \tag{10.2}$$

and those of C_2 can be written as

$$x^\alpha = x^\alpha(t) + \eta^\alpha(t), \tag{10.3}$$

where η^α is a small **connecting vector** which connects points on the two curves with equal values of t. Since the particles have unit mass, the

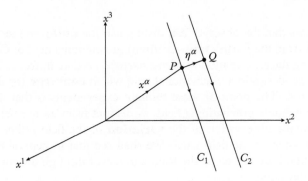

Fig. 10.5 Freely falling gravitational test particles at time t.

equation of motion of the first particle, by (9.5) and (9.6), can be written in the tensor form

$$\ddot{x}^\alpha = -\partial^\alpha \phi, \qquad (10.4)$$

where a dot denotes differentiation with respect to time and

$$\partial^\alpha \phi = \delta^{\alpha\beta} \partial_\beta \phi = \left(\frac{\partial \phi}{\partial x}, \frac{\partial \phi}{\partial y}, \frac{\partial \phi}{\partial z} \right) = (\text{grad}\phi)_P. \qquad (10.5)$$

Similarly, the equation of motion of the second particle is

$$\ddot{x}^\alpha + \ddot{\eta}^\alpha = -(\partial^\alpha \phi)_Q. \qquad (10.6)$$

Since η_α is small, we may expand the term on the right-hand side by Taylor's theorem (exercise), to obtain

$$-(\partial^\alpha \phi)_Q = -(\partial^\alpha \phi)_P - \left(\eta^\beta \partial_\beta \partial^\alpha \phi \right)_P, \qquad (10.7)$$

to first order. Subtracting (10.4) from (10.6), we get

$$\ddot{\eta}^\alpha = -\eta^\beta \partial_\beta \partial^\alpha \phi. \qquad (10.8)$$

If we define the tensor $K^\alpha{}_\beta$ by

$$K^\alpha{}_\beta := \partial^\alpha \partial_\beta \phi, \qquad (10.9)$$

then the equation of motion (10.8) of the connecting vector η_α, which we call the **Newtonian equation of deviation**, becomes

$$\ddot{\eta}^\alpha + K^\alpha{}_\beta \eta^\beta = 0. \qquad (10.10)$$

Note that $K_{\alpha\beta} = \partial_\alpha \partial_\beta \phi$ is symmetric. This equation is intimately connected with the Newtonian field equations in empty space, namely, Laplace's equation (4.6), which can be written (exercise)

$$K^\alpha{}_\alpha = 0. \qquad (10.11)$$

In other words, **the tensor $K_{\alpha\beta}$ is symmetric and trace-free**. We now search for a relativistic generalization of these equations.

10.3 The equation of geodesic deviation

Following the axioms of §8.5, we assume that free test particles in general relativity travel on timelike geodesics. We therefore consider a 2-surface S ruled by a **congruence of timelike geodesics**, that is, a family of geodesics such that exactly one of the curves goes through every point of S. The parametric equation of S is given by

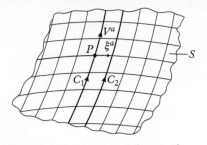

Fig. 10.6 The vectors V^a and ξ^a at a point P in S.

$$x^a = x^a(\tau, \ \sigma), \tag{10.12}$$

where τ is the **proper time** along the geodesics, and σ labels distinct geodesics. We define two vector fields on S by

$$V^a = \frac{\partial x^a}{\partial \tau}, \tag{10.13}$$

and

$$\xi^a = \frac{\partial x^a}{\partial \sigma}. \tag{10.14}$$

Then V^a is the **tangent vector** to the timelike geodesic at each point, and ξ^a is a **connecting vector** connecting two neighbouring curves in the congruence (Fig. 10.6). The commutator of V^a and ξ^a satisfies

$$
\begin{aligned}
[V, \xi]^a &= V^b \partial_b \xi^a - \xi^b \partial_b V^a \\
&= \frac{\partial x^b}{\partial \tau} \frac{\partial}{\partial x^b} \left(\frac{\partial x^a}{\partial \sigma} \right) - \frac{\partial x^b}{\partial \sigma} \frac{\partial}{\partial x^b} \left(\frac{\partial x^a}{\partial \tau} \right) \\
&= \frac{\partial^2 x^a}{\partial \tau \partial \sigma} - \frac{\partial^2 x^a}{\partial \sigma \partial \tau} \\
&= 0
\end{aligned}
\tag{10.15}
$$

since the mixed partial derivatives commute. (It can be shown that the vanishing of the commutator is a necessary and sufficient condition for the vector fields to be **surface-forming**, which means that the congruences generated by the two vectors knit together to form a 2-surface.)

By (6.15), the commutator is also equal to the Lie derivative $\mathrm{L}_V \xi^a$. We now use the result which allows us to replace partial derivatives by covariant derivatives in an expression for a Lie derivative

$$
\begin{aligned}
0 &= \mathrm{L}_V \xi^a \\
&= V^b \partial_b \xi^a - \xi^b \partial_b V^a \\
&= V^b \nabla_b \xi^\alpha - \xi^b \nabla_b V^a \\
&= \nabla_V \xi^a - \nabla_\xi V^a.
\end{aligned}
\tag{10.16}
$$

Taking the covariant derivative of this equation with respect to V^a, we find

$$\nabla_V \nabla_V \xi^a = \nabla_V \nabla_\xi V^a. \tag{10.17}$$

The equation we are seeking derives from the identity (Exercise 6.11)

$$\nabla_X (\nabla_Y Z^a) - \nabla_Y (\nabla_X Z^a) - \nabla_{[X,Y]} Z^a = R^a{}_{bcd} Z^b X^c Y^d. \tag{10.18}$$

If we set $X^a = Z^a = V^a$ and $Y^a = \xi^a$, then the second term on the left vanishes, because V^a is tangent to an affinely parametrized geodesic, and so, by (6.38),

$$\nabla_V V^a = 0. \tag{10.19}$$

The third term vanishes by (10.15), since the covariant derivative of any tensor with respect to the zero tensor is zero. Thus, (10.18) becomes

$$\nabla_V \nabla_\xi V^a - R^a{}_{bcd} V^b V^c \xi^d = 0. \tag{10.20}$$

By definition,

$$\frac{D^2 \xi^a}{D\tau^2} = \nabla_V \nabla_V \xi^a,$$

and so, using (10.17), equation (10.20) becomes the promised **equation of geodesic deviation**

$$\frac{D^2 \xi^a}{D\tau^2} - R^a{}_{bcd} V^b V^c \xi^d = 0. \tag{10.21}$$

The absolute derivative along the curve is the tensorial analogue of the time derivative along the curve in (10.10).

10.4 The vacuum field equations of general relativity

We now want to compare the relativistic equation (10.21) to the Newtonian result (10.10). To do this we define

$$K^a{}_b = R^a{}_{cbd} V^c V^d, \tag{10.22}$$

so that equation (10.21) becomes

$$\frac{D^2 \xi^a}{D\tau^2} + K^a{}_b \xi^b = 0. \tag{10.23}$$

Comparing this with the Newtonian equation (10.10), we see that $K^a{}_b$ defined by (10.22) is the relativistic analogue of the Newtonian quantity $K^\alpha{}_\beta$ defined by (10.11). Continuing with this analogy, we now tentatively suggest that the relativistic version of (10.11) is

$$K^a{}_a = 0. \tag{10.24}$$

From (10.22) this gives

$$R^a{}_{cad} V^c V^d = 0,$$

and thus $R_{ab} V^a V^b = 0$ along any timelike geodesic, where V^a is the tangent to the geodesic. Since at any point P and for any timelike vector V^a

we can find a geodesic through P with this tangent vector, it follows that

$$R_{ab}V^aV^b = 0, \tag{10.25}$$

at any point P and for any timelike vector V^a. Since R_{ab} is symmetric, it follows (exercise) that this is only possible if

$$R_{ab} = 0. \tag{10.26}$$

We therefore take (10.26) as the **vacuum field equations** for general relativity, i.e. the field equations in the absence of matter.

10.5 Freely falling frames

As our first test of these field equations, we want to look at the Newtonian limit of (10.26) and show that it gives Laplace's equation $\nabla^2\phi = 0$ for the Newtonian gravitational potential ϕ. The relationship between Newtonian theory and general relativity is best understood in a **local inertial frame** which is given by a **freely falling frame**. Such a frame consists of a set of four vectors $e_i{}^b$ (where the bold index $\boldsymbol{i} = 0, 1, 2, 3$ labels the vector, and the roman index $b = 0, 1, 2, 3$ gives the components of the vector) defined along γ, the geodesic $x(\tau)^a$ of a freely falling particle parametrized by proper time. We choose

$$e_0{}^b := V^b = \dot{x}^b,$$

to be tangent to the geodesic, and $e_\alpha{}^b$, where $\boldsymbol{\alpha}$ is a bold label running from 1 to 3, to be spacelike vectors which together with e_0 satisfy the following **orthonormality relations**

$$e_0{}^a e_{0a} = -e_1{}^a e_{1a} = -e_2{}^a e_{2a} = -e_3{}^a e_{3a} = 1,$$
$$e_0{}^a e_{1a} = e_0{}^a e_{2a} = e_0{}^a e_{3a} = e_1{}^a e_{2a} = e_1{}^a e_{3a} = e_2{}^a e_{3a} = 0.$$

These can be succinctly summarized as

$$g_{ab}e_i{}^a e_j{}^b = \eta_{ij}, \tag{10.27}$$

where η_{ij} is the Minkowski metric, that is,

$$\eta_{ij} = \mathrm{diag}(1, -1, -1, -1).$$

The four vectors $e_i{}^a$ ($\boldsymbol{ii} = 0, 1, 2, 3$) are said to form a **frame** or **tetrad** (**vierbein**, in German) at P.

Treating $e_i{}^a$ as a 4×4 matrix at P, we can define its inverse (called the **dual frame**) $e^j{}_a$ by requiring

$$e_i{}^a e^j{}_a = \delta_i^j, \tag{10.28}$$

where δ_i^j is the Kronecker delta, or the identity matrix in matrix terms. We have introduced the frame notation merely as a convenience so far, but it turns out that frames possess a powerful formalism of their own (which is outside the scope of this book, but see §20.1). For example, in exactly the same way that we raise and lower **tensor** indices with the metric g_{ab}, we can raise and lower **frame** indices $(i, j, \ldots)$ with the **frame metric** η_{ij}. Let us multiply (10.28) by e^i_b and write it in the form

$$(e^i_b e_i{}^a)e^j_a = e^j_b,$$

from which it should be clear that the quantity in parentheses must be the tensorial Kronecker delta, namely,

$$e^i_b e_i{}^a = \delta^a_b. \tag{10.29}$$

The physical interpretation of the frame is as follows: $e_0{}^a = V^a$ is the 4-velocity of an observer whose world-line is γ, and the three spacelike vectors $e_\alpha{}^a$ are rectangular coordinate vectors (such as the usual Cartesian basis $\boldsymbol{i}, \boldsymbol{j}$, and $\boldsymbol{k}$, for example) at P, where the bold Greek indices run from 1 to 3. So far, the frame has only been defined at P, but we also want the spatial vectors to be non-rotating which requires that

$$\nabla_V e_\alpha{}^b = 0. \tag{10.30}$$

It can be shown that this follows from the physical fact that the spin of a gyroscope is parallely propagated along a geodesic (Schiff 1960, Hartle 2003). Note that, since $e_0{}^b = V^b$, we automatically have

$$\nabla_V e_0^b = \nabla_V V^b = 0. \tag{10.31}$$

Taking (10.30) and (10.31) together, we see that a freely falling frame also satisfies

$$\frac{\mathrm{D}}{\mathrm{D}\tau}\left(e_i{}^a\right) = 0, \tag{10.32}$$

from which we see that the frame is **parallelly transported** along the curve γ.

In the same way as we can get the Cartesian components of a three-dimensional vector by taking the scalar product of it with $\boldsymbol{i}, \boldsymbol{j}$, and $\boldsymbol{k}$, we define the **frame components** of the connecting vector ξ^a by

$$\xi^\alpha = e^\alpha{}_a \xi^a. \tag{10.33}$$

We represent the various quantities schematically in Fig. 10.7. Indeed, we can find the frame components of any tensor by contracting with $e_i{}^b$ and $e^j{}_d$ in order to saturate all the unbold indices and replace them by bold indices. For example, the frame components of a rank 2 tensor T_{ab} are given by

$$T_{ij} = T_{cd} e_i{}^c e_j{}^d.$$

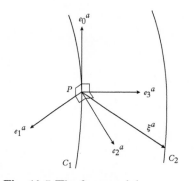

Fig. 10.7 The frame and the connecting vector at P.

To find the equation of geodesic deviation as measured by a freely falling inertial observer, we multiply (10.21) by $e^i{}_a$ and use the fact that $V^a = e_0{}^a$. Taking account of the symmetries of the curvature tensor, this gives

$$e^i{}_a \frac{\mathrm{D}^2 \xi^a}{\mathrm{D}\tau^2} + e^i{}_a R^a{}_{bcd} e_0{}^b e_0{}^d \xi^c = 0. \tag{10.34}$$

Now, by (10.32)

$$e^i{}_a \frac{\mathrm{D}^2 \xi^a}{\mathrm{D}\tau^2} = \frac{\mathrm{D}^2 \left(e^i{}_a \xi^a \right)}{\mathrm{D}\tau^2} = \frac{\mathrm{D}^2 \xi^i}{\mathrm{D}\tau^2}, \tag{10.35}$$

and, using (10.29)

$$e^i{}_a R^a{}_{bcd} e_0{}^b e_0{}^d \xi^c = e^i{}_a R^a{}_{bcd} e_0{}^b e_0{}^d e_j{}^c e^j{}_f \xi^f = R^i{}_{0j0} \xi^j. \tag{10.36}$$

So that the freely falling frame version of (10.21) is

$$\frac{\mathrm{D}^2 \xi^i}{\mathrm{D}\tau^2} + R^i{}_{0j0} \xi^j = 0. \tag{10.37}$$

Note that, because the curvature tensor vanishes if $j = 0$ in the above, we may write this as an equation for the **orthogonal connecting vector**, which is given by $\eta^i = (0, \xi^\alpha)$. The spatial part of this η^α is the precise analogue of the vector η^α of §10.2. In terms of η^α, the spatial part of the equation of geodesic deviation is

$$\frac{\mathrm{D}^2 \eta^\alpha}{\mathrm{D}\tau^2} + K^\alpha{}_\beta \eta^\beta = 0, \tag{10.38}$$

where

$$K^\alpha{}_\beta = R^a{}_{bcd} e^\alpha{}_a e_0{}^b e_\beta{}^c e_0^d = R^\alpha{}_{0\beta0}. \tag{10.39}$$

Equation (10.38) is the spatial part (10.21) as measured in an inertial frame whose Newtonian limit we now calculate to compare with (10.10).

10.6   The Newtonian correspondence

In this section, we consider more precisely the Newtonian limit of a slowly varying weak gravitational field. We shall work in non-relativistic units. In the **Newtonian limit**, we assume that there exists a privileged coordinate system

$$(x^a) = (x^0, x^1, x^2, x^3) = (x^0, x^\alpha) = (ct, x, y, z),$$

in which the metric g_{ab} differs only slightly from the Minkowski metric η_{ab}. Moreover, we assume that the field is produced by bodies whose velocities are small compared with the velocity of light. If v is a typical velocity of the bodies, then we let ε denote a small dimensionless parameter of order v/c, and our basic assumption is

$$g_{ab} = \eta_{ab} + \varepsilon h_{ab} + O(\varepsilon^2), \tag{10.40}$$

where throughout we shall work to lowest order in ε. In time δt, a body moves a distance δx^α with velocity v, i.e.

$$\delta x^\alpha \sim \text{velocity} \times \text{time} \sim v\delta t \sim (v/c)c\delta t \sim \varepsilon \delta x^0,$$

and so

$$\varepsilon/\delta x^\alpha \sim 1/\delta x^0.$$

Then, for any function f, we assume the **slow-motion approximation**

$$\varepsilon \frac{\partial f}{\partial x^\alpha} \sim \frac{\partial f}{\partial x^0}, \tag{10.41}$$

that is, derivatives with respect to x^0 are of order ε times the spatial derivatives. The conditions (10.40) and (10.41) are the starting assumptions for obtaining the Newtonian limit.

We consider the motion of a free test particle moving with a speed of the order of v on a world-line $x^a = x^a(\tau)$ parametrized by the proper time. It travels on a timelike geodesic

$$\frac{\mathrm{d}^2 x^a}{\mathrm{d}\tau^2} + \Gamma^a_{bc} \frac{\mathrm{d}x^b}{\mathrm{d}\tau} \frac{\mathrm{d}x^c}{\mathrm{d}\tau} = 0. \tag{10.42}$$

By definition,

$$\begin{aligned}
c^2 \mathrm{d}\tau^2 &= \mathrm{d}s^2 \\
&= c^2 \mathrm{d}t^2 - \mathrm{d}x^2 - \mathrm{d}y^2 - \mathrm{d}z^2 \\
&= \mathrm{d}t^2(c^2 - v^2) \\
&= c^2 \mathrm{d}t^2(1 - \varepsilon^2),
\end{aligned}$$

and so, taking square roots,

$$\frac{\mathrm{d}\tau}{\mathrm{d}t} = 1 + O(\varepsilon^2). \tag{10.43}$$

Hence, working to lowest order in ε, we can replace τ by t in (10.42). Moreover, from our slow-motion approximation,

$$\mathrm{d}x^\alpha \sim \varepsilon c\,\mathrm{d}t,$$

so that

$$\frac{1}{c}\frac{dx^\alpha}{dt} = O(\varepsilon). \tag{10.44}$$

In addition

$$\Gamma^a_{bc} = \tfrac{1}{2}g^{ad}\left(\partial_c g_{bd} + \partial_b g_{cd} - \partial_d g_{bc}\right)$$
$$= \tfrac{1}{2}\eta^{ad}\varepsilon\left(\partial_c\,h_{bd} + \partial_b\,h_{cd} - \partial_d\,h_{bc}\right) + O(\varepsilon^2), \tag{10.45}$$

so that

$$\Gamma^a_{bc} = O(\varepsilon). \tag{10.46}$$

Since we are interested in the Newtonian limit, we restrict our attention to the spatial part of (10.42), i.e. when $a = \alpha$, and we obtain, by using (10.43) and dividing by c^2

$$0 = \frac{1}{c^2}\frac{d^2 x^\alpha}{dt^2} + \frac{1}{c^2}\Gamma^\alpha_{bc}\frac{dx^b}{dt}\frac{dx^c}{dt}[1 + O(\varepsilon)]$$
$$= \frac{1}{c^2}\frac{d^2 x^\alpha}{dt^2} + \Gamma^\alpha_{00} + 2\Gamma^\alpha_{0\beta}\left(\frac{1}{c}\frac{dx^\beta}{dt}\right) + \Gamma^\alpha_{\beta\gamma}\left(\frac{1}{c}\frac{dx^\beta}{dt}\right)\left(\frac{1}{c}\frac{dx^\gamma}{dt}\right) + O(\varepsilon^2).$$

From (10.44) and (10.46), the third and fourth terms in this equation are $O(\varepsilon^2)$ and $O(\varepsilon^3)$, respectively. From (10.45), the second term is

$$\Gamma^\alpha_{00} = -\tfrac{1}{2}\varepsilon\left(2\frac{\partial h_{0\alpha}}{\partial x^0} - \frac{\partial h_{00}}{\partial x^\alpha}\right)$$
$$= \tfrac{1}{2}\varepsilon\frac{\partial h_{00}}{\partial x^\alpha} + O(\varepsilon^2), \tag{10.47}$$

by the slow motion approximation (10.41). So the spatial part of the geodesic equation can be written

$$\frac{d^2 x^\alpha}{dt^2} = -\tfrac{1}{2}c^2\frac{\partial g_{00}}{\partial x^\alpha}[1 + O(\varepsilon)], \tag{10.48}$$

using (10.40). We compare this with the corresponding Newtonian equation (10.4), namely,

$$\frac{d^2 x^\alpha}{dt^2} = -\frac{\partial\phi}{\partial x^\alpha},$$

where ϕ is the Newtonian gravitational potential. Noting that, at large distances from the sources of the field, $\phi \to 0$ and $g_{00} \to 1$, we conclude

$$g_{00} = 1 + \frac{2\phi}{c^2} + O(v/c). \tag{10.49}$$

This is called the **weak-field limit**.

Let us consider the effect of an infinitesimal coordinate transformation

$$x^a \rightarrow x'^a = x^a + \varepsilon X^a(x),$$

which is consistent with the two assumptions (10.40) and (10.41). Then we find (exercise)

$$g'_{ab} = g_{ab} - \varepsilon \left(\partial_a X_b + \partial_b X_a \right) + O \left(\varepsilon^2 \right), \qquad (10.50)$$

where

$$X_a = \eta_{ab} X^b.$$

To preserve (10.41), we require

$$\frac{\partial X_a}{\partial x^0} \sim \varepsilon \frac{\partial X_a}{\partial x^\alpha},$$

which means from (10.50) that g_{00} is the only component of g_{ab} that does not alter to first order in ε. We have therefore shown that the only component of the metric tensor which is well defined to first order for a slowly varying weak gravitational field is determined to this order by the requirement that the theory should agree with Newtonian theory to this order, and it is given by (10.49). Note that no mention of the field equations has been made in deriving (10.49). It arises purely from assuming geodesic motion and the Newtonian limit as embodied in the equations (10.40) and (10.41).

Having obtained an expression for the weak-field limit, we are now in a position to calculate the Newtonian limit of equation (10.38). We start by looking at the curvature tensor. This consists of derivatives of the connection Γ^a_{bc} and terms quadratic in the connection. Since the connection is $O(\varepsilon)$, we see that the curvature is also $O(\varepsilon)$, with the leading order terms given by the derivative terms. Hence,

$$R^a{}_{bcd} = \Gamma^a_{bd,c} - \Gamma^a_{bc,d} + O(\varepsilon^2).$$

In particular,

$$R^\alpha{}_{0\gamma 0} = \Gamma^\alpha_{00,\gamma} - \Gamma^\alpha_{0\gamma,0} + O(\varepsilon^2).$$

Furthermore, using the slow-motion approximation (10.41) the second term involves a time derivative and so is also $O(\varepsilon^2)$. Thus, ignoring the terms of $O(\varepsilon^2)$ and setting $c = 1$, we have

$$R^\alpha{}_{0\gamma 0} = \Gamma^\alpha_{00,\gamma}$$

$$= \tfrac{1}{2} \varepsilon \frac{\partial}{\partial x^\gamma} \left(\frac{\partial g_{00}}{\partial x^\alpha} \right)$$

$$= \frac{\partial^2 \phi}{\partial x^\alpha \partial x^\gamma}$$

by (10.47) and (10.49). Finally, we note that, for a metric given by (10.40), the components of the frame differ from those of a Minkowski frame by terms of $O(\varepsilon)$ so, to leading order, we may use a standard Minkowski frame whose components coincide with the coordinate vectors. Thus, in the slow-motion weak-field approximation, we find

$$K^{\alpha}{}_{\beta} = R^{\alpha}{}_{0\beta 0} = \frac{\partial^2 \phi}{\partial x^{\alpha} \partial x^{\beta}}. \qquad (10.51)$$

Furthermore, in the slow-motion approximation, $\mathrm{D}/\mathrm{D}t$ is the same as $\mathrm{d}/\mathrm{d}t$, and the spatial components of the equation of geodesic deviation (10.38) reduce to

$$\frac{d^2 \eta^{\alpha}}{dt^2} + \delta^{\alpha\gamma} \frac{\partial^2 \phi}{\partial x^{\gamma} \partial x^{\beta}} = 0,$$

which is identical to the Newtonian equation (10.10).

If we take the trace of $R^a{}_{0b0}$, we get $R^0{}_{000} + R^{\alpha}{}_{0\alpha 0}$. Since the first term vanishes, the trace is just $R^{\alpha}{}_{0\alpha 0} = K^{\alpha}{}_{\alpha}$ which, using (10.51) and the field equation (10.26), gives

$$\delta^{\alpha\beta} \frac{\partial^2 \phi}{\partial x^{\alpha} \partial x^{\beta}} = \nabla^2 \phi = 0,$$

which is just Laplace's equation for the gravitational potential.

We have therefore shown the following result:

In the slow-motion weak field approximation, the relativistic equation of geodesic deviation (10.21) reduces to the Newtonian equation of deviation (10.10), and the relativistic vacuum field equations $R_{ab} = 0$ reduce to the empty-space Newtonian field equations $\nabla^2 \phi = 0$.

10.7   Einstein's route to the field equations of general relativity

Our arrival at the vacuum field equations of general relativity has involved rather a long story. This is not so surprising when you consider that it took Einstein over ten years of endeavour to move from the formulation of the special theory (1905) to a final formulation of the general theory (1916). It might be helpful, therefore, to outline again the main points of the argument.

1. The principle of equivalence reveals that, if we freefall in a gravitational field, then we can eliminate gravity locally and regain special relativity.

2. It also states that, locally, we cannot distinguish a gravitational field from a (uniform accelerating) inertial field and, consequently, we should regard gravitation as an inertial force.

3.   Following special relativity, we assume that free test particles travel on timelike geodesics. Then inertial forces arise in the geodesic equations in the terms involving the metric connection of a flat metric. In order to include the extra effect of gravitation in the metric connection, we generalize the metric to being curved.

4.   The metric then plays the role of the potentials of the theory and, in analogy with Newtonian theory, we seek a set of second-order partial differential equations for the potentials as field equations of the theory. Moreover, by the covariance principle, these equations must be tensorial.

5.   If we now take non-local effects into account; then, a genuine gravitational field can be observed by the variation in the field rather than by an observation of the field itself. This variation causes test particles to travel on timelike geodesics which converge (or diverge), and the convergence is described by the Riemann tensor through the equation of geodesic deviation.

6.   The Riemann tensor is a tensor which involves second partial derivatives of the metric and so we might expect the field equations of the theory to involve the Riemann tensor. The fact that the Newtonian vacuum field equations involve the vanishing of a contracted tensor suggests that we might consider a contraction of the Riemann tensor. There is only one meaningful contraction (why?), namely, the Ricci tensor, and its vanishing is equivalent to the vanishing of the Einstein tensor.

We thus arrive at the equations

$$R_{ab} = 0. \tag{10.52}$$

By Exercise 6.26, the vanishing of the Ricci tensor is equivalent to the vanishing of the Einstein tensor, so that we can write (10.52) in the alternative form

$$G_{ab} = 0. \tag{10.53}$$

Equations (10.52) or (10.53) are the equations which Einstein proposed should serve as the **vacuum field equations of general relativity**. We briefly indicate below why (10.53) is the more natural form when we attempt to generalise Poisson's equation $\nabla^2 \phi = 8\pi G \rho$ in order to include matter sources for the gravitational field.

10.8 The full field equations of general relativity

For completeness, we introduce briefly the full field equations, which hold in the presence of fields other than gravitation. As we shall see, these fields are described by the **energy-momentum tensor** T_{ab}. Now the equivalence of mass and energy from special relativity suggests that **all** forms of energy act as sources for the gravitational field; indeed, this is the content of the weak form of the principle of equivalence P2. We therefore take T_{ab} as a **source** term in the field equations. In special relativity in Minkowski coordinates, the energy-momentum tensor satisfies the conservation equations (see Chapter 12)

$$\partial_b T^{ab} = 0.$$

The principle of minimal gravitational coupling suggests the general relativistic generalization

$$\nabla_b T^{ab} = 0.$$

However, we know that the covariant derivative of the Einstein tensor vanishes through the contracted Bianchi identities (6.87):

$$\nabla_b G^{ab} = 0.$$

The last two equations suggest that the two tensors are proportional, and one can write consistently

$$G^{ab} = \kappa \, T^{ab}, \tag{10.54}$$

where κ is a constant of proportionality called the **coupling constant**. Note that this equation is in line with Mach's principle in the form M1, since the matter (T^{ab}) determines the geometry (G^{ab}) and hence is the source of inertial effects. The constant κ is then determined by the correspondence principle, since this equation must reduce to Poisson's equation (4.5) in the appropriate limit. We shall see in §12.3 that this is given in non-relativistic units by

$$\kappa = 8\pi G/c^4 . \tag{10.55}$$

The equations (10.54) subject to (10.55) constitute the **full field equations** of general relativity. We shall, for the most part, work in **relativistic units**, in which we can take both $c = 1$ and $G = 1$, and then the coupling constant is simply

$$\kappa = 8\pi. \tag{10.56}$$

At this stage, we shall define the **theory of general relativity** to consist of the axioms of special relativity as stated in §8.5 except that I(iii) is now replaced by equation (10.54) subject to (10.55). However, before we consider further the significance of the field equations, we shall look at, in the next chapter, an alternative derivation based on a mathematical principle rather than physical principles, namely, the variational principle, and follow this up with an investigation of the right hand side of (10.54), namely, the energy-momentum tensor.

Exercises

10.1 (§10.2) Taylor's theorem in three dimensions can be written

$$f(\boldsymbol{x} + \boldsymbol{h}) = f(\boldsymbol{x}) + \sum_{1}^{\infty} \frac{(\boldsymbol{h}\cdot\boldsymbol{\nabla})^n}{n!} f(\boldsymbol{x}),$$

where

$$\boldsymbol{x} = x\boldsymbol{i} + y\boldsymbol{j} + z\boldsymbol{k},$$

$$\boldsymbol{h} = h_1\boldsymbol{i} + h_2\boldsymbol{j} + h_3\boldsymbol{k},$$

$$\boldsymbol{\nabla} = \boldsymbol{i}\frac{\partial}{\partial x} + \boldsymbol{j}\frac{\partial}{\partial y} + \boldsymbol{k}\frac{\partial}{\partial z}.$$

Write out the first three terms of the expansion.

10.2 (§10.2) (i) Use Exercise 10.1 to verify (10.7).
(ii) Verify that Laplace's equation can be written in the form (10.11).

10.3 (§10.3) If $V^a = dx^a/d\tau$ is the tangent vector to a timelike geodesic parametrized by the proper time, and ξ^a is an arbitrary vector field, show that
(i) $\nabla_V V^a = 0$,
(ii) $\nabla_V V_a = 0$,
(iii) $V_a \nabla_\xi V^a = 0$,
(iv) $V^a \nabla_\xi V_a = 0$.

10.4 (§10.3) Let $h^a{}_b = \delta^a_b - V^a V_b$ be the projection operator into the space orthogonal to V^a. Show that, if we define the **orthogonal connecting vector** η^a by $\eta^a := h^a{}_b \xi^b$, then (10.21) implies

$$\frac{D^2 \eta^a}{D\tau^2} - R^a{}_{bcd} V^b V^c \eta^d = 0.$$

10.5 (§10.3) Show that a Killing vector X^a satisfies the equation of geodesic deviation

$$\frac{D^2 X^a}{Du^2} - R^a{}_{bcd} \frac{dx^b}{du} \frac{dx^c}{du} X^d = 0$$

along any geodesic $x^a = x^a(u)$. [Hint: use Exercise 7.15.]

10.6 (§10.3) If, at some point P, the symmetric tensor R_{ab} satisfies

$$R_{ab} v^a v^b = 0$$

for an **arbitrary** timelike vector v^a, then deduce that R_{ab} must vanish at P. [Hint: let $v^a = t^a + \lambda s^a$, where $t^a t_a = 1$, $s^a s_a = -1$, $t^a s_a = 0$, $0 \leqslant \lambda < 1$, λ arbitrary, and consider a special coordinate system in which $t^a \overset{*}{=} \delta_0^a$ and $s^a \overset{*}{=} \delta_1^a, \delta_2^a, \delta_3^a$ in turn.]

10.7 (§10.5) Show that, if a frame $e_i{}^a$ is parallelly propagated along C, then so is its dual frame $e^i{}_a$.

10.8 (§10.5) If η^{ij} is the inverse of η_{ij}, then show that

$$g_{ab} = \eta_{ij} e^i{}_a e^j{}_b \quad \text{and} \quad g^{ab} = \eta^{ij} e_i{}^a e_j{}^b.$$

If $(x^a) = (t, r, \theta, \phi)$ and

$$e_0^a = (A^{-1/2}, 0, 0, 0), \quad e_1^a = (0, A^{1/2}, 0, 0),$$
$$e_2^a = (0, 0, 1/r, 0), \quad e_3^a = (0, 0, 0, 1/r\sin\theta),$$

where $A = A(r)$ is an arbitrary function, then find g^{ab}, g_{ab}, and the line element ds^2.

10.9 (§10.6) Write out the argument fully which deduces (10.49) from (10.48).

10.10 (§10.6) Check (10.50). Deduce that g_{00} is the only component not to alter to order ε.

10.11 (§10.7) What principles are used in each of the six steps outlined in §10.7?

10.12 (§10.8) What principles are used in the transition to the full theory?

Further reading

The references for this chapter are similar to those for Chapter 9. The article by Schiff (1960) discusses the motion of gyroscopes in general relativity. This is also discussed in the book by Hartle (1984).

Carroll, S. M. (2004). *Spacetime and Geometry: An Introduction to General Relativity*. Addison Wesley, San Francisco, CA.

Hartle, J. B. (2003). *Gravity: An Introduction to Einstein's General Relativity*. Addison Wesley, San Francisco, CA.

Schutz, B. F. (1985). *A First Course in General Relativity*. Cambridge University Press, Cambridge.

Schiff, L. I. (1960). Motion of a gyroscope according to Einstein's theory of gravitation. *Proceedings of the National Academy of Sciences of the United States of America*, 46(6), 871–82.

Wald, R. M. (1984). *General Relativity*. University of Chicago Press, Chicago, IL.

General relativity from a variational principle

11

11.1 The Palatini equation

Many tensor identities are best derived using the technique of geodesic coordinates, where we choose an arbitrary point P at which $\Gamma^a_{bc} \overset{*}{=} 0$. Then, in particular, covariant derivatives reduce to ordinary derivatives at the point P. The Riemann tensor (6.40) reduces to

$$R^a{}_{bcd} \overset{*}{=} \partial_c \Gamma^a_{bd} - \partial_d \Gamma^a_{bc}. \tag{11.1}$$

We now contemplate a variation of the connection $\Gamma^a{}_{bc}$, to a new connection $\bar{\Gamma}^a_{bc}$

$$\Gamma^a_{bc} \rightarrow \bar{\Gamma}^a_{bc} = \Gamma^a_{bc} + \delta\Gamma^a_{bc}. \tag{11.2}$$

Then $\delta\Gamma^a_{bc}$, being the difference of two connections, is a tensor of type $(1, 2)$. This variation results in a change in the Riemann tensor

$$R^a{}_{bcd} \rightarrow \bar{R}^a{}_{bcd} = R^a{}_{bcd} + \delta R^a{}_{bcd},$$

where

$$\begin{aligned}
\delta R^a{}_{bcd} &\overset{*}{=} \partial_c(\delta\Gamma^a_{bd}) - \partial_d(\delta\Gamma^a_{bc}) \\
&\overset{*}{=} \nabla_c(\delta\Gamma^a_{bd}) - \nabla_d(\delta\Gamma^a_{bc}),
\end{aligned}$$

since partial derivative commutes with variation and is equivalent to co-variant derivative in geodesic coordinates. Now both $\delta R^a{}_{bcd}$, being the difference of two tensors, and the quantities on the right-hand side of the last equation are tensors, and so, by our fundamental result (if a tensor equation holds in one coordinate system it must hold in all coordinate systems), we can deduce the **Palatini equation**

$$\delta R^a{}_{bcd} = \nabla_c(\delta\Gamma^a_{bd}) - \nabla_d(\delta\Gamma^a_{bc}), \tag{11.3}$$

at the point P. Since P is an arbitrary point, the result holds quite generally. Contraction on a and c gives the useful result

$$\delta R_{bd} = \nabla_a(\delta\Gamma^a_{bd}) - \nabla_d(\delta\Gamma^a_{ba}). \tag{11.4}$$

Introducing Einstein's Relativity. Second Edition. Ray d'Inverno and James Vickers, Oxford University Press.
© Ray d'Inverno and James Vickers (2022). DOI: 10.1093/oso/9780198862024.003.0011

11.2 Differential constraints on the field equations

The variational principle proceeds from the specification of a Lagrangian density $\mathcal{L}$, which is assumed to be a functional of the metric g_{ab} and its first and possibly higher derivatives, that is,

$$\mathcal{L} = \mathcal{L}(g_{ab}, \partial_c g_{ab}, \partial_d \partial_c g_{ab}, \ldots). \tag{11.5}$$

$\mathcal{L}$ is required to be a scalar density of weight +1 so that we can form the action integral

$$I = \int_\Omega \mathcal{L} \, d\Omega, \tag{11.6}$$

over some region Ω of the manifold. The principle of stationary action then states that, if we make arbitrary variations of the g_{ab} which vanish on the boundary $\partial\Omega$ of Ω, then I must be stationary. Writing this out formally using the variational notation of Chapter 7, we obtain

$$g_{ab} \to g_{ab} + \delta g_{ab} \;\Rightarrow\; I \to I + \delta I \text{ with } \delta I = 0, \tag{11.7}$$

where

$$\delta I = \int_\Omega \mathcal{L}^{ab} \delta g_{ab} d\Omega, \tag{11.8}$$

and $\mathcal{L}^{ab}$ is the Euler-Lagrange derivative

$$\mathcal{L}^{ab} := \frac{\delta \mathcal{L}}{\delta g_{ab}}. \tag{11.9}$$

The field equations are then

$$\mathcal{L}^{ab} = 0. \tag{11.10}$$

Since δI is the difference between two scalars, it must itself be a scalar, and hence from (11.8) it follows that $\mathcal{L}^{ab}$ is a symmetric tensor density of weight +1. We shall consider the details of the calculation of $\mathcal{L}^{ab}$ in later sections. However, before we do this we shall derive some very important differential constraints on the field equations which hold **whether or not the field equations hold** and which follow simply from the fact that $\mathcal{L}$ **is a density**. In general relativity, these will turn out to be the contracted Bianchi identities.

The idea is to generate a 'variation' in the g_{ab}, which is brought about simply by carrying out a change of coordinates in Ω. Then, since I remains **invariant** it follows that δI must be **identically** zero,

$$\delta I \equiv 0. \tag{11.11}$$

We consider an infinitesimal change of coordinates (7.53) in Ω,

$$x^a \rightarrow x'^a = x^a + \varepsilon X^a(x), \qquad (11.12)$$

where X^a is a smooth vector field which vanishes on the boundary of Ω. Performing a similar calculation to that of §7.7, we find (exercise)

$$\delta g_{ab} = g'_{ab}(x) - g_{ab}(x) = -\mathrm{L}_{\varepsilon X} g_{ab} = -\varepsilon(\nabla_b X_a + \nabla_a X_b). \qquad (11.13)$$

Hence, combining this with (11.8) and (11.11), we obtain

$$0 \equiv \delta I = -2\varepsilon \int_\Omega \mathcal{L}^{ab}(\nabla_b X_a)\mathrm{d}\Omega,$$

since $\mathcal{L}^{ab}$ is symmetric by the definition (11.9). We now use a standard trick, called **integration by parts**, to write the integral as a difference of two terms, namely (check),

$$0 \equiv 2\varepsilon \int_\Omega (\nabla_b \mathcal{L}^{ab}) X_a \mathrm{d}\Omega - 2\varepsilon \int_\Omega \nabla_b[\mathcal{L}^{ab} X_a]\mathrm{d}\Omega. \qquad (11.14)$$

The term in square brackets is a vector density of weight $+1$, and hence by (7.3) its covariant divergence can be replaced by an ordinary divergence. Then the divergence theorem (7.19) gives

$$2\varepsilon \int_\Omega \partial_b[\mathcal{L}^{ab} X_a]\mathrm{d}\Omega = 2\varepsilon \int_{\partial\Omega} \mathcal{L}^{ab} X_a \mathrm{d}S_b, \qquad (11.15)$$

which converts the last term in (11.14) to a surface integral. But, by assumption, X^a vanishes on Ω, and hence this term must vanish. Thus, (11.14) reduces to

$$\int_\Omega (\nabla_b \mathcal{L}^{ab}) X_a \mathrm{d}\Omega \equiv 0, \qquad (11.16)$$

and, since Ω is **arbitrary**, we must conclude (exercise)

$$(\nabla_b \mathcal{L}^{ab}) X_a \equiv 0. \qquad (11.17)$$

Finally, since X^a is arbitrary, we obtain the promised **differential identities**

$$\nabla_b \mathcal{L}^{ab} \equiv 0. \qquad (11.18)$$

11.3 A simple example

Let us use the following notation: a **gothicized** tensor is to represent the corresponding tensor multiplied by $\sqrt{-g}$. Thus, for example,

$$\mathfrak{g}_{ab} = \sqrt{-g}\,g_{ab} \quad \text{and} \quad \mathfrak{T}_{ab} = \sqrt{-g}\,T_{ab}.$$

Then all tensors in gothic type will be tensor densities of weight $+1$.

The simplest scalar density that we can make out of g_{ab} alone is $\sqrt{-g}$ itself, namely,

$$\mathcal{L}(g_{ab}) = \sqrt{-g}, \tag{11.19}$$

where $\sqrt{-g}$ is to be regarded as a functional of the dynamical variable g_{ab}. Recalling (7.8), we write

$$\frac{\partial g}{\partial g_{ab}} = g g^{ab}, \tag{11.20}$$

and so

$$\frac{\partial \sqrt{-g}}{\partial g_{ab}} = \frac{1}{2}\frac{(-g)}{\sqrt{-g}}g^{ab} = \frac{1}{2}\sqrt{-g}g^{ab} = \frac{1}{2}\mathfrak{g}^{ab}, \tag{11.21}$$

from which we deduce that

$$\mathcal{L}^{ab} = \frac{\partial \mathcal{L}}{\partial g_{ab}} = \frac{1}{2}\mathfrak{g}^{ab}.$$

Clearly, $\mathfrak{g}^{ab} = 0$ cannot serve as field equations. The identities (11.18) become

$$\nabla_b \mathfrak{g}^{ab} \equiv 0, \tag{11.22}$$

which is trivially satisfied, since both g_{ab} and $\sqrt{-g}$ have vanishing covariant derivatives by (6.74) and (7.12).

11.4 The Einstein Lagrangian

The Lagrangian (11.19) clearly cannot serve as the Lagrangian of a physical theory. However, it turns out that the next most complicated scalar which can be built out of g_{ab} and its derivatives – and it is very much more complicated – is the curvature scalar R. The resulting Lagrangian,

$$\mathcal{L}_G = \sqrt{-g}R, \tag{11.23}$$

is called the **Einstein Lagrangian**, where the label G denotes that it is the Lagrangian for gravitation. We shall employ the notation of a comma for partial differentiation, otherwise we end up writing terms like $\partial \mathcal{L}/\partial(\partial_c g_{ab})$. Then, explicitly,

$$\mathcal{L}_G = \sqrt{-g}\,g^{cd}R_{cd}$$

$$= \mathfrak{g}^{cd}R^e{}_{ced}$$

$$= \mathfrak{g}^{cd}(\Gamma^e_{cd,e} - \Gamma^e_{ce,d} + \Gamma^f_{cd}\Gamma^e_{fe} - \Gamma^f_{ce}\Gamma^e_{fd})$$

$$= \mathfrak{g}^{cd}\{[\tfrac{1}{2}g^{ef}(g_{cf,d} + g_{df,c} - g_{cd,f})]_{,e}$$

$$- [\tfrac{1}{2}g^{ef}(g_{cf,e} + g_{ef,c} - g_{ce,f})]_{,d}$$

$$+ [\tfrac{1}{2}g^{fh}(g_{ch,d} + g_{dh,c} - g_{cd,h})][\tfrac{1}{2}g^{ei}(g_{fi,e} + g_{ei,f} - g_{fe,i})]$$

$$- [\tfrac{1}{2}g^{fh}(g_{ch,e} + g_{eh,c} - g_{ce,h})][\tfrac{1}{2}g^{ei}(g_{fi,d} + g_{di,f} - g_{fd,i})]\}. \qquad (11.24)$$

We must think of this as a functional of g_{ab} and its first and second derivatives, namely,

$$\mathcal{L}_G = \mathcal{L}_G(g_{ab}, g_{ab,c}, g_{ab,cd}),$$

where we regard g^{ab} and g (and therefore $\mathfrak{g}^{ab}$) as functions of g_{ab}. Note that we could equally well regard $\mathcal{L}_G$ as a functional of one of g^{ab}, $\mathfrak{g}^{ab}$, or $\mathfrak{g}_{ab}$ and their corresponding first and second derivatives. In the case where g_{ab} are the dynamical variables, the Euler-Lagrange derivative is a generalization of (7.36) and becomes

$$\frac{\delta \mathcal{L}_G}{\delta g_{ab}} = \frac{\partial \mathcal{L}_G}{\partial g_{ab}} - \left(\frac{\partial \mathcal{L}_G}{\partial g_{ab,c}}\right)_{,c} + \left(\frac{\partial \mathcal{L}_G}{\partial g_{ab,cd}}\right)_{,cd}. \qquad (11.25)$$

Following the procedure of the last section, we would expect next to calculate actual expressions for each of these terms. For example (exercise),

$$\frac{\delta \mathcal{L}_G}{\delta g_{ab,cd}} = \sqrt{-g}[\tfrac{1}{2}(g^{ac}g^{bd} + g^{ad}g^{bc}) - g^{ab}g^{cd}]. \qquad (11.26)$$

The calculation of the remaining terms, though straightforward, is, unfortunately, absolutely horrendous and we shall not pursue it further. Instead, we will exploit the variational formalism in the next section and show how this indirect approach leads to a more tractable calculation. However, had we proceeded, then we would have found (exercise for the completely dedicated reader!)

$$\mathcal{L}_G^{ab} = \frac{\delta \mathcal{L}_G}{\delta g_{ab}} = -\sqrt{-g}\,G^{ab}, \qquad (11.27)$$

and so the Euler-Lagrange equations lead to the vacuum field equations

$$-\sqrt{-g}\,G^{ab} = 0, \qquad (11.28)$$

that is, the vanishing of the Einstein tensor. In addition the identities (11.18) become

$$\nabla_b[-\sqrt{-g}G^{ab}] \equiv 0 \quad \Rightarrow \quad \nabla_b G^{ab} \equiv 0, \tag{11.29}$$

that is, the contracted Bianchi identities.

11.5 Indirect derivation of the field equations

The approach depends on exploiting the δ notation fully. It can be shown (exercise) that δ behaves much like a derivative when applied to sums, differences, and products. For example, let us see what happens when we apply δ to the tensor δ_c^a. The variation

$$g_{ab} \to g_{ab} + \delta g_{ab},$$

induces a variation in g^{ab}, which we write

$$g^{ab} \to g^{ab} + \delta g^{ab}. \tag{11.30}$$

Then

$$\delta_c^a = g^{ab}g_{bc} \to (g^{ab} + \delta g^{ab})(g_{bc} + \delta g_{bc})$$
$$= \delta_c^a + \delta g^{ab}g_{bc} + g^{ab}\delta g_{bc} + O(\delta^2).$$

But, since δ_c^a is a constant tensor, it cannot change and therefore

$$\delta g^{ab}g_{bc} + g^{ab}\delta g_{bc} = 0, \tag{11.31}$$

to first order, or, multiplying through by g^{cd},

$$\delta g^{ad} = -g^{ab}g^{cd}\delta g_{bc}. \tag{11.32}$$

Compare and contrast this with the corresponding relationship between partial derivatives (7.9).

Starting from I written in the form

$$I = \int_\Omega \mathfrak{g}^{ab}R_{ab}\mathrm{d}\Omega,$$

we carry out a variation and use the Leibniz rule for products, to get

$$\delta I = \int_\Omega (\delta\mathfrak{g}^{ab}R_{ab} + \mathfrak{g}^{ab}\delta R_{ab})\mathrm{d}\Omega. \tag{11.33}$$

We now use the Palatini equation in the form (11.4), so that the second term on the right-hand side becomes

$$\int_\Omega \mathfrak{g}^{ab}\delta R_{ab}\mathrm{d}\Omega = \int_\Omega \mathfrak{g}^{ab}\left[\nabla_c(\delta\Gamma^c_{ab}) - \nabla_b(\delta\Gamma^c_{ac})\right]\mathrm{d}\Omega$$

$$= \int_\Omega \left[\nabla_c(\mathfrak{g}^{ab}\delta\Gamma^c_{ab}) - \nabla_b(\mathfrak{g}^{ab}\delta\Gamma^c_{ac})\right]\mathrm{d}\Omega$$

$$= \int_\Omega \partial_c(\mathfrak{g}^{ab}\delta\Gamma^c_{ab} - \mathfrak{g}^{ac}\delta\Gamma^b_{ab})\mathrm{d}\Omega,$$

since the covariant derivative of $\mathfrak{g}^{ab}$ vanishes identically and the quantities in parentheses are vector densities of weight $+1$. Using the same argument as we did in §11.2, this can be converted to a surface integral by the divergence theorem, which vanishes because the variations are assumed to vanish on the surface of Ω. Hence, (11.33) reduces to

$$\delta I = \int_\Omega R_{ab}\delta\mathfrak{g}^{ab}\mathrm{d}\Omega$$

$$= \int_\Omega R_{ab}\delta[\sqrt{-g}g^{ab}]\mathrm{d}\Omega$$

$$= \int_\Omega [R_{ab}g^{ab}\delta\sqrt{-g} + R_{ab}\sqrt{-g}\delta g^{ab}]\mathrm{d}\Omega$$

$$= \int_\Omega \sqrt{-g}(\tfrac{1}{2}Rg^{cd} - R_{ab}g^{ac}g^{bd})\delta g_{cd}\mathrm{d}\Omega$$

$$= -\int_\Omega \sqrt{-g}(R^{cd} - \tfrac{1}{2}Rg^{cd})\delta g_{cd}\mathrm{d}\Omega$$

$$= \int_\Omega [-\sqrt{-g}G^{ab}]\delta g_{ab}\mathrm{d}\Omega, \tag{11.34}$$

where we have used (11.31) and the result (exercise)

$$\delta\sqrt{-g} = \tfrac{1}{2}\sqrt{-g}g^{ab}\delta g_{ab}. \tag{11.35}$$

Using (11.8), we again get the vacuum field equation in the form (11.28) and the contracted Bianchi identities (11.29) as the corresponding differential constraints on the field equations.

11.6   An equivalent Lagrangian

The resulting field equations are second order in the partial derivatives. This is at first sight rather surprising since by (11.25) we might expect the last term to produce fourth-order equations. However, it turns out, as we have seen in (11.26), that $\partial L_G/\partial g_{ab,cd}$ only involves undifferentiated g_{ab}'s

and $\partial L_{\mathrm{G}}/\partial g_{ab,c}$ only involves once differentiated g_{ab}'s (exercise). In this section, we make the second-order nature of the equations more evident by showing that

$$\mathcal{L}_{\mathrm{G}} = \bar{\mathcal{L}}_{\mathrm{G}} + Q^a{}_{,a}, \tag{11.36}$$

where $\bar{\mathcal{L}}_{\mathrm{G}}$ depends on the metric and its first derivatives only. It can be shown that in applying the variational principle argument to such an equation the divergence term $Q^a{}_{,a}$ can be discarded (by converting to a vanishing surface integral), and hence it follows that $\mathcal{L}_{\mathrm{G}}$ and $\bar{\mathcal{L}}_{\mathrm{G}}$ give rise to the **same** field equations. However, $\bar{\mathcal{L}}_{\mathrm{G}}$ is no longer a scalar density. We sketch the argument below.

The Einstein Lagrangian

$$
\begin{aligned}
\mathcal{L}_{\mathrm{G}} &= \sqrt{-g}R \\
&= \mathfrak{g}^{ab}R_{ab} \\
&= \mathfrak{g}^{ab}(\Gamma^c_{ab,c} - \Gamma^c_{ac,b} + \Gamma^c_{ab}\Gamma^d_{cd} - \Gamma^d_{ac}\Gamma^c_{bd}) \\
&= \mathfrak{g}^{ab}\Gamma^c_{ab,c} - \mathfrak{g}^{ab}\Gamma^c_{ac,b} - \bar{\mathcal{L}}_{\mathrm{G}},
\end{aligned} \tag{11.37}
$$

where

$$\bar{\mathcal{L}}_{\mathrm{G}} = \mathfrak{g}^{ab}(\Gamma^d_{ac}\Gamma^c_{bd} - \Gamma^c_{ab}\Gamma^d_{cd}). \tag{11.38}$$

Integrating the first two terms in (11.37) by parts, we get

$$\mathcal{L}_{\mathrm{G}} = -\mathfrak{g}^{ab}{}_{,c}\Gamma^c_{ab} + \mathfrak{g}^{ab}{}_{,b}\Gamma^c_{ac} - \bar{\mathcal{L}}_{\mathrm{G}} + Q^a{}_{,a}, \tag{11.39}$$

where

$$Q^a = \mathfrak{g}^{bc}\Gamma^a_{bc} - \mathfrak{g}^{ab}\Gamma^c_{bc}. \tag{11.40}$$

From the fact that the covariant derivative of $\mathfrak{g}^{ab}$ vanishes, we find (exercise)

$$\mathfrak{g}^{ab}{}_{,c} = \Gamma^d_{dc}\mathfrak{g}^{ab} - \Gamma^a_{dc}\mathfrak{g}^{db} - \Gamma^b_{dc}\mathfrak{g}^{ad}. \tag{11.41}$$

Substituting in (11.39) and simplifying, we obtain the result (11.36).

Once again, we could consider $\bar{\mathcal{L}}_{\mathrm{G}}$ as a functional of one of g_{ab}, g^{ab}, $\mathfrak{g}_{ab}$, or $\mathfrak{g}^{ab}$ and their corresponding first derivatives. For example, let us choose the $\mathfrak{g}^{ab}$ as the dynamical variables. Then

$$\bar{\mathcal{L}}_{\mathrm{G}} = \bar{\mathcal{L}}_{\mathrm{G}}(\mathfrak{g}^{ab}, \mathfrak{g}^{ab}{}_{,c}),$$

from which it can be shown that

$$\frac{\partial \bar{\mathcal{L}}_G}{\partial \mathfrak{g}^{ab}} = -\Gamma^d_{ac}\Gamma^c_{bd} + \Gamma^c_{ab}\Gamma^d_{cd}, \tag{11.42}$$

and

$$\frac{\partial \bar{\mathcal{L}}_G}{\partial \mathfrak{g}^{ab}_{,c}} = -\Gamma^c_{ab} + \tfrac{1}{2}\delta^c_a \Gamma^d_{bd} + \tfrac{1}{2}\delta^c_b \Gamma^d_{ad}. \tag{11.43}$$

The Euler-Lagrange equations

$$\bar{\mathcal{L}}^{ab}_G = \frac{\partial \bar{\mathcal{L}}_G}{\partial \mathfrak{g}^{ab}} - \left(\frac{\partial \bar{\mathcal{L}}_G}{\partial \mathfrak{g}^{ab}_{,c}}\right)_{,c} = 0, \tag{11.44}$$

then lead to

$$\bar{\mathcal{L}}^{ab}_G = \Gamma^c_{ab,c} - \tfrac{1}{2}\Gamma^d_{bd,a} - \tfrac{1}{2}\Gamma^d_{ad,b} + \Gamma^c_{ab}\Gamma^d_{cd} - \Gamma^d_{ac}\Gamma^c_{bd}. \tag{11.45}$$

If we use the result (exercise)

$$[\ln \sqrt{-g}]_{,a} = \Gamma^d_{ad}, \tag{11.46}$$

then

$$\Gamma^d_{ad,b} = \left[\ln(\sqrt{-g})\right]_{,ab} = \left[\ln(\sqrt{-g})\right]_{,ba} = \Gamma^d_{bd,a},$$

and so (11.45) gives

$$\bar{\mathcal{L}}^{ab}_G = \Gamma^c_{ab,c} - \Gamma^d_{ad,b} + \Gamma^c_{ab}\Gamma^d_{cd} - \Gamma^d_{ac}\Gamma^c_{bd} = R_{ab}.$$

The field equations are correspondingly $R_{ab} = 0$.

11.7   The Palatini approach

The Palatini approach is very elegant and is based on the idea of treating both the metric **and** the connection separately as dynamical variables in the Einstein Lagrangian. To be specific, let us choose $\mathcal{L}_G$ as a functional of $\mathfrak{g}^{ab}$ and a **symmetric** connection Γ^a_{bc} and its derivatives, i.e.

$$\mathcal{L}_G = \mathcal{L}_G(\mathfrak{g}^{ab}, \Gamma^a_{bc}, \Gamma^a_{bc,d}),$$

where

$$\mathcal{L}_G = \mathfrak{g}^{ab} R_{ab}$$
$$= \mathfrak{g}^{ab}(\Gamma^c_{ab,c} - \Gamma^d_{ad,b} + \Gamma^c_{ab}\Gamma^d_{cd} - \Gamma^d_{ac}\Gamma^c_{bd}), \tag{11.47}$$

so that the Ricci tensor depends on Γ^a_{bc} and its derivatives only. Then, if we carry out a variation with respect to g_{ab} only,

$$\delta I = \int_{\Omega} \delta \mathfrak{g}^{ab} R_{ab} d\Omega$$

and the principle of stationary action gives immediately the vacuum field equations $R_{ab} = 0$.

We next carry out a variation with respect to Γ^{a}_{bc} so that

$$\delta I = \int_{\Omega} \mathfrak{g}^{ab} \delta R_{ab} d\Omega$$

$$= \int_{\Omega} \mathfrak{g}^{ab} \left[\nabla_{c}(\delta \Gamma^{c}_{ab}) - \nabla_{b}(\delta \Gamma^{c}_{ac}) \right] d\Omega,$$

by the corollary of the Palatini equation (11.4). Integrating by parts and discarding the divergence term by the usual argument, we get

$$\delta I = \int_{\Omega} \left[\nabla_{b} \mathfrak{g}^{ab} \delta \Gamma^{c}_{ac} - \nabla_{c} \mathfrak{g}^{ab} \delta \Gamma^{c}_{ab} \right] d\Omega$$

$$= \int_{\Omega} \left[(\delta^{b}_{c} \nabla_{d} \mathfrak{g}^{ad} - \nabla_{c} \mathfrak{g}^{ab}) \delta \Gamma^{c}_{ab} \right] d\Omega.$$

Since δI vanishes for arbitrary volumes Ω, the integrand must vanish, i.e.

$$(\delta^{b}_{c} \nabla_{d} \mathfrak{g}^{ad} - \nabla_{c} \mathfrak{g}^{ab}) \delta \Gamma^{c}_{ab}.$$

The variations $\delta \Gamma^{c}_{ab}$ are arbitrary, but **symmetric** in a and b, and so only the symmetric part of the expression in brackets vanishes, i.e.

$$\tfrac{1}{2} \delta^{b}_{c} \nabla_{d} \mathfrak{g}^{ad} + \tfrac{1}{2} \delta^{a}_{c} \nabla_{d} \mathfrak{g}^{bd} - \nabla_{c} \mathfrak{g}^{ab} = 0. \qquad (11.48)$$

Manipulating this equation, one can show in turn (exercise) that the co-variant derivatives of $\mathfrak{g}^{ab}$, $\sqrt{-g}$, g^{ab}, and g_{ab} vanish. Finally, by Exercise 6.21, if

$$\nabla_{c} g_{ab} = 0,$$

and the connection is symmetric, then it follows that Γ^{a}_{bc} is necessarily the metric connection

$$\Gamma^{a}_{bc} = \tfrac{1}{2} g^{ad} (g_{bd,c} + g_{cd,b} - g_{bc,d}).$$

To summarize, the Palatini approach starts from the Einstein Lagrangian (11.47) considered as a functional of a metric and an arbitrary symmetric connection and its derivatives. Variation with respect to the metric produces the vacuum field equations of general relativity, and variation with respect to the connection reveals that the connection is necessarily the metric connection.

11.8   The full field equations

So far, we have been concerned with the vacuum field equations. To obtain the full field equations, we assume that there are other fields present beside the gravitational field, which can be described by an appropriate Lagrangian density $\mathcal{L}_M$ – the matter Lagrangian. The action is then

$$I = \int_\Omega (\mathcal{L}_G + 2\kappa\mathcal{L}_M)\mathrm{d}\Omega, \tag{11.49}$$

where κ is the coupling constant and the reason for the factor of 2 is explained below. Both Lagrangians are to be considered as functionals of the metric and its derivatives, and so, varying with respect to g_{ab} (say), we obtain

$$\frac{\delta\mathcal{L}_G}{\delta g_{ab}} = -\sqrt{-g}\,G^{ab}, \tag{11.50}$$

and

$$\frac{\delta\mathcal{L}_M}{\delta g_{ab}} = \tfrac{1}{2}\sqrt{-g}\,T^{ab}, \tag{11.51}$$

where the latter equation defines the **energy-momentum tensor** T^{ab} for the fields present as being given by

$$T^{ab} = \frac{2}{\sqrt{-g}}\frac{\delta\mathcal{L}_M}{\delta g_{ab}}. \tag{11.52}$$

The reason for the factor of 2 is so that this expression agrees with the so-called canonical energy momentum tensor defined in special relativity (see Exercise 11.13). Calculating the Euler-Lagrange equations for (11.49) and dividing through by $\sqrt{-g}$, the field equations become

$$G^{ab} = \kappa T^{ab}, \tag{11.53}$$

in agreement with (10.54). Note that some authors omit the factor of 2 in both (11.49) and (11.52) which also gives (11.53). In the next chapter, we shall investigate the right-hand side of this equation and look at the definition of the energy-momentum tensor for various important fields.

Exercises

11.1 (§11.2) Show that, under an infinitesimal change of coordinates

$$x^a \to x'^a = x^a + \varepsilon X^a(x),$$

the transformed metric satisfies

$$g'_{ab}(x) - g_{ab} = -\varepsilon(\nabla_b X_a + \nabla_a X_b)$$

to first order in ε.

11.2 (§11.4) Show that

$$\frac{\partial \mathcal{L}_G}{\partial g_{ab,cd}} = \sqrt{-g}\left[\tfrac{1}{2}(g^{ac}g^{bd} + g^{ad}g^{bc}) - g^{ab}g^{cd}\right].$$

11.3 (§11.4) Show that

$$\frac{\partial g^{cd}}{\partial g_{ab}} = -\tfrac{1}{2}(g^{ac}g^{bd} + g^{ad}g^{bc}).$$

11.4 (§11.4) Check that $\partial \mathcal{L}_G/\partial g_{ab,c}$ depends only on g_{ab} and its first derivatives. [Hint: consider (11.24).]

11.5 (§11.5) If y_A are dynamical variables and $L_1 = L_1(y_A)$ and $L_2 = L_2(y_A)$, then show from first principles that
(i) $\delta(\lambda L_1 + \mu L_2) = \lambda \delta L_1 + \mu \delta L_2$, where λ and μ are constants,
(ii) $\delta(L_1 L_2) = L_1 \delta L_2 + L_2 \delta L_1$.

11.6 (§11.5) Show that
(i) $g_{ab}\delta g^{ab} = -g^{ab}\delta g_{ab}$,
(ii) $\delta g = g g^{ab} \delta g_{ab}$ (compare this with (7.9)),
(iii) $\delta\sqrt{-g} = \tfrac{1}{2}\sqrt{-g}g^{ab}\delta g_{ab}$.

11.7 (§11.5) Show that, if we regard $\mathfrak{g}^{ab}$, $\mathfrak{g}_{ab}$, and g^{ab}, respectively, as dynamical variables, then
(i) $\dfrac{\delta \mathcal{L}_G}{\delta \mathfrak{g}^{ab}} = R_{ab}$,

(ii) $\dfrac{\delta \mathcal{L}_G}{\delta \mathfrak{g}_{ab}} = -R^{ab}$,

(iii) $\dfrac{\delta \mathcal{L}_G}{\delta g^{ab}} = \sqrt{-g}G_{ab}$.

What differential constraints do each of these quantities satisfy?

11.8 (§11.5)
(i) If $\int_\Omega \Phi d\Omega = 0$, where Ω is arbitrary, then prove that $\Phi = 0$. [Hint:

choose an arbitrary point P where $\Phi(P) > 0$, say, use continuity to show that there is a region surrounding P where Φ remains positive, deduce that $\int_\Omega \Phi \, d\Omega > 0$ for a suitable Ω, and derive a contradiction; then complete the proof.]

(ii) If $W^a X_a = 0$ where X_a is arbitrary, then show that $W^a = 0$. [Hint: take $X_a \overset{*}{=} (1, 0, 0, 0)$, etc.]

11.9 (§11.6) If the Lagrangians $L(y, y', x)$ and $\bar{L}(y, y', x)$ differ by a divergence, i.e.

$$L = \bar{L} + \frac{\mathrm{d}Q(y, y', x)}{\mathrm{d}x},$$

then show that L and $\bar{L}$ give rise to the same field equation.

11.10 (§11.6)
(i) Establish the results (11.41) and (11.46).
(ii) Establish the result (11.36) for the Einstein Lagrangian.
(iii) Use (11.41) to deduce

$$\mathfrak{g}^{ab}{}_{,c} = -\mathfrak{g}^{ab}\Gamma^c_{bc}.$$

(iv) Use parts (i) and (ii) to establish the result

$$\mathfrak{g}^{ab}{}_{,c}\Gamma^c_{ab} - \mathfrak{g}^{ab}{}_{,b}\Gamma^c_{ac} = -2\bar{\mathcal{L}}_G.$$

(v) Defining L_{ab} by

$$L_{ab} = \Gamma^d_{ac}\Gamma^c_{bd} - \Gamma^c_{ab}\Gamma^d_{dc},$$

so that $\bar{\mathcal{L}}_G = \mathfrak{g}^{ab} L_{ab}$, then show that

$$\mathfrak{g}^{ab}{}_{,c}\delta\Gamma^c_{ab} - \mathfrak{g}^{ab}{}_{,b}\delta\Gamma^a_{ac} = L_{ab}\delta\mathfrak{g}^{ab} - \delta\bar{\mathcal{L}}_G.$$

[Hint: Use parts (i) and (ii) to re-express the LHS and the product law on $\delta\bar{\mathcal{L}}_G$ to re-express the RHS.]

(vi) Take the variation of the equation in part (iv) to establish the result

$$\delta\bar{\mathcal{L}}_G = \left[\tfrac{1}{2}(\delta^c_a\Gamma^d_{bd} - \delta^c_b\Gamma^d_{ad}) - \Gamma^c_{ab}\right]\delta\mathfrak{g}^{ab}{}_{,c} - L_{ab}\delta\mathfrak{g}^{ab},$$

and regarding $\bar{\mathcal{L}}_G = \bar{\mathcal{L}}_G(\mathfrak{g}^{ab}, \mathfrak{g}^{ab}{}_{,c})$ then deduce (11.42) and (11.43). [Hint: Use part (v) and the symmetry of $\delta\mathfrak{g}^{ab}{}_{,c}$ on a and b.]

11.11 (§11.7) Show that, if

$$\tfrac{1}{2}\delta^b_c\nabla_d\mathfrak{g}^{ad} + \tfrac{1}{2}\delta^a_c\nabla_d\mathfrak{g}^{bd} - \nabla_c\mathfrak{g}^{ab} = 0,$$

for an arbitrary symmetric connection, then
(i) $\nabla_c\mathfrak{g}^{ab} = 0$,

(ii) $\nabla_c \sqrt{-g} = 0$,

(iii) $\nabla_c g^{ab} = 0$,

(iv) $\nabla_c g_{ab} = 0$,

and deduce that the connection is necessarily the metric connection.

11.12 (§11.7) Use the variational principle approach to find the field equations of the theory (considered by A. S. Eddington) with Lagrangian

$$\mathcal{L} = \sqrt{-g} R^{abcd} R_{abcd},$$

treating g^{ab} and R^{abcd} as independent variables.

11.13 (§11.8) In Minkowski space, the Lagrangian for a scalar field $\phi(x^a)$ moving in a potential $V(\phi)$ is (using our signature)

$$L = -\tfrac{1}{2} g^{cd} \phi_{,c} \phi_{,d} + V(\phi).$$

(i) The 'canonical' energy-momentum tensor $\Theta^a{}_b$ in Minkowski space for a field with Lagrangian $L(\phi, \partial_a \phi)$ is defined to be given by

$$\Theta^a{}_b := -\frac{\partial L}{\partial(\phi_{,a})} \phi_{,b} + \delta^a_b L.$$

Calculate $\Theta_{ab} = \eta_{ac} \Theta^c{}_a$ for a scalar field and show that

$$\Theta_{00} = \tfrac{1}{2}(\dot{\phi}^2 + |\boldsymbol{\nabla}\phi|^2) + V(\phi),$$

(which is what one would expect for the energy density as the sum of the kinetic and potential energy of the field.)

(ii) Using the the principle of minimal coupling, explain why the Lagrangian density for a scalar field in a curved space-time is

$$\mathcal{L} = [-\tfrac{1}{2} g^{cd} \phi_{,c} \phi_{,d} + V(\phi)]\sqrt{-g}.$$

(iii) Find the energy-momentum tensor T_{ab} for the above Lagrangian density and show that, if we look at this expression in Minkowski space, we find

$$T_{ab} = \Theta_{ab},$$

so that the two definitions agree. [Hint: $T_{ab}\sqrt{-g} = -2\delta\mathcal{L}/\delta g^{ab}$.]

11.14 (§11.8) In calculating the energy-momentum tensor of a field where the Lagrangian involves covariant derivatives, one needs to calculate the variation of the connection. Show that

$$\delta\Gamma^a{}_{bc} = \tfrac{1}{2} g^{ad}[\nabla_c(\delta g_{db}) + \nabla_b(\delta g_{dc}) - \nabla_d(\delta g_{bc})].$$

[Hint: Use the method of geodesic coordinates as used in §11.1.]

Further reading

The approach to the variational principle described here is based on the lovely little book by Schrödinger (1950). A more modern source is the book by Choquet-Bruhat et al. (1977).

Choquet-Bruhat, Y., De Witt-Morette, C., and Dillard-Bleick, M. (1977). *Analysis, Manifolds and Physics*. North-Holland, Amsterdam.

Schrödinger, E. (1950). *Space-time Structure*. Cambridge University Press, Cambridge.

The energy-momentum tensor

12

12.1 Preview

Our programme for this chapter is to look at the three most important energy-momentum tensors in general relativity, namely, the energy-momentum tensors for incoherent matter or dust, a perfect fluid, and the electromagnetic field . In passing, we shall encounter a tensor formulation of Maxwell's equations governing the electromagnetic field. Again, our treatment will not be exhaustive or complete, but will be sufficient for generating the explicit expressions for the three tensors, and these expressions will be essentially all that we require in future chapters. We shall also look at the Newtonian limit of the field equations and discuss the calculation for determining the coupling constant.

12.2 Incoherent matter

We start by considering the simplest kind of matter field, namely, that of **non-interacting incoherent matter**, or **dust**. Such a field may be characterized by two quantities, the **4-velocity** vector field of flow

$$u^a = \frac{\mathrm{d}x^a}{\mathrm{d}\tau},$$

where τ is the proper time along the world-line of a dust particle (Fig. 12.1), and a scalar field

$$\rho_0 = \rho_0(x),$$

describing the **proper density** of the flow, that is, the density which would be measured by an observer moving with the flow (a co-moving observer). The simplest second-rank tensor we can construct from these two quantities is

$$T^{ab} = \rho_0 u^a u^b, \tag{12.1}$$

and this turns out to be the energy-momentum tensor for the matter field.

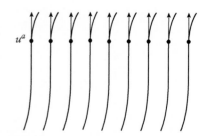

u^a

Fig. 12.1 The world-lines of dust particles.

Introducing Einstein's Relativity. Second Edition. Ray d'Inverno and James Vickers, Oxford University Press.

Let us investigate this tensor in special relativity in Minkowski coordinates. Then, by Exercise 8.10, the 4-velocity is

$$u^a = \gamma(1, \boldsymbol{u}), \tag{12.2}$$

where $\gamma = (1 - u^2)^{-1/2}$. The proper time is defined by

$$
\begin{aligned}
\mathrm{d}\tau^2 &= \mathrm{d}s^2 \\
&= \eta_{ab}\,\mathrm{d}x^a\,\mathrm{d}x^b \\
&= \mathrm{d}t^2 - \mathrm{d}x^2 - \mathrm{d}y^2 - \mathrm{d}z^2 \\
&= \mathrm{d}t^2\left(1 - u^2\right) \\
&= \gamma^{-2}\mathrm{d}t^2.
\end{aligned}
\tag{12.3}
$$

Then the zero–zero component of T^{ab} is

$$T^{00} = \rho_0 \frac{\mathrm{d}x^0}{\mathrm{d}\tau}\frac{\mathrm{d}x^0}{\mathrm{d}\tau} = \rho_0 \frac{\mathrm{d}t^2}{\mathrm{d}\tau^2} = \gamma^2 \rho_0, \tag{12.4}$$

by (12.3). This quantity has a simple physical interpretation. First of all, in special relativity, the mass of a body in motion is greater than its rest mass by a factor γ, by (4.11). In addition, if we consider a moving three-dimensional volume element, then its volume decreases by a factor γ through the Lorentz contraction. Thus, from the point of view of a fixed as opposed to a co-moving observer, the density increases by a factor γ^2. Hence, if a field of material of proper density ρ_0 flows past a fixed observer with velocity $\boldsymbol{u}$, then the observer will measure a density

$$\rho = \gamma^2 \rho_0. \tag{12.5}$$

The component T^{00} may therefore be interpreted as the **relativistic energy density** of the matter field, since the only contribution to the energy of the field is from the motion of the matter (note that this requires a factor of c^2 in the definition (12.1) in non-relativistic units).

The components of T^{ab} can be written, using (12.2) and (12.5), in the form (exercise)

$$T^{ab} = \rho \begin{bmatrix} 1 & u_x & u_y & u_z \\ u_x & u_x^2 & u_x u_y & u_x u_z \\ u_y & u_x u_y & u_y^2 & u_y u_z \\ u_z & u_x u_z & u_y u_z & u_z^2 \end{bmatrix}. \tag{12.6}$$

We now show that the equations governing the force-free motion of a matter field of dust can be written in the following very succinct way

$$\partial_b T^{ab} = 0. \tag{12.7}$$

Using (12.6), in the case when $a = 0$, this equation becomes (exercise)

$$\frac{\partial \rho}{\partial t} + \frac{\partial}{\partial x}(\rho u_x) + \frac{\partial}{\partial y}(\rho u_y) + \frac{\partial}{\partial z}(\rho u_z).$$

This is precisely the classical **equation of continuity**

$$\frac{\partial \rho}{\partial t} + \operatorname{div}(\rho \boldsymbol{u}) = 0. \qquad (12.8)$$

In classical fluid dynamics, this expresses the conservation of matter with density ρ moving with velocity $\boldsymbol{u}$. Since matter is the same as energy in special relativity, it follows that the **conservation of energy** equation for dust is $\partial_b T^{0b} = 0$. The equations corresponding to $a = \alpha$ ($\alpha = 1, 2, 3$) are similarly found to be (exercise)

$$\frac{\partial}{\partial t}(\rho \boldsymbol{u}) + \frac{\partial}{\partial x}(\rho u_x \boldsymbol{u}) + \frac{\partial}{\partial y}(\rho u_y \boldsymbol{u}) + \frac{\partial}{\partial z}(\rho u_z \boldsymbol{u}) = 0.$$

Combining this with (12.8), the equation can be written as (exercise)

$$\rho \left[\frac{\partial \boldsymbol{u}}{\partial t} + (\boldsymbol{u} \cdot \nabla) \boldsymbol{u} \right] = 0. \qquad (12.9)$$

Comparing this with the **Navier–Stokes equation of motion** for a perfect fluid in classical fluid dynamics, namely,

$$\rho \left[\frac{\partial \boldsymbol{u}}{\partial t} + (\boldsymbol{u} \cdot \nabla) \boldsymbol{u} \right] = -\operatorname{grad} p + \rho \boldsymbol{X}, \qquad (12.10)$$

where p is the pressure in the fluid, and $\boldsymbol{X}$ is the body force per unit mass, we see that (12.9) is simply this equation in the absence of pressure and external forces, which is the appropriate equation for dust.

We have seen that the requirement that the energy-momentum tensor has zero divergence in special relativity is equivalent to demanding conservation of energy and conservation of momentum in the matter field – hence the name **energy-momentum** tensor. Moreover, (12.7) is known as the energy-momentum **conservation law**. If we use a non-Minkowskian metric in special relativity, then (12.7) is replaced by its covariant counterpart

$$\nabla_b T^{ab} = 0. \qquad (12.11)$$

We now make the transition to general relativity and once again define the energy-momentum tensor for incoherent matter by (12.1), and, using the principle of minimal gravitational coupling, retain (12.11) as the statement of the conservation law.

12.3 The coupling constant

Before moving on to consider other energy-momentum tensors, we look at the Newtonian limit of the full field equations for incoherent matter, in order to determine the coupling constant. In the weak-field approximation we have

$$g_{ab} = \eta_{ab} + \varepsilon h_{ab} + O(\varepsilon^2). \tag{12.12}$$

Then, as we will see in Chapter 21, if we apply the gauge condition (21.24), then

$$G_{ab} = -\tfrac{1}{2}\varepsilon \Box \left(h_{ab} - \tfrac{1}{2}\eta_{ab}\eta^{cd}h_{cd} \right) + O(\varepsilon^2).$$

so that the full field equations $G_{ab} = \kappa T_{ab}$ become

$$-\tfrac{1}{2}\varepsilon \Box \left(h_{ab} - \tfrac{1}{2}\eta_{ab}\eta^{cd}h_{cd} \right) = \kappa T_{ab} + O(\varepsilon^2). \tag{12.13}$$

Contracting with η^{ab} and applying the slow-motion approximation (10.41), we find (exercise)

$$\tfrac{1}{2}\varepsilon \nabla^2 h_{ab} = \kappa \left(T_{ab} - \tfrac{1}{2}\eta_{ab}\eta^{cd}T_{cd} \right) + O(\varepsilon^2). \tag{12.14}$$

Let us take, as the source of the field, a distribution of dust of **small** proper density ρ_0 moving at low velocity of order v. This assumption means that we neglect terms both of order v/c and $\rho_0 v/c$, and then, by (12.6), in non-relativistic units, the energy-momentum tensor reduces in our privileged coordinate system to

$$T^{ab} = c^2 \rho_0 \, \delta_0^a \, \delta_0^b, \tag{12.15}$$

which, in turn, implies

$$T_{ab} = c^2 \rho_0 \, \delta_a^0 \, \delta_b^0 \quad \text{and} \quad \eta^{cd} T_{cd} = c^2 \rho_0. \tag{12.16}$$

The zero–zero component of the field equations (12.14) then becomes

$$\varepsilon \nabla^2 h_{00} = c^2 \kappa \rho_0 + O(\varepsilon^2). \tag{12.17}$$

But, by (12.12),

$$g_{00} = 1 + \varepsilon h_{00} + O(\varepsilon^2),$$

so that

$$\nabla^2 g_{00} = \varepsilon \nabla^2 h_{00} + O(\varepsilon^2),$$

and, by (10.49),

$$\nabla^2 g_{00} = \nabla^2 \left(\frac{2\phi}{c^2} \right) + O(\varepsilon).$$

Substituting these results in (12.17), we get

$$\nabla^2 \phi = \tfrac{1}{2} c^4 \kappa \rho_0 + O(\varepsilon).$$

Comparing this with Poisson's equation (4.5), namely,

$$\nabla^2 \phi = 4\pi G \rho_0,$$

we obtain the result (10.55), namely,

$$\kappa = 8\pi G/c^4. \tag{12.18}$$

In relativistic units, this reduces to

$$\kappa = 8\pi. \tag{12.19}$$

We have therefore used the correspondence principle with Newtonian theory to obtain the coupling constant κ appearing in the full field equations (10.54).

12.4 Perfect fluid

A **perfect fluid** is characterized by three quantities: a **4-velocity** $u^a = \mathrm{d}x^a/\mathrm{d}\tau$, a **proper density** field $\rho_0 = \rho_0(x)$, and a **scalar pressure** field $p = p(x)$. In the limit as p vanishes, a perfect fluid reduces to incoherent matter. This suggests that we take the energy-momentum tensor for a perfect fluid to be of the form

$$T^{ab} = \rho_0 \, u^a u^b + p S^{ab}, \tag{12.20}$$

for some symmetric tensor S^{ab}. The only second-rank tensors which are associated with the fluid are $u^a u^b$ and the metric g^{ab}, and so the simplest assumption we can make is

$$S^{ab} = \lambda u^a u^b + \mu g^{ab}, \tag{12.21}$$

where λ and μ are constants. Proceeding as we did in the last section, we investigate the conservation law $\partial_b T^{ab} = 0$ in special relativity in Minkowski coordinates and demand that it reduces in an appropriate limit to the continuity equation (12.8) and the Navier–Stokes equation (12.10) in the absence of body forces. This requirement leads to $\lambda = 1$ and $\mu = -1$. Then (12.20) and (12.21) give

$$T^{ab} = (\rho_0 + p) \, u^a u^b - p g^{ab}, \tag{12.22}$$

which we take as the definition for the energy-momentum tensor of a perfect fluid. If we use a non-Minkowskian metric in special relativity, then we again take the covariant form (12.11) for the conservation law.

In the full theory, we also take (12.22) as the definition of a perfect fluid, and (12.11) as the conservation equations.

In addition, p and ρ are related by an **equation of state** governing the particular sort of perfect fluid under consideration. In general, this is an equation of the form $p = p(\rho, T)$, where T is the absolute temperature. However, we shall only be concerned with situations in which T is effectively constant, so that the equation of state reduces to

$$p = p(\rho).$$

12.5   Maxwell's equations

In this section, we wish to reformulate Maxwell's equations for the electromagnetic field in tensorial form. We start by rewriting them in special relativity in Minkowski coordinates. Working in Heaviside–Lorentz units with $c = 1$, we find that **Maxwell's equations in vacuo** for the electromagnetic field split up into two pairs of equations, namely, the **source equations**

$$\operatorname{div} \boldsymbol{E} = \rho, \tag{12.23}$$

$$\operatorname{curl} \boldsymbol{B} - \frac{\partial \boldsymbol{E}}{\partial t} = \boldsymbol{j}, \tag{12.24}$$

and the **internal equations**

$$\operatorname{div} \boldsymbol{B} = 0, \tag{12.25}$$

$$\operatorname{curl} \boldsymbol{E} + \frac{\partial \boldsymbol{B}}{\partial t} = 0, \tag{12.26}$$

where $\boldsymbol{E}$ is the electric field, $\boldsymbol{B}$ is the magnetic induction, ρ is the charge density, and $\boldsymbol{j}$ is the current density. In simple physical terms, (12.23) is the differential form of Gauss's law relating the flux through a closed surface to the enclosed charge, (12.24) is a generalized Ampère's law relating the magnetic field to a flow of current (where the term involving $\boldsymbol{E}$ is Maxwell's displacement current, added in part to produce wave equations for $\boldsymbol{E}$ and $\boldsymbol{B}$), (12.25) is the statement that magnetic monopoles do not exist, and (12.26) is essentially Faraday's law of induction. The quantities ρ and $\boldsymbol{j}$ cannot be prescribed independently because, differentiating (12.23) with respect to t, we get (remembering that $\partial/\partial t$ commutes with $\partial/\partial x$, $\partial/\partial y$, and $\partial/\partial z$)

$$\operatorname{div}\left(\frac{\partial \boldsymbol{E}}{\partial t}\right) = \frac{\partial \rho}{\partial t},$$

and taking the divergence of (12.24) gives

$$-\operatorname{div}\left(\frac{\partial \boldsymbol{E}}{\partial t}\right) = \operatorname{div} \boldsymbol{j}.$$

Thus, ρ and $\boldsymbol{j}$ must satisfy the **equation of continuity**

$$\frac{\partial \rho}{\partial t} + \operatorname{div} \boldsymbol{j} = 0. \tag{12.27}$$

If we interpret $\boldsymbol{j}$ as a convection current, i.e. $\boldsymbol{j} = \rho \boldsymbol{u}$, where $\boldsymbol{u}$ is the velocity field of the material with charge density ρ, then (12.27) is identical to (12.8), the continuity equation of fluid dynamics.

In order to write these equations in tensorial form, we define an anti-symmetric tensor F^{ab}, called the **electromagnetic field tensor** or **Maxwell tensor**, by

$$F^{ab} = \begin{bmatrix} 0 & E_x & E_y & E_z \\ -E_x & 0 & B_z & -B_y \\ -E_y & -B_z & 0 & B_x \\ -E_z & B_y & -B_x & 0 \end{bmatrix}, \tag{12.28}$$

and the **current density** or **source** 4-vector j^a by

$$j^a = (\rho, \boldsymbol{j}). \tag{12.29}$$

Then (exercise) the source equations and internal equations can be written in the form

$$\partial_b F^{ab} = j^a, \tag{12.30}$$

$$\partial_a F_{bc} + \partial_c F_{ab} + \partial_b F_{ca} = 0. \tag{12.31}$$

The anti-symmetry of F_{ab} means that (12.31) can be written more succinctly as

$$\partial_{[a} F_{bc]} = 0. \tag{12.32}$$

The continuity equation (12.27) becomes

$$\partial_a j^a = 0. \tag{12.33}$$

Let us be clear what we have done so far. We have merely shown that, given the definitions (12.28) and (12.29), Maxwell's equations (12.23)–(12.26) can be written **formally** as (12.30) and (12.31). We have treated F^{ab} and j^a as tensors, but the only justification for doing this is knowing their transformation properties under Lorentz transformations. Before the advent of special relativity, their transformation properties were, in fact, unclear. Indeed, from one point of view, it was precisely the desire to make Maxwell's equations Lorentz-covariant that led to the development of special relativity. The approach we shall adopt is to propose (12.28) and (12.29) as a working hypothesis and, from these definitions, work out their transformation properties. The ultimate justification then, as always, lies in comparing the predictions with observation, and there are a host of experiments which support the ansatz.

12.6 Potential formulation of Maxwell's equations

Rather than working with the fields $\boldsymbol{E}$ and $\boldsymbol{B}$ directly, it is usually more convenient to work in terms of the potentials. The **scalar potential** ϕ and the **vector potential** $\boldsymbol{A}$ are defined by

$$\boldsymbol{E} = -\operatorname{grad} \phi - \frac{\partial \boldsymbol{A}}{\partial t}, \tag{12.34}$$

$$\boldsymbol{B} = \operatorname{curl} \boldsymbol{A}. \tag{12.35}$$

If we define the **4-potential** by

$$\phi^a = (\phi, \boldsymbol{A}), \tag{12.36}$$

then we find that (12.34) and (12.35) are equivalent to (exercise)

$$F_{ab} = \partial_b \phi_a - \partial_a \phi_b. \tag{12.37}$$

The 4-potential is not defined uniquely by this equation, since we may perform a **gauge transformation**

$$\phi_a \rightarrow \bar{\phi}_a = \phi_a + \partial_a \psi, \tag{12.38}$$

where ψ is an arbitrary scalar field. Although a gauge transformation alters the potentials, it leaves F_{ab}, and hence $\boldsymbol{E}$ and $\boldsymbol{B}$, unchanged (exercise), and these are the strictly measurable quantities.

In solving particular problems, it is often convenient to reduce the gauge freedom by imposing a constraint on ϕ_a, called a **gauge condition**, which in turn simplifies the problem. For example, an important gauge for discussing electromagnetic radiation is provided by the **Lorentz gauge**

$$\eta^{ab}\phi_{a,b} = \phi^a{}_{,a} = 0. \tag{12.39}$$

Applying this constraint to (12.38), we find that the scalar field ψ is no longer arbitrary but must be a solution of the **wave equation**

$$\Box \psi \equiv \eta^{ab}\psi_{,ab} = 0, \tag{12.40}$$

where $\Box$ is the d'Alembertian operator

$$\Box \equiv \partial_0^2 - \partial_1^2 - \partial_2^2 - \partial_3^2.$$

The definition (12.37) results in the internal equations (12.31) being **automatically** satisfied, that is, they become identities (exercise). The source equations (12.30) become, in terms of the 4-potential,

$$\partial_b \left[\eta^{ac}\eta^{bd} \left(\partial_d \phi_c - \partial_c \phi_d \right) \right] = j^a. \tag{12.41}$$

In the Lorentz gauge, this reduces to (exercise)

$$\Box \phi^a = j^a. \tag{12.42}$$

In source-free regions, j^a vanishes, and this becomes

$$\Box \phi^a = 0, \qquad (12.43)$$

from which it follows that ϕ^a and F^{ab}, and therefore $\boldsymbol{E}$ and $\boldsymbol{B}$, all satisfy wave equations (exercise).

So far, we have restricted our attention to special relativity in Minkowski coordinates. To obtain the covariant formulation, we simply replace ordinary derivatives by covariant derivatives. However, it is not necessary in equations (12.32) and (12.37) because (exercise)

$$\nabla_{[a} F_{bc]} = \partial_{[a} F_{bc]}, \qquad (12.44)$$

and

$$\nabla_{[b} \phi_{a]} = \partial_{[b} \phi_{a]}. \qquad (12.45)$$

The **covariant formulation** of Maxwell's equations **in vacuo** in special relativity is thus

$$\nabla_b F^{ab} = j^a, \qquad (12.46)$$
$$\partial_{[a} F_{bc]} = 0, \qquad (12.47)$$

subject to

$$\nabla_a j^a = 0. \qquad (12.48)$$

In terms of the 4-potential, we still have

$$F_{ab} = \partial_b \phi_a - \partial_a \phi_b. \qquad (12.49)$$

Using the principle of minimal gravitational coupling, we adopt equations (12.46) and (12.47) in general relativity, where, however, the metric is no longer flat but is a solution of the full field equations $G^{ab} = \kappa T^{ab}$, and T^{ab} is the energy-momentum tensor arising from the electromagnetic field – which we now seek.

12.7 The Maxwell energy-momentum tensor

We shall construct the energy-momentum tensor for the electromagnetic field from a variational approach. For simplicity, we shall work **in vacuo** in special relativity in Minkowski coordinates and restrict attention to a source-free region, i.e. a region where j^a vanishes. Consider the Lagrangian for the electromagnetic field defined by

$$\mathcal{L}_E (\phi_a, F_{ab}) = \frac{1}{8\pi} \left[-\tfrac{1}{2} F_{ab} F^{ab} + (\phi_{a,b} - \phi_{b,a}) F^{ab} \right]. \qquad (12.50)$$

Then

$$\frac{\delta \mathcal{L}_E}{\delta \phi_a} = \frac{\partial \mathcal{L}_E}{\partial \phi_a} - \left(\frac{\partial \mathcal{L}_E}{\partial \phi_{a,b}}\right)_{,b}$$

$$= 0 - \frac{1}{8\pi}\left(F^{ab} - F^{ba}\right)_{,b}$$

and the field equations corresponding to a variation with respect to ϕ_a become

$$\left(F^{ab} - F^{ba}\right)_{,b} = 0. \tag{12.51}$$

Similarly,

$$\frac{\delta \mathcal{L}_E}{\delta F_{ab}} = \frac{\partial \mathcal{L}_E}{\partial F_{ab}}$$

$$= \frac{1}{8\pi}\frac{\partial}{\partial F_{ab}}\left[-\tfrac{1}{2}\eta^{ce}\eta^{df}F_{cd}F_{ef} + \eta^{ce}\eta^{df}(\phi_{c,d} - \phi_{d,c})F_{ef}\right]$$

$$= \frac{1}{8\pi}\left[-\tfrac{1}{2}\eta^{ae}\eta^{bf}F_{ef} - \tfrac{1}{2}\eta^{ca}\eta^{db}F_{cd} + \eta^{ca}\eta^{db}(\phi_{c,d} - \phi_{d,c})\right]$$

$$= \frac{1}{8\pi}\eta^{ac}\eta^{bd}\left[-F_{cd} + (\phi_{c,d} - \phi_{d,c})\right],$$

and the field equations corresponding to a variation with respect to F_{ab} become

$$F_{ab} = \phi_{a,b} - \phi_{b,a}. \tag{12.52}$$

This last equation defines F_{ab} in terms of the 4-potential and reveals that F_{ab} is anti-symmetric. The definition also means that the internal equations are satisfied automatically and (12.51) reduces to

$$F^{ab}_{,b} = 0,$$

namely, the source equations (in source-free regions). The result (12.52) also allows us to re-express the Lagrangian as

$$\mathcal{L}_E = \frac{1}{16\pi}\eta^{ac}\eta^{bd}F_{ab}F_{cd}. \tag{12.53}$$

We now make the transition to the full theory and assume that

$$\mathcal{L}_E = \frac{\sqrt{-g}}{16\pi}g^{ac}g^{bd}F_{ab}F_{cd}, \tag{12.54}$$

together with the definition (12.52) of F_{ab} in terms of ϕ_a. The factor $\sqrt{-g}$ is included to ensure that $\mathcal{L}_E$ is a scalar density (note that it reduces to 1 in special relativity in Minkowski coordinates). Then we find (exercise)

$$\frac{\partial \mathcal{L}_E}{\partial g^{ab}} = -\frac{\sqrt{-g}}{8\pi}\left(-g^{cd}F_{ac}F_{bd} + \tfrac{1}{4}g_{ab}F_{cd}F^{cd}\right). \tag{12.55}$$

The analogue of (11.51) for the contravariant metric is

$$\frac{\delta \mathcal{L}_E}{\delta g^{ab}} = -\frac{\sqrt{-g}}{2}T_{ab}. \tag{12.56}$$

These last two equations lead to the definition of the **Maxwell energy-momentum tensor** T_{ab} in source-free regions

$$T_{ab} = \frac{1}{4\pi}\left(-g^{cd}F_{ac}F_{bd} + \tfrac{1}{4}g_{ab}F_{cd}F^{cd}\right). \tag{12.57}$$

From (12.19), we have $\kappa = 8\pi$ in relativistic units. Thus, the full field equations in source-free regions, called the **Einstein–Maxwell equations**, become

$$G_{ab} = -2g^{cd}F_{ac}F_{bd} + \tfrac{1}{2}g_{ab}F_{cd}F^{cd}. \tag{12.58}$$

Let us look at some of the components of T_{ab} in special relativity in Minkowski coordinates. In particular, we find that the **energy density** of the electromagnetic field is given by

$$T_{00} = \frac{1}{8\pi}\left(\boldsymbol{E}^2 + \boldsymbol{B}^2\right), \tag{12.59}$$

which agrees with the usual expression for energy density in electrodynamics. Again, the **momentum density** is

$$(T_{01}, T_{02}, T_{03}) = -\frac{1}{4\pi}\boldsymbol{E}\times\boldsymbol{B}, \tag{12.60}$$

where the vector $\boldsymbol{E}\times\boldsymbol{B}$ is the **Poynting vector** of electrodynamics and represents the momentum density of the electromagnetic field. In addition, it is straightforward to verify that Maxwell's equations imply that T^{ab} is divergenceless, (exercise) i.e.

$$\nabla_b T^{ab} = 0. \tag{12.61}$$

12.8   Other energy-momentum tensors

We have met two methods for obtaining energy-momentum tensors. The first is an **ad hoc** method which constructs likely looking tensors out of the matter and energy fields present and investigates the conservation equations (12.7) in the non-relativistic limit. The second method proceeds from a variational principle formulation and investigates the field equations arising from a proposed Lagrangian. We can construct energy-momentum tensors for other fields or combination of fields using either approach or a combination of them. In particular, we can combine non-interacting fields by superimposing them. For interacting fields, we have to take the interactions into account.

We illustrate this with one example of each procedure. The energy-momentum tensor for a field of charged matter of proper mass density ρ_0 and 4-velocity u^a is (see (12.1) and (12.57))

$$T^{ab} = \rho_0\, u^a u^b + \frac{1}{4\pi}\left(-F^{ac}F^b{}_c + \tfrac{1}{4}g^{ab}F_{cd}F^{cd}\right). \tag{12.62}$$

The conservation equations then express the conservation of energy and the equations of motion for the field. The Lagrangian for an elementary particle of rest mass m_0, for example the π_0-meson, is described by a scalar field $\phi(x)$ given by

$$\mathcal{L}_s = -\frac{\sqrt{-g}}{2}\left(g^{ab}\,\nabla_a\phi\,\nabla_b\phi - m_0{}^2\,\phi^2\right), \tag{12.63}$$

where m_0 is the rest mass of the particle. The energy-momentum tensor is defined by (12.56), and again the conservation equations express the conservation of energy and the equations of motion of the field.

12.9 The dominant energy condition

In general, the components of any tensor in a particular coordinate system do not have an invariant meaning. However, if we choose an invariantly defined frame and look at the **frame components** of the tensor, then these will have physical significance. In the case of the energy-momentum tensor T_{ab}, we choose a frame at a point by looking for solutions of the **eigenvalue** equation

$$T_a{}^b u^a = \lambda u^b,$$

where u^a is the eigenvector corresponding to the eigenvalue λ. This has characteristic equation

$$\det(T_a{}^b - \lambda\delta_a^b) = 0.$$

For all types of standard matter, this equation has real non-zero roots, and the corresponding eigenvectors can be normalized to form a frame $e_i{}^a$ of one timelike and three spacelike vectors. The frame components of T_{ab} are

$$T_{ij} = T_{ab}e_i{}^a e_j{}^b = \mathrm{diag}(\mu, p_1, p_2, p_3),$$

since the matrix is diagonal with the eigenvalues as elements. The eigenvalue μ is called the **energy density**, and $u^a = e_0{}^a$ is the **4-velocity** of the medium. The eigenvalues p_α ($\alpha = 1, 2, 3$) are called the **principal stresses**, and the corresponding eigenvectors $e_\alpha{}^a$, the **principal axes of stress**. An energy-momentum tensor will only represent a physically realistic matter field if the energy density is non-negative and dominates any stresses present. More precisely, all known matter fields satisfy the **dominant energy condition** (Hawking and Ellis 1973)

$$\mu \geqslant 0, \qquad -\mu \leqslant p_\alpha \leqslant \mu. \qquad (12.64)$$

The latter condition can be shown to be equivalent to requiring that the local speed of sound is not greater than the local speed of light.

If, in particular, the three principal stresses are positive and equal to p say, then setting $\mu = \rho_0$, the energy-momentum tensor takes the form of a perfect fluid, (12.22). If the three principal stresses vanish, then the energy-momentum tensor takes the form of dust, (12.1).

Exercises

12.1 (§12.2) Establish (12.6) from (12.1). Show that (12.7) leads to (12.8) and (12.9).

12.2 (§12.3) Derive (12.14) from (12.13) and deduce (12.17).

12.3 (§12.4) Show that the conservation equations for a perfect fluid lead to the equation of continuity and the equation of motion, in special relativity in Minkowski coordinates.

12.4 (§12.5)
(i) Show that Maxwell's equations can be written in the form (12.30) and (12.31), given the definitions (12.28) and (12.29).
(ii) Show that the internal equations can be written in the form (12.32).
(iii) Show that the continuity equation can be written in the form (12.33). Show directly from (12.30) that this equation is an identity.

12.5 (§12.5) Find the transformation properties of E, B, ρ, and j under a boost in the x-direction. [Hint: consider F^{ab} and j^a.]

12.6 (§12.6)
(i) Show that (12.37) is equivalent to (12.34) and (12.35).
(ii) Show that F_{ab} is invariant under a gauge transformation.
(iii) Show that, if F_{ab} is defined in terms of a 4-potential, then the internal equations are automatically satisfied.

12.7 (§12.6) Show that, in an appropriate gauge, Maxwell's equations reduce to $\Box \phi^a = j^a$ in regions where the source 4-vector is non-zero. What remaining gauge freedom is left? Deduce that E and B satisfy the wave equation in source-free regions.

12.8 (§12.6) Check (12.44) and (12.45).

12.9 (§12.7)
(i) Establish (12.55) and (12.57). [Hint: Operate on each side with δ and express each term on the RHS as products with the factor δg^{ab}.]
(ii) Confirm (12.59) and (12.60).
(iii) Show that the conservation equation (12.61) is equivalent to $\partial_b T^{ab} = 0$ and show that this is satisfied by virtue of Maxwell's equations.

12.10 (§12.8) Investigate the conservation equations for the energy-momentum tensor arising from (12.63).

Further reading

All the standard texts listed in the previous chapter have a treatment of the energy-momentum tensor. The book by Hawking and Ellis (1973) has a good treatment of the dominant energy condition as well as the strong and weak energy conditions used in the singularity theorems which we discuss in §20.13. The treatment on the Newtonian limit is based on that of Trautman , Pirani, and Bondi (1964), but see also the book by Hartle (2003).

Hartle, J. B. (2003). *Gravity: An Introduction to Einstein's General Relativity*. Addison Wesley, San Francisco, CA.

Hawking, S. W., and Ellis, G. F. R. (1973). *The Large Scale Structure of Space-Time*. Cambridge University Press, Cambridge.

Trautmann A., Pirani F. A. E., and Bondi, H. (1964). *Lectures on General Relativity*. Brandeis Summer Institute on Theoretical Physics, 1964, vol. 1. Prentice-Hall, Englewood Cliffs, NJ.

The structure of the field equations

13

13.1 Interpretation of the field equations

Before attempting to solve the field equations, we shall consider some of their important physical and mathematical properties in this chapter. The full field equations (in relativistic units) are

$$G_{ab} = 8\pi T_{ab}. \tag{13.1}$$

They can be viewed in three different ways.

1. The field equations are differential equations for determining the metric tensor g_{ab} from a **given energy-momentum tensor** T_{ab}. Here, we are reading the equations from right to left. This is a Machian way of viewing the equations since one specifies a matter distribution and then solves the equations to ascertain the resulting geometry. However, Einstein's equations are not entirely Machian since, without imposing additional conditions, the matter distribution does not determine a **unique** geometry. The most important case of the equations is when $T_{ab} = 0$, in which case we are concerned with finding **vacuum** solutions.

2. The field equations are equations from which the energy-momentum tensor can be read off corresponding to a **given metric tensor** g_{ab}. Here, we are reading the equations from left to right. It was originally thought that this would be a productive way of determining energy-momentum tensors. We simply choose arbitrarily ten functions of the coordinates, namely, the symmetric g_{ab}, and then we can compute G_{ab} and read off T_{ab} from (13.1). However, this rarely turns out to be very useful in practice because the resulting T_{ab} are usually physically unrealistic and violate the dominant energy conditions. In particular, it frequently turns out that the energy density goes negative in some region, which we reject as unphysical because the positive character of energy density dominates gravitation theory.

3. The field equations consist of **ten equations connecting twenty quantities**, namely, the ten components of g_{ab} and the ten components of T_{ab}. Hence, from this point of view, the field equations are to be viewed as constraints on the simultaneous choice of g_{ab} and T_{ab}. For example, when looking at electromagnetism, one solves the Einstein-Maxwell equations (12.58) for the metric given the energy-momentum tensor (12.57), where $F_{ab} := \phi_{a,b} - \phi_{b,a}$ with $F^{ab}_{\;;b} = 0$. This approach is also used when one can partly specify the geometry and the energy-momentum tensor from physical considerations and then the equations are used to try and determine both quantities completely.

Introducing Einstein's Relativity. Second Edition. Ray d'Inverno and James Vickers, Oxford University Press.
© Ray d'Inverno and James Vickers (2022). DOI: 10.1093/oso/9780198862024.003.0012

13.2   Determinacy, non-linearity, and differentiability

Let us consider solving the vacuum field equations

$$G_{ab} = 0, \tag{13.2}$$

for g_{ab}. Then, at first sight, the problem seems well posed: there are ten equations for the ten unknowns g_{ab}. However, the equations are not independent but are connected by four differential constraints through the contracted Bianchi identities

$$\nabla_b G^{ab} \equiv 0. \tag{13.3}$$

So we seem to have a problem of **under-determinacy**, since there are fewer equations than unknowns. However, we cannot expect complete determinacy for any set g_{ab}, since any metric can be transformed with fourfold freedom by a coordinate transformation

$$x^a \rightarrow x'^a = x'^a(x) \quad (a = 0, 1, 2, 3),$$

into an equivalent metric which describes the same geometry, but in different coordinates. From this point of view, we should regard the solutions of Einstein's equations as equivalence classes of space-times possessing metrics which are related by coordinate transformations. In order to work with a particular representative of the equivalence class, we can use the coordinate freedom to impose four conditions on the g_{ab}. These are known as **coordinate conditions** or **gauge conditions**. For example, we could introduce **Gaussian** or **normal coordinates** in which

$$g_{00} \overset{*}{=} 1, \quad g_{0\alpha} \overset{*}{=} 0. \tag{13.4}$$

Then the remaining six unknowns $g_{\alpha\beta}$ can be determined by the six independent equations in (13.2). However, there is rather more to the story, but we postpone its consideration until §13.5. Similar remarks apply to the full theory.

The field equations are very difficult to handle because they are **non-linear**. They do not therefore possess a principle of superposition, that is to say, if you have two solutions of the field equations, then you cannot add them together to obtain a third. Put another way, it means that you cannot analyse a complicated physical problem by breaking it up into simpler constituent parts. The non-linearity reveals itself physically in the following way: the gravitational field produced by some source contains energy and hence, by special relativity, mass, and this mass in turn is itself a source of a gravitational field; that is to say, the gravitational field is coupled to itself. This non-linearity means that the equations are very difficult to solve in general. Indeed, originally Einstein anticipated that one would never be able to find an exact solution of them. It came as something of a surprise when K. Schwarzschild found an exact solution in 1916 shortly after the publication of the theory. However, Schwarzschild's solution arises by making a symmetry assumption, indeed the simplest assumption of all, namely, spherical symmetry. Today there are a large number of solutions in existence, probably in excess of four figures (depending on how you count them). Nearly all of them have been obtained by imposing symmetry conditions or other simplifying assumptions. We discuss the role they play in understanding the behaviour of solutions to Einstein's equations in more detail in §13.10 below.

Ideally, one wants to know what the theory says about physically important situations. In cases where symmetry is absent, or where the symmetry conditions are not strong enough to determine a solution, then recourse has to be made to either numerical or approximation methods. Approximation methods are used in situations where the the gravitational field is **weak** so that some of the terms in Einstein's equations can be ignored. We met an example of using approximation methods in the Newtonian limit of the last chapter. Another situation in which the gravitational field is weak is when we are looking at the gravitational field a long way from an isolated source. From a mathematical viewpoint, the weakness of the gravitational field means that the linear terms in certain equations are more important than the rest. We shall meet a linearized form of the field equations in Chapter 21. Numerical methods are also very important and have played an important role in constructing numerical models of, for example, the gravitational radiation produced by colliding black holes, which have been an essential ingredient in the detection of gravitational waves. The basis of numerical relativity is the so-called **3+1 formalism** which we describe in Chapter 14.

There are important mathematical questions concerning the differentiability of solutions to Einstein's equations. However, we shall not take them into account since we will assume that all our fields are smooth or C^∞, so that they can be differentiated indefinitely. This condition can be weakened considerably; for example, if we assume that the metric is C^2, which means that it can be differentiated twice, then this ensures that the Einstein tensor G_{ab} can be defined and thus the field equations can be constructed. There are other conditions affecting the differentiability which are connected with surfaces of discontinuities that arise in the theory, for example the surface of a material body. One important set of conditions (analogous to the continuity conditions of potential theory) are the **Lichnerowicz conditions**: second and higher derivatives of g_{ab} need not be continuous across a surface of discontinuity S, but g_{ab} and $g_{ab,c}$ must be continuous across S.

13.3 The cosmological term

Einstein was rather sceptical about the full field equations (13.1) and regarded the vacuum field equations (13.2) as more fundamental. However, Einstein considered that even these equations were deficient in that they violated Mach's principle in the form M2, since they admit Minkowski space-time as a solution. This means that a test body in an otherwise empty universe would possess inertial properties (as all bodies do in special relativity) even though there is no matter to produce the inertia. As we pointed out before, a set of partial differential equations possesses large classes of solutions, many of which are unphysical. In order to decide which solutions are realized in nature, one must also prescribe **boundary conditions**. A natural requirement would be to take space-time to be **asymptotically flat** so that the Riemann tensor vanishes at spatial infinity. However, this requirement does not preclude a flat space solution of the vacuum field equations.

Einstein, realizing the need for prescribing appropriate boundary conditions, adopted a different approach. Cosmology, that is, the modelling of the universe, had not really emerged as a separate science prior to general relativity. In as much as there was some generally accepted model of the universe in existence then, it was rather an imprecise one. It suggested that, overall, the universe is **static** (i.e. not undergoing any large-scale motion) and **homogeneous** (i.e. filled uniformly with matter). There are two possible ideas about the spatial extent of the universe, either it is **open** (or **infinite**), in which case it goes on forever in spatial directions, or it is **closed** (**compact** or **finite**), in which case it is bounded in spatial directions. Einstein therefore tried to incorporate a simple model of the universe into the theory and then use this model to prescribe boundary conditions. In particular, he tried to find a static closed solution of the field equations, corresponding to a universe uniformly filled with matter. In so doing, he found he was forced to modify the field equations by introducing an extra term, the **cosmological term** Λg_{ab}, where Λ is a constant called the **cosmological constant**, so that they become (with our sign conventions)

$$G_{ab} - \Lambda g_{ab} = 8\pi T_{ab}. \tag{13.5}$$

Since

$$\nabla_b g^{ab} = 0,$$

we see that (13.5) is consistent with the requirement

$$\nabla_b T^{ab} = 0. \tag{13.6}$$

Using the results of §11.3, the corresponding Lagrangian becomes

$$\mathcal{L} = (R + 2\Lambda)\sqrt{-g} + 2\kappa \mathcal{L}_{\mathbf{M}}. \tag{13.7}$$

Indeed, if, quite generally, we demand that the gravitational field equations should

(1) be generally covariant,

(2) be of second differential order in g_{ab},

(3) involve the energy-momentum tensor T_{ab} linearly,

then it can be shown that the only equation which meets all of these requirements is

$$R_{ab} + \mu R g_{ab} - \Lambda g_{ab} = \kappa T_{ab}, \qquad (13.8)$$

where μ, Λ, and κ are constants. The demand that T_{ab} satisfies the conservation equations (13.6) then leads to $\mu = -\frac{1}{2}$. In fact, it was in the same year as Einstein proposed his equations that the great mathematician Hilbert derived them independently from a variational principle. Of course, they lacked the physical meaningfulness which Einstein had bestowed on them, especially through their reliance on the principle of equivalence.

The full field equations with the cosmological term are Machian in the sense that they no longer admit flat space as a solution. However, shortly after Einstein obtained the static cosmological solution, it was discovered that the universe is not in fact static, but rather is undergoing large-scale expansion, as evidenced by the galactic red shift. Einstein therefore discarded the static solution. At the same time non-static closed solutions of the field equations **without** the cosmological term, corresponding to an expanding distribution of matter, were found. Worse still, from the Machian viewpoint, de Sitter discovered a vacuum solution of the field equations with the cosmological term. These discoveries led Einstein to reject the cosmological term. He did so with some vehemence; he reportedly described his original decision to include it as his 'biggest blunder'. However, despite the fact that the inclusion of the term does not make the theory any more Machian, there is no a priori reason to leave it out. The constant Λ is assumed to be 'very small' in some sense and only of significance on a cosmological scale. Most treatments of cosmology include the term, but it is usually omitted for considerations connected with terrestrial or solar system phenomena and, indeed, we shall neglect it until we come to relativistic cosmology. From the cosmological perspective, rather than regard the term $-\Lambda g_{ab}$ as sitting on the left-hand side of the Einstein equations, one can regard $+\Lambda g_{ab}$ as a source term sitting on the right-hand side of the equations describing the energy-momentum tensor of so-called dark-energy. We will look at this in more detail when we consider relativistic cosmology in Chapter 26. It is also worth noting that it is possible to incorporate a number of **ad hoc** assumptions into Newtonian theory and obtain a cosmological theory which has much in common with relativistic cosmology (see §22.3). In the Newtonian model, if $\Lambda > 0$, then all matter experiences a 'cosmic repulsion', which tends to disperse the matter to spatial infinity. Conversely, $\Lambda < 0$ corresponds to a cosmic attraction. Since all matter experiences the force, it provides, in some sense, a realization of a long-range Machian-type interaction.

13.4 The conservation equations

We have suggested an axiomatic formulation of general relativity which replaces $R^a{}_{bcd} = 0$ by $G_{ab} = 8\pi T_{ab}$ in Axiom I(iii) of §8.5. However, it turns out that, rather surprisingly, the geodesic Axioms II (ii) and II(iii) need not be stated separately in general relativity because it can be shown that they must hold automatically **by virtue of the field equations** themselves. That this is possible can be made plausible by considering more carefully the motion of a test particle or photon in a gravitational field. Strictly speaking, the test particle or photon is itself part of the energy and matter present and so should be contained in the energy-momentum tensor. This tensor, in turn, being the source term in the field equations, determines the geometry of space-time and in particular its geodesic structure. In this sense, the motion of a test particle should somehow be contained in the field equations. In fact, it is coded into the Bianchi identities, since they lead to the requirement that

$$\nabla_b T^{ab} = 0, \tag{13.9}$$

namely, the conservation equations. It is possible to show that these equations specify unique equations of motion for a point particle in a gravitational field and that the ensuing trajectory of that particle is a geodesic of the corresponding metric. The original demonstration of this result was started by Einstein and Grommer, and developed further by Einstein with contributions from Infeld and Hoffman. Their approach rests on treating test particles as singularities of the field and, as a consequence, relies on a special mathematical apparatus which they had to construct to cope adequately with these singularities. The resulting work is both very complicated and voluminous and we will make no attempt to describe it. However, the results were confirmed subsequently by Geroch and Jang, and by Gralla and Wald using more modern mathematical machinery.

There is one neat little calculation which is very suggestive of what happens in essence in the general case. It consists of investigating the equations for a distribution of dust,

$$T^{ab} = \rho_0 u^a u^b.$$

Then the conservation equations (13.9) require

$$\nabla_b[\rho_0 u^a u^b] = 0.$$

The trick is to think of the term in square brackets as being the product of $(\rho_0 u^b)$ and u^a and to apply the Leibniz rule to this product:

$$u^a \nabla_b(\rho_0 u^b) + \rho_0 u^b(\nabla_b u^a) = 0. \tag{13.10}$$

We next contract this equation with u_a and use the result

$$u_a u^a = 1 \quad \Rightarrow \quad u_a(\nabla_b u^a) = 0,$$

which makes the second term vanish, leaving

$$\nabla_b(\rho_0 u^b) = 0.$$

Substituting this result back in (13.10) and dividing by $\rho_0 \neq 0$, we get

$$u^b \nabla_b u^a = 0,$$

which is the condition for u^a to be tangent to a geodesic. In other words, the conservation equations necessitate geodesic motion for the dust particles.

13.5 The Cauchy problem

In this section, we look in some detail at the following mathematical problem:

> Given the metric tensor g_{ab} and its first derivatives at some initial time, then construct the metric which corresponds to a space-time for all future time.

This is the problem of finding the causal development of a physical system from initial data and is a fundamental problem in the theory of partial differential equations. It is known as the **Cauchy problem** or **initial value problem**, or **IVP** for short. For simplicity, we will concentrate on the case of vacuum solutions of the Einstein equations.

We start with a three-dimensional spacelike hypersurface Σ_0 in the manifold, which we can take without loss of generality to be given by $x^0 = 0$. We specify g_{ab} and its first derivatives $g_{ab,c}$ on Σ_0 (Fig. 13.1). However, if we know g_{ab} everywhere on Σ_0, then we know its spacelike derivatives $g_{ab,\alpha}$ everywhere on Σ_0. Hence, it is sufficient to specify the following **initial data** on Σ_0:

$$g_{ab}, \quad g_{ab,0},$$

that is, the metric potentials and their time derivatives. Our problem is then to use the second-order vacuum field equations to try and solve for the second time derivatives $g_{ab,00}$. Let us suppose that we have found some equations for determining $g_{ab,00}$. Then, by repeatedly differentiating these equations with respect to time, we can get all higher time derivatives of g_{ab}.

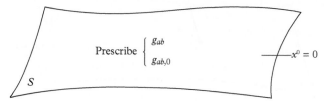

Fig. 13.1 The initial data for the Cauchy IVP.

It follows that, if we assume that g_{ab} is an **analytic function** of x^0, we can develop it in a power series in x^0. More precisely, if P and Q are the points $(0, x_0^\alpha)$ and (x^0, x_0^α) so that Q lies on the x^0-curve passing through P (Fig. 13.2), then, by Taylor's theorem,

$$g_{ab}(Q) = g_{ab}(P) + g_{ab,0}(P)x^0 + \sum_{n=2}^{\infty} \frac{1}{n!} \partial_0^n g_{ab}|_P (x^0)^n. \qquad (13.11)$$

Before considering the more complicated situation of the vacuum Einstein equations, we illustrate the idea by applying the method to determine solutions of the wave equation. The wave equation in Minkowski space may be written as

$$\frac{\partial^2 \phi}{\partial t^2} = \frac{\partial^2 \phi}{\partial x^2} + \frac{\partial^2 \phi}{\partial y^2} + \frac{\partial^2 \phi}{\partial z^2}. \qquad (13.12)$$

Since this is a second-order equation in time, we expect that, if we specify ϕ and $\phi_{,t} := \partial \phi / \partial t$ on some initial hypersurface Σ_0 given by $t = 0$, then there will exist a unique solution of the wave equation satisfying these initial conditions. We now show that this is true in the special case that we have **real analytic** initial data (i.e. both $\phi(0, \boldsymbol{x})$ and $\phi_{,t}(0, \boldsymbol{x})$ are smooth functions with convergent Taylor series).

If we are given ϕ at $t = 0$, then, by differentiating in the hypersurface Σ_0, we know the terms on the right-hand side of (13.12) and hence we know $\phi_{,tt}$ at $t = 0$. Differentiating (13.12) with respect to t, we obtain

$$\frac{\partial^3 \phi}{\partial t^3} = \frac{\partial^3 \phi}{\partial t \partial x^2} + \frac{\partial^3 \phi}{\partial t \partial y^2} + \frac{\partial^3 \phi}{\partial t \partial z^2}, \qquad (13.13)$$

So, if we are given $\phi_{,t}$ on Σ_0, then, by differentiating in the hypersurface again, we obtain the terms on the right-hand side and hence we can find $\phi_{,ttt}$ at $t = 0$.

Similarly, if we differentiate (13.13), we obtain

$$\frac{\partial^4 \phi}{\partial t^4} = \frac{\partial^2 \phi}{\partial t^2 \partial x^2} + \frac{\partial^4 \phi}{\partial t^2 \partial y^2} + \frac{\partial^4 \phi}{\partial t^2 \partial z^2}, \qquad (13.14)$$

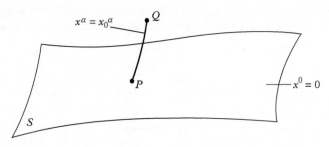

Fig. 13.2 Determining the metric at a later time x^0.

However, since we now know $\phi_{,tt}$ on $t = 0$, we can calculate the right-hand side of (13.14) and read off the value of $\phi_{,tttt}$ at $t = 0$. Carrying on in this way, we can obtain all the t derivatives of ϕ at $t = 0$. We now regard these as the coefficients of a Taylor series expansion in t of $\phi(\boldsymbol{x}, t)$ about $t = 0$. This gives us the power series

$$\phi(t, x) = \sum_{n=0}^{\infty} \left(\frac{\partial^n \phi}{\partial t^n}(0, \boldsymbol{x}) \right) t^n. \tag{13.15}$$

This will be our required solution, provided the series converges. In this simple case, we see that the coefficients of t^{2k} are just the sums of the $2k$ derivatives with respect to x, y, and z of $\phi(0, x, y, z)$ but, since this is real analytic in x, y, and z, we can bound these coefficients by those of a convergent series. For the odd coefficients, a similar result applies, this time using the fact that $\phi_{,t}(0, x, y, z)$ is real analytic.

A very general result which gives existence and uniqueness of solutions to systems of analytic partial differential equations with analytic initial conditions is provided by the following theorem due to Cauchy and Kowalevskya.

Theorem (Cauchy–Kowalevskya)
Let $\Phi^A(t, \boldsymbol{x})$, $A = 1, 2, \ldots, N$, $\boldsymbol{x} \in \mathbb{R}^n$, be functions that satisfy the system of partial differential equations

$$\frac{\partial^2 \Phi^A}{\partial t^2} = F^A \left(t, \boldsymbol{x}, \Phi^B, \frac{\partial \Phi^B}{\partial x^\alpha}, \frac{\partial^2 \Phi^B}{\partial x^\alpha \partial x^\beta}, \frac{\partial^2 \Phi^B}{\partial t \partial x^\alpha} \right), \tag{13.16}$$

where F^A for $A = 1, 2, \ldots N$ are analytic functions of their arguments. Then, given analytic initial data,

$$\Phi^A(0, \boldsymbol{x}) = P^A(\boldsymbol{x}), \qquad \frac{\partial \Phi^A}{\partial t}(0, \boldsymbol{x}) = Q^A(\boldsymbol{x}), \tag{13.17}$$

the initial value problem given by (13.16) has a unique analytic solution in a neighbourhood of $t = 0$.

We do not give the proof of the above theorem but simply remark that the basic idea is to repeatedly differentiate (13.16) and use the initial data to obtain a power series expansion in t for $\Phi^A(t, \boldsymbol{x})$, just as we did for the wave equation. One then uses the analyticity of F^A to construct a simpler scalar equation that we can explicitly solve whose solutions are analytic and bound $|\Phi^A(x, t)|$. Such an equation is said to **majorize** the PDE (13.16). Then, by comparing the Taylor series expansions in t, one concludes that the power series expansion for $\Phi^A(t, \boldsymbol{x})$ converges (for sufficiently small t). Uniqueness follows from the fact that the majorizing solution vanishes for zero initial data.

13.6   Einstein's equations as evolution equations

We now show how Einstein's equations can be regarded as evolution equations. To get an idea of how this works, we introduce a local coordinate system $(x^a) = (x^0, x^\alpha)$, $\alpha = 1, 2, 3$, where x^0 is a timelike coordinate (i.e. $g^{00} > 0$) and x^α are spacelike coordinates which provide coordinates on the spacelike hypersurfaces Σ_t given by $x^0 = t = $ constant.

For simplicity, we will specialize to the vacuum case $R_{ab} = 0$ and split $R_{ab} = 0$ into three equations

$$R_{00} = 0, \tag{13.18}$$

$$R_{0\alpha} = 0, \tag{13.19}$$

$$R_{\alpha\beta} = 0. \tag{13.20}$$

A straightforward calculation (exercise) reveals that the field equations can be written in the following form:

$$R_{00} = -\tfrac{1}{2} g^{\alpha\beta} g_{\alpha\beta,00} + M_{00} = 0, \tag{13.21}$$

$$R_{0\alpha} = \tfrac{1}{2} g^{0\beta} g_{\alpha\beta,00} + M_{0\alpha} = 0, \tag{13.22}$$

$$R_{\alpha\beta} = -\tfrac{1}{2} g^{0\,0} g_{\alpha\beta,0\,0} + M_{\alpha\beta} = 0, \tag{13.23}$$

where the terms involving M can be expressed **solely in terms of the initial data on** Σ_0. This gives rise to two problems of determination:

1. The system (13.21)–(13.23) does not contain $g_{0a,00}$; hence, we have a problem of **under-determination**.

2. The system (13.21)–(13.23) represents ten equations in the six unknowns $g_{\alpha\beta,00}$; hence, we have a problem of **over-determination**. This means that there must be compatibility requirements for the initial data on Σ_0.

We have met Problem 1 before, and it is not unexpected, since it relates to the fourfold freedom of coordinate transformations. Since the evolution of g_{0a} is not constrained by (13.21)–(13.23), we therefore take the bold step of prescribing g_{0a} in a neighbourhood of Σ_0. As we will see below, this amounts to a coordinate or gauge condition and does not affect the geometric or physical content of the equations. For example, one could choose coordinates x^α on Σ_0 and extend these to a neighbourhood of Σ_0 by choosing x^α to be constant along geodesics which have initial vectors as unit normal to the hypersurface, and choosing x^0 to be the corresponding affine parameter. This results in **Gaussian normal coordinates**, where one has $g_{00} \overset{*}{=} 1$ and $g_{0\alpha} \overset{*}{=} 0$. More generally, a choice of g_{0a} amounts to a choice of lapse and shift (see §14.10 for details). Having specified g_{0a}, we see that equation (13.23) has the required form in order to apply the Cauchy–Kowalevskya theorem. So that given analytic initial data $g_{\alpha\beta}(0, x^\gamma)$ and $g_{\alpha\beta,0}(0, x^\gamma)$ on Σ_0, we obtain a unique analytic solution for

$g_{\alpha\beta}(x^0, x^\gamma)$ for x^0 in a neighbourhood of $t = 0$. Combining this with the g_{0a} terms, this determines the metric in a neighbourhood of the initial hypersurface. For this reason, we call the equations $R_{\alpha\beta} = 0$ the **evolution**, **dynamical**, or **main** equations.

However, this all seems too good to be true. We appear to have obtained the metric g_{ab} without using the $R_{0\alpha} = 0$ and $R_{00} = 0$ equations. This is where we need to address Problem 2. If we want our metric to be a solution of the vacuum Einstein equations, we also need (13.21) and (13.22) to be satisfied. Using (13.23) on the initial hypersurface, we can replace the $g_{\alpha\beta,00}$ terms in both these equations by expressions that depend only on the initial data. It therefore transpires that we are not able to choose the initial data freely but must choose it so that the **constraints** $R_{00} = 0$ and $R_{0\alpha} = 0$ are satisfied.

To summarize the situation so far, we chose a gauge by specifying g_{0a} in a neighbourhood of Σ_0 and then chose initial data which satisfy the constraints on Σ_0. Note this is a non-trivial task; see §14.12 for details. We may then use the Cauchy–Kowalevskya theorem to solve the dynamical equation (13.23) and obtain $g_{\alpha\beta}$ in a neighbourhood of Σ_0 which, together with g_{0a}, determines the full metric g_{ab} in a neighbourhood of Σ_0. However, this is not enough for a solution of Einstein's equations. We need the constraints to be satisfied not only for $t = 0$ but at later times as well and it is not at all obvious if this is true. The reassuring answer to this question is that, if the constraints are satisfied on the initial hypersurface Σ_0, they are also satisfied in a neighbourhood of $t = 0$. The reason for this is the contracted Bianchi identities, as we sketch out below. We first show that, if $R_{\alpha\beta} = 0$, then the constraint equations $R_{0\alpha} = 0$ and $R_{00} = 0$ are equivalent to $G_a{}^0 = 0$.

If $R_{\alpha\beta} = 0$, then the scalar curvature is given by

$$R = g^{00}R_{00} + 2g^{0\alpha}R_{0\alpha}. \tag{13.24}$$

So, using $G_a{}^b = R_a{}^b - \frac{1}{2}\delta_a^b R$, we have (exercise)

$$G_0{}^0 = \tfrac{1}{2}g^{00}R_{00}, \tag{13.25}$$

$$G_\alpha{}^0 = g^{00}R_{0\alpha}. \tag{13.26}$$

Since $g^{00} \neq 0$, it immediately follows that, if $R_{\alpha\beta} = 0$, then the constraint equations imply that $G_0{}^0 = 0$ and $G_\alpha{}^0 = 0$. On the other hand, we now show that, if $R_{\alpha\beta} = 0$, then $G_a{}^0 = 0$ implies $G_a{}^b = 0$ and hence, in particular, (13.19) and (13.18) are satisfied. From the definition of $G_a{}^b$, and equation (13.24) for R, we have

$$G_\alpha{}^\beta = g^{0\beta}R_{0\alpha} - \tfrac{1}{2}\delta_\alpha^\beta(g^{00}R_{00} + 2g^{0\lambda}R_{0\lambda}). \tag{13.27}$$

But, by (13.25) and (13.26), $G_a{}^0 = 0$ implies that $R_{0a} = 0$ so that, by the above equation, we must have $G_\alpha{}^\beta = 0$. We also have

$$G_0{}^\beta = R_0{}^\beta = g^{0\beta}R_{00} + g^{\beta\lambda}R_{0\lambda}, \tag{13.28}$$

which again vanishes by virtue of (13.25) and (13.26). Indeed, if we have $R_{\alpha\beta} = 0$ and know $G_a{}^0$, then we may use (13.25) and (13.26) to give R_{00} and $R_{0\alpha}$ and then substitute these in (13.27) and (13.28) and divide by $g^{00} \neq 0$ to obtain the remaining terms (exercise):

$$G_0{}^\beta = \frac{2g^{\beta 0}}{g^{00}} G_0{}^0 + \frac{g^{\beta\lambda}}{g^{00}} G_\lambda{}^0, \qquad (13.29)$$

$$G_\alpha{}^\beta = \frac{g^{0\beta}}{g^{00}} G_\alpha{}^0 - \tfrac{1}{2}\delta_\alpha^\beta \left(2G_0{}^0 + \frac{2g^{0\lambda}}{g^{00}} G_\lambda{}^0 \right). \qquad (13.30)$$

Thus, following Lichnerowicz, we may write the vacuum field equations in the **normal form**

$$R_{\alpha\beta} = 0, \qquad G_a{}^0 = 0,$$

where the first six equations are **evolution** equations for $g_{\alpha\beta,00}$ and the last four equations are **constraint equations** which the initial data must satisfy on Σ_0. This resolves Problem 2.

We now prove a remarkable result

If the constraint equations are satisfied on Σ_0, then they are satisfied for all time, by virtue of the contracted Bianchi identities.

Writing out the contracted Bianchi identity $\nabla_b G_a{}^b = 0$ in coordinates gives

$$\partial_t G_a{}^0 = -\partial_\beta G_a{}^\beta + \Gamma^c_{a0} G_c{}^0 + \Gamma^c_{a\beta} G_c{}^\beta - \Gamma^b_{b0} G_a{}^0 - \Gamma^b_{b\gamma} G_a{}^\gamma. \qquad (13.31)$$

Substituting for $G_0{}^\beta$ and $G_\alpha{}^\beta$ using equations (13.29) and (13.30), we obtain a linear first-order homogeneous system for $G_a{}^0$. Since the constraints are satisfied on the initial surface, we also have $G_a{}^0 = 0$ when $t = 0$. In the analytic case that we are considering here, we may apply the Cauchy–Kowalevskya theorem to show the existence of a unique solution to the above initial value problem. However, since $G_b{}^0$ appears in every term on the right-hand side of the differential equation, we see that $G_b{}^0 \equiv 0$ is a solution, which by uniqueness must be the only solution. We have therefore used the contracted Bianchi identity together with the vanishing of $R_{\alpha\beta}$ to show that, if the constraints are satisfied on the initial hypersurface, they are satisfied at subsequent times. Furthermore, we have also showed that, if $R_{\alpha\beta} = 0$ and the constraints $G_a{}^0$ are satisfied for all t, then Einstein's equations are satisfied.

In summary, we have shown that, if we have a hypersurface Σ_0 with coordinates (x^α) and we want to solve the Cauchy problem, we first need to choose analytic initial data $g_{\alpha\beta}(0)$ and $g_{\alpha\beta,0}(0)$ which satisfy the constraint equations $G_a{}^0 = 0$ on Σ_0. We next need to specify the components of g_{0a} as analytic functions in a neighbourhood of the hypersurface. As we will see in §14.10, this amounts to choosing how we develop the coordinates (x^α) on Σ_0 into coordinates (x^0, x^α) in a neighbourhood of Σ_0.

Having done this, we need to solve the dynamical equations $R_{\alpha\beta} = 0$ for $g_{\alpha\beta}$, which are 6 equations for 6 unknowns. In the case of analytic initial data, we know from the Cauchy–Kowalevskya theorem that there exists a unique analytic solution to these equations. Furthermore, we know that, because of the contracted Bianchi identities for analytic initial data, the constraints are satisfied not just on Σ_0 but in a neighbourhood of the initial hypersurface. By combining $g_{\alpha\beta}$ with g_{0a}, we obtain the space-time metric g_{ab}, which satisfies both the dynamical equations and the constraints and is therefore the required solution to the vacuum Einstein equations. Furthermore, in the given coordinates, this solution is unique.

The above discussion made extensive use of the Cauchy–Kowalevskya theorem, which required analytic coordinate conditions and analytic data. Physically, this is rather a strong restriction, since an analytical function is fully determined by its value and those of its derivatives at a single point. Thus, a knowledge of the function in an arbitrary small region determines its value everywhere. This does not fit in well with our notion of causality in general relativity, where nothing can travel faster than light. Because of this, a major breakthrough in the Cauchy problem was achieved by Choquet-Bruhat in 1952 when she showed the existence and uniqueness of solutions to the vacuum Einstein equations in a small neighbourhood of Σ_0 for smooth (or more precisely at least C^5) initial data. For completeness, we sketch the proof in the next section, which makes use of **harmonic coordinates**.

13.7 Solving Einstein's equations in harmonic coordinates

As we have seen above, obtaining a unique solution of Einstein's equations requires specifying a choice of coordinates. A particularly useful choice is to make use of **harmonic coordinates**, in which the coordinate functions x^a, regarded as scalar fields, satisfy the wave equation

$$\Box x^a := g^{cd}\nabla_c\nabla_d x^a = 0, \quad a = 0,\ldots,3. \tag{13.32}$$

A short calculation (exercise) shows that this is equivalent to the **harmonic gauge condition**

$$H^a := g^{cd}\Gamma^a_{cd} = 0. \tag{13.33}$$

In a general coordinate system, one can show that the Ricci curvature may be written

$$R_{ab} = R^H_{ab} + H_{(a,b)}, \tag{13.34}$$

where

$$R^H_{ab} = -\tfrac{1}{2}g^{cd}g_{ab,cd} + Q_{ab}(g,\partial g), \tag{13.35}$$

and $Q(g, \partial g)$ only depends on g_{ab}, g^{ab}, and $g_{ab,c}$ and contains no second-order derivatives.

So in **harmonic coordinates** the vacuum Einstein equations become the **reduced Einstein equations** given by

$$R_{ab}^H = 0. \tag{13.36}$$

Since the terms involving the highest derivatives are proportional to $g^{cd}\partial_c\partial_d g_{ab}$, the reduced equations give a hyperbolic wave-like equation for g_{ab}. Because of this (and unlike the full Einstein equations), these equations are in a form where one can apply standard PDE theory to show the existence of a unique smooth solution given smooth initial data for g_{ab} and $g_{ab,0}$ on Σ_0. Note that, unlike the previous analytic case, we solve (13.36) for **all the components** of g_{ab}, not just $g_{\alpha\beta}$. Of course, solving the reduced Einstein equations is not the same as solving the full Einstein equations, unless one can also ensure that $H_a = 0$ on M. We now show that one can choose the initial data in such a way that this is true.

To obtain the initial data for (13.36), one first sets $g_{00} = 1$ and $g_{0\alpha} = 0$ on Σ_0 and then solves the constraint equations

$$G_{a0} = G_a{}^0 = 0, \tag{13.37}$$

on Σ_0 to determine a positive definite metric $\gamma_{\alpha\beta} := -g_{\alpha\beta}$ and a symmetric tensor $K_{\alpha\beta} = g_{\alpha\beta,0}$ on Σ_0. To complete the initial data for (13.36), we still need to choose $g_{0b,0}$ on Σ_0 and it turns out (exercise) that one can do this in such a way as to ensure that

$$H_a = 0 \quad \text{on } \Sigma_0. \tag{13.38}$$

Furthermore, the constraint $G_{0a} = 0$ on Σ_0 shows that, if H_a vanishes on Σ_0, then so does $H_{a,0}$. On the other hand, the contracted Bianchi identity $\nabla_a G^a{}_b = 0$ implies that H_a satisfies a wave-like evolution equation, which again can be shown to have unique solutions in the smooth case. Since setting $H_a = 0$ everywhere on M satisfies both the wave-like evolution equation and the initial conditions, by uniqueness it must be **the** solution. Thus, we have shown that we can find initial conditions g_{ab} and $g_{ab,0}$ for (13.36) that satisfy the constraints and ensure that H_a vanishes identically. We now use standard hyperbolic PDE theory to solve $R_{ab}^H = 0$ for g_{ab} and, since H_a vanishes, it follows that we also have $R_{ab} = 0$. Thus, the g_{ab} that we obtain is also a solution of the vacuum Einstein equations. It remains to show that **all** solutions of Einstein's equations can be obtained in this way. We postpone this to the next chapter, where we adopt a more geometrical approach to the Cauchy problem.

A major improvement to this local result was made with the subsequent global existence and uniqueness theorem by Choquet-Bruhat and Geroch. This global theorem shows that, among all the space-times (M, g) which are solutions of the vacuum Einstein equations and such that Σ_0 is an embedded Cauchy surface on which the metric induces the specified initial data, there exists a maximal space-time $(M^\star, g^\star)$ and it is unique. Here the term **maximal** means that any space-time (M, g) that is a solution of the Cauchy problem is isometric to part of $(M^\star, g^\star)$. The questions of existence, uniqueness, and stability (i.e. do 'small' variations of the initial data result in 'small' variations in the solution?), and the extent to which solutions can be developed in general relativity, are deep and complex questions, and are the topics of current research.

13.8 The hole problem

We have, in fact, been somewhat imprecise in setting up the Cauchy problem and in so doing we have covered up something which had originally caused Einstein considerable difficulty. We defined the Cauchy problem as starting with a manifold with no metric on it (a so-called bare manifold), prescribing initial data on a hypersurface in the manifold, and then using the field equations to generate a unique solution for the metric g. However, as we know from the principle of general covariance, we may then apply a coordinate transformation to g and so obtain another solution $\bar{g}$, say. How are the solutions g and $\bar{g}$ related physically?

This question had troubled Einstein and was one of the reasons why, even though the principle of general covariance was formulated in 1907, another eight years were to elapse before the field equations were finally obtained. Einstein raised the question in the form of the 'hole problem'. Suppose that the matter distribution is known everywhere outside of some hole H in the manifold. Then the field equations together with the boundary conditions will enable the metric g to be determined inside H and, in particular, at some point P, say. Now carry out a coordinate transformation which leaves everything outside H fixed, but which (from the active viewpoint) moves points around inside H, for example moving P to P', say (Fig. 13.3). Next, determine afresh the metric $\bar{g}$ in H. Is $\bar{g}$ the same as g? The answer is that, although $\bar{g}$ will in general be functionally different from g (i.e. the components of $\bar{g}$ will involve different functions of its coordinates compared with g), it will still represent the **same** physical solution. How can this be so if the points inside H have moved? The nub of the argument is that the point P in the bare manifold is not distinguished from any other point. It does not become a point with physical meaning (that is, an **event**) until a metric is determined in H. As John Stachel puts it so succinctly, 'no metric, no nothing' (2001). Thus, a physical solution, that is, a **space-time**, consists of a manifold together with a metric. Two space-times are physically equivalent, in other words, give rise to the same gravitational field, if the two metrics can be transformed into each other. Mathematically, we should regard physical solutions as equivalence classes of space-times possessing metrics which are related by coordinate transformations.

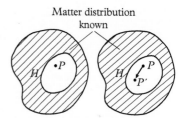

Fig. 13.3 The hole problem.

13.9 The equivalence problem

The question which then arises is, Given two metrics, g and $\bar{g}$, are they in fact the same, that is, does there exist a coordinate transformation transforming one into the other? This is a classic problem in differential geometry, known as the **equivalence problem**, and its classic solution by E. Cartan involves computation and comparison of the tenth covariant derivatives of the Riemann tensors of g and $\bar{g}$.

As one discovers in working out the Riemann tensor, even for something as simple as the Schwarzschild solution (see Exercise 6.32), it is a non-trivial task. It is all too easy to make slips in a longhand calculation. In fact, this task of undertaking large amounts of algebraic calculation has been made much more tractable and less error-prone with the advent of general purpose computer algebra systems, the best known of which include MATHEMATICA, MAPLE, and REDUCE. The system originally most used in general relativity (for which it was specifically designed) is the system SHEEP, together with its extensions CLASSI (for classifying metrics) and STENSOR (for symbolic tensor manipulation). These systems make possible computations which would have been impossible to contemplate undertaking by hand. Even so, they are not capable currently of computing anything like the tenth covariant derivatives of Riemann tensors and so appear to be of little use in the equivalence problem.

The situation has been improved profoundly by the work of A. Karlhede (1980). We will not pursue the details, but in broad outline the Karlhede approach is to classify a geometry by introducing a frame or tetrad, which is defined in stages, such that the Riemann tensor and its covariant derivatives take on a simple or rather **canonical** form at each stage. This is a well-defined procedure leading to a set of invariant quantities characterizing a given geometry. With this approach, the worst case theoretically involves computing the seventh covariant derivative, although, for vacuum solutions of the Einstein equations, it was shown (Ramos and Vickers, 1996) that this can be lowered to the fifth derivative. However, experience in using the algorithm suggests that one rarely needs go beyond the third derivative and often the first derivative is enough. This makes computer calculation a viable proposition. Thus, given two metrics, one first computes their invariant classification. If the two sets are different, then so are the metrics. If they are the same, then there may be a transformation relating them. The problem is then reduced to solving a set of four algebraic equations to determine the transformation. In general, this is non-algorithmic but, in practice, it is often manageable.

13.10 The status of exact solutions

The field equations of general relativity are incredibly difficult to solve. First, they are non-linear so, in particular, you cannot superimpose solutions or break down complex physical situations into its simpler components. But, more immediately, when seen as a set of second-order partial

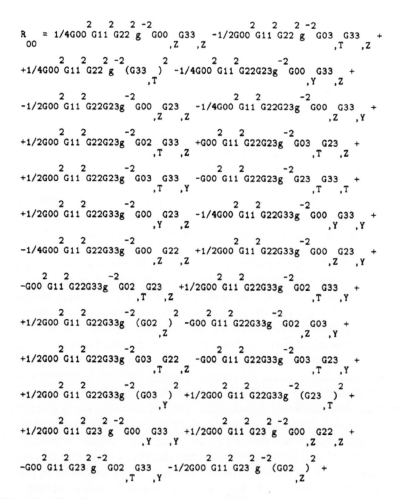

Fig. 13.4 The first twenty-six terms of the Ricci tensor component R_{00} for a general space-time.

differential equations for the metric, they are, indeed, incredibly complicated. One of the authors was the first to write out the equations explicitly using a computer algebra system he had designed called LAM (Lisp Algebraic Manipulator) (see d'Inverno 1980). If we reduce the number of terms by denoting the determinant of the metric as the symbol g (rather than the explicit expression for g in terms of the metric g_{ab}) and write out the components of the metric explicitly as $(G00, G01, \ldots, G33)$ which are all functions of the four coordinates (T, X, Y, Z), then the first twenty-six terms of the components R_{00} are shown in Fig. 13.4. There are of the order of 100,000 terms in the Ricci tensor and, if we were to output the equations in a normal-size font on A4 paper and stack up the paper, then the stack would be of the order of 3 m high! Not surprisingly, Einstein thought it would not be possible to find exact solutions of the field equations. So it came as something of a shock when Schwarzschild produced a vacuum solution in January 1916, a little over a month after the publication of the field equations. Schwarzschild looked for the simplest solution, namely a static spherically symmetric solution. As we shall see, we do not need the assumption of a static space-time because the vacuum field equations force a spherically symmetric solution to be static, again illustrating the special nature of solutions of non-linear partial differential equations. This solution turns out to be the prototype of a massive black hole and was eventually generalized to the Kerr–Newman solution with mass, charge, and spin. Moreover, we have the general theorem that an isolated black hole tends asymptotically in time to the Kerr–Newman solution, so exact solutions have lead to a good understanding of an isolated black hole.

Perhaps more surprisingly, in the following years many thousands of exact solutions were discovered. One way to search for exact solutions is to consider an ansatz of some kind on the functional form of the metric and then try and solve the resulting field equations. This approach has proved to be very fruitful. As an example, Harrison assumed that the metric took on a particular form based on the method of separation of variables and was able to find explicit forms for forty vacuum solutions (Harrison 1955). Using the computer algebra system LAM, it was possible to determine explicitly that the solutions were, indeed, vacuum. This involves a set of calculations which it is estimated would take more than a lifetime to complete by hand - and hence the need for computer algebra systems. Moreover, using the successor system SHEEP (LAM(B) grown up!) and its extension CLASSI, it is possible to provide an invariant geometrical classification of the solutions. This led to the discovery that two of the solutions are in fact the same, i.e. there exists a coordinate transformation transforming one solution into the other. This invariant classification is important for classifying solutions and thereby distinguishing between solutions which are in fact different and not simply exhibited in different coordinate systems. Examples of this are well known: indeed, the Schwarzschild solution itself has apparently been 'discovered' in the literature on some twenty different occasions! The classic book on exact solutions (Kramer et al. 2009) has put the many known exact solutions into an encyclopaedic form.

The fact that there are many known exact solutions would seem like good news. The bad news is that many of them are unphysical in character in that they possess singularities or other unphysical regions and so are unlikely to approximate to real physical situations. It seems to be the nature of non-linear partial differential equations that they throw up these unphysical exact solutions. There are some insightful exact solutions which can be viewed as abstracted away from sources, and exact cosmological solutions have played an important role in cosmology historically. There are many important situations where we would like to have exact solutions but none are know to exist. These include the 2-body problem, interior black hole solutions, and radiation from an isolated source. Since exact solutions for these situations are unavailable, recourse has to be made to approximation theory and numerical relativity, that is, solving Einstein's equations numerically on a computer.

Exercises

13.1 (§13.3) Show that the Lagrangian (13.7) gives rise to the full field equations with cosmological term (13.5).

13.2 (§13.3) Show that, if (13.8) is to be consistent with (13.6), then $\mu = -\frac{1}{2}$.

13.3 (§13.3) Show that the trace of the Maxwell energy-momentum tensor is zero. If $\Lambda = 0$, then what value of μ ensures that both sides of (13.8) are trace-free? Hence, propose an alternative Einstein–Maxwell theory.

13.4 (§13.3) Show that flat space is not a solution of (13.5).

13.5 (§13.4)
(i) Show that the conservation equations for a perfect fluid lead to

$$(\rho_0 + p)u^a \nabla_a u^b + (u^a u^b - g^{ab})\nabla_a p = 0$$

(ii) We suppose that $\rho_0 = \rho_0(p)$ and define the following quantities:

$$f = \exp\left(\int \frac{p}{p + \rho_0(p)}\right),$$
$$C_a = fu_a,$$
$$\Omega_{ab} = \tfrac{1}{2}(\nabla_b C_a - \nabla_a C_b).$$

Deduce that $C^a \Omega_{ab} = 0$.

13.6 (§13.5) If g_{ab} is known everywhere on Σ_0, then establish that $g_{ab,\alpha}$ is known everywhere on S.

13.7 (§13.5) Establish the equations (13.21), (13.22), and (13.23).

13.8 (§13.5) Show that the condition $R_{\alpha\beta} = 0$ leads to equations (13.25) and (13.26). [Hint: use the device of breaking up all latin indices into their zero and Greek consituents, e.g. $g^{0a}R_{0a} = g^{00}R_{00} + g^{0\alpha}R_{0\alpha}$, etc.]

13.9 (§13.5) Derive (13.29) and (13.30).

13.10 (§13.5)
(i) Establish (13.31).
(ii) Show that $G_a{}^0$ satisfies a homogeneous differential equation of the form

$$G_a{}^0{}_{,0} = C^{b\alpha}{}_a G_b{}^0{}_{,\alpha} + D^b{}_a G_b{}^0,$$

where $C^{b\alpha}{}_a$ and $D^b{}_a$ only depend on the metric and its first derivatives.
(iii) Assuming that $G_a{}^0$ is an analytic function of x^0, use (ii) above to develop it in a formal power series in x^0. Show that, if $[G_a{}^0]_S = 0$, then $G_a{}^0 \equiv 0$.

13.11 (§13.7) Show that $\Box x^a = 0$ implies that

$$H^a := g^{cd}\Gamma^a_{cd} = 0.$$

[Hint: for *fixed* index a, the coordinate function x^a is just a scalar field which you can write as $x^{(a)}$ to indicate that a is just a label. It then follows that its derivative $\nabla_b x^{(a)}$ is a covector field (rather than a type $(1,1)$ tensor as might first appear). So establish the result for a fixed a and then relax the restriction.]

13.12 (§13.7) Show that, if $g_{\alpha\beta}$ and $g_{\alpha\beta,0}$ satisfy the constraint equations on Σ_0, one can choose $g_{0b,0}$ on Σ_0 so that

$$H^a = 0 \quad \text{on } \Sigma_0.$$

[Hint: first use $H^0 = 0$ to obtain an expression for $g_{00,0}$ in terms of $g_{\alpha\beta}$ and $g_{\alpha\beta,0}$, then use $H^\alpha = 0$ to find $g_{0\alpha,0}$].

13.13 (§13.10) There are a number of calculations in the book and the ensuing exercises which involve long but straightforward calculations and they would benefit from the use of a computer algebra system. The more important systems include MATHEMATICA, MAPLE, REDUCE, MACSYMA and AXIOM, but there are many others. The optional exercise is to investigate whether you wish and are able to gain access to such a system. There are then the questions of their cost, if any, whether your institution (if you have one) has access to them, how much time and effort is involved in learning to use them, and the associated issue of supporting documentation and help facilities. It is worth mentioning that all the postgraduates currently doing gravity research known to us make use of these systems. The system SHEEP, which we reported on in §13.10 for investigating exact solutions in general relativity, is currently freely available from http://www.maths.qmul.ac.uk/ mm/shp/. See also the Living Reviews article of Malcolm MacCallum "Computer algebra in gravity research" at https://link.springer.com/article/10.1007/s41114-018-0015-6 which has many other links.

Further reading

The treatment of the Cauchy problem is based on that in Adler et al. (1975) and Wald (1984). For a comprehensive treatment of the Cauchy problem, see the book by Ringström (2009). A survey of the use of algebraic computing in general relativity is given in the article by d'Inverno (1980), which appears in the Einstein centenary volume edited by Held

(1980). Most of the known solutions to Einstein's equations can be found in the book by Kramer et al. (2009).

Adler, R., Bazin, M., and Schiffer, M. (1975). *Introduction to General Relativity* (2nd edn). McGraw-Hill, New York, NY.

d'Inverno, R.A. (1980). 'Algebraic computing in general relativity', in Held, A., ed., *General Relativity and Gravitation: One Hundred Years after the Birth of Albert Einstein*, vol. 1. Plenum, New York, NY, 491–537.

Held, A. (ed.) (1980). *General Relativity and Gravitation: One Hundred Years after the Birth of Albert Einstein*, vol. 1. Plenum, New York, NY.

Kramer, D., Stephani, H., Herlt E., and MacCallum, M. A. H. (2009). *Exact Solutions of Einstein's Field Equations* (2nd edn). Cambridge University Press, Cambridge.

Ringström, H. (2009). *The Cauchy Problem in General Relativity*. European Mathematical Society, Zurich.

Wald, R. M. (1984). *General Relativity*. University of Chicago Press, Chicago, IL.

The 3+1 and 2+2 formalisms

14

14.1 The geometry of submanifolds

In the previous chapter, we looked at the Cauchy problem using a special coordinate system. In this chapter, we will look at a more geometrical approach to the problem. We will start by looking at the way in which a hypersurface Σ is embedded in a space-time and the geometric data that one needs to reconstruct the curvature of the space-time, from this. The key quantities are the **induced metric** and the **extrinsic curvature** of the hypersurface Σ. We then go on to look at the way in which one can introduce a time function t and use this to slice up the space-time M by a family of spacelike hypersurfaces, or **foliation**, Σ_t. The geometry of the foliation can be described in terms of the **lapse** function, and the extrinsic curvature can be given an alternative description in terms of the rate of change of the induced metric. The final ingredient is to introduce a timelike vector field, or **fibration**, whose integral curves may be used to identify points on neighbouring, t = constant, hypersurfaces. The geometry of the fibration can then be defined in terms of the **shift** vector. The lapse and shift are freely specifiable and encode the fourfold coordinate freedom in describing the geometry. This machinery allows one to view general relativity from a different perspective as a dynamical theory in which Einstein's equations are encoded in a pair of first-order differential equations which describe the way in which the dynamical variables, the induced metric, and the extrinsic curvature, evolve with the time function t. A knowledge of the induced metric and the lapse and shift may be used to reconstruct the space-time metric and hence the geometry of the space-time. A slightly different form of these equations, called the **ADM formalism**, was derived by Arnowitt, Deser, and Misner (1959) from their Hamiltonian formulation of general relativity and formed the basis for what Wheeler (1963) called **geometrodynamics**. In §14.13 we look at an alternative approach called the $2 + 2$ **formalism**, in which one decomposes space-time into two families of spacelike 2-surfaces. This approach identifies the gravitational degrees of freedom in a geometrically transparent way and is also particularly suited to situations where the initial data is given on a null surface.

Introducing Einstein's Relativity. Second Edition. Ray d'Inverno and James Vickers, Oxford University Press.
© Ray d'Inverno and James Vickers (2022). DOI: 10.1093/oso/9780198862024.003.0014

14.2 The induced metric

Let Σ be a smooth spacelike hypersurface. Then the unit normal n^a to Σ is timelike, may be taken to be future pointing, and satisfies

$$n_a n^a = 1. \tag{14.1}$$

We may use the unit normal n^a to construct the **projection operator**

$$B_b{}^a = \delta_b^a - n_b n^a, \tag{14.2}$$

which projects a space-time vector X^a into the hypersurface to give $\bar{X}^a = B_b{}^a X^b$. We can verify that $\bar{X}^a$ is tangent to the surface since

$$\bar{X}^a n_a = B_b{}^a X^b n_a$$
$$= \delta_b^a X^b n_a - n_b n^a X^b n_a$$
$$= X^b n_b - X^b n_b$$
$$= 0.$$

Similarly, if Y_a is a space-time co-vector, then $\bar{Y}_a = B_a{}^b Y_b$ satisfies $\bar{Y}_a n^a = 0$ (exercise). Both these results also follow directly from the fact that

$$B_a{}^b n_b = 0. \tag{14.3}$$

We also note (exercise) that

$$B_a{}^c B_c{}^b = B_a{}^b, \tag{14.4}$$

as one would expect from a projection operator.

We may use $B_a{}^b$ to project tensor fields on M onto tensor fields on Σ. In particular, we may project the 4-dimensional metric g_{ab} onto Σ to obtain the **induced metric**

$$h_{ab} := B_a{}^c B_b{}^d g_{cd}. \tag{14.5}$$

Using (14.1) and (14.2), we find that

$$h_{ab} = g_{ab} - n_a n_b. \tag{14.6}$$

So, for vectors $\bar{X}^a$ and $\bar{Y}^a$ tangent to Σ, we have

$$h_{ab} \bar{X}^a \bar{Y}^b = g_{ab} \bar{X}^a \bar{Y}^b. \tag{14.7}$$

It is often convenient to introduce coordinates

$$(x^a) = (x^0, x^\alpha) = (t, x^1, x^2, x^3), \tag{14.8}$$

adapted to the hypersurface in which the hypersurface is given by $t = 0$ and (x^α), $\alpha = 1, 2, 3$ are coordinates on the hypersurface. It follows that, in these coordinates, a vector $\bar{X}^a$ tangent to the hypersurface has components $(0, \bar{X}^\alpha)$ so that by (14.7)

$$h_{\alpha\beta}\bar{X}^\alpha\bar{Y}^\beta \overset{*}{=} g_{\alpha\beta}\bar{X}^\alpha\bar{Y}^\beta, \tag{14.9}$$

(where Greek indices are summed over 1 to 3). In other words, in these coordinates, $h_{\alpha\beta}$ are just the spatial components of g_{ab}. With our choice of space-time signature for the metric as $(1, -1, -1, -1)$, we see that $h_{\alpha\beta}$ is a metric with signature $(-1, -1, -1)$. It is therefore convenient to define a positive definite Riemannian metric on Σ by

$$\gamma_{ab} = -h_{ab}. \tag{14.10}$$

14.3   The induced covariant derivative

Let $\bar{X}$ be a vector field on Σ. We now extend this to a vector field X on the whole of M in such a way the the vector field remains the same on Σ. We now consider the projection of the space-time covariant derivative onto the hypersurface:

$$B_a{}^c\nabla_c X^b = B_a{}^c\partial_c X^b + B_a{}^c\Gamma^b_{cd}X^d. \tag{14.11}$$

Since X^b is tangent to Σ and $B_a{}^c\partial_c$ only involves tangential derivatives we see that this is well-defined on Σ and does not depend on how we have extended $\bar{X}$ from Σ to M.

Although $B_a{}^c\nabla_c X^b$ is well-defined, it may not be tangent to Σ. We therefore make a further projection and define the **induced covariant derivative** D_a for vector fields on Σ by

$$D_a X^b := B_d{}^b B_a{}^c \nabla_c X^d. \tag{14.12}$$

In a similar way, if $T^{a_1\cdots a_k}{}_{b_1\cdots b_\ell}$ is a tensor field on Σ, then we define

$$D_e T^{a_1\cdots a_k}{}_{b_1\cdots b_\ell} := B_{c_1}{}^{a_1}\cdots B_{c_k}{}^{a_k} B_{b_1}{}^{d_1}\cdots B_{b_\ell}{}^{d_e} B_e{}^f \nabla_f T^{c_1\cdots c_k}{}_{d_1\cdots d_e}. \tag{14.13}$$

In particular, we find

$$\begin{aligned} D_e h_{ab} &= B_a{}^c B_b{}^d B_e{}^f \nabla_f (g_{cd} - n_c n_d) \\ &= -B_a{}^c B_b{}^d B_e{}^f \nabla_f (n_c n_d) \\ &= -B_a{}^c B_b{}^d n_d B_e{}^f \nabla_f (n_c) - B_a{}^c n_c B_b{}^d B_e{}^f \nabla_f (n_d) \\ &= 0, \end{aligned}$$

by (14.3). It also follows (exercise) from the fact that ∇_a is torsion-free that D_a is also torsion-free. Since D_a is a torsion-free covariant derivative that satisfies $D_a h_{bc} = 0$, it must be the (unique) covariant derivative given by the 3-metric h_{ab}. We have thus shown the following:

> The induced covariant derivative D_a on Σ is the metric covariant derivative of the induced metric h_{ab} on Σ.

If we use coordinates adapted to the hypersurface Σ, we see that $D_a X^b$ only has spatial components which are given by

$$D_\alpha X^\beta \stackrel{*}{=} \partial_\alpha X^\beta + \Gamma^\beta_{\alpha\gamma} X^\gamma. \tag{14.14}$$

On the other hand, since D_α is the metric covariant derivative of $h_{\alpha\beta}$, we have

$$D_\alpha X^\beta \stackrel{*}{=} \partial_\alpha X^\beta + {}^{(3)}\Gamma^\beta_{\alpha\gamma} X^\gamma, \tag{14.15}$$

where ${}^{(3)}\Gamma^\beta_{\alpha\gamma}$ are the Christoffel symbols of $h_{\alpha\beta}$. (Note: these are the same as the connection coefficients of $\gamma_{\alpha\beta} = -h_{\alpha\beta}$, since the Christoffel symbols are quadratic in the metric). Comparing (14.14) and (14.15), we see that, in adapted coordinates,

$$\Gamma^\beta_{\alpha\gamma} \stackrel{*}{=} {}^{(3)}\Gamma^\beta_{\alpha\gamma}. \tag{14.16}$$

Returning to the term $B_a{}^c \nabla_c X^b$ given by (14.11), we now take the normal component by contracting with n_b. This gives a term $n_b B_a{}^c \nabla_c X^b$. Since X^b is tangent to Σ, we have $n_b X^b = 0$, so that

$$n_b \nabla_c X^b + X^b \nabla_c n_b = 0.$$

Hence

$$n_b B_a{}^c \nabla_c X^b = -X^b B_a{}^c \nabla_c n_b \tag{14.17}$$
$$= -X^d B_d{}^b B_a{}^c \nabla_c n_b,$$

where $X^b = B_d{}^b X^d$, since X^b is tangent to Σ.

We now define the tensor K_{ab} by

$$K_{ab} := -B_a{}^c B_b{}^d \nabla_c n_d. \tag{14.18}$$

We call K_{ab} the **extrinsic curvature** of the hypersurface, since it measures the way in which the normal co-vector to the hypersurface bends as it moves about the hypersurface. It follows immediately from the definition that

$$K_{ab} n^a = K_{ab} n^b = 0, \tag{14.19}$$

so that K_{ab} defines a tensor field on Σ. It also follows from the fact that the space-time derivative is torsion-free that

$$K_{ab} = K_{ba}. \tag{14.20}$$

Another useful way of thinking of the extrinsic curvature is as the difference between the space-time covariant derivative and the induced covariant derivative for vector fields X^a and Y^a, which, when restricted to Σ, are tangential. It follows from (14.18) that the tangential part of $Y^b \nabla_b X^a$ is just $Y^b D_b X^a$ and from (14.17) that the normal component is given by $K_{ab} X^a Y^b$. So, by splitting $Y^b \nabla_b X^a$ into its tangential and normal parts on Σ, we obtain the very useful equation

$$Y^b \nabla_b X^a = Y^b D_b X^a + (K_{cd} X^c Y^d) n^a, \tag{14.21}$$

which can be written in coordinate-free notation as

$$\nabla_Y X = D_Y X + K(X, Y) \boldsymbol{n}. \tag{14.22}$$

14.4 The Gauss–Codazzi equations

In the previous section, we showed how the space-time metric g_{ab} and the covariant derivative ∇_a can be used to define the induced metric $h_{ab} = -\gamma_{ab}$ and the extrinsic curvature K_{ab} on Σ. We now show that a knowledge of γ_{ab} and K_{ab} is enough to reconstruct all but one component of the space-time curvature. The first result we will derive in §14.5 below is the **Gauss equation**, which gives an expression for the space-time curvature with all the components projected into the hypersurface

$$B_a{}^e B_b{}^f B_c{}^g B_d{}^h R_{efgh} = {}^{(3)}R_{abcd} + K_{ad} K_{bc} - K_{ac} K_{bd}. \tag{14.23}$$

In adapted coordinates, this is just

$$R_{\alpha\beta\mu\nu} \stackrel{*}{=} {}^{(3)}R_{\alpha\beta\mu\nu} + K_{\alpha\nu} K_{\beta\mu} - K_{\alpha\mu} K_{\beta\nu}. \tag{14.24}$$

The second key equation we will derive in §14.6 is the **Codazzi equation**, which gives an expression for the space-time curvature with three components projected onto the hypersurface and one component contracted with n^a

$$B_a{}^e B_b{}^f B_c{}^g n^h R_{efgh} = D_b K_{ac} - D_a K_{bc}, \tag{14.25}$$

which, in adapted coordinates, is just

$$n^d R_{\alpha\beta\gamma d} \stackrel{*}{=} D_\beta K_{\alpha\gamma} - D_\alpha K_{\beta\gamma}. \tag{14.26}$$

The Gauss and Codazzi equations enable us to obtain all the components of the space-time curvature in terms of $\gamma_{\alpha\beta}$ and $K_{\alpha\beta}$ apart from $R_{\alpha 0 \beta 0}$ (exercise). However, this component requires a knowledge of derivatives of **n** in the normal direction, which depends on the way the space-time is sliced up into a foliation by the constant time surfaces, so we will postpone this to §14.8.

By contracting the Gauss and Codazzi equations, we may obtain expressions for the constraint equation $G_{ab}n^a n^b = 0$ in terms of the intrinsic and extrinsic geometry of Σ. Contracting (14.23) with g^{ac} and using $g^{ac} = h^{ac} + n^a n^c$, we obtain

$$B_b{}^f B_d{}^h R_{fh} - n^e n^g B_b{}^f B_d{}^h R_{efgh} = {}^{(3)}R_{bd} - K^a{}_a K_{bd} + K^a{}_d K_{ba}. \qquad (14.27)$$

Contracting again on b and d gives, after a short calculation (exercise),

$$R - 2n^b n^d R_{bd} = {}^{(3)}R - K^a{}_a K^b{}_b + K^{ab}K_{ab}. \qquad (14.28)$$

On the other hand, $G_{ab} = R_{ab} - \frac{1}{2}g_{ab}R$. Contracting this with $n^a n^b$ gives

$$G_{ab}n^a n^b = R_{ab}n^a n^b - 1/2R$$
$$= -1/2\left({}^{(3)}R - K^a{}_a K^b{}_b + K_{ab}K^{ab}\right). \qquad (14.29)$$

So (in vacuum) the 'constraint' $G_{ab}n^a n^b = 0$ becomes

$${}^{(3)}R - (K^a{}_a)(K^b{}_b) + K_{ab}K^{ab} = 0, \qquad (14.30)$$

or, in adapted coordinates,

$${}^{(3)}R - (K^\alpha{}_\alpha)(K^\beta{}_\beta) + K_{\alpha\beta}K^{\alpha\beta} = 0. \qquad (14.31)$$

Note that, in the above, ${}^{(3)}R$ and contractions of K_{ab} are carried out using $h_{\alpha\beta}$. not $\gamma_{\alpha\beta}$. However, the above equation involves an even number of occurrences of $h_{\alpha\beta}$ and its inverse so that one gets the same answer if one were to use the positive definite metric $\gamma_{\alpha\beta} = -h_{\alpha\beta}$.

Contracting the Codazzi equation (14.25) on a and c, one obtains

$$B_b{}^c n^d R_{cd} = D_b K^a{}_a - D_a K^a{}_b. \qquad (14.32)$$

It then follows that $B_a{}^c n^d G_{cd}$ is given by

$$B_a{}^c n^d G_{cd} = B_a{}^c n^d (R_{cd} - \tfrac{1}{2}g_{cd}R)$$
$$= B_a{}^c n^d R_{cd}$$
$$= D_a K^b{}_b - D_b K^b{}_a \qquad (14.33)$$

by (14.32). So that the 'constraint' $B_a{}^c n^d G_{cd} = 0$ is given by

$$D_b K^b{}_a - D_a K^b{}_b = 0, \qquad (14.34)$$

or, in adapted coordinates,

$$D_\beta K^\beta{}_\alpha - D_\alpha K^\beta{}_\beta = 0. \tag{14.35}$$

We therefore see from (14.31) and (14.35) that the $n^a n^b G_{ab}$ and the $n^a B_c{}^b G_{ab}$ components of the Einstein equations only depend on $\gamma_{\alpha\beta}$, $K_{\alpha\beta}$, and their tangential derivatives so are **constraints** on the intrinsic and extrinsic 3-geometry.

14.5 Calculating the Gauss equation

Let Z^a be a vector field tangent to Σ (i.e. $Z^a n_a = 0$); then, since D_a is the covariant derivative of h_{ab} (and also γ_{ab}), we have

$$D_a D_b Z^c - D_b D_a Z^c = {}^{(3)}R^c{}_{dab}Z^d. \tag{14.36}$$

Now

$$\begin{aligned}
D_a D_b Z^c &= D_a(B_b{}^d B_e{}^c \nabla_d Z^e)\\
&= B_a{}^f B_b{}^g B_k{}^c \nabla_f(B_g{}^d B_e{}^k \nabla_d Z^e)\\
&= B_a{}^f B_b{}^g B_k{}^c B_g{}^d B_e{}^k \nabla_f \nabla_d Z^e\\
&\quad + B_a{}^f B_b{}^g B_k{}^c B_e{}^k (\nabla_f B_g{}^d)(\nabla_d Z^e)\\
&\quad + B_a{}^f B_b{}^g B_k{}^c B_g{}^d (\nabla_f B_e{}^k)(\nabla_d Z^e)\\
&= B_a{}^f B_b{}^d B_e{}^c (\nabla_f \nabla_d Z^e) + B_a{}^f B_b{}^g B_e{}^c (\nabla_f B_g{}^d)(\nabla_d Z^e)\\
&\quad + B_a{}^f B_b{}^d B_k{}^c (\nabla_f B_e{}^k)(\nabla_d Z^e),
\end{aligned}$$

using (14.4). Now

$$\nabla_f B_g{}^d = \nabla_f(\delta^d_g - n_g n^d) = -n_g(\nabla_f n^d) - n^d(\nabla_f n_g), \tag{14.37}$$

so that

$$B_a{}^f B_b{}^g(\nabla_f B_g{}^d) = -B_a{}^f B_b{}^g n_g(\nabla_f n^d) - n^d B_a{}^f B_b{}^g \nabla_f n_g = n^d K_{ab}. \tag{14.38}$$

Similarly, we find using (14.3), (14.4), and (14.18) that

$$B_a{}^f B_k{}^c(\nabla_f B_e{}^k) = n_e K_a{}^c. \tag{14.39}$$

Hence,

$$D_a D_b Z^c = B_a{}^f B_b{}^d B_e{}^c(\nabla_f \nabla_d Z^e) + B_e{}^c K_{ab} n^d(\nabla_d Z^e) + B_b{}^d K_a{}^c n_e(\nabla_d Z^e). \tag{14.40}$$

Now the middle term vanishes when we anti-symmetrize on a and b. Since $n_e Z^e = 0$, we also have $n_e(\nabla_d Z^e) = -Z^e \nabla_d n_e$, so that

$$B_b{}^d n_e(\nabla_d Z^e) = -Z^e B_b{}^d \nabla_d n_e = Z^e K_{eb}. \tag{14.41}$$

Putting all this together, we have

$$D_a D_b Z^c - D_b D_a Z^c = B_a{}^f B_b{}^d B_e{}^c (\nabla_f \nabla_d Z^e - \nabla_d \nabla_f Z^e) + Z^e K_a{}^c K_{eb} - Z^e K_b{}^c K_{ea}, \tag{14.42}$$

which immediately gives

$$^{(3)}R^c{}_{fab} Z^f = B_a{}^f B_b{}^d B_e{}^c R^e{}_{hfd} Z^h + Z^f K_a{}^c K_{fb} - Z^f K_b{}^c K_{fa}. \tag{14.43}$$

We now take Z to be an arbitrary vector field (not necessarily tangent to Σ) and replace Z^h by its tangential component $B_f{}^h Z^f$. Then the above equation remains true and, since Z^h is arbitrary, we have (after some relabelling of indices)

$$^{(3)}R_{abcd} = B_a{}^e B_b{}^f B_c{}^g B_d{}^h R_{efgh} + K_{ac} K_{bd} - K_{ad} K_{bc}, \tag{14.44}$$

which is the **Gauss equation** (14.23).

14.6   Calculating the Codazzi equation

To calculate the Codazzi equation, we differentiate K_{ab}, which from (14.18), is given by

$$K_{ab} = -B_a{}^c B_b{}^d \nabla_c n_d.$$

Since K_{ab} is a tensor tangent to the hypersurface Σ, the 3-dimensional covariant derivative $D_a K_{ab}$ is well-defined. Using the definition of D_a (see (14.13)), we get

$$\begin{aligned}
-D_c K_{ab} &= -B_a{}^e B_b{}^f B_c{}^g \nabla_g K_{ef} \\
&= B_a{}^e B_b{}^f B_c{}^g \nabla_g (B_e{}^j B_f{}^k \nabla_j n_k) \\
&= B_a{}^e B_b{}^f B_c{}^g B_f{}^k (\nabla_g B_e{}^j)(\nabla_j n_k) \\
&\quad + B_a{}^e B_b{}^f B_c{}^g B_e{}^j (\nabla_g B_f{}^k)(\nabla_j n_k) \\
&\quad + B_a{}^e B_b{}^f B_c{}^g B_e{}^j B_f{}^k (\nabla_g \nabla_j n_k) \\
&= B_a{}^e B_b{}^k B_c{}^g (\nabla_g B_e{}^j)(\nabla_j n_k) \\
&\quad + B_a{}^j B_b{}^f B_c{}^g (\nabla_g B_f{}^k)(\nabla_j n_k) \\
&\quad + B_a{}^j B_b{}^k B_c{}^g (\nabla_g \nabla_j n_k),
\end{aligned}$$

using (14.4). Now the first two terms cancel when one anti-symmetrizes on a and c, so that

$$\begin{aligned}
D_a K_{bc} - D_c K_{ba} &= B_a{}^j B_b{}^k B_c{}^g (\nabla_g \nabla_j n_k - \nabla_j \nabla_g n_k) \\
&= -B_a{}^j B_b{}^k B_c{}^g R^\ell{}_{kgj} n_\ell, \tag{14.45}
\end{aligned}$$

which, on relabelling of indices, is just the **Codazzi equation** (14.25).

14.7 The geometry of foliations

A **Cauchy surface** is a spacelike hypersurface Σ such that each endless timelike or null curve intersects Σ exactly once (Fig 14.1). A space-time that possesses a Cauchy surface is called **globally hyperbolic**. The name comes from the fact that the wave equation $g^{ab}\nabla_a\nabla_b\phi = 0$ (a hyperbolic PDE) has a unique globally defined solution on such a space-time. From the point of view of Einstein's equations, the important fact about globally hyperbolic space-times is that the manifold M has topology

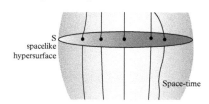

Fig. 14.1 A Cauchy surface.

$$M = \Sigma \times \mathbb{R}, \tag{14.46}$$

so that the manifold can be sliced up into hypersurfaces Σ_t, which are the level surfaces of some time function. We call such a slicing a **foliation**, which we define more precisely below.

By a **foliation** of a manifold, we mean that there exists a smooth scalar field ϕ (our time function)

$$\phi : M \to \mathbb{R}, \tag{14.47}$$

which has non-vanishing gradient and whose level surfaces

$$\Sigma_t = \{x \in M : \phi(x) = t\}, \tag{14.48}$$

give the whole of M, so that

$$M = \bigcup_{t\in\mathbb{R}} \Sigma_t. \tag{14.49}$$

Since ϕ has non-vanishing gradient, it follows that the hypersurfaces Σ_t are non-intersecting, so that each point x on M lies on precisely one hypersurface Σ_t (see Fig. 14.2).

Since we want to consider Einstein's equations, as evolution equations, we will be interested in the case where the hypersurfaces Σ_t are spacelike and so we now specialize to this case. Let n^a be the unit, future-pointing, timelike normal to the constant time surfaces Σ_t. Since we now have a family of hypersurfaces Σ_t rather than a single hypersurface Σ, we may also take derivatives of n^a off the hypersurface Σ_t. This enables us to define a new geometric quantity, $A = \nabla_{\boldsymbol{n}}\boldsymbol{n}$, which in coordinates is given by

$$A^a = n^b\nabla_b n^a. \tag{14.50}$$

As n^a is a unit timelike vector, it can be regarded as tangent to the world-line of an observer. The quantity A^a therefore measures the **acceleration** of such an observer. In the differential geometry literature, A^a is called the **geodesic curvature** of the unit normal curves, since it vanishes when the worldlines are affinely parametrized geodesics. Differentiating the equation $n^a n_a = 1$ and contracting with n^b gives

$$n_a n^b \nabla_b n^a = 0, \tag{14.51}$$

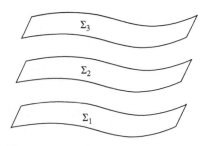

Fig. 14.2 A foliation of space-time by spacelike hypersurfaces Σ_t.

which implies that

$$n_a A^a = 0. \tag{14.52}$$

So that A^a is a spacelike vector tangent to the Σ_t hypersurfaces.

We have seen that we may use geometric quantities defined on the hypersurface, namely the induced metric h_{ab} and the extrinsic curvature K_{ab}, to reconstruct all but one component of the space-time curvature. We now show that A^a is the additional quantity that we need to know on the hypersurface to obtain the missing piece of the space-time curvature. This is given by the **Ricci equation**

$$B_a{}^c n^d B_b{}^e n^f R_{cdef} = \mathbf{L}_{\boldsymbol{n}} K_{ab} + D_a A_b - A_a A_b + K_a{}^c K_{cb}. \tag{14.53}$$

14.8   Derivation of the Ricci equation

Now that we have the definition of the acceleration, we may compute all the components of $\nabla_a n_b$ (not just those in the hypersurface). Then we find (exercise)

$$\nabla_a n_b = -K_{ab} + n_a A_b. \tag{14.54}$$

We are now in a position to compute $\mathbf{L}_{\boldsymbol{n}} K_{ab}$. From the definition of the Lie derivative

$$\mathbf{L}_{\boldsymbol{n}} K_{ab} = n^c \nabla_c K_{ab} + (\nabla_a n^c) K_{cb} + (\nabla_b n^c) K_{ac}. \tag{14.55}$$

Now by (14.54) the last two terms can be written

$$(\nabla_a n^c) K_{cb} + (\nabla_b n^c) K_{ac} = (-K_a{}^c + n_a A^c) K_{cb} + (-K_b{}^c + n_b A^c) K_{ac}. \tag{14.56}$$

Again using (14.54), we may write the first term as (exercise)

$$n^c \nabla_c K_{ab} = -n^c (\nabla_c \nabla_a n_b) + A_a A_b + n_a n^c (\nabla_c A_b). \tag{14.57}$$

Now

$$
\begin{aligned}
n_a n^c (\nabla_c A_b) &= (\delta_a^c - B_a{}^c) \nabla_c A_b \\
&= \nabla_a A_b - B_a{}^c (\nabla_c A_b) \\
&= \nabla_a A_b - D_a A_b - (K_a{}^c A_c) n_b.
\end{aligned} \tag{14.58}
$$

Also using (14.52) and (14.54), we obtain

$$
\begin{aligned}
\nabla_a A_b &= \nabla_a (n^c \nabla_c n_b) \\
&= (\nabla_a n^c)(\nabla_c n_b) + n^c \nabla_a \nabla_c n_b \\
&= (-K_a{}^c + n_a A^c)(-K_{cb} + n_c A_b) + n^c \nabla_a \nabla_c n_b \\
&= K_a{}^c K_{cb} - n_a A^c K_{cb} + n^c (\nabla_a \nabla_c n_b).
\end{aligned} \tag{14.59}
$$

Combining (14.57), (14.58), and (14.59), we obtain for the first term in (14.55)

$$n^c \nabla_c K_{ab} = -n^c(\nabla_c\nabla_a - \nabla_a\nabla_c)n_b + A_a A_b - D_a A_b + K_a{}^c K_{cb}$$
$$- n_a A^c K_{cb} - n_b A^c K_{ca}. \tag{14.60}$$

Using $n^c(\nabla_c\nabla_a - \nabla_a\nabla_c)n_b = -n^c n^d R_{dbca}$ and substituting for (14.56) and (14.60) in (14.55) gives

$$\mathbf{L}_n K_{ab} = n^c n^d R_{dbca} + A_a A_b - D_a A_b - K_a{}^c K_{cb}. \tag{14.61}$$

Note that, because of the symmetries of the Riemann tensor, contracting $n^c n^d R_{dbca}$ with n^a or n^b gives zero and hence $n^c n^d R_{dbca}$ is a tensor in the hypersurface. Equation (14.61) can therefore be written as

$$B_a{}^c n^d B_b{}^e n^f R_{cdef} = \mathbf{L}_n K_{ab} + D_a A_b - A_a A_b + K_a{}^c K_{cb}, \tag{14.62}$$

which is just the Ricci equation. In adapted coordinates, this is

$$n^c n^d R_{\alpha c\beta d} = \mathbf{L}_n K_{\alpha\beta} + D_\alpha A_\beta - A_\alpha A_\beta + K_\alpha{}^\gamma K_{\gamma\beta}. \tag{14.63}$$

14.9 The lapse function

Since the gradient of ϕ is orthogonal to the level surfaces, this corresponds to the case where $w^a = g^{ab}\phi_{,b}$ is a timelike vector (i.e. $g_{ab}w^a w^b > 0$), which, without loss of generality, we may assume to be future pointing. In such a case, ϕ is called a **time function** and we will write the scalar field as $t(x)$ rather than $\phi(x)$. Since w^a is normal to the constant time hypersurfaces, it must be proportional to the future-pointing normal n^a to Σ_t. We may therefore write

$$n^a = N w^a, \tag{14.64}$$

for some scalar function N which is called the **lapse** of the foliation. Since both w^a and n^a are future pointing, the lapse is positive, so that (check)

$$N = \frac{1}{\sqrt{g^{ab} t_{,a} t_{,b}}}. \tag{14.65}$$

Note that, in the numerical relativity literature, where it is common to use the alternative $(-, +, +, +)$ signature, a time function that increases to the future results in a past-pointing $w^a = g^{ab}\nabla_b T$, so that, to ensure that N is positive, n^a is taken to be equal to $-Nw^a$.

Since n^a is a unit timelike vector field, it can be regarded as tangent to the worldline of some observer. We call such an observer an **Eulerian observer** for the foliation. The constant time hypersurfaces Σ_t are then orthogonal to the worldlines of such an observer and consist of events that are considered simultaneous from their point of view.

Let x^a be an event on the hypersurface Σ_t and let $\tilde{x}^a$ be an event at a small proper time $\delta\tau$ later on the worldline of an Eulerian observer (Fig 14.3). Since n^a is a unit vector tangent to the worldline and $\delta\tau$ is small, we may write

$$\tilde{x}^a = x^a + (\delta\tau)n^a. \tag{14.66}$$

To find out what hypersurface $\tilde{x}^a$ is on, we calculate $t(\tilde{x}^a)$

$$
\begin{aligned}
t(\tilde{x}^a) &= t(x^a + (\delta\tau)n^a) \\
&= t(x^a) + \left(\frac{\partial t}{\partial x^b}\right)(\delta\tau)n^b + O\left((\delta\tau)^2\right) \\
&\simeq t(x^a) + (\delta\tau)w_b n^b \\
&= t(x^a) + (\delta\tau)/N,
\end{aligned}
$$

so that $\tilde{x}^a$ lies on the hypersurface $\Sigma_{t+\delta t}$, where $\delta t = \delta\tau/N$. In other words, $\delta\tau = N\delta t$, so that $N \simeq \delta\tau/\delta t$ and in the limit we have

$$N = \frac{\mathrm{d}\tau}{\mathrm{d}t}. \tag{14.67}$$

This tells us that the lapse is just the rate of change of proper time with respect to coordinate time for an Eulerian observer.

Since we now have a family of spacelike hypersurfaces Σ_t rather than a single hypersurface, we can calculate how the induced metric changes in going from one hypersurface to the next. The rate of change of h_{ab} with respect to proper time τ as measured by an Eulerian observer is given geometrically by $L_n h_{ab}$. Now

$$L_{\boldsymbol{n}} h_{ab} = L_{\boldsymbol{n}}(g_{ab} - n_a n_b). \tag{14.68}$$

Calculating the first term gives

$$
\begin{aligned}
L_{\boldsymbol{n}} g_{ab} &= n^c(\nabla_c g_{ab}) + (\nabla_a n^c)g_{cb} + (\nabla_b n^c)g_{ac} \\
&= \nabla_a n_b + \nabla_b n_a, \tag{14.69}
\end{aligned}
$$

since the covariant derivative of the metric vanishes, while the second term gives (exercise)

$$L_{\boldsymbol{n}}(n_a n_b) = A_a n_b + A_b n_a. \tag{14.70}$$

Hence, substituting for (14.69) and (14.70) in (14.68) gives

$$L_{\boldsymbol{n}} h_{ab} = (\nabla_a n_b - n_a A_b) + (\nabla_b n_a - n_b A_a) = -2K_{ab}, \tag{14.71}$$

by (14.54). So, in the case of a foliation, we have

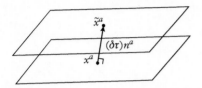

Fig. 14.3 Two hypersurfaces for an Eulerian observer.

$$K_{ab} = -\tfrac{1}{2}\mathrm{L}_{\boldsymbol{n}}h_{ab}. \tag{14.72}$$

Note that, in the numerical relativity literature, this is sometimes taken as the definition of the extrinsic curvature but it relies on the existence of a foliation, while the definition we gave in §14.3 only requires a single hypersurface.

If we introduce coordinates (x^{α}) on some initial hypersurface Σ_0, then we may use the worldlines of the Eulerian observers to give space-time coordinates (t, x^{α}) in a neighbourhood of Σ_0 by requiring that their worldlines are given by $x^{\alpha} = $ constant, $\alpha = 1, 2, 3$. Since in these coordinates the geometric version of $\partial/\partial t$ is $\mathrm{L}_{N\boldsymbol{n}}$, we may write (14.72) as (exercise)

$$\dot{\gamma}_{\alpha\beta} \overset{*}{=} 2NK_{\alpha\beta}, \tag{14.73}$$

where $\gamma_{\alpha\beta} = -h_{\alpha\beta}$ and a dot indicates a derivative with respect to the coordinate time t rather than the proper time τ.

We may also use the lapse to give an expression for the acceleration vector A^b (exercise)

$$A_a = -\frac{1}{N}D_a N. \tag{14.74}$$

Hence,

$$A_a = -D_a(\ln N). \tag{14.75}$$

We can use this to simplify (14.61) since

$$D_a A_b - A_a A_b = -\frac{1}{N}D_a D_b N + \frac{1}{N^2}(D_a N)(D_b N) - \frac{1}{N^2}(D_a N)(D_b N)$$

$$= -\frac{1}{N}D_a D_b N.$$

Substituting this into (14.61) gives

$$\mathrm{L}_{\boldsymbol{n}}K_{ab} = n^c n^d R_{dbca} - K_a{}^c K_{cb} + \frac{1}{N}D_a D_b N - 2K_{ac}K_b{}^c. \tag{14.76}$$

We may then use the contracted Gauss equation (14.27) to replace the term $n^c n^d R_{dbca}$ by the space-time Ricci tensor. This gives an evolution equation for the extrinsic curvature in terms of quantities defined in the hypersurface

$$\mathrm{L}_{\boldsymbol{n}}K_{ab} = B_a{}^c B_b{}^d R_{cd} - {}^{(3)}R_{ab} + K_c{}^c K_{ab} + \frac{1}{N}D_a D_b N - 2K_{ac}K_b{}^c. \tag{14.77}$$

14.10   The 3+1 decomposition of the metric

In the previous section, we introduced observers who move along world-lines that are normal to the foliation and used them to construct adapted coordinates (t, x^α). However, although this is geometrically convenient, it is both physically and mathematically a restriction. We therefore now consider an arbitrary congruence of curves which are nowhere tangent to the leaves of the foliation Σ_t and with the property that every point on M lies on precisely one such curve. This is called a **fibration** in differential geometry. Since the curves are nowhere tangent to the constant time surfaces, we may parameterize these curves by t and define T^a to be the corresponding tangent vector. Such a vector field is called a **rigging vector**. We may now project T^a into components normal and tangential to Σ_t and write

$$T^a = \alpha n^a + \beta^a, \tag{14.78}$$

where $\alpha = T^a n_a$ and $\beta^a = B_b{}^a T^b$ (Fig. 14.4). Note that, by the definition of T^a, moving the hypersurface Σ_t forward by $\delta t\, T^a$ results in the hypersurface $\Sigma_{t+\delta t}$, so that α is just the lapse

$$\alpha = N = \frac{1}{\sqrt{g^{ab} t_{,a} t_{,b}}}. \tag{14.79}$$

The quantity β^a is called the **shift**. By definition, it is tangent to the hypersurfaces so that

$$\beta^a n_a = 0. \tag{14.80}$$

Writing equation (14.78) in the form

$$n^a = \frac{1}{N}(T^a - \beta^a), \tag{14.81}$$

and recalling that $g^{ab} = h^{ab} + n^a n^b$, we find

$$g^{ab} = \frac{1}{N^2}(T^a - \beta^a)(T^b - \beta^b) + h^{ab}. \tag{14.82}$$

We now introduce coordinates adapted to the foliation and rigging as follows. Let (x^α), $\alpha = 1, 2, 3$, be coordinates on some initial slice Σ_0. Then we may extend these to coordinates (x^0, x^α) on M by defining the curves in the fibration to be given, in these coordinates, by $x^0 = t$, $x^\alpha =$ constant, $\alpha = 1, 2, 3$.

It then follows that, in these coordinates,

$$T = \frac{\partial}{\partial t}, \tag{14.83}$$

so that

$$T^a \stackrel{*}{=} \delta^a_0. \tag{14.84}$$

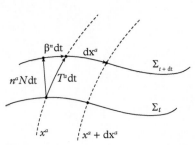

Fig. 14.4 A fibration showing the rigging vector T^a decomposed into the lapse and shift.

Since β^a and h^{ab} have purely spatial components given by β^μ and $h^{\mu\nu}$, it follows that

$$\beta^a \overset{*}{=} \delta^a_\mu \beta^\mu, \tag{14.85}$$

$$h^{ab} \overset{*}{=} \delta^a_\mu \delta^a_\nu h^{\mu\nu}. \tag{14.86}$$

Substituting these into (14.82) gives

$$g^{ab} = \begin{pmatrix} 1/N^2 & -\beta^\mu/N \\ -\beta^\mu/N & -\gamma^{\mu\nu} + \beta^\mu\beta^\nu/N^2 \end{pmatrix}, \tag{14.87}$$

where $\gamma^{\mu\nu} := -h^{\mu\nu}$.

We note from taking the determinant of g^{ab} that $\det g^{ab} = -\det\gamma^{\mu\nu}/N^2$ so that $\det(\gamma^{\mu\nu}) = -N^2\det(g^{ab})$ and $\gamma^{\mu\nu}$ is invertible. If we now define $\gamma_{\mu\nu}$ by

$$\gamma_{\mu\nu}\gamma^{\nu\sigma} = \delta^\sigma_\mu \tag{14.88}$$

and use $\gamma^{\mu\nu}$ and $\gamma_{\mu\nu}$ to raise and lower Greek indices, then, inverting (14.87) using $g_{ab}g^{bc} = \delta^c_a$, we obtain

$$g_{ab} = \begin{pmatrix} N^2 + \beta^\sigma\beta_\sigma & -\beta_\mu \\ -\beta_\mu & -\gamma_{\mu\nu} \end{pmatrix}. \tag{14.89}$$

Hence we find the line element (using our choice of metric signature) is given by

$$ds^2 = (N^2 + \beta^\sigma\beta_\sigma)dt^2 - 2\beta_\mu dt dx^\mu - \gamma_{\mu\nu}dx^\mu dx^\nu. \tag{14.90}$$

Notice that, in these coordinates, $g_{\alpha\beta} = -\gamma_{\alpha\beta}$ (where the minus sign has been introduced to make $\gamma_{\alpha\beta}$ a positive definite Riemannian metric). Note, however, that $g^{\alpha\beta} \neq -\gamma^{\alpha\beta}$ unless we choose coordinates in which the shift vanishes.

14.11   The 3+1 decomposition of the vacuum Einstein equations

So far, we have adopted a purely geometric approach in which we have given a decomposition of the metric relative to a foliation of a space-time by spacelike hypersurfaces transvected by a timelike fibration. The extrinsic curvature of a given $\Sigma(t)$ also describes the manner in which the hypersurface is embedded in the enveloping 4-geometry.

An alternative 3+1 viewpoint regards general relativity as a dynamical theory in which space time is comprised of the 'time history' of a spacelike hypersurface $\Sigma(t)$ regarded as an 'instant of time'. The geometry of $\Sigma(t)$ is described by the 3-metric $\gamma_{\mu\nu}$ while the rate of change of $\gamma_{\mu\nu}$ gives an expression for the extrinsic curvature of $\Sigma(t)$. These together with the acceleration enable one to reconstruct the 4-metric g_{ab}, which gives the geometry of the space-time. There are two types of variables: the four functions comprising the lapse α and shift β are kinematical and are freely specifiable, since they embody the fourfold coordinate freedom of general relativity. The six functions comprising the 3-metric $\gamma_{\mu\nu}$ are the dynamical variables. We now show how these dynamical equations can be can be formulated as a pair of partial differential equations which are first order in time.

The first step is to split the ten vacuum Einstein equations $G_{ab} = 0$ into three parts. The first is a part where we project in the direction normal to the hypersurface

$$n^a n^b G_{ab} = 0. \tag{14.91}$$

Using the twice-contracted Gauss equation (14.29), we have shown this can be written

$$^{(3)}R - (K^a{}_a)(K^b{}_b) + K_{ab}K^{ab} = 0. \tag{14.92}$$

This only depends on the intrinsic and extrinsic geometry of the hypersurface and, in adapted coordinates, is given by

$$^{(3)}R - (K^\alpha{}_\alpha)(K^\beta{}_\beta) + K_{\alpha\beta}K^{\alpha\beta} = 0, \tag{14.93}$$

where $^{(3)}R$ is the scalar curvature of $\gamma_{\alpha\beta}$ and the Greek indices in the above equation are raised using $\gamma^{\alpha\beta}$.

The second part involves projecting one component normal to the hypersurface and one component into the hypersurface

$$B_a{}^c n^d G_{cd} = 0. \tag{14.94}$$

Using the contracted Codazzi equation (14.33), we have shown that this can be written

$$D_b K^b{}_a - D_a K^b{}_b = 0. \tag{14.95}$$

Again, this only depends on the intrinsic and extrinsic geometry of the hypersurface and, in adapted coordinates, is given by

$$D_\beta K^\beta{}_\alpha - D_\alpha K^\beta{}_\beta = 0, \tag{14.96}$$

where D_α is the metric covariant derivative of $\gamma_{\alpha\beta}$ and the Greek indices are raised with $\gamma^{\alpha\beta}$.

The third part involves projecting both components of the Einstein equations into the hypersurface. However, rather than work with the projection of $G_{ab} = 0$, it is mathematically more convenient to work with the projection of $R_{ab} = 0$, which gives

$$B_a{}^c B_b{}^d R_{cd} = 0. \tag{14.97}$$

As previously shown in §13.6, equations (14.91), (14.94), and (14.97) are equivalent to $G_{ab} = 0$ so there is no problem in using (14.97) rather than $B_a{}^c B_b{}^d G_{cd} = 0$. We have already shown in (14.77) that the Ricci equation gives rise to a propagation equation for K_{ab}

$$L_n K_{ab} = -{}^{(3)}R_{ab} + K_c{}^c K_{ab} - 2K_{ac}K^c{}_b + \frac{1}{N}D_a D_b N. \qquad (14.98)$$

We also have a propagation equation for h_{ab}

$$L_n h_{ab} = -2K_{ab}, \qquad (14.99)$$

which together with (14.98) may be regarded as a second-order evolution equation for the 3-metric $\gamma_{ab} = -h_{ab}$. However, to use these equations in computations, we need to write them in the form of partial differential equations rather than in the geometrical form given above. The key to doing this is to note that the Lie derivative with respect to the vector field T^a is just the partial derivative $\partial/\partial t$ in the adapted coordinates. We therefore look at modified versions of (14.98) and (14.99) which involve L_T rather than L_n. We start with equation (14.99). Using the fact that $T^a = Nn^a + \beta^a$, we have

$$L_T h_{ab} = L_{Nn} h_{ab} + L_\beta h_{ab}. \qquad (14.100)$$

Now for the first term we have

$$
\begin{aligned}
L_{Nn} h_{ab} &= Nn^c(\nabla_c h_{ab}) + (\nabla_a(Nn^c))h_{cb} + (\nabla_b(Nn^c))h_{ac} \\
&= Nn^c(\nabla_c h_{ab}) + N(\nabla_a n^c)h_{cb} + n^c(\nabla_a N)h_{cb} + N(\nabla_b n^c)h_{ac} \\
&\quad + n^c(\nabla_b N)h_{ac} \\
&= Nn^c(\nabla_c h_{ab}) + N(\nabla_a n^c)h_{cb} + N(\nabla_b n^c)h_{ac} \\
&= NL_n h_{ab} \\
&= -2NK_{ab},
\end{aligned}
$$

using $n^a h_{ab} = 0$ from (14.5). For the $L_\beta h_{ab}$ term, we expand the Lie derivative using the 3-dimensional covariant derivative D_a and use the fact that $D_c h_{ab} = 0$. This gives

$$
\begin{aligned}
L_\beta h_{ab} &= \beta^c D_c h_{ab} + (D_a \beta^c)h_{cb} + (D_b \beta^c)h_{ac} \\
&= D_a \beta_b + D_b \beta_a.
\end{aligned}
$$

Combining these results, we have

$$L_T h_{ab} = -2NK_{ab} + D_a \beta_b + D_b \beta_a. \qquad (14.101)$$

In the adapted coordinates, this is

$$\frac{\partial \gamma_{\mu\nu}}{\partial t} \stackrel{*}{=} 2NK_{\mu\nu} - D_\mu \beta_\nu - D_\nu \beta_\mu. \qquad (14.102)$$

We thus obtain an expression for $K_{\mu\nu}$ in terms of derivatives of hypersurface quantities

$$K_{\mu\nu} = \frac{1}{2N}\left(\frac{\partial\gamma_{\mu\nu}}{\partial t} + D_\mu\beta_\nu + D_\nu\beta_\mu\right).\tag{14.103}$$

We now compute $L_T K_{ab}$. This is given by

$$L_{Nn+\beta}K_{ab} = L_{Nn}K_{ab} + L_\beta K_{ab}.\tag{14.104}$$

Now

$$\begin{aligned}
L_{Nn}K_{ab} &= Nn^c\nabla_c K_{ab} + (\nabla_a(Nn^c))K_{cb} + (\nabla_b(Nn^c))K_{ac}\\
&= Nn^c\nabla_c K_{ab} + N(\nabla_a n^c)K_{cb} + n^c(\nabla_a N)K_{cb} + N(\nabla_b n^c)K_{ac}\\
&\quad + n^c(\nabla_b N)K_{ac}\\
&= Nn^c\nabla_c K_{ab} + N(\nabla_a n^c)K_{cb} + N(\nabla_b n^c)K_{ac}\\
&= NL_n K_{ab}\\
&= D_a D_b N - N\left({}^{(3)}R_{ab} + 2K_{ac}K^c{}_b - K^c{}_c K_{ab}\right),
\end{aligned}$$

using (14.19) and (14.98). For the $L_\beta K_{ab}$ term, we have

$$L_\beta K_{ab} = \beta^c D_c K_{ab} + (D_a\beta^c)K_{cb} + (D_b\beta^c)K_{ac}.\tag{14.105}$$

Combining these, we have

$$\begin{aligned}
L_T K_{ab} = {}& D_a D_b N + \beta^c D_c K_{ab} + (D_a\beta^c)K_{cb} + (D_b\beta^c)K_{ac}\\
& - N\left({}^{(3)}R_{ab} + 2K_{ac}K^c{}_b - K^c{}_c K_{ab}\right),
\end{aligned}\tag{14.106}$$

which in the adapted coordinates becomes

$$\begin{aligned}
\frac{\partial K_{\mu\nu}}{\partial t} \overset{*}{=} {}& D_\mu D_\nu N + \beta^\rho D_\rho K_{\mu\nu} + (D_\mu\beta^\rho)K_{\rho\nu} + (D_\nu\beta^\rho)K_{\mu\rho}\\
& - N\left({}^{(3)}R_{\mu\nu} + 2K_{\mu\rho}K^\rho{}_\nu - K^\rho{}_\rho K_{\mu\nu}\right).
\end{aligned}\tag{14.107}$$

This gives the time derivative of $K_{\mu\nu}$ in terms of quantities in the hypersurface. If one now uses (14.103) to substitute for $K_{\mu\nu}$ in the above, one sees that this gives a second-order evolution equation for $\gamma_{\mu\nu}$.

Notice that, with this 3+1 decomposition of Einstein's equations, we do not have evolution equations for the lapse N or the shift β^a; instead, these are 'gauge' quantities that can be thought of as defining the coordinate system rather than as dynamical variables that describe the geometry of space-time. Indeed, the lapse and shift encode essentially the same information as g_{0a}, which, as we saw from §13.6, is part of the data we need to specify to find an analytic solution.

The procedure for solving the Cauchy problem for the vacuum Einstein equations is therefore as follows. One first fixes a coordinate system x^α on Σ_0 and finds initial data $\gamma_{\mu\nu}(0)$ and $K_{\mu\nu}(0)$ that solve the constraint equations (14.93) and (14.96). Note that this is a non-trivial task (see brief discussion below). One then specifies the lapse and shift, which extends the coordinates x^α to give coordinates x^a on a neighbourhood of Σ_0. This enables one to evolve $\gamma_{\mu\nu}$ and $K_{\mu\nu}$ forward in time, for some interval $0 \leqslant t \leqslant C$, as a pair of first-order evolution equations, using (14.102), (14.107), and the initial data. The Bianchi identities show that, if the constraints are initially satisfied, they are satisfied at future times so we now have $\gamma_{\mu\nu}(t)$ and $K_{\mu\nu}(t)$ which satisfy (14.91), (14.94), and (14.97) for $0 \leqslant t \leqslant C$ and hence satisfy $G_{ab} = 0$ on that interval.

Writing the Einstein equations in this way was first derived by Darmois, as early as 1927, in the special case $N = 1$ and $\beta^a = 0$ i.e. these are the so-called **Gaussian normal coordinates** corresponding to the motion of Eulerian observers with the foliation parametrized by proper time. The case $N \neq 1$, but still with $\beta^a = 0$, was considered by Lichnerowicz in 1939 and the general case with arbitrary lapse and shift by Choquet-Bruhat in 1948. A slightly different form, with K_{ab} replaced by the 'momentum conjugate to γ_{ab}', namely $p^{ab} := \sqrt{\gamma}(K\gamma^{ab} - K^{ab})$, was derived by Arnowitt, Deser, and Misner from their Hamiltonian formulation of general relativity in 1959. This is the origin of the term Hamiltonian constraint and momentum constraint for equations (14.30) and (14.34) and explains why this description of the Einstein equations as a first-order system is sometimes called the **ADM formalism**.

In the previous chapter, we outlined the proof of the existence of solutions to the coordinate version of the Cauchy problem in the analytic case by means of the Cauchy-Kowalevskya theorem. A similar strategy in the analytic case also works with the above formulation of the equations, as shown by Darmois and Lichnerowicz. However, from the physical point of view, asking for analytic initial data seems an unreasonable requirement. Unfortunately, it is much harder to prove existence and uniqueness with non-analytic initial data but the above formalism is the basis for writing Einstein's equations as a well-posed initial value problem for which one can apply standard theorems from the theory of partial differential equations.

14.12 The 3+1 equations and numerical relativity

In this section, we will review the current status of the 3+1 formalism without considering the calculational details. The 3+1 approach has been used extensively in **numerical relativity**, that is, solving Einstein's equations numerically on a computer, and has played a key role in our current understanding of gravitational collapse and gravitational waves.

We begin by discussing how to solve the constraint equations. Unlike the case for the rest of the book we will mostly just be quoting results from the literature in this section. One of the most useful ways of solving this is the **conformal approach** of York and O'Murchadha. The key idea here is to introduce conformal scalings so that the constraint equations are cast into a set of four quasilinear elliptic partial differential equations for four gravitational 'potentials'. This idea facilitates both theoretical analysis as well as providing a numerical technique. We start by introducing a conformal factor ψ and write the 3-metric $\gamma_{\mu\nu}$ in the form

$$\gamma_{\mu\nu} = \psi^4 \hat{\gamma}_{\mu\nu}. \tag{14.108}$$

The conformal factor is one of the 'potentials' which will be fixed by the Hamiltonian constraint (14.30). Then, among other things, the scalar curvature transforms as

$$R = \psi^{-4}\hat{R} - 8\psi^{-5}(\hat{D}^\mu \hat{D}_\mu)\psi. \tag{14.109}$$

We then perform a so-called **transverse traceless** decomposition of the extrinsic curvature tensor, which introduces three additional 'potentials' X^μ which will be fixed by the momentum constraints. Defining the trace-free part of the extrinsic curvature by

$$A^{\mu\nu} = K^{\mu\nu} - \frac{1}{3}\gamma^{\mu\nu}K, \tag{14.110}$$

then the choice

$$A^{\mu\nu} = \psi^{-10}\hat{A}^{\mu\nu}, \tag{14.111}$$

results in the property

$$D_\nu A^{\mu\nu} = \psi^{-10}\hat{D}_\nu \hat{A}^{\mu\nu}. \tag{14.112}$$

As with any traceless symmetric tensor, $A^{\mu\nu}$ can be decomposed into a part $\hat{A}_{TT}^{\mu\nu}$ with vanishing divergence and trace, and another trace-free part which can be obtained from differentiating a vector potential W^ν, namely

$$\hat{A}^{\mu\nu} = \hat{A}_{TT}^{\mu\nu} + (\hat{\ell}W)^{\mu\nu}, \tag{14.113}$$

where

$$(\hat{\ell}W)^{\mu\nu} = \hat{D}^\mu W^\nu + \hat{D}^\nu W^\mu - \frac{2}{3}\hat{\gamma}^{\mu\nu}\hat{D}_\sigma W^\sigma \tag{14.114}$$

and the TT (Transverse-Traceless) part $\hat{A}_{TT}^{\mu\nu}$ satisfies

$$\hat{D}_\nu \hat{A}_{TT}^{\mu\nu} = 0. \tag{14.115}$$

In practice, it will generally be inconvenient to give the freely specifiable part of the conformally scaled extrinsic curvature in terms of a transverse-traceless tensor. So we 'reverse decompose' $\hat{A}_{TT}^{\mu\nu}$ as

$$\hat{A}_{TT}^{\mu\nu} = \hat{T}^{\mu\nu} - (\hat{\ell}V)^{\mu\nu}, \tag{14.116}$$

where the traceless, symmetric tensor $\hat{T}^{\mu\nu}$ is freely specifiable and V^ν is another vector field. Then

$$\hat{A}^{\mu\nu} = \hat{T}^{\mu\nu} + (\hat{\ell}W)^{\mu\nu} - (\hat{\ell}V)^{\mu\nu}$$

$$= \hat{T}^{\mu\nu} + (\hat{\ell}X)^{\mu\nu}, \quad \text{where } X = W - V.$$

The Hamiltonian and momentum constraints become, in this approach,

$$\hat{\Delta}\psi := (\hat{D}^\nu \hat{D}_\nu)\psi = \frac{1}{8}\hat{R}\psi + \frac{1}{12}K^2\psi^5 - \frac{1}{8}\left(\hat{T}^{\mu\nu} + (\hat{\ell}X)^{\mu\nu}\right)^2 \psi^{-7}, \tag{14.117}$$

$$(\hat{\Delta}_\ell)^\mu := \hat{D}_\nu(\hat{\ell}X)^{\mu\nu} = -\hat{D}_\nu \hat{T}^{\mu\nu} + \frac{2}{3}\psi^6 \hat{D}^\mu K. \tag{14.118}$$

These equations are a set of four quasilinear, coupled elliptic PDEs for the four gravitational potentials $\{\psi, X^\nu\}$. So, to summarize, the procedure is:

- freely specify $\{\hat{\gamma}^{\mu\nu}, K, \hat{T}^{\mu\nu}\}$,
- solve the constraints for the potentials $\{\psi, X^\nu\}$,
- construct physical initial data using

$$\gamma_{\mu\nu} = \psi^4 \hat{\gamma}_{\mu\nu}, \tag{14.119}$$

$$K^{\mu\nu} = \left(\hat{T}^{\mu\nu} + (\hat{\ell}X)^{\mu\nu}\right)\psi^{-10} + \frac{1}{3}K\psi^{-4}\hat{\gamma}^{\mu\nu}. \tag{14.120}$$

A particularly simple choice for solving the constraints is:

- introduce Cartesian coordinates (x, y, z) and take the conformal metric to be flat i.e. $\hat{\gamma}_{\mu\nu} = \text{diag}(1, 1, 1)$,
- choose the initial slice to be maximal i.e. $K = K_\mu{}^\mu = 0$,
- choose a minimal radiation condition i.e. $\hat{T}^{\mu\nu} = 0$.

With the above choices, the constraints become

$$\hat{\Delta}\psi = -18(\hat{\ell}X)^{\mu\nu}(\hat{\ell}X)_{\mu\nu}\psi^{-7}, \tag{14.121}$$

$$(\hat{\Delta}_\ell X)^\mu = 0. \tag{14.122}$$

This greatly simplifies the problem since the momentum constraint is decoupled from the Hamiltonian constraint and is linear. It has been possible this way to even find analytic solutions of the momentum constraints, for example, corresponding to one or more black holes with freely specified linear and angular momentum.

The second issue we need to address is the choice of **gauge conditions** for the choice of lapse and shift. There are two requirements which motivate the choice of a particular gauge:

- The avoidance of both coordinate and physical singularities: The latter can be avoided by slowing down the evolution of the spatial region near the singularity: This is controlled by the lapse N.
- Making the Einstein evolution equations as simple as possible, so that a numerical solution is not unduly complicated: This is often controlled by the shift β^a. For example, a good choice can lead to several of the components of $\gamma_{\mu\nu}$ vanishing, which reduces the size of expressions such as that for $^{(3)}R$.

Choices for the lapse are as follows:

Geodesic slicing. If $N = 1$ is combined with $\beta^a = 0$ then we are using Eulerian observers who are freely falling. The spatial hypersurfaces are geodesically parallel. This slicing is singularity seeking.

Lagrangian slicing. In spherical symmetry, we can use $NU^0 = 1$ combined with $\beta^\mu = 0$, where U^0 is the time-component of the fluid 4-velocity. Then $U_\mu = 0$ and fluid world-lines are orthogonal to the spatial hypersurfaces (which is possible because there is no vorticity). Since $\beta^\mu = 0$, the coordinates follow the matter. The fluid world-lines will focus towards any singularity and so this slicing is again singularity seeking.

Maximal slicing. We can avoid the focusing of world-lines towards singularities by choosing N in such a way that $K = K^\alpha{}_\alpha$ remains zero on each hypersurface if it is zero on the initial one. Physically, K measures the expansion of a congruence of world-lines normal to the foliation. Substituting both $K = 0$ and $\partial K/\partial t = 0$ into the evolution equations gives

$$\gamma^{\mu\nu} D_\mu D_\nu \alpha - \alpha^{(3)} R = 4\pi(S - 3\rho). \tag{14.123}$$

This is an elliptic equation for α to be solved on each slice $\Sigma(t)$, which means that it can be computationally expensive.

Choices for the shift are as follows:

Eulerian gauge. We simply set $\beta^\mu = 0$ so that the coordinate congruence is normal to the foliation. Early stellar collapse codes used this gauge.

Lagrangian gauge. For space-times containing matter, we can set $\beta^\mu = U^\mu/U^0$, where U^a is the fluid 4-velocity. Thus, the coordinate congruence coincides with the congruence of the fluid world-lines. For one-dimensional flows, this is a convenient choice but, in two or three dimensions, where vorticity may be present, the coordinate grid can become severely distorted, leading to a loss of accuracy.

Isothermal and radial gauge. A particular choice of β^μ can simplify Einstein's equations by making certain components of the 3-metric zero. Three conditions on β^μ enable three components to be eliminated, for example giving a diagonal line element

$$d\sigma^2 = \gamma_{xx}dx^2 + \gamma_{yy}dy^2 + \gamma_{zz}dz^2. \tag{14.124}$$

In an isothermal gauge, $\gamma_{r\theta} = 0$ and $\gamma^{rr} = \gamma^{\theta\theta}$. In a radial gauge, $\gamma_{r\theta} = \gamma_{r\phi} = 0$ and $\gamma_{\theta\theta}\gamma_{\phi\phi} - \gamma_{\theta\phi}^2 = r^4 \sin^2\theta$, and the metric has the form

$$d\sigma^2 = A^2 dr^2 + r^2 B^{-2} d\theta^2 + r^2 B^2 (\sin\theta d\phi + \xi d\theta)^2, \tag{14.125}$$

where A, B, and ξ are metric functions to be determined. This is a particularly useful gauge for gravitational radiation in asymptotically flat space-times.

14.13 The 2+2 and characteristic approaches

There are two major limitations of the 3+1 approach. The first is that the initial data is not freely specifiable, but must satisfy the constraints. The conformal approach is a powerful technique for achieving this but does not reveal what the freely specifiable initial data i.e. the true **gravitational degrees of freedom** are in clear geometric terms.

Let us do some counting in the 3+1 regime. Restricting attention to the vacuum case, we start off with ten field unknowns, namely the components of the 4-dimensional metric, and ten vacuum field equations. However, four of these unknowns may be prescribed arbitrarily because of the fourfold coordinate field freedom, leaving six components of the 4-metric freely specifiable. Moreover, the field equations are not independent but satisfy four differential constraints, namely the contracted Bianchi identities. The Lichnerowicz lemma reveals that, if the constraints are satisfied initially and the evolution equations hold generally, then the constraints are satisfied for all time by virtue of the contracted Bianchi identities. Then, as we have seen in the 3+1 first-order formulation of the initial value problem, we end up needing to specify on an initial slice the six components of the 3-metric $\gamma_{\mu\nu}$ together with the six variables $K_{\mu\nu}$ subject to the four constraints. So this leaves eight variables freely specifiable. However, there exists a threefold coordinate freedom within the initial slice. This can be used, for example, to specify three of the $\gamma_{\mu\nu}$. This leaves five variables free. Finally, there is a condition which describes the embedding of the initial slice into the 4-geometry. This is a little harder to see, but we have already met examples of conditions like this such as maximal slicing $K = 0$. The point is that relationships like these are constraints between the $\gamma_{\mu\nu}$ and the $K_{\mu\nu}$ which encode the nature of the embedding. Such a constraint finally reduces the number of freely specifiable data to 4. These can be thought of, from a Lagrangian point of view, as being two q's and two $\dot{q}$' that is, two pieces of information encoded in the metric and two pieces in its time derivative (or, equivalently, two pieces of information in the extrinsic curvature). It is in this sense that we say the gravitational field has two dynamical degrees of freedom. But what are they explicitly in the 3+1 case? Moreover, why are there six evolution equations rather than the two you would expect for a system with two degrees of freedom? The 2+2 approach answers these questions in a transparent way.

The second problem is that the 3 + 1 approach fails if the foliation becomes null and, furthermore, null foliations are important in their own right, as we shall see when discussing gravitational radiation in Chapter 23.

The basis of the 2+2 approach is to decompose space-time into two families of spacelike 2-surfaces. We can view this as a constructive procedure in which an initial 2-dimensional submanifold S is chosen in a bare manifold, together with two vector fields $\boldsymbol{n}_0$ and $\boldsymbol{n}_1$ which transvect the submanifold everywhere (Fig. 14.5). The two vector fields can then be used to drag the initial 2-surface out into two foliations of 3-surfaces. The character of these 3-surfaces will depend in turn on the character of the two vector fields. The most important cases are when at least one of the vector fields is taken to be null. For example, if both vector fields are null, we obtain a double-null foliation (indicated schematically in Fig. 14.6) or, if one is null and the other is timelike, we obtain a null-timelike foliation (Fig. 14.7).

The most elegant way of proceeding is to introduce a formalism which is manifestly covariant and which uses projection operators and Lie derivatives associated with the two vector fields. The resulting formalism is called the **2+2 formalism** (d'Inverno and Smallwood 1980). When the vector fields are of a particular geometric character, then this can be refined further into a 2+(1+1) formalism. Finally, in analogy to the conformal approach of the last chapter, one extracts a conformal factor from the spacelike 2-geometries to isolate the gravitational degrees of freedom.

The resulting formalism leads to a number of advantages. First of all, it identifies the two gravitational degrees of freedom in an explicit geometrical way as residing in the conformal 2-geometry (d'Inverno and Stachel 1978). Secondly, the data is unconstrained and satisfies two dynamical equations which are simply ordinary differential equations along the vector fields. Most importantly, the formalism applies to situations where the foliation either is null or becomes null. Such initial value problems are called null or **characteristic initial value problems**. They are the natural vehicle for studying gravitational radiation problems (since gravitational radiation propagates along null geodesics), asymptotics of isolated systems (since future and past null infinity are null hypersurfaces), and problems in cosmology (since we gain information about the universe along our past null cone). From a calculational viewpoint,

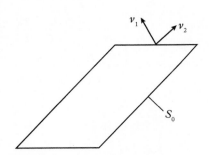

Fig. 14.5 2-dimensional manifold and two transvecting submanifolds.

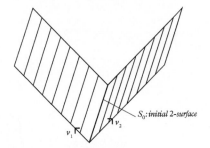

Fig. 14.6 Double null foliation.

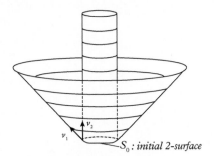

Fig. 14.7 Null timelike foliation.

this formalism allows null infinity to be incorporated into the calculational domain and so allows one to define gravitational radiation in an
unambiguous manner.

We shall not develop the 2+2 formalism in the same detail as we did
with the 3+1 formalism in the previous section. This is partly because it is
rather complicated looking at first sight. However, much of the procedure
is analogous to that of the 3+1 decomposition, and largely rests on the
use of projection operators and Lie derivatives. The only new entities are
tensors which essentially encode 2-dimensional Lie derivatives. However,
we shall look in detail at the 2+2 decomposition of the metric so we can
compare it with the 3+1 case.

14.14 The 2+2 metric decomposition

In the 3+1 formalism, we slice up space-time into spacelike hypersurfaces
which are the level surfaces of a time function. In the 2+2 formalism, we
instead slice up space-time using a 2-parameter family of 2-surfaces. We
can think of these 2-surfaces as being obtained from the intersection of
two hypersurfaces Σ^0 and Σ^1, which may be defined as the level surfaces
of two scalar functions ϕ^0 and ϕ^1. We then define

$$\Sigma^0(u) = \{x \in M : \phi^0(x^a) = u = \text{constant}\},$$

$$\Sigma^1(v) = \{x \in M : \phi^1(x^a) = v = \text{constant}\},$$

where we use the bold numbers **0** and **1** to label the hypersurfaces and
associated geometric quantities. We now assume that these hypersurfaces
intersect to define a family of 2-surfaces by

$$\mathcal{S}(u,v) = \Sigma^0(u) \cap \Sigma^1(v),$$

and restrict attention to the case when $\mathcal{S}(u,v)$ is spacelike. At each point
on $\mathcal{S}(u,v)$, we may define the set of tangent directions $\{S\}$ and a set of
directions orthogonal to $\{S\}$ which we call $\{T\}$ (see Fig. 14.8). We say
that $\{T\}$ is **integrable** if we can find a 2-surface $\mathcal{T}$ such that the vectors
in $\{T\}$ are all tangent to $\mathcal{T}$. However, as we now show, in general $\{T\}$ will
not be integrable.

We first use ϕ^0 and ϕ^1 to define co-vectors normal to Σ^0 and Σ^1,
respectively, by

$$n_a^0 = \frac{\partial \phi^0}{\partial x^a} \quad \text{and} \quad n_a^1 = \frac{\partial \phi^1}{\partial x^a} \tag{14.126}$$

We then use these to define a pair of vectors n_0^a and n_1^a which satisfy

$$n_A^a n_a^B = \delta_B^A, \text{ where } A, B = 0, 1. \tag{14.127}$$

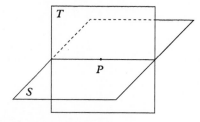

Fig. 14.8 The timelike 2-space $\{T\}$ orthogonal to $\{S\}$ at P.

These vectors together span $\{T\}$ and are called a **vector dyad** for $\{T\}$.
The condition for a pair of vectors to be **surface forming** is that the Lie
bracket vanishes. However,

$$[\boldsymbol{n}_0, \boldsymbol{n}_1] \neq 0 \quad \text{in general,} \tag{14.128}$$

so that $\{T\}$ does not form an integrable distribution. If, however, the Lie bracket vanishes, then $\{T\}$ forms a 2-dimensional subspace of M and is said to be **holonomic**. We use the $\boldsymbol{n}_A$ to define a 2×2 matrix of scalars N_{AB} by

$$N_{AB} = g_{ab} n_A^a n_B^b, \tag{14.129}$$

where the bold letters $\boldsymbol{A}$ and $\boldsymbol{B}$ etc range over 0 and 1. Because $\boldsymbol{n}_0$ and $\boldsymbol{n}_1$ are independent, the matrix N_{AB} is invertible, with inverse denoted N^{AB}. We may use N_{AB} and N^{AB} to relate n_a^A and n_B^b since

$$n_A^a = g^{ab} N_{AB} n_b^B, \tag{14.130}$$

and

$$n_a^A = g_{ab} N^{AB} n_B^b. \tag{14.131}$$

We define projection operators into $\{S\}$ and $\{T\}$ by

$$B_b^a = \delta_b^a - n_A^a n_b^A, \tag{14.132}$$

$$T_b^a = n_A^a n_b^A, \tag{14.133}$$

where $\boldsymbol{A}$ is summed over 0 and 1. The 2-metric induced on $\mathcal{S}$ is given by the projection

$$^2 g_{ab} = B_a^c B_b^d g_{cd} = B_{ad} B_b^d = B_{ab}.$$

Similarly, the 2-metric induced on $\{T\}$ is given by the projection

$$h_{ab} = T_a^c T_b^d g_{cd} = T_{ad} T_b^d = T_{ab}.$$

Since the components of h_{ab} lie in $\{T\}$, we may use $\boldsymbol{n}_0$ and $\boldsymbol{n}_1$ to give the dyad components of h_{ab} by defining $h_{AB} := h_{ab} n_A^a n_B^b$. It then follows that

$$h_{AB} = h_{ab} n_A^a n_B^b = g_{ab} n_A^a n_B^b = N_{AB}.$$

so that N_{AB} are just the dyad components of the orthogonal metric h_{ab}. In particular, the elements N_{00} and N_{11} define the lapses of $\mathcal{S}$ in $\{\Sigma_0\}$ and $\{\Sigma_1\}$, respectively.

We now choose a pair of vectors E_A^a which connect neighbouring 2-surfaces in $\{S\}$. We choose them such that

$$n_a^A E_B^a = \delta_B^A, \tag{14.134}$$

which defines E_A^a up to an arbitrary shift vector b_A^a, i.e.

$$E_A^a = n_A^a + b_A^a, \tag{14.135}$$

with

$$n_a^A b_B^a = 0. \tag{14.136}$$

Although, in general, the vectors n_A^a do not commute, it is always possible to choose the b_A^a so that

$$[\boldsymbol{E}_0, \boldsymbol{E}_1] = 0. \tag{14.137}$$

Thus, each E_A^a is tangent to a congruence of curves in Σ_A parametrized by $\phi^A(x^a)$. We may therefore choose coordinates such that $\phi^0(x^a) = x^0$, $\phi^1(x^a) = x^1$, with x^2 and x^3 being constant along the congruence of curves.

In these coordinates,

$$E_0 = \frac{\partial}{\partial x^0}, \qquad E_1 = \frac{\partial}{\partial x^1},$$

so that

$$n_0 = E_0 - b_0 = (1, 0, -b_0{}^i),$$
$$n_1 = E_1 - b_1 = (0, 1, -b_1{}^i),$$

where i ranges over 2 and 3. This results in the 2+2 decomposition of the contravariant metric

$$g^{ab} = \begin{pmatrix} N^{AB} & -N^{AB} b_B^i \\ -N^{AB} b_B^i & {}^2g^{ij} + N^{AB} b_A^i b_B^j \end{pmatrix}, \tag{14.138}$$

where the indices i and j range over 2 and 3. The contravariant metric has components

$$g_{ab} = \begin{pmatrix} N_{AB} + {}^2g_{ij} b_A^i b_B^j & {}^2g_{ij} b_A^j \\ {}^2g_{ij} b_A^j & {}^2g_{ij} \end{pmatrix}. \tag{14.139}$$

Compare and contrast (14.138) and (14.139) with (14.87), and (14.89), respectively. Note that, in the 2+2 case, the lapse function becomes a 2×2 lapse matrix and there are two shift vectors

$$\boldsymbol{b}_0 = b_0^a \partial/\partial x^a \quad \text{and} \quad \boldsymbol{b}_1 = b_1^a \partial/\partial x^a. \tag{14.140}$$

In the 2+2 formalism, the next procedure is to extract the conformal factor γ given by

$$\gamma = |{}^2g_{ij}|, \tag{14.141}$$

and define the **conformal 2-structure** ${}^2\bar{g}_{ij}$ by

$$ {}^2\bar{g}_{ij} = (\gamma^{-2}) \, {}^2g_{ij}. \tag{14.142}$$

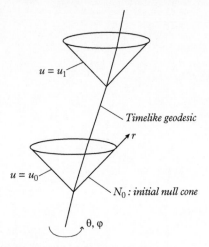

$u = u_1$

Timelike geodesic

r

$u = u_0$

N_0 : *initial null cone*

θ, φ

Fig. 14.9 Bondi type coordinates.

An analysis of the field equations goes to show that the two gravitational degrees of freedom may be chosen to lie in the conformal 2-structure ${}^2\bar{g}_{ij}$. An attractive feature of this formalism is that, if we determine the Euler-Lagrange equations generated by these two degrees of freedom for the Einstein action, then they turn out to be precisely the two dynamical Einstein equations. We will not pursue the matter further here but will consider a particular application of the 2+2 formalism to describe Bondi's radiating metric in Chapter 23 (see Fig. 14.9).

Exercises

14.1 (§14.2) Show that, if Y_a is a space-time co-vector, then $\bar{Y}_a = B_a{}^b Y_b$ satisfies $\bar{Y}_a n^a = 0$.

14.2 (§14.2) Show that the projection operator $B_a{}^b$ satisfies $B_a{}^c B_c{}^b = B_a{}^b$.

14.3 (§14.3)
(i) Show that, if the vector fields X^a and Y^a are both tangent to Σ, then so is the Lie bracket $[X, Y]$.
(ii) If D_a is the induced covariant derivative on Σ, then show that

$$X^a D_a Y^b - Y^a D_a X_b = B_d{}^b (X^a \nabla_a Y^d - Y^a \nabla_a X^d).$$

(iii) Deduce that

$$X^a D_a Y^b - Y^a D_a X^b = X^a \nabla_a Y^b - Y^a \nabla_a X^b,$$

and explain why this shows that D_a is torsion-free.

14.4 (§14.3) Calculate $\gamma_{\alpha\beta}$ and $K_{\alpha\beta}$ for the $t = 0$ slice of the spherically symmetric metric (considered in §15.4) given by

$$ds^2 = e^\nu dt^2 - e^\lambda dr^2 - r^2(d\theta^2 + \sin^2\theta d\phi^2),$$

where $\nu = \nu(t, r)$ and $\lambda = \lambda(t, r)$.

14.5 (§14.4) Show that, by using the Gauss and Codazzi equations, one can obtain expressions for all the components of the space-time curvature in terms of $\gamma_{\alpha\beta}$ and $K_{\alpha\beta}$ apart from $R_{\alpha 0 \beta 0}$. [Hint: use $n^a = (e^{\frac{\nu}{2}}, 0, 0, 0)$.].

14.6 (§14.4) Establish equation (14.28).

14.7 (§14.8)
(i) Show that $\nabla_a n_b = -K_{ab} + n_a A_b$. [Hint: start from $\nabla_a n_b = \delta_a^c \delta_b^d \nabla_c n_d$

and write δ_a^b in terms of the projection operator $B_a{}^b$].
(ii) Show that

$$n^c \nabla_c K_{ab} = -n^c (\nabla_c \nabla_a n_b) + A_a A_b + n_a n^c \nabla_c A_b.$$

14.8 (§14.9) Show that the Lie derivative L_n applied to the normal covector n_a satisfies

$$L_n(n_a n_b) = A_a n_b + A_b n_a.$$

14.9 (§14.9) Show that in coordinates adapted to an Eulerian observer one may write (14.72) as

$$\dot{\gamma}_{\alpha\beta} \overset{*}{=} 2NK_{\alpha\beta},$$

where $\gamma_{\alpha\beta} = -h_{\alpha\beta}$ and a dot indicates a derivative with respect to the coordinate time t. [Hint: use $n = (1/N)\partial/\partial t$.].

14.10 (§14.9) Show that one may write the acceleration as

$$A_a = -D_a(\ln N),$$

where N is the lapse

14.11 (§14.9) Use equation (14.76) together with (14.27) to establish (14.77).

14.12 (§14.14)
(i) Use the definition of the projection operator in (14.132) to show that the 2-metric on S is given in terms of the 4-metric by

$$^2 g^{ab} = g^{ab} - N^{AB} n_A^a n_B^b,$$

where $A, B = 0, 1$.
(ii) Use (14.135) to write n_A^a in terms of E_A^α and b_A^a.
(iii) Use adapted coordinates in which

$$E_0 \overset{*}{=} (1,0,0,0), \quad E_1 \overset{*}{=} (0,1,0,0), \quad b_0 \overset{*}{=} (0,0,b_0^i), \quad b_1 \overset{*}{=} (0,0,b_1^i),$$

to establish the 2 + 2 decomposition of the contravariant metric given in (14.138).
(iv) Confirm the form of the 2 + 2 covariant metric given in (14.139).
[Hint: use block multiplication of the matrices in (14.138) and (14.139) to show that $g_{ab} g^{bc} = \delta_a^c$.]

Further reading

There are many standard treatments of the 3+1 formalism in general relativity; see, for example, the book by Smarr (1979). Our approach follows that of Gourgoulhon (2012), who is careful to look at the intrinsic and extrinsic geometry of hypersurfaces before going on to look at foliations and finally introducing the lapse and shift. The 3+1 formalism plays a key role in numerical relativity by converting the Cauchy problem into a set of first-order PDEs. However, many of the treatments of the 3+1 formalism aimed at numerical relativity go straight to the lapse and shift description. Also note that in numerical relativity it is usual to take a metric to have the signature $(-, +, +, +)$ rather than $(+, -, -, -)$, which accounts for some sign differences in some of the equations.

The 3+1 formalism is not well adapted to looking at the 'characteristic initial data problem' in which one specifies the initial data on a null hypersurface rather than a Cauchy surface. It turns out that the 2+2 formalism is particularly well adapted to this. The approach we follow here is based on the summary in the article by d'Invero (1996). For a detailed description, see the article by d'Inverno and Smallwood (1980) and, for an alternative treatment, see the paper by Brady et al. (1996).

Brady, P. R., Droz, S., Israel, W., and Morsink, S. M. (1996). Covariant double null dynamics: (2+2) splitting of the Einstein equations. *Classical and Quantum Gravity*, *13*(8), 2211.

d'Inverno, R. A. (1996). 'Numerical computing in general relativity', in Hall, G. S., and Pulham, J. R., eds, *General Relativity*. Institute of Physics Publishing, London, 331–76.

d'Inverno, R. A., and Smallwood, J. (1980). Covariant 2+2 formulation of the initial-value problem in general relativity. *Physical Review D, 22*(6), 1233.

d'Inverno, R. A., and Stachel, J. (1978). Conformal two-structure as the gravitational degrees of freedom in general relativity. *Journal of Mathematical Physics, 19*(12) 2447–60.

Gourgoulhon, E. (2012). *The 3+1 Formalism in General Relativity*. Lecture Notes in Physics. Springer, Berlin.

Smarr, L. (1979). *Sources of Gravitational Radiation*. Cambridge University Press, Cambridge.

The Schwarzschild solution

15

15.1 Stationary solutions

We now turn our attention to solving the vacuum field equations in the simplest case, namely, that of spherical symmetry. As a preliminary, in the next two sections, we make clear the distinction between stationary and static solutions. In simple terms, a solution is stationary if it is time-independent. This does not mean that the solution is in no way evolutionary, but simply that the time does not enter into it explicitly. On the other hand, the stronger requirement that a solution is static means that it cannot be evolutionary. In such a case, nothing would change if at any time we ran time backwards, i.e. static means time-symmetric about any origin of time. Think of the motion of gas in a pipe (Fig. 15.1). If it is being pumped by some time-dependent device, then the motion will be non-stationary. If the gas travels with constant velocity at each point in the pipe, then the motion is stationary. If the gas velocity is zero everywhere, then the system is static.

A metric will be stationary if there exists a special coordinate system in which the metric is visibly time-independent, i.e.

$$\frac{\partial g_{ab}}{\partial x^0} \overset{*}{=} 0, \tag{15.1}$$

where x^0 is a timelike coordinate. Of course, in an arbitrary coordinate system, the metric will probably depend explicitly on all the coordinates; so we need to make the statement (15.1) coordinate-independent. If we define a vector field

$$X^a \overset{*}{=} \delta_0^a, \tag{15.2}$$

in the special coordinate system, then

$$L_X g_{ab} = X^c g_{ab,c} + g_{ac} X^c{}_{,b} + g_{bc} X^c{}_{,a}$$
$$\overset{*}{=} \delta_0^c g_{ab,c} = g_{ab,0} = 0,$$

by (15.1). However, $L_X g_{ab}$ is a tensor so if it vanishes in one coordinate system it vanishes in all coordinate systems. Hence, it follows that X^a is a **Killing vector field**. Conversely, given a **timelike** Killing vector field X^a, then there always exists a coordinate system which is **adapted** to the

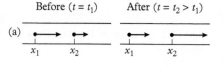
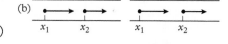

Fig. 15.1 Two gas particles in a pipe in (a) non-stationary, (b) stationary, and (c) static flow.

Introducing Einstein's Relativity. Second Edition. Ray d'Inverno and James Vickers, Oxford University Press.
© Ray d'Inverno and James Vickers (2022). DOI: 10.1093/oso/9780198862024.003.0015

Killing vector field, that is, in which (15.2) holds, and then

$$0 = \mathrm{L}_X g_{ab} \overset{*}{=} g_{ab,0},$$

and so the metric is stationary. We have therefore established the coordinate-independent definition:

> A space-time is said to be **stationary** if and only if it admits a timelike Killing vector field.

15.2　Hypersurface-orthogonal vector fields

In order to discuss static solutions in a coordinate-independent way, we need to introduce the concept of a hypersurface-orthogonal vector field, which we do in this section. We start with the equation of a **family** of hypersurfaces given by

$$f(x^a) = \mu, \tag{15.3}$$

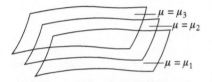

Fig. 15.2 A family of hypersurfaces labelled by μ.

where different members of the family correspond to different values of μ (Fig. 15.2). Let P be a point on S and let T^a be a tangent vector to S at P. Then we can find a curve $x^a(s)$ lying in S such that $x^a(0) = P$ and $\dot{x}^a(0) = T^a$. Since S is given by $f(x^a) = \mu = \text{constant}$, we have

$$f(x^a(s)) = \text{constant}.$$

Differentiating this with respect to s by the function of a function rule, we have

$$\frac{\partial f}{\partial x^a} \frac{\mathrm{d}x^a}{\mathrm{d}s} = 0, \tag{15.4}$$

at P. If we define the **covariant vector field** n_a to the family of hypersurfaces by

$$n_a := \frac{\partial f}{\partial x^a}, \tag{15.5}$$

then (15.4) becomes

$$n_a T^a = g_{ab} n^a T^b = 0,$$

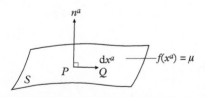

Fig. 15.3 The normal vector field n^a at the point P.

at P. Since T^a is an arbitrary tangent vector to S, this tells us that n^a is orthogonal to the tangent space of S and is therefore known as the **normal vector field** to S at P (Fig. 15.3). Any other vector field X^a is said to be **hypersurface-orthogonal** if it is everywhere orthogonal to the family of hypersurfaces, in which case it must be proportional to n^a everywhere, i.e.

$$X^a = \lambda(x) n^a, \tag{15.6}$$

for some proportionality factor λ, which in general will vary from point to point.

Then the integral curves of X^a are orthogonal to the family of hypersurfaces (Fig. 15.4). From (15.6) and (15.5), the hypersurface-orthogonal condition can also be written

$$X_a = \lambda f_{,a},$$ (15.7)

and so

$$X_a \partial_b X_c = \lambda f_{,a} \lambda_{,b} f_{,c} + \lambda^2 f_{,a} f_{,cb}.$$

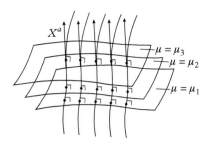

Fig. 15.4 A hypersurface-orthogonal vector field X^a.

Taking the totally anti-symmetric part of this equation and noting that the first term on the right is symmetric in a and c and the second term is symmetric in b and c, we see that their totally anti-symmetric parts vanish, and hence

$$X_{[a} \partial_b X_{c]} = 0.$$ (15.8)

This equation is unchanged if we replace the ordinary derivative by a covariant derivative (exercise), namely,

$$X_{[a} \nabla_b X_{c]} = 0.$$ (15.9)

We have shown that any hypersurface-orthogonal vector field satisfies (15.9). We shall now establish a partial converse, namely, any non-null Killing vector field satisfying (15.9) is necessarily hypersurface orthogonal. Since X^a is a Killing vector, it satisfies (7.55), namely,

$$L_X g_{ab} = \nabla_b X_a + \nabla_a X_b = 0.$$

It follows that interchanging indices on the covariant derivative of X_a introduces a minus sign:

$$\nabla_a X_b = -\nabla_b X_a.$$ (15.10)

Using this, the six terms in (15.9) reduce to three terms:

$$X_a \nabla_b X_c + X_c \nabla_a X_b + X_b \nabla_c X_a = 0.$$

Contracting with X^c and writing $X^2 = X^a X_a$, we get

$$X_a X^c \nabla_b X_c + X^2 \nabla_a X_b + X_b X^c \nabla_c X_a = 0,$$

or, using (15.10),

$$X_a X^c \nabla_b X_c + X^2 \nabla_a X_b - X_b X^c \nabla_a X_c = 0.$$ (15.11)

Interchanging the raised and lowered indices for the dummy index c (why can we do this?) and using (15.10) on the middle term, this becomes

$$X_a X_c \nabla_b X^c - X^2 \nabla_b X_a - X_b X_c \nabla_a X^c = 0.$$ (15.12)

Adding (15.11) and (15.12), we get

$$X_a \nabla_b X^2 - X_b \nabla_a X^2 + X^2 (\nabla_a X_b - \nabla_b X_a) = 0,$$

or, since X^2 is a scalar field and the terms in the parentheses involving the connection vanish (see (12.45)),

$$X_a \partial_b X^2 - X_b \partial_a X^2 + X^2 (\partial_a X_b - \partial_b X_a) = 0.$$

We write this in the form

$$X^2 \partial_a X_b - X_b \partial_a X^2 = X^2 \partial_b X_a - X_a \partial_b X^2,$$

or, equivalently, dividing by X^4,

$$\partial_a \left(\frac{X_b}{X^2} \right) = \partial_b \left(\frac{X_a}{X^2} \right), \tag{15.13}$$

since X^a is non-null by assumption and so $X^2 \neq 0$. This last equation requires that the term in parentheses be a gradient of some scalar field, f, say, i.e.

$$\frac{X_a}{X^2} = f_{,a}, \tag{15.14}$$

and so, finally,

$$X_a = X^2 f_{,a}. \tag{15.15}$$

This is the hypersurface-orthogonal condition (15.7) with $\lambda = X^2$.

15.3 Static solutions

If a solution is stationary, then, in an adapted coordinate system, the metric will be time-independent but the line element will still in general contain cross terms in $\mathrm{d}x^0 \mathrm{d}x^\alpha$. If, in addition, the metric is static, we would expect these cross terms to be absent for the following reason. Consider the interval between two events (x^0, x^1, x^2, x^3) and $(x^0 + \mathrm{d}x^0, x^1 + \mathrm{d}x^1, x^2, x^3)$ in our special coordinate system. Then

$$\mathrm{d}s^2 \overset{*}{=} g_{00}(\mathrm{d}x^0)^2 + 2g_{01}\mathrm{d}x^0 \mathrm{d}x^1 + g_{11}(\mathrm{d}x^1)^2, \tag{15.16}$$

where all the g_{ab} depend on x^α only (why?). Under a time reversal

$$x^0 \to x'^0 = -x^0, \tag{15.17}$$

the g_{ab} remain unchanged, but $\mathrm{d}s^2$ becomes

$$\mathrm{d}s^2 \overset{*}{=} g_{00}(\mathrm{d}x^0)^2 - 2g_{01}\mathrm{d}x^0 \mathrm{d}x^1 + g_{11}(\mathrm{d}x^1)^2. \tag{15.18}$$

The assumption that the solution is static, means that ds^2 is invariant under a time reversal about any origin of time, and so, equating (15.16) and (15.18), we find that g_{01} vanishes. Similarly, g_{02} and g_{03} must vanish, and so we have shown that there are no cross terms $dx^0 dx^\alpha$ in the line element in the special coordinate system. Thus, for a **static** space-time, there exists a coordinate system such that x^0 is a timelike coordinate and the metric takes the form

$$ds^2 \overset{*}{=} g_{00}(x^\gamma)(dx^0)^2 + g_{\alpha\beta}(x^\gamma)dx^\alpha dx^\beta, \qquad \alpha, \beta = 1, 2, 3. \quad (15.19)$$

Let us investigate the hypersurface-orthogonal condition (15.9) in a stationary space-time. We have shown that (15.9) implies (15.15) so that, in a coordinate system adapted to the timelike Killing vector field, that is, $X^a \overset{*}{=} \delta_0^a$, then

$$X_a = g_{ab}X^b \overset{*}{=} g_{ab}\delta_0^b = g_{0a},$$

and

$$X^2 = X_a X^a \overset{*}{=} g_{0a}\delta_0^a = g_{00}.$$

So (15.15) gives

$$g_{0a} \overset{*}{=} g_{00} f_{,a}, \qquad (15.20)$$

for some scalar field f. When $a = 0$, this produces $f_{,0} \overset{*}{=} 1$, and so integration gives

$$f \overset{*}{=} x^0 + h(x^\alpha),$$

where h is an arbitrary function of the spacelike coordinates only. Consider the coordinate transformation defined by

$$x^0 \rightarrow x'^0 = x^0 + h(x^\alpha), \quad x^\alpha \rightarrow x'^\alpha = x^\alpha. \quad (15.21)$$

Then we find, in the new coordinate system (exercise),

$$X'^a \overset{*}{=} \delta_0^a, \qquad (15.22)$$

$$g'_{ab,0} \overset{*}{=} 0, \qquad (15.23)$$

$$g'_{00} \overset{*}{=} g_{00}, \qquad (15.24)$$

$$g'_{0\alpha} \overset{*}{=} 0. \qquad (15.25)$$

The last equation reveals that there are no cross terms in $dx^0 dx^\alpha$ and so the solution is static. Conversely, if there exists a coordinate system in which the metric takes the form (15.19), one can show (exercise) that $X^a \overset{*}{=} \delta_0^a$ is a hypersurface-orthogonal timelike Killing vector. We have therefore established the following coordinate-free definition of a static space-time.

A space-time is said to be **static** if and only if it admits a hypersurface-orthogonal timelike Killing vector field.

Moreover, we have established the following important result.

In a static space-time, there exists a coordinate system adapted to the timelike Killing vector field in which the metric is time-independent and no cross terms appear in the line element involving the time, i.e. the metric takes the form (15.19).

It can be shown (exercise) that there still exists the coordinate freedom

$$x^0 \to x'^0 = Ax^0 + B, \quad x^\alpha \to x'^\alpha = h'^\alpha(x^\beta), \tag{15.26}$$

where A and B are constants and the functions h'^α are arbitrary. If the boundary conditions require $g_{00} \to 1$ at spatial infinity, then this requires $A = \pm 1$. Neglecting time reversal, then this fixes A to be 1, and so we have defined a time coordinate, called **world time**, which is defined to within an unimportant additive constant. Thus, in a static space-time, we have regained the old Newtonian idea of an absolute time in the sense that the manifold can be sliced up in a well-defined way into hypersurfaces $t = $ constant (Fig. 15.5). Then there exist a privileged class of observers who measure world time and hence can agree on events being simultaneous. The corresponding coordinates are Gaussian, since $g_{0a} \overset{*}{=} \delta_{0a}$.

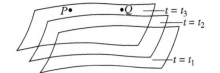

Fig. 15.5 Two 'simultaneous' events in world time.

15.4 Spherically symmetric solutions

Spherical symmetry can be defined rigorously using the notion of isometry. In §7.7 we defined an isometry $\phi : x^a \to \tilde{x}^a$ to be a map such that the metric satisfied (7.52). It then follows that the inverse map $\phi^{-1} : \tilde{x}^a \to x^a$ is an isometry (exercise). If we have a second isometry, $\psi : \tilde{x}^a \to \hat{x}^a$, then the composition of the two, $\psi \circ \phi : x^a \to \hat{x}^a$, is also an isometry (exercise). As a result of this, the set of isometries of a metric forms a group, called the **isometry group**. A space-time is spherically symmetric if the metric remains invariant under a spatial rotation. More precisely:

A space-time is said to be **spherically symmetric** if and only if its isometry group contains a subgroup isomorphic to the rotation group in three dimensions, and the orbits of this group are topologically 2-spheres S^2.

In particular, since there must be isometries that generate rotations, there exist three linearly independent spacelike Killing vector fields X^a which correspond to rotations about the x, y, and z axes and therefore satisfy

$$[X^1, X^2] = X^3, \quad [X^2, X^3] = X^1, \quad [X^3, X^1] = X^2. \tag{15.27}$$

Then (see Exercise 7.4 and Exercise 8.5) there exists a coordinate system in which the Killing vectors take on a standard form, as expressed in the following result.

> In a spherically symmetric space-time, there exists a coordinate system (x^a) (called Cartesian) in which the Killing fields X^a are of the form
>
> $$X^0 \overset{*}{=} 0,$$
>
> $$X^\alpha \overset{*}{=} \omega^\alpha{}_\beta x^\beta, \quad \omega_{\alpha\beta} = -\omega_{\beta\alpha}.$$

The quantity $\omega_{\alpha\beta}$ depends on three parameters which specify three space-like rotations. These results then lead to a canonical form for the line element. The calculation is rather detailed, so we shall proceed in a different manner and present a heuristic argument for determining the form of the line element.

Intuitively, spherical symmetry means that there exists a privileged point, called the origin O, such that the system is invariant under spatial rotations about O. Then, if we fix the time and consider a point P a distance a from O, the spatial rotations will result in P sweeping out a 2-sphere centred on O. We can then introduce an axial coordinate ϕ and an azimuthal coordinate θ on the sphere in the usual way. Dropping a perpendicular from P to the equatorial plane $z = 0$ at Q, then ϕ is the angle which OQ makes with the positive x-axis, and θ is the angle which OP makes with the positive z-axis (Fig. 15.6). All points on the 2-sphere will be covered by the coordinate ranges

$$0 \leqslant \theta \leqslant \pi, \tag{15.28}$$

$$-\pi < \phi \leqslant \pi. \tag{15.29}$$

Moreover, the line element of the 2-sphere is (exercise)

$$\mathrm{d}s^2 = a^2(\mathrm{d}\theta^2 + \sin^2\theta\,\mathrm{d}\phi^2). \tag{15.30}$$

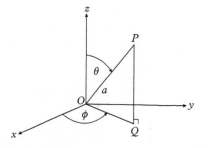

Fig. 15.6 The standard spherical coordinates θ and ϕ.

It is then natural to assume that, in four dimensions, we can augment θ and ϕ with an arbitrary timelike coordinate t and some radial-type parameter r, so that the line element reduces to the form (15.30) on a 2-sphere t = constant, r = constant. Spherical symmetry requires that the line element does not vary when θ and ϕ are varied, so that θ and ϕ only occur in the line element in the form $(\mathrm{d}\theta^2 + \sin^2\theta\,\mathrm{d}\phi^2)$. Moreover, using an argument analogous to the one we used at the beginning of §15.3, there can be no cross terms in $\mathrm{d}\theta$ or $\mathrm{d}\phi$ (exercise) because the metric must be invariant separately under the reflections

$$\theta \to \theta' = \pi - \theta, \tag{15.31}$$

and

$$\phi \to \phi' = -\phi. \tag{15.32}$$

Our starting ansatz, then, for a spherically symmetric space-time is that there exists a special coordinate system

$$(x^a) = (x^0, x^1, x^2, x^3) = (t, r, \theta, \phi),$$

in which the line element has the form

$$ds^2 \stackrel{*}{=} A dt^2 - 2B dt\, dr - C dr^2 - D(d\theta^2 + \sin^2\theta\, d\phi^2), \qquad (15.33)$$

where A, B, C, and D are as yet undetermined functions of t and r, i.e.

$$A = A(t, r), \quad B = B(t, r), \quad C = C(t, r), \quad D = D(t, r).$$

If we introduce a new radial coordinate by the transformation

$$r \rightarrow r' = D^{\frac{1}{2}},$$

then (15.33) becomes

$$ds^2 = A'(t, r') dt^2 - 2B'(t, r') dt\, dr' - C'(t, r') dr'^2 - r'^2 (d\theta^2 + \sin^2\theta\, d\phi^2). \qquad (15.34)$$

Consider the differential

$$A'(t, r') dt - B'(t, r')\, dr'.$$

The theory of ordinary differential equations tells us that we can always multiply this by an integrating factor, $I = I(t, r')$, say, which makes it a perfect differential. We use this result to define a new time coordinate t' by requiring

$$dt' = I(t, r')[A'(t, r') dt - B'(t, r') dr'].$$

Squaring, we obtain

$$dt'^2 = I^2(A'^2 dt^2 - 2A'B' dt\, dr' + B'^2 dr'^2),$$

and so

$$A' dt^2 - 2B' dt\, dr' = A'^{-1} I^{-2} dt'^2 - A'^{-1} B'^2 dr'^2,$$

and the line element (15.34) becomes

$$ds^2 = A'^{-1} I^{-2} dt'^2 - (C' + A'^{-1} B'^2) dr'^2 - r'^2 (d\theta^2 + \sin^2\theta\, d\phi^2).$$

Defining two new functions ν and λ by

$$A'^{-1} I^{-2} = e^\nu \qquad (15.35)$$

and

$$C' + A'^{-1}B'^2 = e^{\lambda}, \qquad (15.36)$$

and dropping the primes, we finally obtain the form

$$ds^2 = e^{\nu}dt^2 - e^{\lambda}dr^2 - r^2(d\theta^2 + \sin^2\theta \, d\phi^2), \qquad (15.37)$$

where

$$\nu = \nu(t, r), \quad \lambda = \lambda(t, r).$$

The definitions of ν and λ in (15.35) and (15.36) are given in terms of exponentials, which, since they are always positive, guarantees that the signature of the metric is -2. In fact, there are rigorous arguments which confirm that the most general spherically symmetric line element in four dimensions (with signature -2) can be written in the canonical form (15.37).

15.5 The Schwarzschild solution

We now use Einstein's vacuum field equations to determine the unknown functions ν and λ in (15.37). The covariant metric is

$$g_{ab} = \text{diag}(e^{\nu}, \, -e^{\lambda}, \, -r^2, \, -r^2\sin^2\theta), \qquad (15.38)$$

and, since the metric is diagonal, its contravariant form is

$$g^{ab} = \text{diag}(e^{-\nu}, \, -e^{-\lambda}, \, -r^{-2}, \, -r^{-2}\sin^{-2}\theta). \qquad (15.39)$$

If we denote derivatives with respect to t and r by dot and prime, respectively, then, by Exercise 6.32(v), the non-vanishing components of the mixed Einstein tensor are

$$G_0{}^0 = e^{-\lambda}\left(\frac{\lambda'}{r} - \frac{1}{r^2}\right) + \frac{1}{r^2}, \qquad (15.40)$$

$$G_0{}^1 = -e^{-\lambda}r^{-1}\dot{\lambda} = -e^{\lambda-\nu}G_1{}^0, \qquad (15.41)$$

$$G_1{}^1 = -e^{-\lambda}\left(\frac{\nu'}{r} + \frac{1}{r^2}\right) + \frac{1}{r^2}, \qquad (15.42)$$

$$G_2{}^2 = G_3{}^3 = \frac{1}{2}e^{-\lambda}\left(\frac{\nu'\lambda'}{2} + \frac{\lambda'}{r} - \frac{\nu'}{r} - \frac{\nu'^2}{2} - \nu''\right)$$

$$+ \frac{1}{2}e^{-\nu}\left(\ddot{\lambda} + \frac{\dot{\lambda}^2}{2} - \frac{\dot{\lambda}\dot{\nu}}{2}\right). \qquad (15.43)$$

The contracted Bianchi identities reveal that equation (15.43) vanishes **automatically** if the equations (15.40), (15.41), and (15.42) all vanish (exercise). Hence, there are three independent equations to solve, namely,

$$e^{-\lambda}\left(\frac{\lambda'}{r} - \frac{1}{r^2}\right) + \frac{1}{r^2} = 0, \tag{15.44}$$

$$e^{-\lambda}\left(\frac{\nu'}{r} + \frac{1}{r^2}\right) - \frac{1}{r^2} = 0, \tag{15.45}$$

$$\dot{\lambda} = 0. \tag{15.46}$$

Adding (15.44) and (15.45), we get

$$\lambda' + \nu' = 0,$$

and integration gives

$$\lambda + \nu = h(t), \tag{15.47}$$

where $h(t)$ is an arbitrary function of integration. Here, λ is purely a function of r by (15.46), and so (15.44) is simply an **ordinary** differential equation, which we write

$$e^{-\lambda} - re^{-\lambda}\lambda' = 1,$$

or, equivalently,

$$(re^{-\lambda})' = 1.$$

Integrating, we get

$$re^{-\lambda} = r + \text{constant}.$$

Choosing the constant of integration to be $-2m$, for later convenience, we then obtain

$$e^{\lambda} = (1 - 2m/r)^{-1}. \tag{15.48}$$

It then follows from (15.47) that

$$e^{\nu} = e^{h(t)}e^{-\lambda} = e^{h(t)}(1 - 2m/r). \tag{15.49}$$

So, at this stage, the metric has been reduced to

$$g_{ab} = \text{diag}\left[e^{h(t)}(1 - 2m/r), -(1 - 2m/r)^{-1}, -r^2, -r^2\sin^2\theta\right]. \tag{15.50}$$

The final stage is to eliminate $h(t)$. This is done by transforming to a new time coordinate t', i.e. $t \rightarrow t'$, where t' is determined by the relation

$$t' = \int_c^t e^{h(u)/2}du, \tag{15.51}$$

where c is an arbitrary constant. Then the only component of the metric which changes is (exercise)

$$g'_{00} = (1 - 2m/r).$$

Dropping primes, we have shown that it is always possible to find a coordinate system in which the most general spherically symmetric solution of the vacuum field equations is

$$ds^2 = (1 - 2m/r)\,dt^2 - (1 - 2m/r)^{-1}dr^2 - r^2(d\theta^2 + \sin^2\theta\,d\phi^2). \quad (15.52)$$

This is the famous **Schwarzschild line element**.

15.6 Properties of the Schwarzschild solution

We restrict attention to the **exterior** region $r > 2m$, where the coordinates t and r are timelike and spacelike, respectively (see §17.1). It is immediate from (15.52) that $g_{ab,0} \stackrel{*}{=} 0$, and so the solution is **stationary**. Moreover, the coordinates are **adapted** to the Killing vector field $X^a \stackrel{*}{=} \delta^a_0$. Since

$$X_a = g_{ab}X^b \stackrel{*}{=} g_{ab}\,\delta^b_0 = g_{0a} = g_{00}\,\delta^0_a = (1 - 2m/r, 0, 0, 0),$$

we see that X^a is **hypersurface-orthogonal**, that is, $X_a = \lambda f_{,a}$, with

$$\lambda = X^2 \stackrel{*}{=} g_{00} \quad \text{and} \quad f(x^a) \stackrel{*}{=} t = \text{constant}.$$

Alternatively, we can check (exercise) that

$$X_{[a}\partial_b X_{c]} = 0. \quad (15.53)$$

Thus, the timelike Killing vector field X^a is hypersurface orthogonal to the family of hypersurfaces $t = $ constant, and hence the solution is **static** and t is a **world time**. Alternatively, it is immediate from (15.52) that the solution is **time-symmetric**, since it is invariant under the time reflection $t \to t' = -t$, and **time translation invariant**, since it is invariant under the transformation $t \to t' = t + $ constant, and so again it is static (see Exercise 15.1). We have thus proved the following somewhat unexpected result.

Birkhoff's theorem: A spherically symmetric vacuum solution in the exterior region is necessarily static.

This is unexpected because, in Newtonian theory, spherical symmetry has nothing to do with time dependence. This highlights the special character of non-linear partial differential equations and the solutions they admit. In particular, Birkhoff's theorem implies that, if a spherically symmetric source like a star changes its shape, but does so always remaining

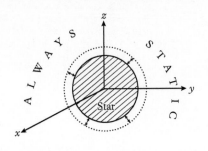

Fig. 15.7 A pulsating spherical star cannot emit gravitational waves.

spherically symmetric, then it cannot propagate any disturbances into the surrounding space. Looking ahead, this means that a pulsating spherically symmetric star cannot emit gravitational waves (Fig. 15.7). If a spherically symmetric source is restricted to the region $r \leqslant a$ for some $a > 2m$, then the solution for $r > a$ must be the Schwarzschild solution or, to give it its full name, the Schwarzschild **exterior** solution. However, the converse is not true: a source which gives rise to an exterior Schwarzschild solution is **not** necessarily spherically symmetric. Some counter examples are known. Thus, in general, a source need not inherit the symmetry of its external field.

If we take the limit of (15.52) as $r \to \infty$, then we obtain the flat space metric of special relativity in spherical polar coordinates, namely,

$$ds^2 = dt^2 - dr^2 - r^2(d\theta^2 + \sin^2\theta \, d\phi^2). \tag{15.54}$$

We have therefore shown that a spherically symmetric vacuum solution is necessarily **asymptotically flat**. Some authors obtain the Schwarzschild solution from the starting assumptions that the solution is spherically symmetric, static, and asymptotically flat. However, as we have seen, there is no need to adopt these last two assumptions a priori, because the field equations force them on you. Let us attempt an interpretation of the constant m appearing in the solution, by considering the Newtonian limit. A point mass M situated at the origin O in Newtonian theory gives rise to a potential $\phi = -GM/r$. Inserting this into the weak-field limit (10.49) gives

$$g_{00} \simeq 1 + 2\phi/c^2 = 1 - 2GM/(c^2 r),$$

and, comparing this with (15.52), we see that

$$m = GM/c^2 \tag{15.55}$$

in non-relativistic units. In other words, if we interpret the Schwarzschild solution as due to a particle situated at the origin, then the constant m is simply the mass of the particle in relativistic units. It is clear from (15.52) that m has the dimensions of length, which is consistent with the dimensions given by (15.55). It is sometimes known as the **geometric mass**. We postpone a discussion of the coordinate ranges and the interpretation of the coordinates until Chapter 17. We end this section by summarizing the properties we have met. The exterior Schwarzschild solution:

(1) is spherically symmetric;

(2) is stationary;

(3) has coordinates adapted to the timelike Killing vector field X^a;

(4) is static $\Leftrightarrow$ is time-symmetric and time-translation invariant,
$\Leftrightarrow$ has a hypersurface-orthogonal timelike Killing vector field X^a;

(5) is asymptotically flat;

(6) has geometric mass $m = GMc^{-2}$.

15.7 Isotropic coordinates

In this section, we seek an alternative set of coordinates in which the time slices $t = $ constant are as close as we can get them to Euclidean 3-space. More specifically, we attempt to write the line element in the form

$$ds^2 = A(r)\,dt^2 - B(r)\,d\sigma^2,$$

where $d\sigma^2$ is the line element of Euclidean 3-space, namely,

$$d\sigma^2 = dx^2 + dy^2 + dz^2,$$

in Cartesian coordinates or, equivalently,

$$d\sigma^2 = dr^2 + r^2 d\theta^2 + r^2 \sin^2 \theta\, d\phi^2,$$

in spherical polar coordinates. In this form, the metric in a slice $t = $ constant is **conformal** to the metric of Euclidean 3-space, and hence, in particular, angles between vectors and ratios of lengths are the same for each metric (see Exercise 6.28).

We consider a transformation in which the coordinates θ, ϕ, and t remain unchanged while

$$r \rightarrow \rho = \rho(r), \tag{15.56}$$

so that ρ is some other radial coordinate, and we attempt to put the solution in the form

$$ds^2 = (1 - 2m/r)\,dt^2 - [\lambda(\rho)]^2[d\rho^2 + \rho^2(d\theta^2 + \sin^2 \theta\, d\phi^2)]. \tag{15.57}$$

We could consider how (15.52) transforms under the transformation (15.56), but it is easier to proceed as follows. Comparing (15.57) with (15.52), the coefficients of $d\theta^2 + \sin^2 \theta\, d\phi^2$ must be equal, which requires

$$r^2 = \lambda^2 \rho^2. \tag{15.58}$$

Equating the two radial elements produces

$$(1 - 2m/r)^{-1} dr^2 = \lambda^2 d\rho^2. \tag{15.59}$$

Eliminating λ and taking square roots, we find

$$\frac{dr}{(r^2 - 2mr)^{\frac{1}{2}}} = \pm\frac{d\rho}{\rho}. \tag{15.60}$$

This is an ordinary differential equation in which the variables are separated. Since we require $\rho \to \infty$ as $r \to \infty$, we take the positive sign, and by integration we find (exercise)

$$r = \rho(1 + \tfrac{1}{2}m/\rho)^2, \tag{15.61}$$

and so, from (15.58),

$$\lambda^2 = (1 + \tfrac{1}{2}m/\rho)^4. \tag{15.62}$$

Using (15.61) to eliminate r, we find that the Schwarzschild solution can be written in the following **isotropic form**

$$ds^2 = \frac{(1 - \tfrac{1}{2}m/\rho)^2}{(1 + \tfrac{1}{2}m/\rho)^2}dt^2 - (1 + \tfrac{1}{2}m/\rho)^4[d\rho^2 + \rho^2(d\theta^2 + \sin^2\theta\,d\phi^2)]. \tag{15.63}$$

15.8 The Schwarzschild interior solution

We end this chapter with a brief discussion of static spherically symmetric perfect fluid solutions of Einstein's equations that can serve as an interior solution to the Schwarzschild solution. As before, the assumption of spherical symmetry leads to the metric (15.37). However, since we are now assuming that the solution is static, there is no t dependence so we now have

$$\nu = \nu(r), \qquad \lambda = \lambda(r).$$

The matter is a perfect fluid with energy-momentum tensor given by (12.22) so that

$$T^{ab} = (\rho + p)u^a u^b - pg^{ab}, \tag{15.64}$$

where u^a is the 4-velocity of the fluid. Since the solution is static, u^a must be proportional to the timelike Killing vector $\xi^a = (1, 0, 0, 0)$ and, since it is a 4-velocity, it must satisfy $u^a u_a = 1$. Hence,

$$u^a = \frac{\xi^a}{(\xi^b \xi_b)^{1/2}}. \tag{15.65}$$

It follows from this (exercise) that

$$T_0{}^0 = \rho, \tag{15.66}$$

and

$$T_i^j = -p\delta_i^j, \quad i,j = 1,\dots,3. \tag{15.67}$$

The expression for $G_0{}^0$ is still given by (15.40) so that this component of Einstein's equations gives

$$8\pi\rho = e^{-\lambda}\left(\frac{\lambda'}{r} - \frac{1}{r^2}\right) + \frac{1}{r^2},$$

$$= \frac{1}{r^2}\left(1 - \frac{d}{dr}(re^{-\lambda})\right). \qquad (15.68)$$

We now define the function $m(r)$, which in the Newtonian case would be the total mass inside the radius r, by

$$m(r) = 4\pi\int_0^r \rho(s)s^2 ds. \qquad (15.69)$$

Then multiplying (15.68) by r^2, rearranging, and integrating, we obtain (exercise)

$$e^{-\lambda(r)} = 1 - \frac{2m(r)}{r}. \qquad (15.70)$$

The $G_1{}^1$ component of Einstein's equations gives

$$8\pi p = e^{-\lambda}\left(\frac{\nu'}{r} + \frac{1}{r^2}\right) - \frac{1}{r^2}, \qquad (15.71)$$

which, using (15.70), simplifies to (exercise)

$$\frac{d\nu}{dr} = \frac{1}{r}\left(1 - \frac{2m(r)}{r}\right)^{-1}\left(\frac{2m(r)}{r} + 8\pi r^2 p\right). \qquad (15.72)$$

We now use the fact that $\nabla_b T_a{}^b = 0$ to deduce (exercise) that

$$\frac{dp}{dr} = -\frac{1}{2}(\rho + p)\frac{d\nu}{dr}. \qquad (15.73)$$

Eliminating ν' from (15.72) and (15.73) and rearranging gives

$$\frac{dp}{dr} = -\frac{1}{r^2}(\rho + p)(m(r) + 4\pi r^3 p)\left(1 - \frac{2m(r)}{r}\right)^{-1}. \qquad (15.74)$$

This is called the **Tolman–Oppenheimer–Volkoff equation**, or **TOV** equation for short, which describes the behaviour of a static spherically symmetric perfect fluid. If $\rho(r)$ is a given function of r, then one can, in principle, obtain expressions for the pressure and metric coefficients. A particularly simple case is when the density is constant so $\rho = \rho_0 = \text{constant}$. This yields the so-called **Schwarzschild interior solution**. If the radius of the star is R, then the total mass is $M = \frac{4}{3}\pi R^3\rho_0$, and $m(r)$ is given by (exercise)

$$m(r) = \begin{cases} Mr^3/R^3 & r \leqslant R, \\ M & r > R. \end{cases} \qquad (15.75)$$

Then (15.74) can be integrated with the boundary condition $p(R) = 0$ (so that the pressure vanishes at the edge of the star) to give the pressure in the interior $(r < R)$ as (exercise)

$$p(r) = \rho_0 \left[\frac{(1 - 2M/R)^{1/2} - (1 - 2Mr^2/R^3)^{1/2}}{(1 - 2Mr^2/R^3)^{1/2} - 3(1 - 2M/R)^{1/2}} \right]. \qquad (15.76)$$

Then $\lambda(r)$ is given by (15.70) while $\nu(r)$ is given by integrating (15.72) to give

$$e^{\nu(r)} = \begin{cases} \left[\tfrac{3}{2}(1 - 2M/r)^{1/2} - \tfrac{1}{2}(1 - 2Mr^2/R^3)^{1/2} \right]^2 & r \leqslant R, \\ 1 - 2M/r & r > R. \end{cases} \qquad (15.77)$$

The above equations give the pressure and metric functions for the Schwarzschild interior solution.

However, it is important to note that, apart from the simple examples, like the constant density case illustrated above, it is not usual to be able to specify the density as a given function of r. Instead, one has an **equation of state** giving a functional relationship between the pressure and the density so that

$$p = p(\rho).$$

Differentiating this gives

$$\frac{\mathrm{d}p}{\mathrm{d}r} = \frac{\mathrm{d}p}{\mathrm{d}\rho} \frac{\mathrm{d}\rho}{\mathrm{d}r},$$

which may be substituted into (15.74) to give a (complicated) non-linear equation for the density. In general this cannot be solved in closed form, and numerical methods must be used to obtain the solution of the TOV equation.

Exercises

15.1 (§15.1) A system is time-symmetric if it is invariant under

$$t \rightarrow t' = -t.$$

Give an example of a non-stationary time-symmetric system. Show that, if a time-symmetric system is also time-translation invariant, i.e. invariant under

$$t \rightarrow t' = t + \text{constant},$$

then the system is static. Deduce that a stationary time-symmetric system is necessarily static.

15.2 (§15.1) Show that, if g_{ab} is stationary, then there exists a privileged coordinate system (t, x^α) in which the Killing vector field X reduces to $X = \partial/\partial t$ with $X(g_{ab}) = 0$. Show that X generates a time-translation invariance

$$t \to t' = t + \text{constant}.$$

15.3 (§15.2)
(i) Take the differential of (15.3) to confirm (15.4).
(ii) Show that (15.9) is equivalent to (15.8).
(iii) Check that (15.14) is consistent with (15.13).

15.4 (§15.3)
(i) Establish (15.22)–(15.25) under the transformation (15.21).
(ii) Show that there still remains the coordinate freedom (15.26).

15.5 (§15.3) Show that, if there exists a coordinate system in which the metric takes the form (15.19), then $X^a \overset{*}{=} \delta_0^a$ is a hypersurface-orthogonal timelike Killing vector.

15.6 (§15.4) Show that the composition of two isometries is an isometry.

15.7 (§15.4) Consider a point P on a 2-sphere of radius a centred at the origin. Find the distance P travels under an increase of coordinates
(i) $\theta \to \theta + \mathrm{d}\theta$,
(ii) $\phi \to \phi = \mathrm{d}\phi$.

Use Pythagoras' theorem to obtain the line element (15.30) for a 2-sphere.

15.8 (§15.4) Show that a spherically symmetric line element cannot possess cross terms in $\mathrm{d}\theta$ and $\mathrm{d}\phi$ because the metric must be invariant under the reflections (15.31) and (15.32). [Hint: assume that all the metric components g_{ab} $(a, b \neq 0)$ and $g_{33} \sin^{-2}\theta$ do not depend on θ or ϕ.]

15.9 (§15.5) Show that, if (15.40), (15.41), and (15.42) vanish, then so does (15.43), by the contracted Bianchi identities.

15.10 (§15.5) Show that, under the transformation to a new time coordinate t' given by (15.51), the line element (15.50) is transformed into the form (15.52), where primes have been dropped in (15.52).

15.11 (§15.6) Check that (15.53) holds for the Schwarzschild line element where X^a is the timelike Killing vector field.

15.12 (§15.6) Find the dimensions of the gravitational constant G. [Hint: use (4.4) and Newton's second law.] Use (15.55) to show that m has the dimensions of a length.

15.13 (§15.6) Find the non-zero components of R_{abcd} for the Schwarzschild solution.

15.14 (§15.7)
(i) Show that (15.60) taken with the positive sign integrates to give (15.61).
(ii) Use (15.57), (15.61), and (15.62) to derive (15.63).

15.15 (§15.7) Consider (15.63) in the weak-field limit $m \ll \rho$ to show that $g_{00} \simeq 1 - 2m/\rho$ and confirm (15.55).

15.16 (§15.7) Which of the six properties listed at the end of §15.6 still hold for the isotropic form of the Schwarzschild line element?

15.17 (§15.7) Confirm by direct computation that the isotropic form of the Schwarzschild solution

$$ds^2 = \frac{(1 - m/2\rho)^2}{(1 + m/2\rho)^2} dt^2 - (1 + m/2\rho)^4 \left[dx^2 + dy^2 + dz^2\right],$$

where

$$r = (x^2 + y^2 + z^2)^{1/2} = \rho(1 + m/2\rho)^2$$

admits the Killing vector fields

$$\frac{\partial}{\partial t}, \quad x\frac{\partial}{\partial y} - y\frac{\partial}{\partial x}, \quad y\frac{\partial}{\partial z} - z\frac{\partial}{\partial y}, \quad z\frac{\partial}{\partial x} - x\frac{\partial}{\partial z}.$$

[Hint: This is a fairly long calculation and you will need to compute in turn $\partial r/\partial x$, $dr/d\rho$ and hence $\partial\rho/\partial x$. Then use the symmetry in x, y, and z.] Find all their commutators.

15.18 (§15.8)
(i) Verify that, for a static perfect fluid, the energy-momentum tensor satisfies (15.66) and (15.67).
(ii) Show that (15.68) leads to (15.70).
(iii) Show that $\nabla_b T_a{}^b = 0$ implies (15.73). [Hint: you need only consider the case $a = 1$ and then use the results of Exercise 6.32 (ii).]
(iv) Use (15.72) and (15.73) to obtain the TOV equation (15.74).
(v) Integrate the TOV equation for the constant density case, $\rho = 3M/4\pi R^3$ for $0 \leqslant r \leqslant R$ and $\rho = 0$ for $r > R$, to obtain (15.76). Calculate $\lambda(r)$ and $\nu(r)$, and show this gives the interior Schwarzschild solution for $r \leqslant R$ and the Schwarzschild exterior solution for $r > R$.

Further reading

All the general relativity textbooks listed below contain material on the Schwarzschild exterior solution. However, we have only sketched the details of an interior source for the Schwarzschild solution. A simple treatment of this is given in the book by Hughston and Tod (1990).

Carroll, S. M. (2004). *Spacetime and Geometry: An Introduction to General Relativity*. Addison Wesley, San Francisco, CA.

Hartle, J. B. (2003). *Gravity: An Introduction to Einstein's General Relativity*. Addison Wesley, San Francisco, CA.

Schutz, B. F. (1985). *A First Course in General Relativity*. Cambridge University Press, Cambridge.

Wald, R. M. (1984). *General Relativity*. University of Chicago Press, Chicago, IL.

Hughston, L. P., and Tod, K. P. (1990). *An Introduction to General Relativity*. Cambridge University Press, Cambridge.

Classical experimental tests of general relativity

16

16.1 Introduction

In this chapter, we shall consider various experimental tests of general relativity. In particular we will focus on the tests of general relativity that can be carried out in the solar system. However, these tests are really tests of general relativity in the weak-field regime, in which the gravitational effects are not significantly different from the corresponding Newtonian ones. We will show that, in this regime, it is possible to introduce a scheme called the **parametrized-post-Newtonian** (PPN) framework in which one can compare general relativity with alternative gravitational theories. We will see that by using this framework for weak gravitational fields, the predictions of general relativity, are confirmed with an error of just a few tenths of a per cent. However, some of the most interesting predictions of general relativity, such as the the structure of the early universe, the existence of black holes, and the emission of gravitational waves by colliding stars, involve strong gravitational fields so cannot be described through a perturbation of Newtonian gravity. Indirect evidence for the existence of gravitational waves comes from monitoring binary pulsars, such as PSR 1913+16, whose orbits decay due to the emission of gravitational radiation. Direct evidence for the existence of gravitational waves comes from the measurement of gravitational waves by two LIGO detectors in 2015, where the signal matched the numerical predictions of the merger of two black holes. This event, which lasted only a few seconds, was the most powerful astronomical event ever observed. We will describe in more detail the experimental evidence for the existence of black holes and gravitational radiation in Parts D and E.

Historically, the first tests of the theory were the three so-called classical tests of general relativity, namely, the precession of the perihelion of Mercury, the bending of light, and the gravitational redshift. These tests were augmented subsequently by a fourth classical test, the delay of a light signal in a gravitational field. The test of gravitational redshift was originally thought to be a direct test of general relativity, since it makes use of the Schwarzschild solution but it was soon realized that it is really just a test of the **weak equivalence principle**. However, since the equivalence principle is such a fundamental part of general relativity, high-precision measurements of gravitational redshift are important in confirming the foundations of the theory. Furthermore, taken together with some tests of special relativity showing local Lorentz invariance and other experiments showing local position invariance, one is able to provide strong experimental evidence for the **Einstein equivalence principle**

Introducing Einstein's Relativity. Second Edition. Ray d'Inverno and James Vickers, Oxford University Press.
© Ray d'Inverno and James Vickers (2022). DOI: 10.1093/oso/9780198862024.003.0016

(see §16.4), which allows one to deduce that gravity can be described through the curvature of space-time endowed with a symmetric metric. This paves the way for using the other classical tests to compare general relativity with other metric theories using the PPN framework. There have probably been at least a score of alternative relativistic theories of gravitation proposed since the advent of general relativity. However, use of the PPN formalism shows that the predictions of all these theories in terms of the solar system tests must all be extremely close to those of general relativity. This taken with the observations of gravitational radiation strongly supports general relativity as being the best and simplest classical theory that we have. We end with a brief chronology of the main experimental or observational events connected with general relativity.

16.2 Gravitational red shift

Since it is so central to the underpinning of the theory, and was the first test suggested by Einstein, we begin by considering gravitational redshift. As we have said, at first it was thought that this was a direct test of general relativity since it employed the Schwarzschild solution, but it is now clear that any relativistic theory of gravitation consistent with the principle of equivalence will predict a redshift. We outline below a thought experiment which leads directly to the existence of a gravitational redshift. Consider an endless chain running between the Earth and the Sun, carrying buckets containing atoms in an excited state on one side and an equal number of atoms in the ground state on the other side (Fig. 16.1). Since the excited atoms possess greater energy, they must have greater mass (using $E = mc^2$). They will be heavier than the ground-state atoms and so, by the principle of equivalence, they will fall towards the Sun, whose gravitational field predominates. Suppose we have a device which returns an atom arriving at the Sun to its ground state, collects the emitted energy radiated in a mirror, and reflects it back to the Earth, where it is used to excite an incoming atom in the ground state. Then the rotating chain will run on indefinitely. In this way, we have constructed a **perpetuum mobile**, or perpetual-motion machine. Such a device contradicts the principle of conservation of energy, the cornerstone of physics, and so something must be wrong with the argument. It breaks down because the radiation arriving at the Earth is not sufficiently energetic to excite the incoming ground-state atom. In other words, it gets downgraded climbing up the gravitational field: the radiation has been shifted to the **red**.

We shall next obtain a quantitative expression for the redshift in the special case of a general **static** space-time. The coordinates are taken to be

$$(x^a) = (x^0, x^\alpha),$$

where x^0 is the world time and x^α are spatial coordinates. We consider two observers carrying ideal atomic clocks whose world-lines are $x^\alpha = x_1^\alpha = $ constant and $x^\alpha = x_2^\alpha = $ constant, respectively (see Fig. 16.2). Let the first observer possess an atomic system which is sending out

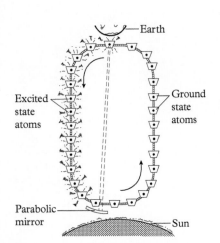

Fig. 16.1 A gravitational perpetuum mobile?

Earth

Excited state atoms

Ground state atoms

Parabolic mirror

Sun

radiation to the second observer. We denote the time separation between successive wave crests as measured by the first clock by $d\tau$ in terms of proper time and by dx_1^0 in terms of coordinate time. It follows from the definition of proper time that

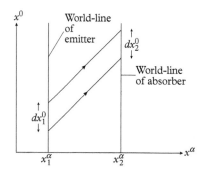

$$d\tau^2 = g_{ab}\left(x_1^\alpha\right) dx_1^a \, dx_1^b = g_{00}\left(x_1^\alpha\right)\left(dx_1^0\right)^2, \tag{16.1}$$

since g_{ab} can only depend on the spatial coordinates. Let the corresponding interval of reception recorded by the second observer be $k d\tau$ in proper time and dx_2^0 in coordinate time. Then, similarly,

Fig. 16.2 Emission and reception of successive wave crests of a signal.

$$(k \, d\tau)^2 = g_{00}\left(x_2^\alpha\right)\left(dx_2^0\right)^2. \tag{16.2}$$

However, the assumption that the space-time is static means that

$$dx_1^0 = dx_2^0, \tag{16.3}$$

because otherwise there would be a build-up or depletion of wave crests between the two observers, in violation of the static assumption. Dividing (16.1) and (16.2), we find

$$k = \left(\frac{g_{00}\left(x_2^\alpha\right)}{g_{00}\left(x_1^\alpha\right)}\right)^{\frac{1}{2}}. \tag{16.4}$$

The factor k records how many times the second clock has ticked between the reception of the two wave crests. It follows that, if the atomic system has characteristic frequency ν_1, then the second observer will measure a frequency for the first clock of ν_2, where

$$\nu_2 = \frac{\nu_1}{k} = \nu_1\left(\frac{g_{00}\left(x_1^\alpha\right)}{g_{00}\left(x_2^\alpha\right)}\right)^{\frac{1}{2}}. \tag{16.5}$$

Then, in particular,

$$g_{00}\left(x_1^\alpha\right) < g_{00}\left(x_2^\alpha\right) \quad \Rightarrow \quad \nu_2 < \nu_1, \tag{16.6}$$

which means that the frequency is shifted to the red. We define the fractional **frequency shift** to be

$$\frac{\Delta\nu}{\nu} = \frac{\nu_2 - \nu_1}{\nu_1}, \tag{16.7}$$

which, in the case of the weak-field limit (10.49), namely,

$$g_{00} \simeq 1 + 2\phi/c^2,$$

gives (exercise)

$$\frac{\Delta \nu}{\nu} \simeq \frac{\phi_1 - \phi_2}{c^2}. \tag{16.8}$$

Note that we have obtained this expression **without** recourse to the field equations. In the special case of the Schwarzschild solution, this becomes, in non-relativistic units,

$$\frac{\Delta \nu}{\nu} \simeq -\frac{GM}{c^2}\left(\frac{1}{r_1} - \frac{1}{r_2}\right). \tag{16.9}$$

Then

$$r_1 < r_2 \quad \Rightarrow \quad \Delta \nu < 0, \tag{16.10}$$

and so the frequency is shifted to the red.

If we take r_1 to be the observed radius of the Sun, and r_2 the radius of the Earth's orbit (Fig. 16.3), then (neglecting the Earth's gravitational field)

$$\Delta \nu / \nu \simeq -2.12 \times 10^{-6}. \tag{16.11}$$

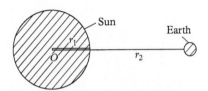

Fig. 16.3 Observation of the redshift of atoms near the Sun's edge.

Observations of the Sun's spectra near its edge give results of this order, but there is great difficulty in interpreting the results generally because of lack of knowledge of the detailed structure of the Sun and the solar atmosphere. Similar remarks hold about white dwarfs, which, because of their small radii compared with their masses, have a more pronounced shift.

Since there are difficulties associated with astronomical measurements of the gravitational redshift, there has been interest in the possibility of a terrestrial test. This is a difficult task because the expected shift over a vertical distance of 30 m, say, is only of the order of 10^{-15}. Fortunately, the discovery of the Mössbauer effect in 1958 gave a method of producing and detecting gamma rays which are monochromatic to one part in 10^{12}, and so makes a terrestrial test feasible. Pound and Rebka carried out such a test in 1960. They placed a gamma ray emitter at the bottom of a vertical 22 m tower with an absorber at the top. Gamma rays emitted at the bottom then suffered a gravitational redshift climbing up the Earth's gravitational field to the top of the tower and were therefore less favourably absorbed. By moving the emitter upwards at a small measured velocity, a compensating Doppler shift was produced which allowed the rays to be resonantly absorbed. The experimental result gave 0.997 ± 0.009 times the predicted shift of 4.92×10^{-15}, that is, an agreement of better than 1%. Other experiments since 1960 have measured the change in the rate of atomic clocks transported on aircraft, rockets, and satellites; these have produced agreement with the theoretical predictions to about the same order of accuracy. One example being the shift experienced by radio signals from the space probe Voyager I in its flight past Saturn in 1980. The accuracy was increased by two more orders of magnitude over the 1960 result in 1976 when a hydrogen maser clock was flown on a

Scout rocket to an altitude of some 10,000 km and compared to a similar clock on the ground; this showed that the differences between the theoretical and measured values for the redshift were less than one part in 2×10^{-4}. It is intriguing to note that the length of the Scout rocket was almost exactly the same as the height of the Jefferson Physical Laboratory tower at Harvard University used for the 1960 experiment.

Although the time differences due to gravitational redshift are small, it turns out that taking them into account is vital for the accuracy of the **Global Positioning System** (GPS). The basic idea is that a GPS satellite sends out a microwave signal encoded with the time and position of the satellite (as measured in an inertial frame located at the Earth's centre). An observer then measures the time at which they receive the signal and, using the fact that the speed of light is c, they can work out their distance from the satellite at the time the signal was emitted. By doing this for four or more satellites, the observer's position can be fixed in space and time even if the observer does not have an accurate clock. However, in order to make accurate measurements of position, the measurements need to be corrected for relativistic effects. The main corrections that are needed are (i) to account for the special relativistic time dilation due to the fact that the satellite is moving relative to the inertial frame of the Earth and (ii) the gravitational redshift due to the fact that the satellite is further from the centre of the Earth than the observer. Both these effects mean that the signals will be received at a slower rate than the rate they were emitted. The GPS satellites orbit at a radius of about $R_S = 2.7 \times 10^4$ km, which gives a radial speed of about 4 km/s and this results in a gamma factor of $\gamma \simeq 8 \times 10^{-9}$. On the other hand, the fractional correction for the gravitational redshift is by (16.9)

$$\frac{GM_e}{c^2 R_S} \simeq 16 \times 10^{-9}. \tag{16.12}$$

Thus, the effect of gravitational redshift is about twice that of the special relativistic time dilation. Although neither of these effects seem very large, they account for a time difference of approximately one nanosecond or a distance of 30 cm. These errors will accumulate over time and without taking into account the relativistic corrections quickly result in GPS errors of tens of metres.

16.3 The Eötvös experiment

We have seen that the gravitational redshift is essentially a test of the principle of equivalence. Since the principle of equivalence is so central to general relativity, we mention briefly here the important Eötvös torsion balance experiment, which tests the equivalence of gravitational and inertial mass. The experiment grew out of the much earlier work of Newton and Bessel using pendula. The Eötvös experiment consists of two objects of different composition connected by a rod of length ℓ and suspended horizontally by a fine wire (Fig. 16.4). If the gravitational acceleration of

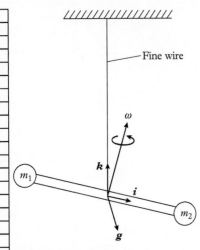

Fig. 16.4 The Eötvös torsion balance.

the two masses is different, then it can be shown that there will be a torque N on the wire with

$$|N| = \eta l (g \times k) i, \qquad (16.13)$$

where g is the gravitational acceleration, i and k are unit vectors along the rod and the wire, and η is a limit on the difference in acceleration called the Eötvös ratio. If the apparatus is rotated with angular velocity ω, then the torque will be modulated with period $2\pi/\omega$. In the original experiment of Baron von Eötvös around the beginning of the last century, g was the gravitational acceleration due to the Earth and the apparatus was rotated about the direction of the wire. Eötvös found a limit on η of $|\eta| < 5 \times 10^{-9}$.

The experiment has been repeated and improved by Dicke at Princeton and Braginski at Moscow. In their experiments, g was due to the Sun, and the rotation of the Earth provided the modulation of the torque. The torque was determined by measuring the force required to keep the rod in place in the Princeton experiment and gave a result $|\eta| < 10^{-11}$. In the Moscow experiment, the torque was determined by measuring the torsional motion of the rod and produced $|\eta| < 10^{-12}$, one of the most accurate results in physics. These results have been confirmed by Su et al. (1994).

16.4   The Einstein equivalence principle

In Chapter 9, we considered the strong and weak forms of the equivalence principle. However, in the context of the experimental tests of general relativity, it is useful to introduce a version called the **Einstein equivalence principle**, which in some sense lies between the weak and strong versions, and for which there is very good experimental evidence. The significance of the principle is that it can be shown to lead to the conclusion that gravity can be described geometrically as a metric theory, which enables one to use the classical tests to compare general relativity with alternative metric theories of gravity.

The **Einstein equivalence principle** states the following:

1. The trajectory of a freely moving 'test' particle is independent of its internal structure and composition (i.e the weak equivalence principle is valid).

2. The outcome of any local non-gravitational experiment is independent of the velocity of the freely-falling reference frame in which it is performed.

3. The outcome of any local non-gravitational experiment is independent of where and when in the universe it is performed.

Mathematically, point (1) is just one version of the **weak equivalence principle**, point (2) amounts to **local Lorentz invariance**, while point (3) is **local space-time position invariance** in both space and time. In the previous sections, we outlined strong experimental evidence for the validity of the weak equivalence principle, using both measurements of gravitational redshift and Eötvös torsion balance experiments. Evidence for local Lorentz invariance comes from the many high-precision tests of special relativity. These include a refined version of the Michelson–Morley experiment, measurements of the independence of the speed of light of the velocity of the sources using binary X-ray sources and high-energy pions as well as experiments demonstrating the isotropy of the speed of light. Evidence of local position independence in space comes from measurements of atomic clocks on rockets and satellites. Evidence for local position independence in time comes from the measurement of spectral lines in distant galaxies (where the light was emitted in the distant past) as well as comparing the gravitational redshift of the Earth-bound clocks relative to the highly stable millisecond pulsar PSR 1937+21 (Will 2006).

The Einstein equivalence principle is of crucial importance because it is possible to argue from this that gravitation must be a 'curved space-time phenomenon', in which the effects of gravity are equivalent to those of living in a curved space-time. Details of this assertion are given in the book by Will (1993), but the argument is based on the fact that, if the principle is valid, then, in local freely falling frames, the laws governing experiments must be independent of the velocity of the frame (local Lorentz invariance), with constant values for the various fundamental constants in order to be independent of location. The only laws we know of that fulfil these requirements are those that are compatible with special relativity. Furthermore, according to the equivalence principle, in local freely falling frames, test bodies appear to be unaccelerated, in other words they move on straight lines; but such 'locally straight' lines simply correspond to 'geodesics' in a curved space-time. Thus, the concepts of inertial observer and Lorentz invariance together imply the use of a Lorentz signature metric to describe the kinematics.

As a consequence of these arguments, theories of gravity that satisfy the Einstein equivalence principle must also satisfy the postulates of **metric theories of gravity**, which are as follows:

1. Space-time is endowed with a symmetric metric.
2. The trajectories of freely falling test bodies are geodesics of that metric.
3. In local freely falling reference frames, the non-gravitational laws of physics are those written in the language of special relativity.

We will see in §16.9 that the two other 'classical' tests, the advance of the perihelion and the bending of light, together with the Shapiro time delay test, enable one to compare the predictions of general relativity with other metric theories of gravity – a test general relativity passes with flying colours.

16.5 Classical Kepler motion

Before considering the motion of a test particle in the Schwarzschild metric, we first review the classical Kepler problem, namely, the motion of a test particle in the gravitational field of a massive body, before considering its general relativistic counterpart. It starts from the assumption that a particle of mass m moves under the influence of an inverse square law force whose centre of attraction is at the origin O, that is,

$$F = -m\frac{\mu}{r^2}\hat{r},$$ (16.14)

where μ is a constant. Then Newton's second law is

$$m\ddot{r} = -m\frac{\mu}{r^2}\hat{r}.$$ (16.15)

The **angular momentum** of m is defined as

$$L = r \times m\dot{r},$$ (16.16)

and so

$$\frac{dL}{dt} = \dot{r} \times m\dot{r} + r \times m\ddot{r}$$

$$= r \times \left(-m\frac{\mu}{r^2}\hat{r}\right)$$

$$= 0,$$

where the cross products of $\dot{r}$ with itself and r with $\hat{r}$ both vanish because the vectors are parallel. Hence, the angular momentum is conserved and

$$L = mh,$$ (16.17)

where h is a constant vector. Assuming $h \neq 0$, it follows from (16.16) that r is always perpendicular to h, and so the particle is restricted to move in a plane. If we introduce plane polar coordinates (R, ϕ), then the equation of motion (16.15) becomes

$$\left(\ddot{R} - R\dot{\phi}^2\right)\hat{R} + \frac{1}{R}\frac{d}{dt}\left(R^2\dot{\phi}\right)\hat{\phi} = -\frac{\mu}{R^2}\hat{R}.$$ (16.18)

Taking the scalar product with $\hat{\phi}$ throughout and integrating produces

$$R^2\dot{\phi} = h,$$ (16.19)

which is conservation of angular momentum again, where h is the magnitude of the angular momentum per unit mass. Taking the scalar product with $\hat{R}$ throughout (16.18) gives

$$\ddot{R} - R\dot{\phi}^2 = -\mu/R^2.$$ (16.20)

We are interested in obtaining the equation of the orbit of the particle, which in plane polar coordinates is

$$R = R(\phi).$$ (16.21)

If we introduce the new variable $u = R^{-1}$, then this can also be written as $u = u(\phi)$. Using the function of a function rule, we find

$$\dot{R} = \frac{dR}{dt} = \frac{d}{dt}\left(\frac{1}{u}\right) = -\frac{1}{u^2}\frac{du}{d\phi}\frac{d\phi}{dt} = -\frac{1}{u^2}hu^2\frac{du}{d\phi} = -h\frac{du}{d\phi}$$

by (16.19). Similarly (exercise),

$$\ddot{R} = -h^2u^2\frac{d^2u}{d\phi^2}, \qquad (16.22)$$

and so (16.20) becomes **Binet's equation**

$$\frac{d^2u}{d\phi^2} + u = \frac{\mu}{h^2}. \qquad (16.23)$$

Binet's equation is the orbital differential equation for the particle, and has solution (exercise)

$$u = \frac{\mu}{h^2} + C\cos(\phi - \phi_0), \qquad (16.24)$$

where C and ϕ_0 are constants. This can be written in terms of R as (exercise)

$$\ell/R = 1 + e\cos(\phi - \phi_0), \qquad (16.25)$$

where $\ell = h^2/\mu$ and $e = Ch^2/\mu$. This is the polar equation of a conic section in which ℓ (semi-latus rectum) determines the scale, e (eccentricity) the shape, and ϕ_0 the orientation (relative to the x-axis). In particular, if $0 < e < 1$, then the conic is an ellipse (Fig. 16.5), and the point of nearest approach to the origin is called the **perihelion**.

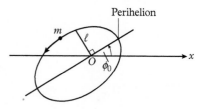

Fig. 16.5 Kepler motion in an ellipse.

The motion of a test particle in the field of a massive body is called the **one-body problem**. We shall establish the classic result that in Newtonian theory the **two-body problem** of two point masses moving under their mutual gravitational attraction can be reduced to a one-body problem. Consider two masses m_1 and m_2 with position vectors $\mathbf{r}_1$ and $\mathbf{r}_2$, respectively (Fig. 16.6). Define the position vector of m_1 (say) relative to m_2 by

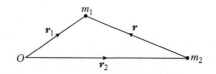

Fig. 16.6 The two-body problem.

$$\mathbf{r} = \mathbf{r}_1 - \mathbf{r}_2.$$

If $\mathbf{F}_{12}$ is the force on m_1 due to m_2, and $\mathbf{F}_{21}$ is the force on m_2 due to m_1, then, by Newton's third law,

$$\mathbf{F}_{21} = -\mathbf{F}_{12}. \qquad (16.26)$$

Using Newton's second law, (16.26), and Newton's universal law of gravitation (4.4), we obtain

$$\mathbf{F}_{12} = m_1\ddot{\mathbf{r}}_1 = -m_2\ddot{\mathbf{r}}_2 = -\frac{Gm_1m_2}{r^2}\hat{\mathbf{r}},$$

and so

$$\ddot{\boldsymbol{r}} = \ddot{\boldsymbol{r}}_1 - \ddot{\boldsymbol{r}}_2 = -\frac{Gm_2}{r^2}\hat{\boldsymbol{r}} - \frac{Gm_1}{r^2}\hat{\boldsymbol{r}} = -\frac{G(m_1 + m_2)}{r^2}\hat{\boldsymbol{r}}.$$

We find, finally, that the equation of motion can be written as

$$\boldsymbol{F}_{12} = m\ddot{\boldsymbol{r}} = -m\frac{\mu}{r^2}\hat{\boldsymbol{r}}, \tag{16.27}$$

where m, the **reduced mass**, is given by

$$m = m_1 m_2/(m_1 + m_2) \tag{16.28}$$

and

$$\mu = G(m_1 + m_2). \tag{16.29}$$

Comparing (16.27) with (16.15), we see that this is the one-body problem we discussed earlier. In the simplest model of planetary motion, we take m_2 to be the mass of the sun, and m_1 to be the mass of the planet. Then, suitably interpreted (see Exercise 16.6), the motion of a planet is again a Kepler ellipse.

16.6 Advance of the perihelion of Mercury

We now look at the one-body problem in general relativity. We assume that the central massive body produces a spherically symmetric gravitational field. The appropriate solution in general relativity is then the Schwarzschild solution. Moreover, a test particle moves on a timelike geodesic, and so we begin by studying some of the geodesics of the Schwarzschild solution. The simplest approach is to employ the variational method of §7.6. Letting a dot denote differentiation with respect to proper time τ, we then find, for timelike geodesics,

$$2K = (1 - 2m/r)\dot{t}^2 - (1 - 2m/r)^{-1}\dot{r}^2 - r^2\dot{\theta}^2 - r^2\sin^2\theta\,\dot{\phi}^2 = 1. \tag{16.30}$$

We next work out the Euler-Lagrange equations. It turns out to be sufficient to restrict attention to the three simplest equations, which are given when $a = 0, 2, 3$ in (7.47) and which are

$$\frac{\mathrm{d}}{\mathrm{d}\tau}\left[(1 - 2m/r)\,\dot{t}\right] = 0, \tag{16.31}$$

$$\frac{\mathrm{d}}{\mathrm{d}\tau}\left(r^2\,\dot{\theta}\right) - r^2 \sin\theta \cos\theta\,\dot{\phi}^2 = 0, \tag{16.32}$$

$$\frac{\mathrm{d}}{\mathrm{d}\tau}\left(r^2 \sin^2\theta\,\dot{\phi}\right) = 0. \tag{16.33}$$

This is because we need four differential equations to determine our four unknowns, namely,

$$t = t(\tau), r = r(\tau), \theta = \theta(\tau), \phi = \phi(\tau).$$

However, (16.30) is itself an integral of the motion and so, together with (16.31)–(16.33), provides the four equations needed. We have seen in Newtonian theory that the corresponding motion is confined to a plane. Let us see if this is still true in general relativity. Consider a particle with initial position in Schwarzschild coordinates given by $x^a(0)$ and initial velocity $\dot{x}^a(0)$. Then, since the Schwarzschild solution is spherically symmetric, we can, without loss of generality, choose the coordinates such that $\theta(0) = \pi/2$ and $\dot{\theta}(0) = 0$. It then follows from (16.32) that $\ddot{\theta}(0) = 0$. Differentiating (16.32), we can show that all higher derivatives of θ must vanish as well, and hence it follows that the motion is in the equatorial plane (why?). Then, setting $\theta = \pi/2$ in (16.33), this can be integrated directly to give

$$r^2\dot{\phi} = h, \tag{16.34}$$

where h is a constant. This is conservation of angular momentum (compare with (16.19)) and note that, in the equatorial plane, the spherical polar coordinate r is the same as the plane polar coordinate R. Similarly, (16.31) gives

$$(1 - 2m/r)\,\dot{t} = k, \tag{16.35}$$

where k is a constant. Substituting for $\dot{\phi}$ and $\dot{t}$ in (16.30), we obtain

$$k^2(1 - 2m/r)^{-1} - (1 - 2m/r)^{-1}\dot{r}^2 - r^2\dot{\phi}^2 = 1. \tag{16.36}$$

We proceed as we did in the classical theory and set $u = r^{-1}$, which leads to

$$\dot{r} = -h\frac{\mathrm{d}u}{\mathrm{d}\phi}$$

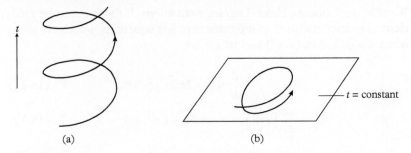

Fig. 16.7 Motion of a test particle (a) in space-time and (b) projected onto $t =$ constant.

Then, using (16.34), we find (16.36) becomes

$$\left(\frac{\mathrm{d}u}{\mathrm{d}\phi}\right)^2 + u^2 = \frac{k^2 - 1}{h^2} + \frac{2m}{h^2}u + 2mu^3. \tag{16.37}$$

This is a first-order differential equation for determining the orbit of a test particle or, more precisely, the trajectory of the test body projected into a slice $t =$ constant (Fig. 16.7). It can be integrated directly by using elliptic functions. We shall use an approximation method to solve it.

Differentiating (16.37), we obtain the second-order equation

$$\frac{\mathrm{d}^2u}{\mathrm{d}\phi^2} + u = \frac{m}{h^2} + 3mu^2. \tag{16.38}$$

This is the relativistic version of Binet's equation (16.23) and differs from the Newtonian result by the presence of the last term. For planetary orbits, this last term is comparatively small, because the ratio of the two terms on the right-hand side of (16.38) is given in relativistic units by $3h^2/r^2$, which for planetary orbits in the solar system is small. On this assumption, we may solve the equation approximately by a perturbation method. The first step is to write down equation (16.38) in a dimensionless form, where we note that, in the relativistic units we are using, m, r, and h all have the dimensions of length. For the Newtonian equation, we have seen that $u \sim m/h^2$ so that we define the dimensionless variable $\tilde{u} = h^2u/m$, which in the Newtonian case is approximately one. Writing (16.38) in terms of $\tilde{u}$, we obtain

$$\frac{\mathrm{d}^2\tilde{u}}{\mathrm{d}\phi^2} + \tilde{u} = 1 + \frac{3m^2}{h^2}\tilde{u}^2.$$

We now introduce the dimensionless quantity

$$\varepsilon = 3m^2/h^2, \tag{16.39}$$

and write the differential equation as

$$\frac{\mathrm{d}^2\tilde{u}}{\mathrm{d}\phi^2} + \tilde{u} = 1 + \varepsilon\tilde{u}^2. \tag{16.40}$$

For a planetary orbit, ε is very small, for example $\varepsilon \sim 10^{-7}$ for Mercury, and, since $\tilde{u} \sim 1$ in the Newtonian case (when $\varepsilon = 0$), the additional perturbation term on the right-hand side will produce an $O(\varepsilon)$ change in the solution. We may therefore assume that the equation has a solution of the form

$$\tilde{u} = \tilde{u}_0 + \varepsilon \tilde{u}_1 + O(\varepsilon^2). \tag{16.41}$$

Substituting in (16.40), we find

$$\tilde{u}_0'' + \tilde{u}_0 - 1 + \varepsilon \left(\tilde{u}_1'' + \tilde{u}_1 - u_0^2\right) + O(\varepsilon^2) = 0. \tag{16.42}$$

If we equate the coefficients of different powers of ε to zero, then the zeroth-order solution u_0 is the usual conic section (16.24)

$$\tilde{u}_0 = (1 + e \cos \phi),$$

where, for convenience, we have taken $\phi_0 = 0$. The first-order equation is

$$\tilde{u}_1'' + \tilde{u}_1 = \tilde{u}_0^2, \tag{16.43}$$

and so, substituting for u_0, we get

$$\begin{aligned}
\tilde{u}_1'' + \tilde{u}_1 &= (1 + e \cos \phi)^2 \\
&= (1 + 2e \cos \phi + e^2 \cos^2 \phi) \\
&= (1 + \tfrac{1}{2}e^2) + 2e \cos \phi + \tfrac{1}{2}e^2 \cos 2\phi.
\end{aligned}$$

If we try a particular solution of the form

$$\tilde{u}_1 = A + B\phi \sin \phi + C \cos 2\phi, \tag{16.44}$$

then we find (exercise)

$$A = (1 + \tfrac{1}{2}e^2), B = e, C = -\frac{e^2}{6}. \tag{16.45}$$

Thus, the general solution of (16.40) is

$$\tilde{u} = (1 + e \cos \phi) + \varepsilon \left[1 + e\phi \sin \phi + e^2(\tfrac{1}{2} - \tfrac{1}{6} \cos 2\phi)\right]. \tag{16.46}$$

The most important correction to u_0 is the term involving $e\phi \sin \phi$, because, after each revolution, it gets larger and larger. If we neglect the other corrections and multiply by m/h^2 to obtain u, this gives

$$u \simeq \frac{m}{h^2}[1 + e \cos \phi + \varepsilon e\phi \sin \phi],$$

or

$$u \simeq \frac{m}{h^2} \left\{ 1 + e \cos \left[\phi \left(1 - \varepsilon \right) \right] \right\}, \qquad (16.47)$$

again neglecting terms of order ε^2 (check). Thus, the orbit of the test body is only approximately an ellipse. The orbit is still periodic, but no longer of period 2π; rather, it is of period

$$\frac{2\pi}{1 - \varepsilon} \simeq 2\pi \left(1 + \varepsilon \right). \qquad (16.48)$$

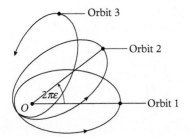

Orbit 3

Orbit 2

Orbit 1

$2\pi\varepsilon$

O

Fig. 16.8 Precession of the perihelion.

In simple intuitive terms, a planet will travel in an ellipse but the axis of the ellipse will rotate, moving on by an amount $\Delta\phi = 2\pi\varepsilon$ between two points of closest approach (Fig. 16.8). This is the famous **precession of the perihelion**. Using the fact that, in the Newtonian case, $h^2 = ma(1 - e^2)$, where a is the semi-major axis of the ellipse, we get the formula

$$\Delta\phi \simeq \frac{6\pi GM}{c^2 a(1 - e^2)}, \qquad (16.49)$$

where we have used (15.55) to write m in non-relativistic units. Furthermore, we may also eliminate M and write this entirely in terms of the orbital parameters as (exercise)

$$\Delta\phi \simeq \frac{24\pi^3 a^2}{c^2 T^2 \left(1 - e^2 \right)}, \qquad (16.50)$$

where T is the period of the orbit.

Now, in fact, in Newtonian theory, there is also an advance of the perihelion. This is because the planetary system is not a two-body system but rather an n-body system, and all the other planets produce a perturbation effect on the motion of one particular planet (rather similar in effect to the perturbation in (16.38)). For example, the planet Jupiter produces a measurable perturbation because its mass is relatively large, being about 0.1% of that of the Sun. Mercury has an orbit with high eccentricity and small period (see (16.50)) and the perihelion position can be accurately determined by observation. Before general relativity, there was a discrepancy between the classical prediction and the observed shift of some 43 seconds of arc per century. Even though this is a very small difference, it is very significant on an astrophysical scale and represents about a hundred times the probable observational error. This discrepancy had worried astronomers since the middle of the 19th century. In fact, in an attempt to explain the discrepancy, it was suggested that there existed another planet, which was given the name Vulcan, whose orbit was inside the orbit of Mercury. (Indeed, there is a famous incident of its reported 'observation' by a French astronomer.) However, Vulcan does not exist, and general relativity appears to explain the discrepancy, since it gives a theoretical prediction of 42.98 seconds of arc per century. This compares very well with the 2013 observation which constrains the anomalous precession to be 42.98 ± 5 seconds of arc per century (Lo et al. 2013). The perihelion shift has also been measured for binary pulsar systems, with that for PSR 1913+16 amounting to $4.2°$ per year, in agreement with the predictions of general relativity. We will analyse the behaviour of binary pulsars in more detail in Chapter 21. The agreement of the residual perihelion precession with the other planets is not so marked because their observed precessions are very small and some of the observational data involved is not sufficiently accurate. One exception is a measurement in 1971 of the residual precession of the minor planet Icarus, which once again is in good agreement with the predicted values of general relativity (Table 16.1).

Table 16.1 Predicted and observed values of residual perihelion precession.

Planet	GR prediction	Observed
Mercury	43.0	43.1 ± 0.5
Venus	8.6	8.4 ± 4.8
Earth	3.8	5.0 ± 1.2
Icarus	10.3	9.8 ± 0.8

16.7 Bending of light

We next consider the case of the trajectory of a light ray in a spherically symmetric gravitational field. The calculation is essentially the same as that given in the last section, except that a light ray travels on a **null** geodesic and so a dot now denotes differentiation with respect to an affine parameter, and the right-hand side of (16.30) is zero. The analogue of (16.38) is easily found to be (exercise)

$$\frac{d^2 u}{d\phi^2} + u = 3mu^2. \tag{16.51}$$

In the limit of **special relativity**, m vanishes and the equation becomes

$$\frac{d^2 u}{d\phi^2} + u = 0, \tag{16.52}$$

the general solution of which can be written in the form

$$u = \frac{1}{D} \sin(\phi - \phi_0), \tag{16.53}$$

where D is a constant. This is the equation of a straight line (exercise) as ϕ goes from ϕ_0 to $\phi_0 + \pi$, where D is the distance of closest approach to the origin. For convenience, without loss of generality, we may assume that the coordinates have been chosen so that $\phi_0 = 0$. The straight-line motion (Fig. 16.9) is the same as is predicted by Newtonian theory.

The equation of a light ray in Schwarzschild space-time (16.51) can again be thought of as a perturbation of the classical equation (16.52). As before, the first step is to write down the differential equation in dimensionless form by introducing the dimensionless variable $\tilde{u} = Du$, which, in the Newtonian case, is approximately one. Writing (16.51) in terms of $\tilde{u}$, we obtain

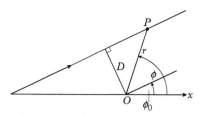

Fig. 16.9 Straight-line motion of a light ray in special relativity.

$$\frac{d^2 \tilde{u}}{d\phi^2} + \tilde{u} = \frac{3m}{D}\tilde{u}^2. $$

We now introduce the dimensionless quantity

$$\varepsilon = \frac{3m}{D}, \tag{16.54}$$

and write the differential equation as

$$\frac{\mathrm{d}^2 \tilde{u}}{\mathrm{d}\phi^2} + \tilde{u} = \varepsilon \tilde{u}^2. \tag{16.55}$$

Then, for a light ray grazing the surface of the Sun, $3m/D \sim 10^{-6}$, so we may regard ε as a small parameter and, since $\tilde{u} \sim 1$ in the unperturbed case, we can expand the solution to (16.55) in powers of ε and look for a solution of the form

$$\tilde{u} = \tilde{u}_0 + \varepsilon \tilde{u}_1 + O(\varepsilon^2). \tag{16.56}$$

Using the fact that, by (16.53), $\tilde{u}_0 = \sin \phi$, substituting into (16.55), and equating powers of ε, we get

$$\tilde{u}_1'' + \tilde{u}_1 = \tilde{u}_0^2 = \sin^2 \phi. \tag{16.57}$$

This has $(1 + C\cos\phi + \cos^2\phi)/3$ as solution (exercise), where C is an arbitrary constant of integration. Substituting in (16.56) and dividing by D, we see that u is given by

$$u = \frac{\sin \phi}{D} + \frac{m\left(1 + C\cos\phi + \cos^2\phi\right)}{D^2} + O(\varepsilon^2). \tag{16.58}$$

Since m/D is small, this is clearly a perturbation from straight-line motion. We are interested in determining the angle of deflection, $\Delta\phi$, for a light ray in the presence of a spherically symmetric source, such as the Sun. A long way from the source, $r \to \infty$ and hence $u \to 0$, which requires the right-hand side of (16.58) to vanish. Let us take the values of ϕ for which $r \to \infty$, that is, the angles of the asymptotes, to be $-\varepsilon_1$ and $\pi + \varepsilon_2$, respectively, as shown in Fig. 16.10. Using the small-angle formulae for ε_1 and ε_2, we get

$$-\frac{\varepsilon_1}{D} + \frac{m}{D^2}(2 + C) = 0, \qquad -\frac{\varepsilon_2}{D} + \frac{m}{D^2}(2 - C) = 0.$$

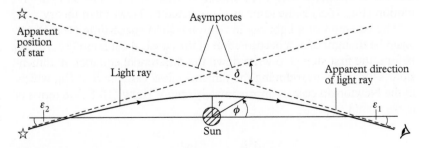

Fig. 16.10 Deflection of light in a gravitational field.

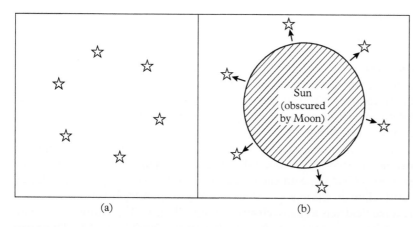

Fig. 16.11 Position of stars in a field (a) when the Sun is absent and (b) during a total eclipse.

Adding, we find

$$\Delta\phi = \varepsilon_1 + \varepsilon_2 = 4m/D\,, \qquad (16.59)$$

or, in non-relativistic units,

$$\Delta\phi = 4GM/c^2 D\,. \qquad (16.60)$$

The deflection predicted for a light ray which just grazes the Sun is 1.75 seconds of arc. Attempts have been made to measure this deflection at a time of total eclipse when the light from the Sun is blocked out by the Moon, so that the apparent position of the stars can be recorded. Then, if photographs of a star field in the vicinity of the Sun at a time of total eclipse are compared with photographs of the same region of the sky taken at a time when the Sun is not present, they reveal that the stars appear to move out radially because of light deflection (Fig. 16.11).

The first expedition to record a total eclipse was one in 1919 under the leadership of Sir Arthur Eddington. The fact that this took place shortly after the end of World War I (and, moreover, that the expedition was led by an English scientist attempting to confirm a theory of a German scientist) caught the imagination of a war-weary world. When Eddington reported that the observations confirmed Einstein's theory, Einstein became something of a celebrity, and the newspapers of the day carried popular articles attempting to explain how we now lived in a curved four-dimensional world. Einstein was so convinced that his theory was right that he reportedly remarked that he would have been sorry for God if the observations had disagreed with the theory. In fact, it is now believed that the observations were not as clear cut as they then seemed, because of problems associated with the solar corona, systematic errors, and photographic emulsions. There have been over ten attempts to make eclipse measurements, and the results have varied markedly from 0.7 to 1.55 times the Einstein prediction. The 2017 eclipse measurement is regarded as one of the most accurate and agreed with the predicted value to about 0.01%, with a measurement error of less than 3%. With the advent of large radio telescopes and the discovery of pointlike sources called **quasars** (quasi-stellar objects), which emit huge amounts of electro magnetic radiation, the deflection can now be measured using long-baseline interferometry when such a source passes close to the sun (Lebach et al. 1995), which gives excellent agreement with the predictions of general relativity (see §16.9 for details).

If one considers a family of curves representing light rays coming in parallel to each other from a distant source, then the presence of a massive object like the Sun causes the light rays to converge and produce a caustic line on the axis $\phi = 0$. In this way, a spherically symmetric gravitational field acts as a **gravitational lens** (Fig. 16.12). Moreover, distant point-like sources can produce double images (see Fig. 16.13). There was considerable interest in 1980 when astronomers first reported the identification of what was previously considered two distinct quasars (known as 0957 + 561A, B) separated by 6 seconds of arc. The evidence is that there is a galaxy, roughly a quarter of the way from us to the quasar, which is the principal component of a gravitational lens. With the advent of the Hubble space telescope, there have been numerous observations of gravitational lensing (see Fig 16.14). An important feature of gravitational lensing, unlike optical lenses, is that it is **achromatic**, that is, the deflection is independent of the frequency of the light. Also, since gravitational lensing only depends on the mass, it can be used as a method for detecting matter in the universe whether it is visible or not, which is an important issue in cosmology (see Chapter 26). The topic of gravitational lensing is now an important one in modern astronomy (see e.g. Perlick 2004 for more details on this.)

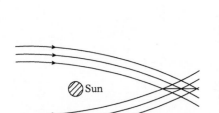

Fig. 16.12 The gravitational lens effect of a Schwarzschild field.

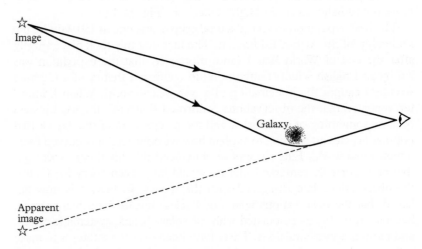

Fig. 16.13 Schematic representation of the double-image effect of the gravitational lens.

Fig. 16.14 Photograph of gravitational lensing taken from the Hubble space telescope.

16.8 Time delay of light

A fourth test which may also be considered a classical test of general relativity was proposed by Shapiro in 1964. The idea is to use radar methods to measure the time travel of a light signal in a gravitational field. Because space-time is curved in the presence of a gravitational field, this travel time is greater than it would be in flat space, and the difference can be tested experimentally.

We begin by considering the path of a light ray in the equatorial plane $\theta = \pi/2$ in Schwarzschild space-time, where, using (15.52),

$$2K = (1 - 2m/r)\,\dot{t}^2 - (1 - 2m/r)^{-1}\,\dot{r}^2 - r^2\,\dot{\phi}^2 = 0, \qquad (16.61)$$

and the dot denotes differentiation with respect to an affine parameter along the ray. To find the travel time of a light ray, we need to eliminate ϕ in terms of r and so obtain a differential equation for dt/dr. The Euler–Lagrange equation for ϕ gives conservation of angular momentum (see (16.34))

$$r^2 \dot{\phi} = h, \qquad (16.62)$$

where h is a constant, and the Euler–Lagrange equation for t is (see (16.35))

$$(1 - 2m/r)\dot{t} = k, \qquad (16.63)$$

where k is a constant. If we let $r = D$ denote the point of closest approach to the Sun (see Fig. 16.15), then, since r is increasing on either side of this it follows that

$$\dot{r} = 0 \text{ when } r = D. \tag{16.64}$$

Substituting this result into (16.61), we get, using (16.62), (16.63), and (16.64) (exercise),

$$h^2/k^2 = D^2/(1 - 2m/D). \tag{16.65}$$

From (16.62) and (16.63), we also get

$$\left(\frac{d\phi}{dt}\right)^2 = \left(\frac{\dot{\phi}}{\dot{t}}\right)^2 = \frac{D^2(1 - 2m/r)^2}{r^4(1 - 2m/D)}, \tag{16.66}$$

which we can use to eliminate $\dot{\phi}$ in (16.61) giving (exercise)

$$\left(\frac{dr}{dt}\right)^2 = (1 - 2m/r)^2 \left[1 - \frac{D^2(1 - 2m/r)}{r^2(1 - 2m/D)}\right]. \tag{16.67}$$

Taking square roots, we get

$$dt = \pm(1 - 2m/r)^{-1} \left[1 - \frac{D^2(1 - 2m/r)}{r^2(1 - 2m/D)}\right]^{-1/2} dr, \tag{16.68}$$

and, using Taylor's theorem to expand the right-hand side in powers of m, we find to order m^2 (exercise)

$$dt = \pm \frac{r}{(r^2 - D^2)^{1/2}} \left(1 + \frac{2m}{r} + \frac{mD}{r(r + D)}\right) dr. \tag{16.69}$$

If we denote the time a light ray travels from the planet to the closest approach to the Sun as $f(D_P)$ then choosing the positive sign (since $r > D$), we get

$$f(D_P) = \int_D^{D_P} dt = \int_D^{D_P} \frac{r}{(r^2 - D^2)^{1/2}} \left(1 + \frac{2m}{r} + \frac{mD}{r(r + D)}\right) dr, \tag{16.70}$$

and the three terms can be integrated directly to give (exercise)

$$f(D_P) = \sqrt{D_P^2 - D^2} + 2m \ln\left(\frac{D_P + \sqrt{D_P^2 - D^2}}{D}\right) + m\left(\frac{D_P - D}{D_P + D}\right)^{1/2}. \tag{16.71}$$

A similar calculation gives the time taken for a light ray to travel from the closest approach to the Sun to the Earth as given by the above formula with D_E replacing D_P. The total time T for a light ray to travel from the Earth to a planet and back is therefore

$$T = 2[f(D_P) + f(D_E)], \tag{16.72}$$

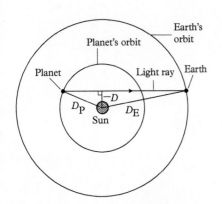

Fig. 16.15 A light ray travelling from a planet to the Earth in the Sun's gravitational field.

which gives finally (exercise),

$$T = 2\left[\left(D_P^2 - D^2\right)^{1/2} + \left(D_E^2 - D^2\right)^{1/2}\right]$$
$$+ \; 4m\ln\left\{\left[\left(D_P^2 - D^2\right)^{1/2} + D_P\right]\left[\left(D_E^2 - D^2\right)^{1/2} + D_E\right]/D^2\right\}$$
$$+ \; 2m\left[\left(\frac{D_P - D}{D_P + D}\right)^{1/2} + \left(\frac{D_E - D}{D_E + D}\right)^{1/2}\right], \qquad (16.73)$$

where D is the closest approach to the Sun, D_P is the planet's orbit radius, and D_E is the Earth's orbit radius (see Fig. 16.15). The first term in square brackets in (16.73) represents the flat space result (as should be clear from the figure and also by setting $m = 0$). The actual experiment consists of sending a pulse from the Earth to the planet and back again. The biggest effect is when D is close to the solar radius D_S. In this case, the excess delay ΔT compared to the special relativistic result is given to a good approximation by (exercise)

$$\Delta T \simeq 4m\left[\ln\left(\frac{4D_P D_E}{D_S^2}\right) + 1\right], \qquad (16.74)$$

or in non-relativistic units

$$\Delta T \simeq \frac{4GM}{c^3}\left[\ln\left(\frac{4D_P D_E}{D_S^2}\right) + 1\right]. \qquad (16.75)$$

The experimental verification of the delay consists in sending pulsed radar signals from the Earth to Venus and Mercury and timing the echoes as the positions of the Earth and the planet change relative to the Sun. For Venus, the measured delay is about 200 μs, which gives an agreement with the theoretical prediction of better than 5%. A more accurate measurement of ΔT was carried out with the Viking mission to Mars in 1976, which verified the general relativistic prediction with an accuracy of better than 1%.

16.9 The PPN parameters

The purpose of the original classical tests was to compare the predictions of general relativity with those of Newtonian gravity. However, the results of the 1919 eclipse measurement of the bending of light were not immediately accepted by other scientists. In his analysis of the 1922 solar eclipse, Eddington replaced the Schwarzschild solution by a more general spherical symmetric metric with a parameter γ which is related to the deflection of light (see equation (16.81)) whose value is 1 for general relativity. This enabled him to give a more quantitative analysis of the results of the 1922 eclipse measurements, which led to them being more generally accepted. This approach was later refined by Robertson and Schiff, who introduced

a second parameter β that, in a certain sense, measures the non-linearity of the theory. The **parametrized-post-Newtonian** (PPN) that we describe below generalizes the work of Eddington, Robertson, and Schiff and provides a systematic way of comparing the predictions of general relativity with a wide class of alternative gravitational theories.

As discussed earlier, general relativity cannot be considered a completely Machian theory and, in an attempt to produce a relativistic theory of gravitation which better incorporated Mach's principle, Brans and Dicke proposed an alternative theory in 1961. We shall not discuss the details of it here except to say that it is motivated in part by the idea of treating the Newtonian constant G as a function of epoch (time), rather than a constant as in general relativity. The resulting theory has an adjustable parameter in it called ω and, if, for suitable boundary conditions, we allow $\omega \to \infty$, then the theory corresponds to general relativity. Indeed, over the years, a number of alternative relativistic tensorial theories of gravity have been proposed. In discussing how the predictions of these alternative theories compare with general relativity, it is useful to have a framework in which the corrections to Newtonian theory provided by these theories can be compared.

As we have seen, there is considerable experimental evidence supporting the Einstein equivalence principle. Mathematically, this supports the conclusion that the only viable theories of gravity are metric theories, or possibly theories that are metric, apart from very weak or short-range non-metric couplings (as in string theory). We will therefore restrict attention to metric theories of gravity. For such theories, the PPN framework provides a useful way of describing the corrections to Newtonian gravity as measured by the experimental tests. There are a total of ten PPN parameters, which are related to the ten independent components of the metric g_{ab}. However, four of these are related to violations of conservation of momentum, while another four are related to the existence of preferred frames or locations, and none of these eight parameters are relevant to the 'classical' experimental tests of general relativity. There are also other experiments which show that all of these eight parameters are extremely small. We will therefore concentrate on two parameters, γ and β, which we describe below and which do influence the results of the classical tests (Will 1993, 2014).

It follows from (15.37) that the most general **static** spherically symmetric metric can be put into the form

$$ds^2 = A(r)dt^2 - B(r)dt^2 - r^2(d\theta^2 + \sin^2\theta d\phi^2). \qquad (16.76)$$

However, if we work in non-relativistic units, we must replace dt by $c dt$, and r by a dimensionless quantity. Assuming that the only physical parameter that determines the geometry of the gravitational field of the star is the mass M, then the only dimensionless quantity that we can construct from r using M, G, and c is

$$\tilde{r} = \frac{c^2 r}{GM}. \qquad (16.77)$$

So, in non-relativistic units, (16.76) becomes

$$ds^2 = A(\tilde{r})c^2 dt^2 - B(\tilde{r})dr^2 - r^2(d\theta^2 + \sin^2\theta d\phi^2). \qquad (16.78)$$

We know from the weak-field limit that to have agreement with Newtonian theory we must have

$$A(\tilde{r}) \simeq 1 - \frac{2}{\tilde{r}} = 1 - \frac{2GM}{c^2 r}.$$

We get the **post-Newtonian** correction by expanding $A(\tilde{r})$ in further inverse powers of $\tilde{r}$.

$$A(\tilde{r}) = 1 - \frac{2}{\tilde{r}} + \frac{A_2}{\tilde{r}^2} + O(\tilde{r}^{-3}).$$

Similarly, in order to obtain the Newtonian limit of Euclidean space when $c \to \infty$, we must have $B(\tilde{r}) \to 1$ as $c \to \infty$, or equivalently as $1/\tilde{r} \to 0$. The post-Newtonian term for B is obtained by expanding it in powers of $1/\tilde{r}$. This gives

$$B(\tilde{r}) = 1 + \frac{B_1}{\tilde{r}} + O(\tilde{r}^{-2})$$

for some constant B_1. The constants A_2 and B_1 are related to the PPN parameters β and γ by (exercise)

$$A_2 = 2(\beta - \gamma), \qquad B_1 = 2\gamma$$

where the PPN parameters β and γ are defined by (16.79) and come from a similar expansion of the isotropic form of the spherically symmetric metric. The leading order terms given by

$$ds^2 = \left(1 - \frac{2GM}{c^2\rho} + 2\beta\left(\frac{GM}{c^2\rho}\right)^2\right)$$
$$- \left(1 - 2\gamma\frac{GM}{c^2\rho}\right)\left[d\rho^2 + \rho^2(d\theta^2 + \sin^2\theta d\phi^2)\right]. \qquad (16.79)$$

The parameter β in some sense measures the 'non-linearity of gravity' while the parameter γ measures the 'space-time curvature produced, per unit rest mass' (Will, 2014)

Using the same perturbation methods as described earlier in the chapter, one can now calculate (exercise) the values of the classical tests using the metric (16.78) and one finds the following:

1. The precession $\Delta\phi_P$ of a planet per orbit is

$$\Delta\phi_P \simeq \frac{1}{3}(2 + 2\gamma - \beta)\frac{6\pi GM}{c^2 a(1 - e^2)}. \qquad (16.80)$$

2. The angle of deflection $\Delta\phi_L$ of a light ray is

$$\Delta\phi_L \simeq \left(\frac{1+\gamma}{2}\right)\left(\frac{4GM}{c^2D}\right). \tag{16.81}$$

3. The excess time delay in the Shapiro test is

$$\Delta T \simeq \left(\frac{1+\gamma}{2}\right)\frac{4GM}{c^3}\left[\ln\left(\frac{4D_E D_P}{D_S}\right)+1\right]. \tag{16.82}$$

This enables one to use the experimental tests to measure the values of β and γ and compare these with the general relativistic values.

The most accurate value of γ comes from measuring the time delay and gives

$$\gamma = 1.000 \pm 0.002 \,.$$

Using this value of γ in (16.80), the most accurately measured values of the advance of the perihelion of Mercury give a value of

$$\beta = 1.000 \pm 0.003 \,.$$

Note, however, that, in making these calculations, we have assumed that the Sun is spherical. However, this is not quite true, since the rotation of the Sun results in it being **oblate** so that it is shorter along the axis of revolution compared to the equatorial radius, due to centrifugal effects. However, accurate measurements of this by Brown et al. (1989) show that these distortions are too small to influence the bounds on β given above.

In conclusion, we see that the solar system tests described in this chapter confirm the predictions of general relativity with an accuracy of only a few tenths of a per cent.

16.10 A chronology of experimental and observational events

We end our considerations of experimental relativity with a brief chronology of the more important experimental and observational events which relate to general relativity.

1919 Eclipse expedition

1922 Eötvös torsion balance experiments
1922 Eclipse expedition

1929 Eclipse expedition

1936 Eclipse expedition

1947 Eclipse expedition

1953 Eclipse expedition

1954 Measurement of red-shift in spectrum of a white dwarf

1960 Hughs–Drever mass-anisotropy experiments
      Pound–Rebka gravitational red-shift experiment

1962 Princeton Eötvös experiments

1965 Discovery of 3 K cosmic microwave background radiation

1966 Reported detection of solar oblateness

1967 Discovery of pulsars

1968 Planetary radar measurements of time delay
      First radio deflection measurements

1970 Cygnus X1: first black hole candidate
      Mariners 6 and 7 time-delay measurements

1971 Measurement of Shaprio time delay

1972 Moscow Eötvös experiments

1973 Eclipse expedition

1974 Discovery of binary pulsar

1976 Rocket gravitational red-shift experiment
      Mariner 9 and Viking time-delay results

1978 Measurement of orbit-period decrease in the Hulse—Taylor binary
      pulsar

1979 Scout rocket maser clock red-shift measurements

1980 Discovery of gravitational lens

2003 Measurement of orbit-period decrease in double pulsar

2004 Frame-dragging measured by Gravity Probe B

2016 Observation of gravitational waves by advanced LIGO

2018 Accurate measurement of gravitational lensing by Hubble Space
      Telescope

2019 Observation of black hole by Event Horizon Telescope

16.11 Rubber-sheet geometry

We end our considerations of general relativity with the description of a
simple model which may help in understanding the theory. Although it is
not in any sense a quantitative model, it has some features in common with
general relativity and, in particular, it illustrates the way that curvature of
space-time can lead to the bending of light. The model consists of an open
box with a sheet of rubber stretched tightly over it. If a marble is then projected across the sheet, then it will move (approximately) in a straight-line
with constant velocity. This simulates **flat** space or special relativity, with
the marble's path corresponding to the straight line geodesic motion of
special relativity (Fig. 16.16). Next, a weight is placed on the centre of
the sheet, causing the rubber to become curved. If the marble is now projected correctly, it will be seen to orbit the central weight. This simulates
general relativity, where a central mass curves up space-time in its vicinity in such a way that a particle with suitable initial conditions will orbit

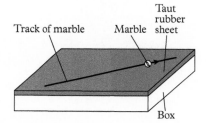

Fig. 16.16 Simulation of straight-line
geodesic motion in special relativity.

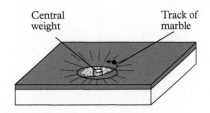

Central
weight

Track of
marble

Fig. 16.17 Simulation of precessing
elliptical motion in general relativity.

the mass. The orbiting marble is performing the 'straightest' motion possible on the curved rubber sheet, or, more precisely, it is travelling on a geodesic of the sheet. Moreover, if the marble is projected carefully, it can be seen to be travelling on an elliptically shaped orbit which, owing to frictional effects between the marble and the rubber sheet, precesses about the central weight in analogy to a planetary orbit (Fig. 16.17).

We can relate this model better to the full theory if we consider an embedding diagram of the Schwarzschild solution in a slice $t = $ constant and in the equatorial plane $\theta = \pi/2$. The line element then reduces to

$$ds^2 = (1 - 2m/r)^{-1}dr^2 + r^2 \, d\phi^2. \qquad (16.83)$$

The curved geometry of this two-dimensional surface is best understood if it is embedded in the flat geometry of a three-dimensional Euclidean manifold. This is depicted in Fig. 16.18, where the distance between two neighbouring points (r, ϕ) and $(r + dr, \phi + d\phi)$ defined by (16.83) is correctly represented. However, distances measured off the curved surface have no direct physical meaning, nor do points off the curved surface; only the curved 2-surface has meaning. If we fill in the interior of the Schwarzschild solution for $r \leqslant r_0$ $(r_0 > 2m)$, then this represents the gravitational field due to a spherical star, and the embedding diagram looks like Fig. 16.19. The surface depicted in Fig. 16.19 is similar in nature to the curved surface of the rubber sheet in Fig. 16.17. This embedding diagram also helps us to understand the phenomenon of **light bending** (Fig. 16.20).

Although these diagrams are helpful in providing some insight into the idea of a curved space-time, they need to be used with caution. For example, the actual deflection of light is twice that suggested by Fig. 16.20 because the light travel takes place in **space-time** rather than space. What they do show, however, is how mass curves up space (actually, space-time) in its vicinity and how free particles and photons travel in the straightest lines possible, namely, on the geodesics of the curved space. As Wheeler puts it so succinctly, 'Space-time tells matter how to move;

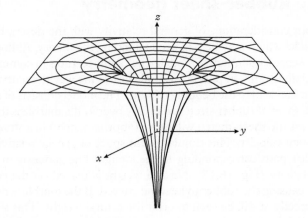

Fig. 16.18 Schwarzschild solution ($t = $ constant, $\theta = \pi/2$) embedded in Euclidean 3-space.

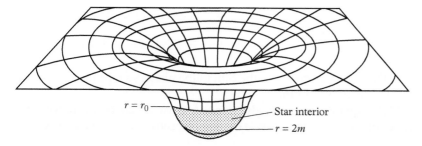

$r = r_0$ — Star interior
$r = 2m$

Fig. 16.19 Embedded geometry exterior and interior to a spherical star.

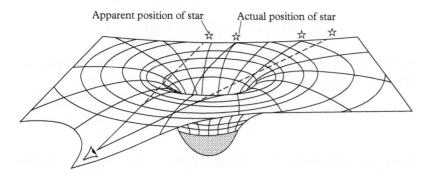

Apparent position of star   Actual position of star

Fig. 16.20 Depiction of light bending in the gravitational field of a star.

matter tells space-time how to curve'. The model also explains how the influence of the central mass is communicated to free particles and photons. This is very different from the action-at-a-distance theory of Newtonian gravitation, where a central mass communicates its influence on a distant particle in a rather mysterious or at least unexplained way. Moreover, if the central mass changes in any way in Newtonian theory, then its influence is altered at all distant points **instantaneously**. In general relativity, any change in the mass of the central source will spread out like a ripple in the rubber-sheet geometry, travelling with the speed of light. This leads to the beginnings of understanding how gravitational waves are generated, which we shall consider further in Part E.

Exercises

16.1 (§16.2) Show that (16.5) leads to (16.8) in the weak-field limit. Deduce (16.9) for the Schwarzschild solution.

16.2 (§16.5) Show that (16.16) and (16.17) lead immediately to (16.19) if $h \neq 0$. What is the motion if $h = 0$?

16.3 (§16.5) Establish the result (16.22), Binet's equation (16.23), and its solution (16.24) and (16.25).

16.4 (§16.5) Establish Kepler's laws of planetary motion for the one-body problem, namely,

K1: Each planet moves about the Sun in an ellipse, with the Sun at one focus.

K2: The radius vector from the Sun to the planet sweeps out equal areas in equal intervals of time.

K3: The squares of the periods τ of any two planets are proportional to the cubes of the semi-major axes a of their respective orbits, that is, $\tau \propto a^{3/2}$.

16.5 (§16.5) Show that the total energy E for the one-body problem can be written in terms of (R, ϕ) as

$$E = \tfrac{1}{2} m (\dot{R}^2 + R^2 \dot{\phi}^2) - m\mu/R.$$

Express this in terms of (u, ϕ) and use (16.25) to identify the parameters as

$$\ell = h^2/\mu, e = (1 + 2Eh^2/m\mu^2)^{1/2}.$$

16.6 (§16.5) Establish (16.27) subject to (16.28) and (16.29) for the two-body problem.

16.7 (§16.5) Define the centre of mass $\mathbf{R}$ by

$$\mathbf{R} = \frac{m_1 \mathbf{r}_1 + m_2 \mathbf{r}_2}{m_1 + m_2},$$

for the two-body problem and deduce that it moves with constant velocity. Transform to an inertial frame S' in which the centre of mass is at rest and situated at the origin O' of the frame S'. Define position vectors $\mathbf{r}_1$ and $\mathbf{r}_2$ of m_1 and m_2 relative to O', and hence describe the motion of m_1 and m_2 relative to O'. How are Kepler's laws modified in the case of the two-body problem? Show that, in particular,

$$\tau \simeq 2\pi a^{3/2} (Gm_{\mathrm{sun}})^{-1/2}.$$

16.8 (§16.6) Establish the Euler–Lagrange equations (16.31)–(16.33). Write down the equation corresponding to $a = 1$ and confirm that (16.31)–(16.33) are the three simplest Euler-Lagrange equations.

16.9 (§16.6) Derive (16.37) and deduce (16.38) from it. What do the equations become in special relativity?

16.10 (§16.6) Show that (16.44) subject to (16.45) is a particular solution of (16.43). Hence establish (16.47).

16.11 (§16.6) Establish the result (16.50). [Hint: replace t by ct in (15.52) and use (15.55) and Exercise 16.7.]

16.12 (§16.6)
(i) Show that the equation for a test particle orbiting in the equatorial plane of the Schwarzschild solution can be written

$$\tfrac{1}{2}\dot{r}^2 + V(r) = k^2/2,$$

where

$$V(r) = \tfrac{1}{2}(1 - 2m/r)(1 + h^2/r^2),$$

h is given by (16.34), and k is given by (16.35).
(ii) Show that $V(r)$ has turning points at

$$r_\pm = \frac{h^2}{2m}\left[1 \pm \sqrt{1 - 12\left(\frac{m}{h}\right)^2}\right],$$

and that r_- is a maxima and r_+ is a minima.
(iii) Take the time derivative of the first equation in (i) to obtain an equation for $\ddot{r}$ and show that $r = r_0 = $ constant implies that $V'(r_0) = 0$. By considering the value of $V'(r)$ near r_+ and r_-, deduce that **stable** circular orbits are only possible when $r = r_+$. Hence show that the **innermost stable circular orbit** (ISCO) occurs when $h^2 = 12m^2$ and at a radius of $r = 6m$.

16.13 (§16.7) Use the method of §16.6 to show that, for a light ray,

$$k^2(1 - 2m/r)^{-1} - (1 - 2m/r)^{-1}\dot{r}^2 - r^2\dot{\phi}^2 = 0.$$

Hence derive equation(16.51).

16.14 (§16.7) Show that (16.53) is the general solution of (16.52) and interpret (16.53) geometrically. Hence establish (16.58) as the approximate solution of (16.51).

16.15 (§16.7) Show that, in the Schwarzschild metric, it is possible for a photon to travel in a circular orbit of radius $r = 3m$ (which is called the **photon sphere**). [Hint: It is enough to consider the case of orbits in the equatorial plane $\theta = \pi/2$.]

16.16 (§16.8)
(i) Establish (16.65).
(ii) Establish (16.67) and deduce (16.69) to order m^2.
(iii) Integrate the three terms in the integrand of (16.70) to establish (16.71) and deduce (16.73). [Hint: use the substitution $r = D\cosh u$ to integrate the first two terms.]

16.17 (§16.8)
(i) Show that (16.73) leads to (16.74).
(ii) Use a dimensional argument to establish (16.75).

16.18 (§16.9) Assuming a power series expansion of r in terms of ρ of the form

$$r = \rho + d + O(\rho^{-1}).$$

where d is a constant then, by comparing coefficients in the spherical symmetric solution (16.76) and the isotropic form (16.79), show that

$$A_2 = 2(\beta - \gamma), \qquad B_1 = 2\gamma.$$

16.19 (§16.9) Show that for the metric (16.76) the formula for the bending of light is given by (16.81).

Further reading

A good discussion of the experimental tests of general relativity is included in the textbook by Hartle (2003), which also discusses gravitational lensing in more detail than we do. A complete but more advanced treatment of the experimental tests is contained in the book and the *Living Reviews in Relativity* article by Will (2014). There is also a comprehensive *Living Reviews in Relativity* article on gravitational lensing by Perlick (2004). The quote 'Space-time tells matter how to move; matter tells space-time how to curve' comes from Wheeler's autobiography *Geons, Black Holes and Quantum Foam* (Wheeler and Ford, 2000), and a shorter version of it is on page 5 of the classic text by Misner, Thorne, and Wheeler (1973).

Hartle J. B. (2003) *Gravity: An Introduction to Einstein's General Relativity*. Addison Wesley, San Francisco, CA.

Misner, C. W., Thorne, K. S., and Wheeler, J. A. (1973). *Gravitation*. Freeman, San Francisco, CA.

Perlick, V. (2004). Gravitational lensing from a spacetime perspective. *Living Reviews in Relativity*, 7, 9.

Wheeler, J. A., and Ford, K. (2000). *Geons, Black Holes, and Quantum Foam: A Life in Physics*. W. W. Norton & Co., New York, NY.

Will, C. M. (1993). *Theory and Experiment in Gravitational Physics* (revised edn). Cambridge University Press, Cambridge.

Will, C. M. (2014). The confrontation between general relativity and experiment. *Living Reviews in Relativity*, 17, 4.

Part D
Black Holes

Non-rotating black holes 17

17.1 Characterization of coordinates

In this chapter, we are going to make an effort to understand the
Schwarzschild vacuum solution. The solution (15.52) is exhibited in a
particular coordinate system. In general, if we wish to write down a solution of the field equations, then we need to do so in some particular
coordinate system. But what, if any, is the significance of any particular coordinate system? For example, take the Schwarzschild solution and
apply as complicated a coordinate transformation as you can imagine, la-
belling the new coordinates x'^a. Now suppose you had been given this
solution and were asked to interpret the solution and identify the coordinates x'^a. The solution will, of course, still satisfy the vacuum field
equations, but there is likely to be little or no geometrical significance attached to the coordinates x'^a. For example, one cannot just set $x'^0 = t$,
say, and interpret t as a 'time' parameter. As a trivial illustration of this,
consider the transformation

$$x'^0 = \theta, \quad x'^1 = r, \quad x'^2 = t, \quad x'^3 = \phi.$$

One thing we can do, however, is establish whether the coordinate
hypersurface

$$x^{(a)} = \text{constant}, \tag{17.1}$$

(where the parentheses enclosing the label a mean that it is to be regarded
as fixed) is timelike, null, or spacelike at a point. The normal co-vector
field to (17.1) is given by

$$n_b = \frac{\partial x^{(a)}}{\partial x^b} = \delta_b^{(a)}.$$

So that the normal vector is

$$n^c = g^{cb} n_b = g^{cb} \delta_b^{(a)} = g^{c(a)},$$

which has magnitude squared given by

$$n^2 = n^c n_c = g^{c(a)} \delta_c^{(a)} = g^{(a)(a)} \quad \text{(not summed)}.$$

Introducing Einstein's Relativity. Second Edition. Ray d'Inverno and James Vickers, Oxford University Press.
© Ray d'Inverno and James Vickers (2022). DOI: 10.1093/oso/9780198862024.003.0017

Hence the hypersurface (17.1) at P is **timelike**, **null**, or **spacelike**, depending on whether $g^{(a)(a)}$ is > 0, $= 0$, or < 0. At any point where the coordinate system is regular, the coordinate hypersurfaces may have any character, but the four normal vector fields $n^b_{(a)}$ must be linearly independent. Thus, for example, the hypersurfaces could all be null, timelike, or spacelike, or any combination of these. We shall be meeting the three most common situations where the four coordinates consist of:

    1 timelike, 3 spacelike;
    1 null, 3 spacelike;
    2 null, 2 spacelike.

Although a metric may be displayed in any coordinate system, if it possesses symmetries, then there will exist preferred coordinates adapted to the symmetries. We have already seen in Chapter 15 that, if a solution possesses a Killing vector field, then the coordinates may be adapted to the Killing vector field. If a solution possesses more than one Killing vector field, then the coordinates can be adapted to each of them as long as the Killing vector fields commute, that is, their Lie brackets vanish. If they do not commute, then the story is more complicated, but none the less the symmetries can be used to tie down the possible coordinate systems.

With these ideas in mind, let us look at the Schwarzschild solution in the form (15.52) to see if we can characterize the coordinates (t, r, θ, ϕ). First of all, since

$$g^{00} = \left(1 - \frac{2m}{r}\right)^{-1}, \quad g^{11} = -\left(1 - \frac{2m}{r}\right), \quad g^{22} = -\frac{1}{r^2},$$

$$g^{33} = -\frac{1}{r^2 \sin^2 \theta}, \tag{17.2}$$

it follows that $x^0 = t$ is timelike and $x^1 = r$ is spacelike, as long as $r > 2m$ and both $x^2 = \theta$ and $x^3 = \phi$ are spacelike. Next, since the metric is independent of t and there are no cross terms in dt, it follows that the solution is static and t is the invariantly defined world time of §15.3. The coordinate r is a radial parameter which has the property that the 2-sphere $t = $ constant, $r = $ constant, has the standard line element

$$ds^2 = -r^2(d\theta^2 + \sin^2 \theta d\phi^2),$$

from which it follows that the surface area of the 2-sphere is $4\pi r^2$. This would fail to be the case if we had chosen a different radial parameter, such as the isotropic coordinate ρ in (15.63). Then, finally, θ and ϕ are the usual spherical polar angular coordinates on the 2-spheres, which are invariantly defined by the spherical symmetry and are unique up to rotations. In short, the Schwarzschild coordinates (t, r, θ, ϕ) are canonical coordinates defined invariantly by the symmetries present.

17.2 Singularities

We now turn to another problem associated with coordinates, that is, the fact that, in general, a coordinate system only covers a portion of the manifold. Thus, for example, the Schwarzschild coordinates do not cover the axis $\theta = 0$, $\theta = \pi$, because the line element becomes degenerate there and the metric ceases to be of rank 4. This degeneracy could be removed by introducing Cartesian coordinates (x, y, z), where, as usual,

$$x = r \sin \theta \cos \phi, \quad y = r \sin \theta \sin \phi, \quad z = r \cos \theta.$$

Such points are called **coordinate singularities** because they reflect deficiencies in the coordinate system used and are therefore **removable**. There are two other values of the coordinates for which the Schwarzschild solution is degenerate, namely, $r = 2m$ and $r = 0$. The value $r = 2m$ is known as the **Schwarzschild radius**. The hypersurface $r = 2m$ again turns out to be a removable coordinate singularity. This is indicated by the fact that the Riemann tensor scalar invariant

$$R_{abcd}R^{abcd} = 48m^2 r^{-6},$$

is finite at $r = 2m$. Since it is a scalar, its value remains the same in all coordinate systems. By the same token, this invariant blows up at the origin $r = 0$. The singularity at the origin is indeed irremovable and is variously called an **intrinsic**, **curvature**, **physical**, **essential**, or **real** singularity. Notice also by (16.4) that, since g_{00} vanishes at the Schwarzschild radius, the surface $r = 2m$ is a surface of infinite redshift. We shall pursue this later.

The normal interpretation of the Schwarzschild solution is as a vacuum solution exterior to some spherical body of radius $a > 2m$ (Fig. 17.1). A different metric would describe the body itself for $r < a$, and would then correspond to some distribution of matter resulting in a non-zero energy-momentum tensor. As we saw in §15.8, Schwarzschild obtained a spherically symmetric static perfect fluid solution known as the **interior** Schwarzschild solution. Our programme in this chapter will be to investigate the Schwarzschild vacuum solution abstracted away from any source for all values of r. In such a case, it should be clear from (17.2) that $r = 2m$ is a null hypersurface dividing the manifold into two disconnected components:

    I. $2m < r < \infty$,
    II. $0 < r < 2m$.

Inside the region II the coordinates t and r reverse their character, with t now being spacelike and r timelike. However, as regions I and II, as given above, are disconnected, we cannot regard them as a single space-time but need to treat them for the moment as separate.

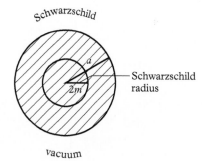

Fig. 17.1 Standard interpretation of the Schwarzschild exterior solution.

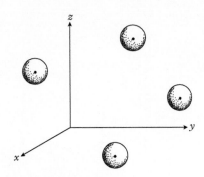

Fig. 17.2 Spatial diagram of Minkowski space-time.

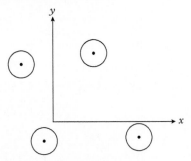

Fig. 17.3 Spatial diagram of Minkowski space time (one spatial dimension suppressed).

17.3 Spatial and space-time diagrams

The main technique we shall use to help interpret the solution is to investigate its local future light cone structure. A local **light cone** at the point P is defined to be the set of null directions at P and so the **future light cone** spans the set of possible directions of light rays emerging from P. Light rays are null geodesics along which

$$ds^2 = g_{ab}dx^a dx^b = 0,$$

so that light rays are tangent to the light cone at every point on the ray. The light cone structure puts constraints on the possible histories of an observer, since an observer moves on a timelike world-line whose direction at any point must lie within the future light cone at the point. Various diagrams will help us in trying to understand the nature of the solution.

In a purely **spatial diagram**, we shall be interested in what happens at various points in the manifold at two successive intervals of time, t_1 and t_2, say. At time t_1, a light flash is emitted from each point of interest and the spatial diagram indicates where the wave fronts of these flashes have reached at time t_2. This is illustrated in Fig. 17.2 for Minkowski space-time. In this figure, the light from each point will form a spherical wave front centred on the point. If there are symmetries present, it may be sufficient to consider what happens if we suppress one spatial dimension. For example, Fig. 17.2 becomes Fig. 17.3 in the plane $z = 0$, say, and the spheres now become circles.

In a **space-time diagram**, we are interested in the history of these light flashes. Suppose we take successive 'snapshots' of the wave fronts emanating from some point P at instants t_1, t_2, t_3, and so on (Fig. 17.4). The idea in a space-time diagram is to stack these pictures up in time. Since this would involve a four-dimensional picture – and there are enough problems in drawing three-dimensional pictures in two dimensions – we suppress one spatial dimension and, as in Chapter 2, we draw the time axis vertically. To be specific, let us restrict attention to the plane $z = 0$ and then the wave fronts will become circles (which will appear as ellipses in the diagram to take some account of perspective) lying on the future light cone through P (Fig. 17.5). In the same way, we can include the past light cone, which can be thought of as an imploding wave front. Again, it will often be sufficient to consider a space-time diagram with two spatial dimensions suppressed (Fig. 17.6). In a **curved** space-time, the

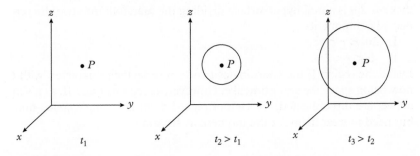

Fig. 17.4 Light flash from a point at three successive times.

curvature manifests itself in space-time diagrams through the light cones being squashed or opened out and tipped or tilted in various ways, as we shall see below.

17.4 Space-time diagram in Schwarzschild coordinates

We first consider the class of **radial null geodesics** defined by requiring

$$\mathrm{d}s^2 = \dot{\theta} = \dot{\phi} = 0. \tag{17.3}$$

Then, using our variational principle approach, we have

$$2K = (1 - 2m/r)\dot{t}^2 - (1 - 2m/r)^{-1}\dot{r}^2 = 0, \tag{17.4}$$

where a dot denotes differentiation with respect to an affine parameter u along the null geodesic. The Euler-Lagrange equation (7.47) corresponding to x^0 is

$$\frac{\mathrm{d}}{\mathrm{d}u}\left[(1 - 2m/r)\dot{t}\right] = 0,$$

which integrates to give

$$(1 - 2m/r)\dot{t} = k, \tag{17.5}$$

where k is a constant. Substituting in (17.4) we find

$$\dot{r}^2 = k^2, \tag{17.6}$$

or

$$\dot{r} = \pm k, \tag{17.7}$$

from which it follows that r is an affine parameter (exercise). Rather than find the parametric equation of these curves, let us look directly for their equation in the form $t = t(r)$. Then

$$\frac{\mathrm{d}t}{\mathrm{d}r} = \frac{\mathrm{d}t/\mathrm{d}u}{\mathrm{d}r/\mathrm{d}u} = \frac{\dot{t}}{\dot{r}}, \tag{17.8}$$

which can be found from (17.5) and (17.7). Taking the positive sign in (17.7), we get

$$\frac{\mathrm{d}t}{\mathrm{d}r} = \frac{r}{r - 2m}, \tag{17.9}$$

which can be integrated, to give (exercise)

$$t = r + 2m\ln|r - 2m| + \text{constant.} \tag{17.10}$$

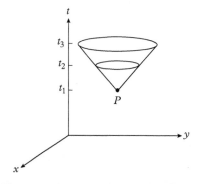

Fig. 17.5 Space-time diagram of light flash (one spatial dimension suppressed).

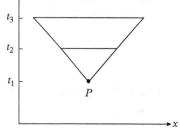

Fig. 17.6 Space-time diagram of light flash (two spatial dimensions suppressed).

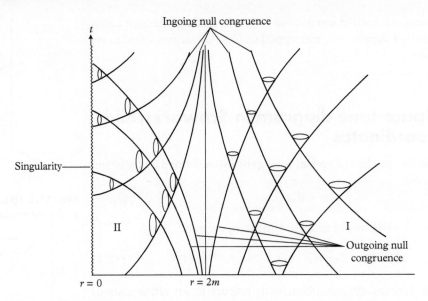

Fig. 17.7 Schwarzschild solution in Schwarzschild coordinates (two dimensions suppressed).

In the region I, by (17.9),

$$r > 2m \quad \Rightarrow \quad \frac{\mathrm{d}r}{\mathrm{d}t} > 0,$$

so that r increases as t increases. We therefore define the curves (17.10) to be a congruence of **outgoing** radial null geodesics. Similarly, the negative sign gives the congruence of **ingoing** radial null geodesics

$$t = -(r + 2m \ln |r - 2m| + \text{constant}). \qquad (17.11)$$

Notice that, under the transformation $t \to -t$, ingoing and outgoing geodesics get interchanged, as we would expect.

We can now use these congruences to draw a space-time diagram (Fig. 17.7) of the Schwarzschild solution in Schwarzschild coordinates with two dimensions suppressed (exercise). The space-time diagram is drawn for some fixed θ and ϕ. Since the diagram will be the same for all θ and ϕ, we should think of each point (t, r) in the diagram as representing a 2-sphere of area $4\pi r^2$. Notice that, as $r \to \infty$, the null geodesics make angles of 45° with the coordinate axes as in flat space in relativistic units, which we should expect since the solution is asymptotically flat. The local light cones tip over in region II, because the coordinates t and r reverse their character. For example, the line $t = \text{constant}$ is a timelike line in region II and so must lie within the local light cone. An observer in region II cannot stay at rest, that is, at a constant value of r, but is forced to move in towards the intrinsic singularity at $r = 0$. This diagram seems to suggest that an observer in region I moving in towards the origin would take an infinite amount of time to reach the Schwarzschild radius

$r = 2m$. Equally, the diagram suggests that the same is true for an in-going light ray. However, we need to remember that regions I and II are really distinct space-times and it turns out that this space-time diagram is misleading, as we shall see.

17.5   A radially infalling particle

Let us consider the path of a radially infalling free particle. It will move on a timelike geodesic given by the equations (exercise)

$$(1 - 2m/r)\dot{t} = k, \qquad (17.12)$$

$$(1 - 2m/r)\dot{t}^2 - (1 - 2m/r)^{-1}\dot{r}^2 = 1, \qquad (17.13)$$

where a dot now denotes differentiation with respect to τ, the proper time along the world-line of the particle. Different choices of the constant k correspond to different initial conditions. Let us make the choice $k = 1$, which corresponds to dropping in a particle from infinity with zero initial velocity (exercise), so that, for large r, we have $\dot{t} \simeq 1$, that is, asymptotically $t \simeq \tau$. Then (17.12) and (17.13) give

$$\left(\frac{d\tau}{dr}\right)^2 = \frac{r}{2m}. \qquad (17.14)$$

Taking the negative square root (why?) and integrating, we find (exercise)

$$\tau - \tau_0 = \frac{2}{3(2m)^{1/2}}(r_0^{3/2} - r^{3/2}), \qquad (17.15)$$

where the particle is at r_0 at proper time τ_0. This is, perhaps rather surprisingly, precisely the same as the classical Newtonian result. No singular behaviour occurs at the Schwarzschild radius, and the body falls continuously to $r = 0$ in a finite proper time.

If, instead, we describe the motion in terms of the Schwarzschild coordinate time t, then

$$\frac{dt}{dr} = \frac{\dot{t}}{\dot{r}} = -\left(\frac{r}{2m}\right)^{1/2}\left(1 - \frac{2m}{r}\right)^{-1}. \qquad (17.16)$$

Integrating, we obtain (exercise)

$$t - t_0 = -\frac{2}{3(2m)^{1/2}}(r^{3/2} - r_0^{3/2} + 6mr^{1/2} - 6mr_0^{1/2})$$

$$+ 2m \ln \frac{[r^{1/2} + (2m)^{1/2}][r_0^{1/2} - (2m)^{1/2}]}{[r_0^{1/2} + (2m)^{1/2}][r^{1/2} - (2m)^{1/2}]}. \qquad (17.17)$$

For situations where r_0 and r are much larger than $2m$, the results (17.15) and (17.17) are approximately the same, as we should expect. If, however, r is very near to $2m$, then we find (exercise)

$$r - 2m = (r_0 - 2m)e^{-(t-t_0)/2m}, \qquad (17.18)$$

from which it is clear that

$$t \to \infty \quad \Rightarrow r - 2m \to 0,$$

so that $r = 2m$ is approached but never passed. The two situations are illustrated in Fig. 17.8.

The coordinate t is useful and physically meaningful asymptotically at large r since it corresponds to the proper time measured by an observer at rest far away from the origin. From the point of view of such an observer, it takes an infinite amount of time for a test body to reach $r = 2m$. However, as we have seen, from the point of view of the test body itself, it reaches both $r = 2m$ and $r = 0$ in finite proper time. Clearly, then, the Schwarzschild time coordinate t is inappropriate for describing this motion. Moreover, the coordinate system goes wrong at $r = 2m$, as is evident from the behaviour of the line element there. In the next section, we shall introduce a new time coordinate which is adapted to radial infall, and in the process we shall remove the coordinate singularity at $r = 2m$.

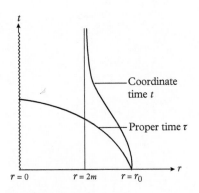

Fig. 17.8 Radially infalling particle in times τ and t.

17.6 Eddington-Finkelstein coordinates

The idea is very simple: we change to a new time coordinate in region I in which the ingoing radial null geodesics become straight lines. It follows immediately from (17.10) that, for $r > 2m$, the appropriate change is given by

$$t \to \bar{t} = t + 2m \ln(r - 2m), \qquad (17.19)$$

because, in the new $(\bar{t}, r, \theta, \phi)$ coordinate system, (17.11) becomes

$$\bar{t} = -r + \text{constant}, \qquad (17.20)$$

which is a straight line making an angle of $-45°$ with the r-axis. Differentiating (17.19), we get

$$\mathrm{d}\bar{t} = \mathrm{d}t + \frac{2m}{r - 2m}\mathrm{d}r, \qquad (17.21)$$

and, substituting for $\mathrm{d}t$ in the Schwarzschild line element (15.52), we find the **Eddington-Finkelstein** form (exercise)

$$\mathrm{d}s^2 = \left(1 - \frac{2m}{r}\right)\mathrm{d}\bar{t}^2 - \frac{4m}{r}\mathrm{d}\bar{t}\mathrm{d}r - \left(1 + \frac{2m}{r}\right)\mathrm{d}r^2 - r^2(\mathrm{d}\theta^2 + \sin^2\theta\mathrm{d}\phi^2).$$

$$(17.22)$$

This solution is now regular and invertible at $r = 2m$ (exercise); indeed, it is regular for the whole range $0 < r < 2m$ so that we can use this form of the metric to extend the coordinate range from $2m < r < \infty$ to $0 < r < \infty$

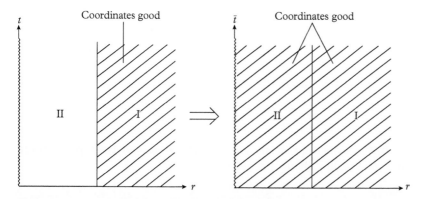

Fig. 17.9 Analytic extension of the Schwarzschild solution.

and obtain a bigger vacuum solution. The process is rather reminiscent of analytically continuing a function in complex analysis and, because of this, (17.22) is called an **analytic extension** of (15.52) (see Fig. 17.9).

One could object that the coordinate transformation (17.19) cannot be used at $r = 2m$ because it becomes singular. However, (17.19) is just a convenient device to get us from (15.52) to (17.22). Our starting point is really the two line elements (15.52) and (17.22). Given these solutions, we then ask the question, What is the largest range of the coordinates for which each solution is regular? The answer is the patch $2m < r < \infty$ (together with, of course, $-\infty < t < \infty$, $0 \leqslant \theta \leqslant \pi$, and $-\pi < \phi \leqslant \pi$, apart from the usual problem with the coordinates on the axis $\theta = 0, \pi$) for (15.52) and the patch $0 < r < \infty$ for (17.22). In the **overlap** region ($2m < r < \infty$), the two solutions are related by using (17.19) in this region, and hence they must represent the **same** solution in this region. Put another way, the region $2m < r < \infty$ of (17.22) is **isometric** to region I of (15.52). It also turns out that the region $0 < r < 2m$ of (17.22) is isometric to region II of (15.52) but we cannot apply (17.19) in the whole region $0 < r < \infty$ since this transformation is not defined for $r \leqslant 2m$. Instead, in the region $r < 2m$, we need to use the transformation $\bar{t} = t + 2m \ln(2m - r)$, which is well defined.

In summary we have started with the Schwarzschild metric (15.52) given in Schwarzschild coordinates for the region $r < 2m < \infty$ (region I) and introduced new Eddington-Finkelstein coordinates given by (17.19) to write it in the form (17.22) in order to remove the coordinate singularity at $r = 2m$. Looking at the metric in these coordinates, we are able to **extend** the metric (17.19) to a vacuum solution of Einstein's equation defined on a larger connected manifold with $0 < r < \infty$. The $r > 2m$ part of the larger solution is isometric to the exterior Schwarzschild solution using (17.19), and the part $0 < r < 2m$ is isometric to region II of (15.52) but using a **different** transformation.

In making the extension using (17.19), we note that the solution in Eddington-Finkelstein coordinates is no longer time symmetric. We can obtain a time-reversed solution by introducing a different time coordinate,

$$t \rightarrow t^* = t - 2m\ln(r - 2m),$$

which straightens out **outgoing** radial null geodesics and results in a **different extension**.

We can write (17.22) in a simpler form by introducing a null coordinate

$$v = \bar{t} + r, \tag{17.23}$$

which, for historical reasons, is called an **advanced time parameter**. The resulting line element is (exercise)

$$ds^2 = (1 - 2m/r)dv^2 - 2dv\,dr - r^2(d\theta^2 + \sin^2\theta d\phi^2). \tag{17.24}$$

It is then easy to show that the congruence of ingoing radial null geodesics is given by $v = $ constant, which should be evident from (17.20). The space-time diagram for the Schwarzschild solution in Eddington-Finkelstein coordinates is given in Fig. 17.10. As before, the light cones open out to 45° cones as $r \rightarrow \infty$. The left-hand edge of the light cones are all at $-45°$ to the r-axis. The right-hand edge starts at 45° to the r-axis at infinity and tips up as r decreases, becoming vertical at $r = 2m$, and tipping inwards for $r < 2m$. Notice that, at $r = 2m$, radially outgoing photons 'stay where they are'. We can get a three-dimensional picture (in

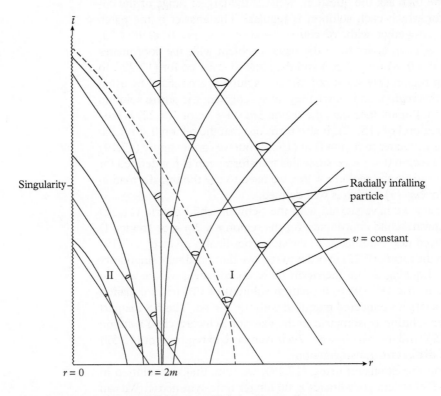

Fig. 17.10 Schwarzschild solution in advanced Eddington-Finkelstein coordinates.

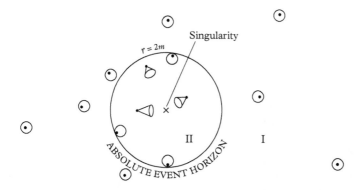

Fig. 17.11 Spatial diagram of the Schwarzschild solution in advanced Eddington-Finkelstein coordinates.

the equatorial plane $\theta = 0$, say) by rotating Fig. 17.10 about the $\bar{t}$-axis. Figure 17.10 now illustrates correctly what happens to a radially infalling particle.

17.7 Event horizons

Figure 17.10 suppresses the angular information in the Schwarzschild solution. This can best be depicted in the equatorial plane in a spatial diagram, as shown in Fig. 17.11. A long way from the origin, the spatial picture is similar to the special relativity picture (Fig. 17.3). As we move close to the origin, the spherical wave fronts are attracted inwards, so that the points from which they emanate are no longer at the centre. This becomes more marked until, on the surface $r = 2m$, only radial outgoing photons stay where they are, whereas all the rest are dragged inwards. In region II, all photons, even radially 'outgoing' ones, are dragged inwards towards the singularity.

It is clear from this picture that the surface $r = 2m$ acts as a **one-way membrane**, letting future-directed timelike and null curves cross only from the outside (region I) to the inside (region II). Moreover, no future-directed null or timelike curve can escape from region II to region I. The surface $r = 2m$ is called an **event horizon** because it represents the boundary of all events which can be observed in principle by an external inertial observer. The situation is reminiscent of the event horizons of hyperbolic motions in §3.8. However, they were observer dependent. The Schwarzschild event horizon is **absolute**, since it seals off all internal events from **every** external observer.

If, instead, we use the null coordinate

$$w = t^* - r, \qquad (17.25)$$

called a **retarded time parameter**, then the line element becomes

$$ds^2 = (1 - 2m/r)dw^2 + 2dw\,dr - r^2(d\theta^2 + \sin^2\theta d\phi^2). \qquad (17.26)$$

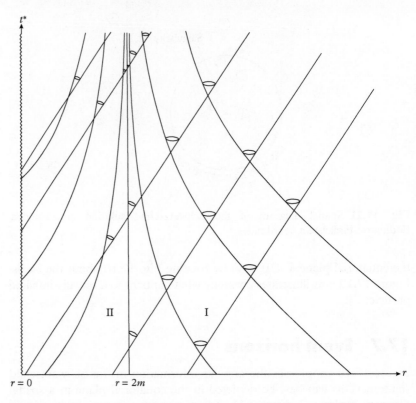

Fig. 17.12 The Schwarzschild solution in retarded Eddington-Finkelstein coordinates.

This solution is again regular for $0 < r < \infty$ and corresponds to the time reversal of the advanced Eddington-Finkelstein solution (17.22) (Fig. 17.12) and for this reason is sometimes called a **white hole**. For $r > 2m$, this again is just the exterior Schwarzschild solution but, for $0 < r < 2m$, this is a **different extension** to that given by (17.24). The surface $r = 2m$ is again a null surface which acts as a one-way membrane. However, this time it acts in the other direction of time, letting only past-directed timelike or null curves cross from the outside to the inside.

17.8 Black holes

The theory of stellar evolution tells us that stars whose masses are of the order of the Sun's mass can reach a final equilibrium state as a white dwarf or a neutron star. But, for much larger masses, no such equilibrium is possible, and in such a case the star will contract to such an extent that the gravitational effects will overcome the internal pressure and stresses which will not be able to halt further contraction. General relativity predicts that a spherically symmetric star will necessarily contract

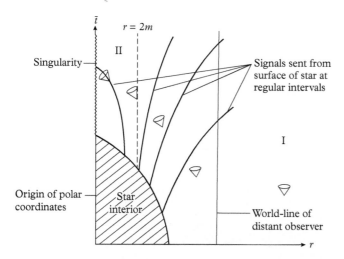

Fig. 17.13 Gravitational collapse (two spatial dimensions suppressed).

until all matter contained in the star arrives at a singularity at the centre of symmetry.

We imagine a situation in which the collapse of a spherically symmetric non-rotating star takes place and continues until the surface of the star approaches its Schwarzschild radius. To get an idea of the magnitude of the Schwarzschild radius, we note that the Schwarzschild radius for the Earth is about 1 cm and that of the Sun is 3 km. As long as the star remains spherically symmetric, its external field remains that given by the Schwarzschild vacuum solution. Figure 17.13 is a two-dimensional space-time diagram of the gravitational collapse, where the Schwarzschild vacuum solution is taken to be in Eddington-Finkelstein coordinates. As is clear from the diagram, an observer can follow a collapsing star through its Schwarzschild radius. If signals are sent out from an observer on the surface of the star at regular intervals according to that observer's clock, then as the surface of the star reaches the Schwarzschild radius, a distant observer will receive these signals with an ever-increasing time gap between them. The signal at $r = 2m$ will never escape from $r = 2m$, and all successive signals will ultimately be dragged back to the singularity at the centre. In fact, no matter how long the distant observer waits, it will only be possible to see the surface of the star as it was just before it plunged through the Schwarzschild radius. In practice, however, the distant observer would soon see nothing of the star's surface, since the observed intensity would die off very fast owing to the infinite redshift at the Schwarzschild radius. The star would quickly fade from view, leaving behind a 'black hole' in space, waiting to gobble up anything which ventured too close.

For completeness, we conclude this section with a three-dimensional space-time diagram of gravitational collapse (Fig. 17.14), which is obtained essentially by rotating Fig. 17.13 about the $\bar{t}$-axis.

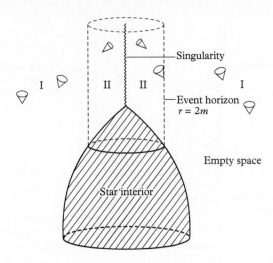

Fig. 17.14 Gravitational collapse (one spatial dimension suppressed).

17.9 A Newtonian argument

The idea of a black hole, in the restricted sense of the gravitational field of a star being so strong that light cannot escape to distant regions, is in fact a consequence of Newtonian theory, if we adopt a particle theory of light. Consider a particle of mass m moving away radially from a spherically symmetric distribution of matter of radius R, uniform density ρ, and total mass M (Fig. 17.15). If the particle possesses a velocity v at a distance r from the centre, then conservation of energy E gives

$$E = \text{kinetic energy} + \text{potential energy}$$
$$= \tfrac{1}{2}mv^2 - GMm/r \,. \tag{17.27}$$

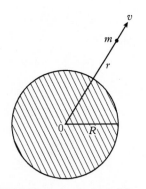

Fig. 17.15 Escape velocity in Newtonian gravitation.

We define the **escape velocity** v_0 to be the velocity at the surface of the distribution of matter which enables the particle to escape to infinity with zero velocity. This requires $v \to 0$ as $r \to \infty$, which by (17.27) results in $E = 0$. Solving for v, we find $v^2 = 2GM/r$, and hence the escape velocity is

$$v_0^2 = 2GM/R. \tag{17.28}$$

Then, if a particle has a radial velocity less than v_0 at the surface, it will eventually be pulled back by the gravitational attraction of the distribution. If light has velocity c, then it will just escape to infinity if it is related to the mass and radius of the distribution by

$$c^2 = 2GM/R\,. \tag{17.29}$$

Thus, if the mass M were increased (keeping the radius constant) or, equivalently, the radius R decreased (keeping the mass constant), then it follows that light could no longer escape. This was recognized by Laplace in 1798 who pointed out that a body of about the same density as the Sun but 250 times its radius would prevent light from escaping. Note that the limiting condition (17.29) in terms of the radius R is

$$R = 2GM/c^2 , \qquad (17.30)$$

or $R = 2m$ in relativistic units, which is the Schwarzschild radius.

17.10   Tidal forces in a black hole

Consider a distribution of non-interacting particles falling freely towards the Earth in Newtonian theory, where initially the distribution is spherical (see Exercise 9.6). Each particle moves on a straight line through the centre of the Earth, but those nearer the Earth fall faster because the gravitational attraction is stronger. The sphere no longer remains a sphere but is distorted into an ellipsoid with the same volume (Fig. 17.16). Thus, the gravitation produces a **tidal force** in the sphere of particles. The tidal effect results in an elongation of the distribution in the direction of motion and a compression of the distribution in transverse directions. The same effect occurs in a body falling towards a spherical object in general relativity but, if the object is a black hole, the effect becomes infinite as the singularity is reached. We can gain some idea of this by considering the equation of geodesic deviation (see (10.38) and (10.39)) in the form

$$\frac{D^2 \eta^\alpha}{D\tau^2} + R^a{}_{bcd}\, e^\alpha{}_a e_0{}^b e_\beta{}^c e_0^d \eta^\beta = 0, \qquad (17.31)$$

for the spacelike components of the orthogonal connecting vector η^α connecting two neighbouring particles in freefall. Let the frame $e_i{}^a$ be defined in Schwarzschild coordinates as

$$e_0{}^a \overset{*}{=} (1 - 2m/r)^{-\frac{1}{2}} (1, 0, 0, 0), \qquad (17.32)$$

$$e_1{}^a \overset{*}{=} (1 - 2m/r)^{\frac{1}{2}} (0, 1, 0, 0), \qquad (17.33)$$

$$e_2{}^a \overset{*}{=} r^{-1} (0, 0, 1, 0), \qquad (17.34)$$

$$e_3{}^a \overset{*}{=} (r \sin \theta)^{-1} (0, 0, 0, 1), \qquad (17.35)$$

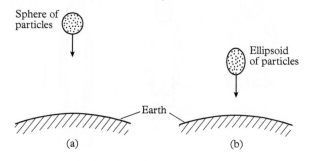

Fig. 17.16 Newtonian tidal force: (a) before; (b) after.

and let us denote the components of η^α by

$$\eta^\alpha = (\eta^1, \eta^2, \eta^3) = (\eta^r, \eta^\theta, \eta^\phi). \tag{17.36}$$

Then (17.31) reduces in Schwarzschild space-time in the above frame to the equations (exercise)

$$\frac{\mathrm{D}^2 \eta^r}{\mathrm{D}\tau^2} = +\frac{2m}{r^3} \eta^r, \tag{17.37}$$

$$\frac{\mathrm{D}^2 \eta^\theta}{\mathrm{D}\tau^2} = -\frac{m}{r^3} \eta^\theta, \tag{17.38}$$

$$\frac{\mathrm{D}^2 \eta^\phi}{\mathrm{D}\tau^2} = -\frac{m}{r^3} \eta^\phi. \tag{17.39}$$

The positive sign in (17.37) indicates a tension or stretching in the radial direction, and the negative signs in (17.38) and (17.39) indicate a pressure or compression in the transverse directions (see Misner, Thorne, and Wheeler 1973 for further details). Moreover, the equations reveal that the effect becomes infinite at the singularity $r = 0$.

Consider an intrepid astronaut falling feet first into a black hole (Fig. 17.17). The astronaut's feet are attracted to the centre by an infinitely mounting gravitational force, while the astronaut's head is accelerated downward by a smaller though ever-rising force. The difference between the two forces becomes greater and greater as the astronaut reaches the centre, where the difference becomes infinite. At the same time as the head-foot stretching, the astronaut is pulled by the gravitational field into regions with ever-decreasing circumference and so the astronaut is squashed on all sides. Again the squashing becomes infinite at the centre. Indeed, not only do the tidal effects tear the astronaut to pieces, but the very atoms of which the astronaut is composed must ultimately share the same fate!

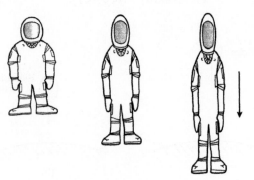

Fig. 17.17 Successive times in the astronaut's fall.

17.11   Observational evidence for black holes

Observing a black hole directly is impossible, unless one were lucky enough to see a star disappear. However, it is certainly possible to infer the existence of a black hole through its gravitational effects on its surroundings. There is now a wealth of such evidence for the existence of black holes that we briefly describe below. For more details, see Cardoso and Pani (2019).

The first such evidence comes from X-ray binaries; these are double stars with one standard star and a second, compact, invisible component. By studying the motion of the standard star, one can deduce the mass of the invisible partner and, if it is much larger than the maximum mass a star can have without collapsing, then it cannot be a neutron star and is a black hole candidate. The black hole will suck matter from its visible partner, forming an accretion disc, and the hot inner regions will produce intense bursts of X-rays formed by synchrotron radiation shortly before the spiralling matter disappears down the hole (Fig. 17.18). It was the discovery in 1971 of the rapid variations of the X-ray source Cygnus X1 by telescopes aboard the Uhuru satellite that provided the first evidence of the likely existence of black holes. The visible component is a supergiant star, and detailed study of the X-rays led to the conclusion that the unseen body is a compact object with a mass in excess of nine solar masses. Since the maximum masses of white dwarfs and neutron stars are believed to be approximately 1.4 and 4 solar masses, respectively, then the simplest conclusion is that the object is a black hole. Since 1971, a number of other black hole candidates have been found in X-ray binaries.

A more direct confirmation of the existence of black holes comes from the observation of gravitational radiation. In September 2015, LIGO

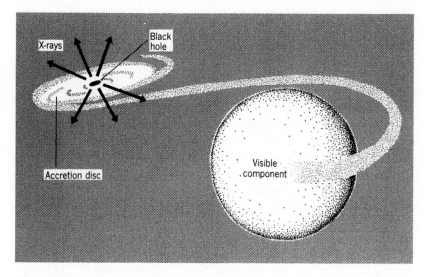

Fig. 17.18 A binary star with one visible and one black hole component.

measured gravitational waves consistent with the theoretical predictions for the radiation produced by the merger of two black holes of about 36 and 29 solar masses, respectively. Since then, many more gravitational wave events have been observed which are completely in line with the predictions of general relativity and the existence of black holes. We will say more about this in Chapter 21.

Although black holes were originally thought of as coming from the collapse of individual stars, there is now considerable evidence that many galaxies have supermassive black holes at their centre. Astronomers use the term 'active galaxy' to describe galaxies with unusual characteristics, such as unusual spectral line emission and very strong radio emission. Theoretical and observational studies have shown that the activity in these active galactic nuclei (AGNs) may be explained by the presence of supermassive black holes, which can be millions of times more massive than stellar ones. The models of these AGNs consist of a central black hole that may be millions or billions of times more massive than the Sun. In particular, by observing the proper motions of stars near the centre of our own Milky Way, one can deduce that there lies at the centre an object with a mass of about 2.6 million solar masses and a radius of at most 2^{10} km. Although the upper limit of the radius is larger than the Schwarzschild radius, this is strongly indicative of the existence of a black hole, as there are no other plausible scenarios for confining so much invisible mass to such a small volume.

Further direct evidence for the existence of supermassive black holes at the centre of galaxies comes from the Event Horizon Telescope. This is a large optical telescope array sufficient to observe objects the size of the event horizon of a supermassive black hole. The idea is to look at the image of the accretion disc of an AGN and compare this with the substantially distorted image due to the extreme gravitation lensing of the black hole as predicted by general relativity. Following observation of a potential black hole in the center of Messier 87 in 2017, the Event Horizon Telescope project spent two years analysing the data and in 2019 released an image that agreed very well with the predictions of general relativity.

17.12 Theoretical status of black holes

When considering black holes at the theoretical level, there is the objection that the solution is too special in being spherically symmetric. For example, no account has been taken of charge or rotation. In Chapters 18 and 19, we shall consider the Reissner-Nordström and Kerr solutions, which deal with charged and rotating black holes, respectively. We shall see that, although the story changes in detail, the chief characteristics of a black hole, namely, the existence of absolute event horizons and singularities, persist. The next objection is that asymmetries have been excluded. It is not surprising, it can be argued, that, if all the matter is moving in radially towards the centre, then it will ultimately result in a singularity there. However, perturbations of the Schwarzschild solution have been considered and appear to suggest that all asymmetries are eventually radiated away and that, asymptotically in time, the system settles down to a Schwarzschild black hole (Fig. 17.19).

Another objection relates to the particular set of field equations used, namely, those of general relativity. However, Penrose and Hawking have managed to prove some remarkable theorems, the so-called **singularity theorems**, which suggest that many of the qualitative features of this collapse picture remain in a more general situation. Their results do not depend on having spherical symmetry or the particular field equations of general relativity, but on much weaker assumptions such as a metric theory of gravity and the consequent curvature of space-time (as implied by the Einstein equivalence principle), relativistic causality, and energy conditions (see §20.13 for more details). The theorems prove that, with these very reasonable assumptions, as a result of the gravitational collapse of a star, there exist geodesics which come to an end, that is to say, that cannot be extended any further. This is usually taken to mean that they are ending on a singularity. Quite where the singularities are located and what their structure is like are issues which these theorems do not directly address. Of course, even these very weak assumptions may not apply in extremely strong gravitational fields. It could be possible, for example, that such fields result in violations of the energy conditions or failures of causality. The general belief, however, is that the theorems provide strong evidence that singularities are, in fact, generic features of relativistic theories of gravitation.

There is another problem which has not yet been resolved. In order to discuss in detail the stability of a collapse situation, we need to understand what is going on inside the star. That is, we need realistic **interior** solutions which can be matched on to the known exterior solutions. However, all attempts at finding a realistic interior Kerr solution, and there have been many of them, appear to have failed. This is somewhat disturbing, because the attempts seem to suggest that the matching cannot be done. Were we to have an interior solution, it is conceivable that the motion might be unstable, leading finally to fragmentation rather than collapse. Finally, we point out that gravitational collapse deals with situations of high densities and that these are really the province of quantum theory. It seems likely that a classical theory like general relativity might be modified profoundly by quantum effects. Indeed, some theories of quantum gravity suggest that the collapse is halted before a singularity is reached and a bounce takes place. However, Penrose has pointed out that we do not need high densities to create event horizons. Since the radius R of the event horizon is proportional to the mass m, and the volume is proportional to R^3, and hence m^3, we see that the average density of a black hole is proportional to $1/m^2$. Hence, perhaps counter intuitively, the larger the black hole the smaller the density. In particular, the average density of the supermassive black hole that is thought to exist at the centre of our galaxy is of the order of 10% of the density of water. Furthermore, the gravitational field at the event horizon for such a supermassive black hole produces a tidal force comparable to that on the surface of the Earth so this does not involve any extreme physics either.

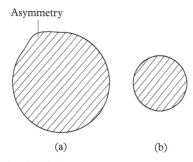

Asymmetry

(a)　　　　(b)

Fig. 17.19 Asymmetry radiated away: (a) before; (b) eventually.

Exercises

17.1 (§17.1) Interpret the solution

$$ds^2 = \left(1 - \frac{2m}{\phi}\right) d\theta^2 - \left(1 - \frac{2m}{\phi}\right)^{-1} d\phi^2 - \phi^2 dt^2 - \phi^2 \sin^2 t \, dr^2.$$

17.2 (§17.1) Apply the transformations

$$t = \bar{t}^2 \bar{r}, \quad r = \bar{r}\cos\bar{\theta} + 2m,$$
$$\theta = \sin^{-1}(\bar{r}\bar{\theta}), \quad \phi = \cos(\bar{\phi}\bar{t})$$

to the Schwarzschild line element (15.52) and find the coefficient of $d\bar{t}^2$.

17.3 (§17.1) What is the character of the coordinates of
(i) (t, ρ, z, ϕ) in

$$ds^2 = \rho^{-2m} dt^2 - \rho^{-2m}[\rho^{2m^2}(d\rho^2 + dz^2) + \rho^2 d\phi^2];$$

(ii) (u, r, x, y) in

$$ds^2 = x^2 du^2 - 2du dr + 4rx^{-1} du dx - r^2 dx^2 - x^2 dy^2.$$

17.4 (§17.1) Dingle's metric is the most general diagonal metric

$$ds^2 = A dt^2 - B dx^2 - C dy^2 - D dz^2,$$

where A, B, C, and D are functions of all four coordinates. What does this solution become if $\partial/\partial x$, $\partial/\partial y$, and $\partial/\partial z$ are commuting vector fields and the solution is adapted to these Killing vector fields?

17.5 (§17.2) Write the Schwarzschild line element (14.47) in coordinates (t, x, y, z) where x, y, and z are defined by

$$x = r\sin\theta\cos\phi, \quad y = r\sin\theta\sin\phi, \quad z = r\cos\theta.$$

17.6 (§17.3) Draw a two-dimensional space-time diagram of null geodesics in special relativity. Draw the world-line of an observer moving into the origin and out again.

17.7 (§17.4) Integrate (17.6). Deduce that r is an affine parameter. Integrate (17.9) to obtain (17.10).

17.8 (§17.4) Confirm Fig. 17.7 by first drawing the graphs of
(i) $y = \ln x \ (x > 0)$,
(ii) $y = \ln|x|$,

(iii) $y = 2m \ln |x|$,

(iv) $y = x + 2m \ln |x|$,

in turn, translating the y-axis to $x = 2m$, and then drawing the graphs of

(v) $y = x - 2m + 2m \ln |x - 2m|$ $(x > 0)$,

(vi) $y = x + 2m \ln |x - 2m| + c$ $(x > 0)$,

for different values of the constant c. What is the slope of the radial null geodesics at $r = 0$?

17.9 (§17.5) Establish (17.12) and (17.13) for the equations of a radially infalling particle. Show that the choice $k = 1$ corresponds to the particle having zero velocity at spatial infinity $(r = \infty)$.

17.10 (§17.5) Integrate (17.14) to obtain (17.15). Show that this is the same result as that for a particle falling radially from r_0 to r in Newtonian theory under the influence of a point particle situated at the origin of mass M, where the particle has zero velocity at infinity.

17.11 (§17.5) Integrate (17.16) to obtain (17.17).

17.12 (§17.5) If r is near $2m$, set $\varepsilon = 1 - r/2m$ and show that the dominant term in (17.16) is $1/\varepsilon$. Hence deduce (17.18).

17.13 (§17.6) Show that (17.19) transforms the Schwarzschild line element (15.52) into the form (17.22). Use (17.23) to express the resulting line element in the form (17.24).

17.14 (§17.6) Calculate the contravariant form g^{ab} of the Eddington-Finkelstein metric (17.22).

17.15 (§17.6) Draw the Schwarzschild solution in advanced Eddington–Finkelstein coordinates with one spatial dimension suppressed in the equatorial plane $\theta = \pi/2$. (Hint: rotate Fig. 17.10 about the $\bar{t}$-axis.)

17.16 (§17.7) Show that (17.25) leads to the form (17.26). Find the equations for radial null geodesics and establish Fig. 17.12.

17.17 (§17.8) Draw the white hole analogue of Fig. 17.13 and describe its appearance to an external observer.

17.18 (§17.10) Show that (17.32)–(17.35) defines an orthonormal frame in Schwarzschild space-time. Show that the spatial part of the equation of geodesic deviation leads to (17.37)–(17.39). [Hint: Use the results of Exercise 15.13.] Give a qualitative argument which reveals that η^r increases without bound as $r \to 0$.

Further reading

The main reference for black holes is the book by Hawking and Ellis (1973). A summary of the observational evidence for black holes is surveyed in the *Living Reviews in Relativity* article by Cardoso and Pani (2019). We also give a link to the results of the Event Horizon Telescope.

Cardoso, V., and Pani, P. (2019). Testing the nature of dark compact objects: A status report. *Living Reviews in Relativity*, 22, 4.

Hawking, S. W., and Ellis, G. F. R. (1973). *The Large Scale Structure of Space-Time*. Cambridge University Press, Cambridge.

The Event Horizon Telescope Collaboration, Akiyama, K., Alberdi, A., et al. (2019). First M87 event horizon telescope results. I. The shadow of the supermassive black hole. *Astrophysical Journal Letters*, 875, 1.

Maximal extension and conformal compactification

18

18.1 Maximal analytic extensions

We saw in the last chapter that the Schwarzschild solution for $2m < r < \infty$ can be extended either into the advanced Eddington-Finkelstein solution (17.24) or the retarded Eddington-Finkelstein solution (17.26), where $0 < r < \infty$. That this is possible is indicated by the fact that a radial timelike geodesic can be extended through $r = 2m$ down to $r = 0$. The question naturally arises, Is it possible to extend these solutions further?

We need to make this question more precise, which we do by introducing a couple of definitions. A manifold endowed with an affine or metric geometry is said to be **maximal** if every geodesic emanating from an arbitrary point of the manifold either can be extended to infinite values of the affine parameter along the geodesic in both directions or terminates at an intrinsic singularity (see §17.2). If, in particular, all geodesics emanating from any point can be extended to infinite values of the affine parameters in both directions, the manifold is said to be **geodesically complete**. Clearly, a geodesically complete manifold is maximal, but the converse is not true in general. Minkowski space-time provides a trivial example of a geodesically complete manifold. Neither the Schwarzschild nor the Eddington-Finkelstein advanced or retarded extensions is in fact maximal. However, Kruskal has found the maximal analytic extension of the Schwarzschild solution and, moreover, this extension is unique. The Kruskal solution, although maximal, is again not complete because of the existence of intrinsic singularities. The Kruskal solution can be obtained by simultaneously straightening out both incoming and outgoing radial null geodesics. We shall sketch the original procedure of Kruskal in the next section.

18.2 The Kruskal solution

We start by introducing both an advanced null coordinate v and a retarded null coordinate w, in which case, in the coordinates (v, w, θ, ϕ), the Schwarzschild line element becomes (exercise)

$$ds^2 = (1 - 2m/r)\, dvdw - r^2 \left(d\theta^2 + \sin^2\theta\, d\phi^2 \right), \qquad (18.1)$$

Introducing Einstein's Relativity. Second Edition. Ray d'Inverno and James Vickers, Oxford University Press.
© Ray d'Inverno and James Vickers (2022). DOI: 10.1093/oso/9780198862024.003.0018

where $r(v, w)$ is a function of the coordinates v and w and is determined implicitly by

$$\tfrac{1}{2}(v - w) = r + 2m \ln(r - 2m). \tag{18.2}$$

Note that, using (17.23) and (17.25), we may write v and w more symmetrically as

$$v = t + r^*, \quad w = t - r^*, \tag{18.3}$$

where r^* is defined for $r > 2m$ by

$$r^* = r + 2m \ln(r - 2m), \tag{18.4}$$

and is called the **tortoise radial coordinate** by some authors.

The 2-space $\theta = $ constant, $\phi = $ constant, has metric

$$ds^2 = (1 - 2m/r)\, dv dw, \tag{18.5}$$

and hence by the second theorem in §6.13 must be conformally flat. To make this evident, we define

$$\tilde{t} = \tfrac{1}{2}(v + w), \quad \tilde{x} = \tfrac{1}{2}(v - w),$$

and then (18.5) becomes

$$ds^2 = (1 - 2m/r)(d\tilde{t}^2 - d\tilde{x}^2).$$

The most general coordinate transformation which leaves the 2-space (18.5) expressed in such conformally flat double null coordinates is

$$v \to v' = v'(v), \quad w \to w' = w'(w),$$

where v' and w' are arbitrary functions, which leads to

$$ds^2 = (1 - 2m/r)\frac{dv}{dv'}\frac{dw}{dw'}dv'dw'.$$

Introducing

$$t' = \tfrac{1}{2}(v' + w'), \quad x' = \tfrac{1}{2}(v' - w'),$$

we can write (18.5) in the general form

$$ds^2 = F^2(t', x')(dt'^2 - dx'^2).$$

A particular choice of v' and w' will then determine the precise form of the line element.

The choice which Kruskal made was

$$v' = \exp(v/4m), \tag{18.6}$$

$$w' = -\exp(-w/4m). \tag{18.7}$$

The radial coordinate r is to be considered a function of t' and x' determined implicitly by the equation

$$t'^2 - x'^2 = -(r - 2m)\exp(r/2m), \tag{18.8}$$

and F is given by

$$F^2 = \frac{16m^2}{r}\exp(-r/2m).$$

Then the line element becomes

$$ds^2 = \frac{16m^2}{r}\exp\left(-\frac{r}{2m}\right)dt'^2 - \frac{16m^2}{r}\exp\left(-\frac{r}{2m}\right)dx'^2$$
$$- r^2\left(d\theta^2 + \sin^2\theta d\phi^2\right), \tag{18.9}$$

where $r(x', t')$ is given by (18.8).

A two-dimensional space-time diagram of the Kruskal solution is shown in Fig. 18.1. As we indicated, all the light cones are now 45° cones and the incoming and outgoing radial null geodesics are straight lines. Figure 18.1 shows a radial timelike geodesic which starts from $(r = 4m, t' = 0)$ and falls into the event horizon $r = 2m$, ending up on the future singularity at $r = 0$. The figure includes some of the signals sent out from this geodesic and illustrates the trapped nature of the signals sent inside the event horizon. Notice from (18.8), which is quadratic in t' and x', that one value of r determines two hypersurfaces. In two dimensions, the space-time is bounded by two hyperbolae representing the intrinsic singularity at $r = 0$. They are termed the **past** singularity and the **future** singularity, respectively. The future singularity is **spacelike** and hence unavoidable in region II. The asymptotes of the hyperbolae represent the event horizons corresponding to $r = 2m$. These asymptotes divide the space-time into four regions labelled I, II, I', and II'. Regions I and II correspond to the advanced Eddington-Finkelstein solution (see Fig. 17.10), with region I corresponding to the Schwarzschild solution for $r > 2m$, and region II corresponding to the black hole solution. regions I and II' correspond to the retarded Eddington-Finkelstein solution (see Fig. 17.12), with region II' corresponding to the white hole solution. What is surprising is that there is a new region called I' which is geometrically identical to the asymptotically flat exterior Schwarzschild solution region I. The topology connecting I and I' is rather complicated and we consider it next.

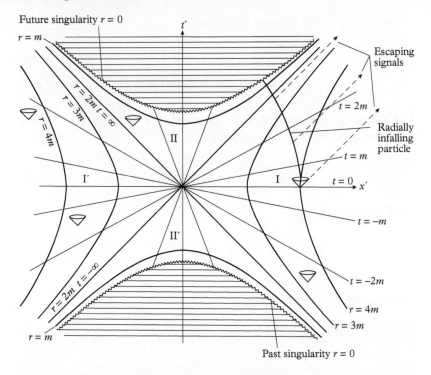

Fig. 18.1 Space-time diagram of the Kruskal solution.

18.3 The Einstein-Rosen bridge

Remember that each point in the diagram represents a 2-sphere. We can gain some intuitive idea of the overall four-dimensional structure if we consider first the submanifold $t' = 0$. Then, from (18.9), the line element induced on this hypersurface is given by

$$ds^2 = -F^2 \, dx'^2 - r^2 \left(d\theta^2 + \sin^2 \theta \, d\phi^2 \right). \tag{18.10}$$

Setting $t' = 0$ in (18.8), we see that, as we move along the x'-axis from $+\infty$ to $-\infty$, the value of r decreases to a minimum $2m$ at $x' = 0$ and then increases again as x' goes to $-\infty$. We can draw a cross-section of this manifold corresponding to the equatorial plane $\theta = \pi/2$, in which case (18.10) reduces further to

$$ds^2 = -\left(F^2 \, dx'^2 + r^2 \, d\phi^2 \right). \tag{18.11}$$

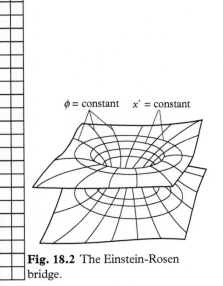

Fig. 18.2 The Einstein-Rosen bridge.

To interpret this, we consider a two-dimensional surface possessing this line element embedded in a flat three-dimensional space. The surface appears as in Fig. 18.2. Thus, at $t' = 0$, the Kruskal manifold can be thought of as being formed by two distinct but identical asymptotically flat Schwarzschild manifolds joined at the 'throat' $r = 2m$. As t' increases, the same qualitative picture holds but the throat narrows down, the universes

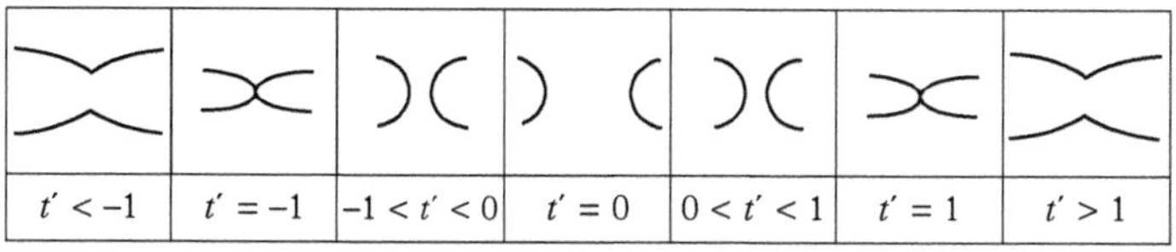

$t' < -1$	$t' = -1$	$-1 < t' < 0$	$t' = 0$	$0 < t' < 1$	$t' = 1$	$t' > 1$

Fig. 18.3 Time evolution of the Einstein-Rosen bridge.

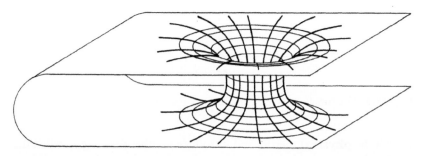

Fig. 18.4 A Schwarzschild wormhole.

joining at a value of $r < 2m$. At $t' = 1$, the throat pinches off completely and the two universes touch at the singularity $r = 0$. For larger values of t', the two universes, each containing a singularity at $r = 0$, are completely separate. The Kruskal solution is time-symmetric with respect to t', and so the same thing happens if we run time backwards from $t' = 0$. The full time evolution is shown schematically in Fig. 18.3, where each diagram should be rotated about the central vertical axis to get the two-dimensional picture analogous to that shown in Fig. 18.2.

The intriguing question of whether or not the mathematical procedure for extending the solution which results in the 'new universe' I' has any physical significance is still an open one. Although Einstein's equations fix the local geometry of space-time, they do not fix its global geometry or its topology. In Fig. 18.4, we see an embedding of the slice $t' = $ constant which is geometrically identical but topologically different. This embedding leads to a Schwarzschild 'wormhole' which connects two distant regions of a single asymptotically flat universe. We shall not pursue the idea further.

Although Fig. 18.1 is very informative, it does not indicate what happens to points at 'infinity'. We shall see that the process of conformal compactification allows us to investigate the structure of these points and leads to another picture called a **Penrose diagram**.

18.4 Penrose diagram for Minkowski space-time

We shall introduce the idea of a Penrose diagram by first of all considering the procedure for Minkowski space-time. This will provide a prototype for other solutions. The essential idea is to start off with a metric g_{ab}, which

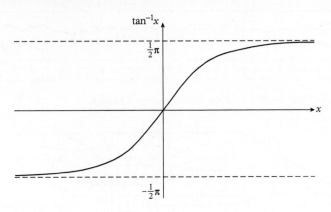

Fig. 18.5 The function $\tan^{-1}(x)$ maps $(-\infty, \infty)$ onto $(-\pi/2, \pi/2)$.

we call the **physical** metric, and introduce another metric $\bar{g}_{ab}$, called the **unphysical** metric, which is conformally related to g_{ab}, that is,

$$\bar{g}_{ab} = \Omega^2 g_{ab}, \tag{18.12}$$

where Ω is the conformal factor. Then, by a suitable choice of Ω^2, it may be possible to 'bring in' the points at infinity to a finite position and hence study the causal structure of infinity. As we found in Exercise 6.29, the null geodesics of conformally related metrics are the same. The null geodesics determine the light cones, which in turn define the causal structure. The essential idea for bringing in the points at infinity is to use coordinate transformations involving functions like $\tan^{-1}(x)$, which, for example, maps the infinite interval $(-\infty, \infty)$ onto the finite interval $(-\pi/2, \pi/2)$ (Fig. 18.5).

We introduce double null coordinates defined by

$$v = t + r, \tag{18.13}$$

$$w = t - r, \tag{18.14}$$

in which case the line element of Minkowski space-time becomes (exercise)

$$ds^2 = dv\, dw - \tfrac{1}{4}(v - w)^2 \left(d\theta^2 + \sin^2\theta\, d\phi^2\right). \tag{18.15}$$

From (18.13) and (18.14), it follows that $r = \tfrac{1}{2}(v - w)$, and so the co-ordinate range $(-\infty < t < \infty,\ 0 \leqslant r < \infty)$ becomes $(-\infty < v < \infty,\ -\infty < w < \infty)$, with the requirement

$$r \geqslant 0 \quad \Rightarrow \quad v - w \geqslant 0 \quad \Rightarrow \quad v \geqslant w. \tag{18.16}$$

The space-time diagram for Minkowski space-time is shown in Fig. 18.6. We next define new coordinates p and q by

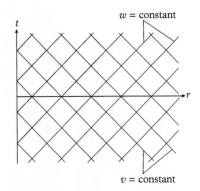

Fig. 18.6 Space-time diagram of Minkowski space-time.

$$p = \tan^{-1}v, \tag{18.17}$$

$$q = \tan^{-1}w, \tag{18.18}$$

with the coordinate ranges $-\frac{1}{2}\pi < p < \frac{1}{2}\pi$ and $-\frac{1}{2}\pi < q < \frac{1}{2}\pi$, and whereby (18.16)

$$p \geqslant q. \tag{18.19}$$

Then (18.15) becomes (exercise)

$$ds^2 = \tfrac{1}{4}\sec^2 p \sec^2 q \left[4dp\,dq - \sin^2{(p-q)}\left(d\theta^2 + \sin^2\theta\,d\phi^2\right)\right], \tag{18.20}$$

and the line element of the unphysical metric is

$$d\bar{s}^2 = \bar{g}_{ab}\,d\bar{x}^a\,d\bar{x}^b = 4dp\,dq - \sin^2{(p-q)}\left(d\theta^2 + \sin^2\theta\,d\phi^2\right), \tag{18.21}$$

with the conformal factor

$$\Omega^{-2} = \tfrac{1}{4}\sec^2 p \sec^2 q.$$

Finally, we introduce the coordinates

$$t' = p + q, \tag{18.22}$$

$$r' = p - q, \tag{18.23}$$

where the coordinate range is

$$-\pi < t' + r' < \pi, \tag{18.24}$$

$$-\pi < t' - r' < \pi, \tag{18.25}$$

$$r' \geqslant 0, \tag{18.26}$$

the last condition resulting from (18.19). The unphysical line element is now

$$d\bar{s}^2 = dt'^2 - dr'^2 - \sin^2 r' \left(d\theta^2 + \sin^2\theta\,d\phi^2\right) \tag{18.27}$$

subject to the coordinate range (18.24)–(18.26).

The line element (18.27) is that of the **Einstein static universe**, which we introduced in §13.3 and which we shall meet in more detail in Part F. The topology of this solution is **cylindrical**, with the time coordinate running along the generators of the cylinder. A cross-section of the cylinder, $t' = $ constant, has the topology of a 3-sphere S^3. Then the coordinate range of the manifold is

$$-\infty < t' < \infty, \quad 0 \leqslant r' \leqslant \pi, \quad 0 \leqslant \theta \leqslant \pi, \quad -\pi < \phi \leqslant \pi, \tag{18.28}$$

where $r' = 0, \pi$ and $\theta = 0, \pi$ are coordinate singularities. We shall discuss this further in Part F, but for the moment it is sufficient to think of an S^3 as a three-dimensional generalization of a 2-sphere S^2. In fact, the Einstein static universe can be embedded as the cylinder

$$x^2 + y^2 + z^2 + w^2 = 1$$

in a five-dimensional flat space of signature -3 with line element

$$\mathrm{d}s^2 = \mathrm{d}t^2 - \mathrm{d}x^2 - \mathrm{d}y^2 - \mathrm{d}z^2 - \mathrm{d}w^2.$$

(Suppressing two dimensions, this is the more familiar equation of a cylinder, namely, $x^2 + y^2 = 1$, in a three-dimensional space, but with a Minkowski-type geometry $\mathrm{d}s^2 = -\mathrm{d}t^2 + \mathrm{d}x^2 + \mathrm{d}y^2$.) The Einstein static universe then has line element (18.27) and coordinate range (18.28). We have shown that Minkowski space-time is conformal to that part of the Einstein static universe defined by the coordinate range (18.24)–(18.26). This is depicted in Fig. 18.7. The coordinate range (18.24)–(18.26) defines the diamond-shape region of the cylinder indicated. Thus, the whole of Minkowski space-time has been shrunk or compacted into this finite region. The process is called **conformal compactification** and the region is called **compactified Minkowski space-time**. The boundary of this region represents the conformal structure of infinity for Minkowski space-time. In terms of the coordinates p and q, it consists of the following:

a null surface $p = \frac{1}{2}\pi$ called $\mathscr{I}^+$,
a null surface $q = -\frac{1}{2}\pi$ called $\mathscr{I}^-$,
a point $(p = \frac{1}{2}\pi, q = \frac{1}{2}\pi)$ called i^+,
a point $(p = \frac{1}{2}\pi, q = -\frac{1}{2}\pi)$ called i^0,
a point $(p = -\frac{1}{2}\pi, q = -\frac{1}{2}\pi)$ called i^-,

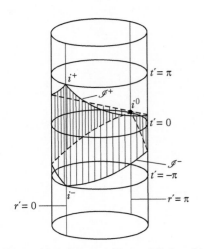

Fig.   18.7 Compactified   Minkowski space-time (two dimensions suppressed).

where $\mathscr{I}$ is pronounced 'scri' – short for script i. Then it can be shown that all timelike geodesics originate at i^- and terminate at i^+. Similarly, null geodesics originate at points of $\mathscr{I}^-$ and end at points of $\mathscr{I}^+$, while spacelike geodesics both originate and end at i^0 (but these rules are not satisfied by non-geodesic curves). Thus, one may regard i^+ and i^- as representing **future** and **past timelike infinity**, $\mathscr{I}^+$ and $\mathscr{I}^-$ as representing **future** and **past null infinity**, and i^0 as representing **spacelike infinity**. This is illustrated in Fig. 18.8.

A **Penrose diagram** is a space-time diagram of a conformally compactified space-time. The Penrose diagram for Minkowski space-time is shown in Fig. 18.9. The diagram shows the curves $r = $ constant which correspond to the histories of 2-spheres $r = $ constant, and the curves $t = $ constant which correspond to timelike slices. Ingoing and outgoing radial null geodesics are represented by the straight lines $p = $ constant and $q = $ constant, making angles $-45°$ and $45°$, respectively. A large class of asymptotically flat space-times, which Penrose calls **simple** space-times, can be analysed in a similar manner.

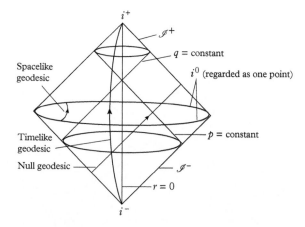

Fig. 18.8 Origin and termination of geodesics in compactified Minkowski space-time (one dimension suppressed).

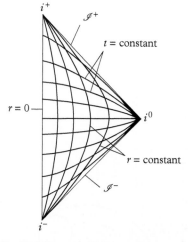

Fig. 18.9 Penrose diagram of Minkowski space-time (two dimensions suppressed).

18.5 Penrose diagram for the Kruskal solution

The conformal compactification of the Kruskal solution may be obtained by defining new advanced and retarded null coordinates in terms of the null coordinates v' and w' of §18.2

$$v'' = \tan^{-1}\left[v'/(2m)^{\frac{1}{2}}\right], \quad w'' = \tan^{-1}\left[w'/(2m)^{\frac{1}{2}}\right],$$

for the coordinate range

$$-\tfrac{1}{2}\pi < v'' < \tfrac{1}{2}\pi,$$
$$-\tfrac{1}{2}\pi < w'' < \tfrac{1}{2}\pi,$$
$$-\pi < v'' + w'' < \pi.$$

We omit the calculational details and simply present the Penrose diagram in Fig. 18.10. Again, null geodesics and light cones have angles $\pm 45°$ in the figure. Both regions I and I′ have their own future, past, and null infinities. For any point outside $r = 2m$, an outward radial null geodesic ends up at $\mathscr{I}^+$ but an inward radial null geodesic ends up at the future singularity. For any point lying inside $r = 2m$, both outward and inward radial null geodesics end up on the future singularity.

We now take into account the fact that each point in the diagram represents a 2-sphere. Consider a 2-sphere S_0 situated in region I which is illuminated at some time. Then the photons at each point of S_0 move out in a 2-sphere and the envelope of these 2-spheres is again the two 2-spheres S_1 and S_2 as shown in Fig. 18.11. The area of S_2 will be greater than S_0, which in turn will have a greater area than S_1. However, if S_0 lies in region II, both wave fronts are imploding and the areas of S_1 and S_2

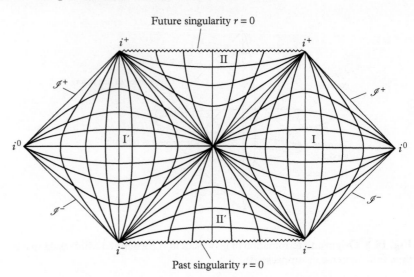

Fig. 18.10 Penrose diagram of the Kruskal solution.

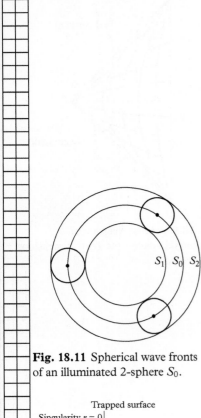

Fig. 18.11 Spherical wave fronts of an illuminated 2-sphere S_0.

Fig. 18.12 Penrose diagram of spherically symmetric gravitational collapse.

will both be less than S_0. Such a **closed** two-surface in which the light rays **converge in both directions** is called a **trapped surface**. Similarly, each point in region II$'$ represents a time-reversed trapped surface. It turns out that it is precisely the existence of trapped surfaces which lead in the singularity theorems to the existence of singularities. Note that Hawking and Ellis call such a surface a "closed trapped surface" but we prefer to use Penrose's original terminology as there is not really such a thing as an "open trapped surface", since the convergence condition on its own does not lead to trapping. For example the surface formed by the intersection of two past null cones in Minkowski space satisfies the convergence condition but is certainly not trapped.

In Fig. 18.12, we show the Penrose diagram for a collapsing spherical star (compare with Figs. 17.13 and 17.14).

Exercises

18.1 (§18.2) Show that Schwarzschild space-time can be written in the form (18.1) subject to (18.2) in double null coordinates. [Hint: use (17.23) and (17.25)].

18.2 (§18.2) Show that (18.6) and (18.7) lead to the form (18.9) subject to (18.8).

18.3 (§18.2) Show that radial null geodesics make angles of $\pm 45°$ with the x'-axis in the Kruskal space-time diagram.

18.4 (§18.2) Where can observers from Universes I and I$'$ meet in the Kruskal solution? What is their ultimate fate?

18.5 (§18.4) Show that Minkowski space-time takes the form (18.15) in double null coordinates.

18.6 (§18.4) Show that Minkowski space-time takes the form (18.20) under the coordinate transformations (18.17) and (18.18).

18.7 (§18.4) Draw a diagram of the region in the (t', r')-plane described by the inequalities (18.24) and (18.25). What subregion satisfies (18.26) as well?

18.8 (§18.4) Write down the transformation from the usual Minkowski coordinates (t, r) to (t', r') given in (18.22) and (18.23). Find the equations for the curves $t = $ constant and $r = $ constant in terms of t' and r' and draw them in the Penrose diagram of Minkowski space-time.

18.9 (§18.5) Draw the analogue of Fig. 18.11 for a trapped surface. Draw the corresponding figure for a 2-sphere in region II$'$.

18.10 (§18.5) Consider the transition from Fig. 18.10 to Fig. 18.12. What has happened to regions I$'$ and II$'$?

Further reading

Again, the main source for this chapter is the book by Hawking and Ellis (1973).

Hawking, S. W., and Ellis, G. F. R. (1973). *The Large Scale Structure of Space-Time*. Cambridge University Press, Cambridge.

Charged black holes

19

19.1 The field of a charged mass point

In this chapter, we shall obtain and investigate the Reissner-Nordström solution for a **charged** mass point. The importance of this solution is that its structure is in many ways similar to that of the more complicated Kerr solution describing rotating black holes, which we shall meet in the next chapter. The approach we adopt is to look for a static, asymptotically flat. spherically symmetric solution of the Einstein-Maxwell field equations. The Einstein-Maxwell equations are

$$G_{ab} = 8\pi T_{ab}, \tag{19.1}$$

where T_{ab} is the Maxwell energy-momentum tensor, which in source-free regions is given by (12.57). In Exercise 13.3, we saw that this tensor is trace-free, which, by (19.1), implies that the Ricci scalar vanishes (exercise). We can therefore also work with the equivalent equations to (19.1), namely,

$$R_{ab} = 8\pi T_{ab}. \tag{19.2}$$

In addition, the Maxwell tensor F_{ab} must satisfy Maxwell's equations in source-free regions

$$\nabla_b F^{ab} = 0, \tag{19.3}$$

$$\partial_{[a} F_{bc]} = 0. \tag{19.4}$$

The assumption of spherical symmetry means that we can introduce coordinates (t, r, θ, ϕ) in which the line element reduces to the canonical form (15.37), namely,

$$ds^2 = e^\nu dt^2 - e^\lambda dr^2 - r^2(d\theta^2 + \sin^2\theta d\phi^2), \tag{19.5}$$

where ν and λ are functions of t and r. If we next impose the condition that the solution is static, then this requires that ν and λ are functions of r only, namely,

$$\nu = \nu(r), \qquad \lambda = \lambda(r). \tag{19.6}$$

The assumption that the field is due to a charged particle, which we take to be situated at the origin of coordinates, means that the line element and the

Introducing Einstein's Relativity. Second Edition. Ray d'Inverno and James Vickers, Oxford University Press.
© Ray d'Inverno and James Vickers (2022). DOI: 10.1093/oso/9780198862024.003.0019

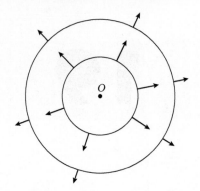

Fig. 19.1 Radial electrostatic field of a charged point particle.

Maxwell tensor will become singular there. Moreover, the charged particle will give rise to an electrostatic field which is purely radial (Fig. 19.1). This means that the Maxwell tensor must take on the form (exercise)

$$F_{ab} = E(r) \begin{bmatrix} 0 & -1 & 0 & 0 \\ 1 & 0 & 0 & 0 \\ 0 & 0 & 0 & 0 \\ 0 & 0 & 0 & 0 \end{bmatrix}. \tag{19.7}$$

Plugging the assumptions (19.5)–(19.7) in (19.3) and (19.4), we find (exercise) that (19.4) is satisfied automatically and (19.3) reduces to one equation, namely,

$$(e^{-\frac{1}{2}(\nu+\lambda)} r^2 E)' = 0, \tag{19.8}$$

where the prime indicates differentiation with respect to r. This integrates to give

$$E = e^{\frac{1}{2}(\nu+\lambda)} \varepsilon / r^2, \tag{19.9}$$

where ε is a constant of integration. Our assumption that the solution is asymptotically flat requires

$$\nu, \lambda \to 0 \quad \text{as} \quad r \to \infty, \tag{19.10}$$

and so $E \sim \varepsilon/r^2$ asymptotically. This latter result is exactly the same as the classical result for the electric field of a point particle of charge ε situated at the origin. We therefore interpret ε as the **charge** of the particle.

We now use (19.5) to (19.9) together with (12.57) to compute the Maxwell energy momentum tensor T_{ab}. Plugging this into the field equations (19.2), we find that the 00 and 11 equations lead to

$$\lambda' + \nu' = 0, \tag{19.11}$$

which by (19.10) results in $\lambda = -\nu$. The 22 equation is the one remaining independent equation and it leads to

$$(re^\nu)' = 1 - \varepsilon^2/r^2 , \tag{19.12}$$

which integrates immediately to give

$$e^\nu = 1 - 2m/r + \varepsilon^2/r^2 , \tag{19.13}$$

where m is a constant of integration. We have finally obtained the **Reissner-Nordström solution**

$$ds^2 = \left(1 - \frac{2m}{r} + \frac{\varepsilon^2}{r^2}\right) dt^2 - \left(1 - \frac{2m}{r} + \frac{\varepsilon^2}{r^2}\right)^{-1} dr^2 - r^2(d\theta^2 + \sin^2\theta\, d\phi^2).$$
$$(19.14)$$

When $\varepsilon = 0$, this reduces to the Schwarzschild line element (15.52), and so we again identify m as the **geometric mass**. In deriving this solution, we have, in addition to assuming spherical symmetry, also assumed the solution is static and asymptotically flat. In fact, as in the case of the Schwarzschild solution, it is **not** necessary to adopt these last two assumptions: they are forced on you. The full calculation is similar to the Schwarzschild case but rather longer, which is why we have omitted it. There is therefore an analogue to Birkhoff's theorem.

Theorem: A spherically symmetric exterior solution of the Einstein-Maxwell field equations is necessarily static.

19.2   Intrinsic and coordinate singularities

Consider the coefficients

$$g_{00} = -(g_{11})^{-1} = 1 - 2m/r + \varepsilon^2/r^2 = Q(r)/r^2,$$

where

$$Q(r) = r^2 - 2mr + \varepsilon^2. \tag{19.15}$$

The discriminant of the quadratic Q is

$$\Delta = m^2 - \varepsilon^2,$$

and, if this is negative, i.e. $\varepsilon^2 > m^2$, the quadratic has no real roots and is positive for all values of r. Hence, it follows that the line element (19.14) is non-singular for all values of r except at the origin $r = 0$. The solution possesses an intrinsic singularity at $r = 0$ – as can be shown by calculating the Riemann invariant $R^{abcd}R_{abcd}$ – which is not surprising, since this is where the point charge producing the field is located. The more interesting case occurs when, $\varepsilon^2 \leqslant m^2$, for then the metric has singularities when Q vanishes, namely, at $r = r_+$ and $r = r_-$, where

$$r_{\pm} = m \pm (m^2 - \varepsilon^2)^{\frac{1}{2}}. \tag{19.16}$$

In Fig. 19.2, we plot g_{00} in the case $\varepsilon^2 < m^2$ and compare it with the Schwarzschild coefficient $_S g_{00} = 1 - 2m/r$.

The line element (19.14) is regular in the regions:

I. $r_+ < r < \infty$,
II. $r_- < r < r_+$,
III. $0 < r < r_-$.

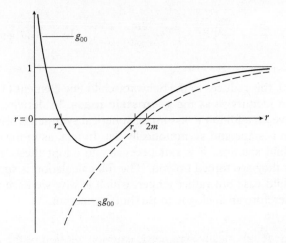

Fig. 19.2 Graphs of g_{00} for Reissner-Nordström and Schwarzschild solutions.

If $\varepsilon^2 = m^2$, then only the regions I and III exist. The regions are separated by the null hypersurfaces $r = r_+$ and $r = r_-$. The situation at $r = r_+$ is rather similar to the Schwarzschild case at $r = 2m$. The coordinates t and r are timelike and spacelike, respectively, in the regions I and III, but interchange their character in region II. Thus, regions I and III are static, but region II is not. As in the case of the Schwarzschild solution, these coordinates suggest that the regions I, II, and III appear totally disconnected because the light cones have totally different orientations on either side of the null hypersurfaces $r = r_\pm$. We will not pursue the structure of the solution in these coordinates further, but rather proceed as we did with the Schwarzschild solution and look for the analogue of the Eddington-Finkelstein coordinates.

19.3   Space-time diagram of the Reissner-Nordström solution

In the next two sections, we restrict our attention to the important case $\varepsilon^2 < m^2$. We first find the equation for the congruence of ingoing radial null geodesics (exercise). Then, defining for $r > r_+$ the new time coordinate

$$\bar{t} = t + \frac{r_+^2}{r_+ - r_-}\ln(r - r_+) - \frac{r_-^2}{r_+ - r_-}\ln(r - r_-), \qquad (19.17)$$

the line element takes on the form (exercise)

$$ds^2 = (1 - f)d\bar{t}^2 - 2f d\bar{t}dr - (1 + f)dr^2 - r^2(d\theta^2 + \sin^2\theta d\phi^2), \quad (19.18)$$

where, for convenience, we define

$$f = 1 - g_{00} = 2m/r - \varepsilon^2/r^2. \tag{19.19}$$

This form of the metric is regular for all positive values of r and again has an intrinsic singularity at $r = 0$. The conditions for radial null geodesics are

$$\dot{\theta} = \dot{\phi} = \mathrm{d}s^2 = 0. \tag{19.20}$$

These lead to (exercise) the ingoing family of null geodesics

$$\bar{t} + r = \text{constant}, \tag{19.21}$$

and the outgoing family whose differential equation is

$$\frac{\mathrm{d}\bar{t}}{\mathrm{d}r} = \frac{1+f}{1-f}. \tag{19.22}$$

We do not, in fact, need to solve this equation exactly since our aim is to draw a space-time diagram, in which case it is sufficient to use the equation to obtain qualitative information about the slope for different values of r. The graphs of $1+f$ and $1-f$ are shown in Fig. 19.3. At infinity, f vanishes and so the slope is 45°, as we would expect for an asymptotically flat solution. As we come in from infinity, $1+f$ increases and $1-f$ decreases and so the slope increases until, at $r = r_+$, $1 - f$ vanishes and the slope becomes infinite. In region II, the slope increases from $-\infty$ at $r = r_+$ to some maximum negative value at $r = \varepsilon^2/m$, and then decreases again to $-\infty$ as r approaches r_-. In region III, the slope decreases from $+\infty$ to 1, where the graphs cross, and continues decreasing to zero, where the graph $1 + f$ crosses the r-axis. The slope then decreases through negative values until it reaches -1 at the origin. With this information, we can draw the space-time diagram in Fig. 19.4. It is clear from the light cones at $r = r_+$ that no light signal can escape from region II to region I. Thus, the surface $r = r_+$ is an **event horizon**. In region II, the light cones are inclined towards the singularity $r = 0$, and hence any particle entering region II will move necessarily towards the centre until it either crosses $r = r_-$ or reaches it asymptotically. In the region III, the light cones are no

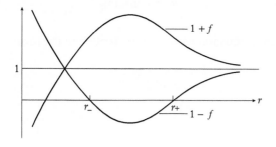

Fig. 19.3 Graphs of the functions $1 + f$ and $1 - f$.

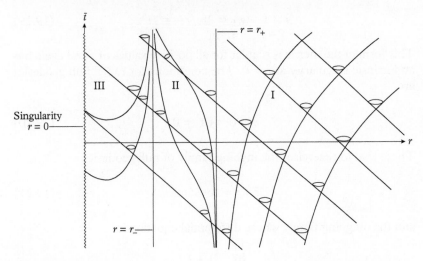

Fig. 19.4 Reissner-Nordström solution ($\varepsilon^2 < m^2$) in advanced Eddington-Finkelstein-type coordinates.

longer inclined towards the centre and consequently particles need not fall into the singularity. In fact, the opposite occurs in that neutral particles **cannot** reach the singularity, as we shall next show.

19.4 Neutral particles in Reissner-Nordström space-time

To consider the motion of a neutral test particle , we shall investigate a radial timelike geodesic, the conditions for which are

$$\dot{\theta} = \dot{\phi} = 0, \quad \dot{s}^2 = 1, \tag{19.23}$$

where dot denotes differentiation with respect to the proper time τ. Defining the covariant 4-velocity $u_a = g_{ab}\mathrm{d}x^b/\mathrm{d}\tau$, we find that the geodesic equations lead to a first integral of the motion (exercise)

$$u_0 = \text{constant}, \tag{19.24}$$

and a remaining equation which can be written in the form

$$\dot{r}^2 = A, \tag{19.25}$$

where

$$A = u_0^2 - g_{00}. \tag{19.26}$$

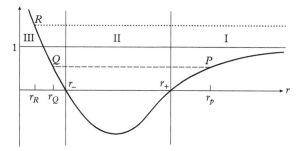

Fig. 19.5 Bounded and unbounded motion of a neutral particle.

We can investigate what qualitative forms of motion are possible by plotting the curve of g_{00} against r and drawing in lines parallel to the r-axis a distance u_0^2 from it, for various values of u_0^2 (Fig. 19.5). Consider first the case when $u_0^2 < 1$. Then the line intersects the graph of g_{00} at two points P and Q in regions I and III, respectively. At these points, A vanishes by (19.26), and (19.25) then shows that $\dot{r} = 0$. Moreover, from (19.25), the left-hand side of the equation is positive, from which it follows that A must be positive. Therefore, by (19.26), motion is only possible when $u_0^2 \geqslant g_{00}$. It follows that the motion is bounded between the two values $r = r_P$ and $r = r_Q$. Similar arguments show that, if $u_0^2 > 1$, then unbounded motion is possible, but there is a minimum distance of approach $r = r_R$ in region III. Thus, the point charge at the origin produces a potential barrier, which means that a neutral free particle can only approach within a certain distance before being repelled.

According to Fig. 19.4, once a particle is in region III, it cannot cross $r = r_-$ but can only reach it asymptotically. However, it can be shown that if a particle reaches $r = r_-$ then it does so in **finite** proper time. The diagram is misleading in exactly the same way as the Schwarzschild diagram in Schwarzschild coordinates (Fig. 17.7) is misleading in describing what happens to a radially infalling free particle in region I. Thus, the manifold described by the line element (19.18) is not maximal and needs extending in analogy with the Kruskal case.

19.5   Penrose diagrams of the maximal analytic extensions

In this section, we again restrict attention to the physically important case in which $\varepsilon^2 < m^2$. Then, following the example of the Kruskal extension of Schwarzschild described in §18.2, we start by introducing double null coordinates in the exterior region $r > r_+$ given by

$$v = t + r^*, \qquad w = t - r^*, \qquad (19.27)$$

where

$$r^* = r + \frac{r_+^2}{r_+ - r_-} \ln(r - r_+) - \frac{r_-^2}{r_+ - r_-} \ln(r - r_-). \qquad (19.28)$$

Then we find the line element (19.18) takes on the double null form (exercise)

$$ds^2 = \left(1 - \frac{2m}{r} + \frac{\varepsilon^2}{r^2} \right) dv\, dw - r^2 (d\theta^2 + \sin^2\theta d\phi^2), \qquad (19.29)$$

where $r(v, w)$ is defined implicitly using $\frac{1}{2}(v - w) = r^*$ and (19.28).

We now define new coordinates

$$v' = \exp\left(\frac{r_+ - r_-}{2r_+^2} v \right), \qquad w' = -\exp\left(\frac{r_- - r_+}{2r_+^2} w \right), \qquad (19.30)$$

which transforms the line element into the form

$$ds^2 = \frac{4r_+^4 (r - r_-)^{1 + r_-^2/r_+^2}}{r^2 (r_+ - r_-)^2} \exp\left(\frac{r_- - r_+}{r_+^2} r \right) dv' dw'$$
$$- r^2 (d\theta^2 + \sin^2\theta d\phi^2), \qquad (19.31)$$

where $r(v', w')$ is now defined implicitly by

$$v'w' = -\exp\left(\frac{r_+ - r_-}{2r_+^2} r \right) (r - r_+)^{1/2} (r - r_-)^{-r_-^2/2r_+^2}. \qquad (19.32)$$

This line element is the analogue of the Kruskal solution and represents the maximal analytic extension of the Reissner-Nordström solution for $\varepsilon^2 < m^2$.

The Penrose diagram for this maximal extension is obtained by setting

$$v'' = \arctan v', \qquad w'' = \arctan w' \qquad (19.33)$$

and is shown in Fig. 19.6.

This time, the maximal extension gives rise to an infinity of 'new universes'. There are an infinite number of asymptotically flat regions I where $r > r_+$. These are connected by intermediate regions II and III where $r_- < r < r_+$ and $0 < r < r_-$, respectively. Region III possesses an intrinsic singularity at $r = 0$ but, unlike the Kruskal solution, it is **timelike** and so can be avoided by a future-directed timelike curve from region I which crosses $r = r_+$. A timelike curve is drawn in Fig. 19.6 which starts in a particular region I, passes through regions II, III, and II and re-emerges into another asymptotically flat region I. This gives rise to the highly speculative possibility that it may be possible to travel to other universes by passing through the 'wormholes' produced by charges. Unfortunately, it would seem as though it would not be possible to return. However, there is the possibility of identifying regions I (giving rise to a more complicated topology), so that a particle could then re-emerge from the black hole through the horizon $r = r_+$. Whether or not the particle emerges into the same part or a different part of the universe will depend on how

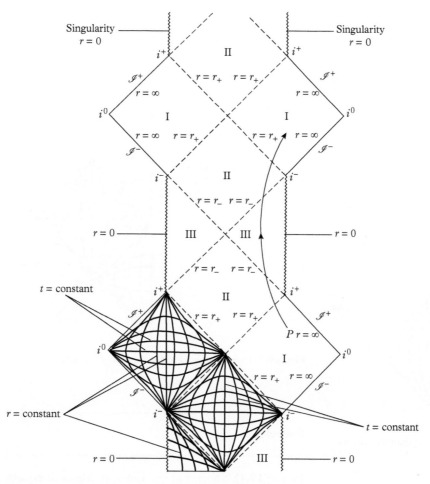

Fig. 19.6 Penrose diagram for maximal analytic extension ($\varepsilon^2 < m^2$).

the identification is made. A particle crossing the event horizon $r = r_+$ would appear to suffer an infinite red shift to an observer who remains in region I. In region II, each point represents a trapped surface. The extended solution possesses a very bizarre property in that any observer crossing the surface $r = r_-$ would see the whole of the remaining history of the asymptotically flat region I in a finite time! Objects crossing into this region would therefore be infinitely blue-shifted, which suggests that the surface $r = r_-$ would be unstable to small perturbations. This is a topic of current research. The line element (19.31) has a coordinate singularity at $r = r_-$. It is therefore necessary to introduce new null coordinates (in fact, an infinity of such coordinates) in order to 'patch' the manifold together. We shall not pursue this further.

The case $\varepsilon^2 = m^2$ can be extended similarly and the Penrose diagram is shown in Fig. 19.7. The remaining case $\varepsilon^2 > m^2$ is inextendible and the Penrose diagram is shown in Fig. 19.8.

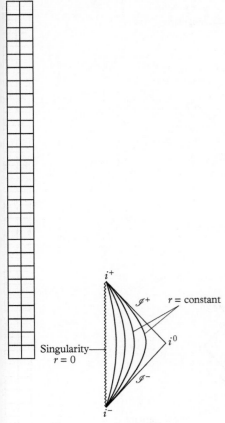

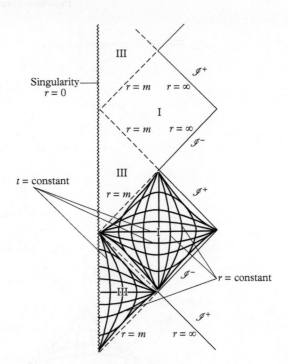

Fig. 19.7 Penrose diagram for the case $\varepsilon^2 = m^2$.

Fig. 19.8 Penrose diagram for the case $\varepsilon^2 > m^2$.

Exercises

19.1 (§19.1) Show that the Einstein–Maxwell equations can be written in the equivalent form (19.2) in source-free regions.

19.2 (§19.1) Given the definition (12.28) of the Maxwell tensor in Minkowski coordinates (t, x, y, z), find its components in spherical polar coordinates (t, r, θ, ϕ). Hence confirm the ansatz (19.7).

19.3 (§19.1) Show that the assumptions (19.5)–(19.7) lead to the result that (19.4) is satisfied automatically and (19.3) reduces to (19.8). [Hint: You may need to use equation (7.10)]

19.4 (§19.1) Use (19.5)–(19.8) and (12.57) to compute the energy-momentum tensor T_{ab}, in terms of ν, λ and ε. Show that the two independent Einstein-Maxwell field equations are (19.11) and (19.12). Hence obtain the Reissner-Nordström solution.

19.5 (§19.2) Establish Fig. 19.2.

19.6 (§19.2) Establish the character of the coordinates t and r in (19.14) for $\varepsilon^2 < m^2$ in the regions I, II, and III. Find the surfaces of infinite red shift.

19.7 (§19.2) Draw a retarded Eddington-Finkelstein space-time diagram for the Reissner–Nordström solution.

19.8 (§19.3) Find the equation for the congruence of ingoing radial null geodesics for the line element (19.14) in the case $\varepsilon^2 < m^2$.

19.9 (§19.3) Draw a space-time diagram for the Reissner–Nordström solution in the coordinates of (19.14) for $\varepsilon^2 < m^2$. What happens to the diagram when $\varepsilon^2 = m^2$?

19.10 (§19.3) Show that (19.17) transforms (19.14) into the form (19.18). Show that the two families of radial null geodesics are given by (19.21) and (19.22).

19.11 (§19.3) Show that the transformations

$$\bar{t} = t + \frac{r_+^2}{r_+ - r_-} \ln(r_+ - r) - \frac{r_-^2}{r_+ - r_-} \ln(r - r_-)$$

for $r_- < r < r_+$,

$$\bar{t} = t + \frac{r_+^2}{r_+ - r_-} \ln(r_+ - r) - \frac{r_-^2}{r_+ - r_-} \ln(r_- - r)$$

for $0 < r < r_-$, transform (19.14) for $\varepsilon^2 < m^2$ into the form (19.18).

19.12 (§19.3) Show that the transformations

$$\bar{t} = t + m\ln(r - m)^2 - \frac{m^2}{r - m}, \quad \text{if } \varepsilon^2 = m^2,$$

$$\bar{t} = t + m\ln(r^2 - 2mr - \varepsilon^2) + \frac{2m^2 - \varepsilon^2}{(\varepsilon^2 - m^2)^{1/2}} \tan^{-1} \frac{r - m}{(\varepsilon^2 - m^2)^{1/2}},$$
$$\text{if } \varepsilon^2 > m^2.$$

transform (19.14) into the form (19.18).

19.13 (§19.3) Find the advanced Eddington–Finkelstein form of the Reissner-Nordström solution.

19.14 (§19.3) Consider the graphs of $1 + f$ and $1 - f$ in Fig. 19.3. Where is the slope $d\bar{t}/dr$ a maximum in region II? Where is the slope zero in region III? What is the slope at the origin?

19.15 (§19.4) Show that the equation for a radial timelike geodesic for the solution (19.14) in the case $\varepsilon^2 < m^2$ leads to (19.24) and (19.25).

19.16 (§19.5) Show that the transformation (19.27) transforms the line element (19.18) into the form (19.29). Show that the transformation (19.30) transforms (19.29) into the form (19.31) subject to (19.32).

19.17 (§19.5) Consider the world-line of the observer in Fig. 19.6 emanating from the point P, and the histories of all timelike geodesics in the region I containing P. Hence show that the observer will see all the remaining history of these geodesics as the horizon $r = r_-$ is crossed.

Further reading

As before, the main source for this chapter is Hawking and Ellis (1973).

Hawking, S. W., and Ellis, G. F. R. (1973). *The Large Scale Structure of Space-Time*. Cambridge University Press, Cambridge.

Rotating black holes 20

20.1 Null tetrads

In this chapter, we shall investigate the Kerr solution, which describes rotating black holes. It turns out to be a rather long process to solve Einstein's vacuum equations directly for the Kerr solution. We shall, instead, describe a 'trick' of Newman and Janis for obtaining the Kerr solution from the Schwarzschild solution. This same trick can then be applied to the Reissner-Nordström solution to obtain the Kerr–Newman solution, the most general solution for a charged rotating black hole. In order to discuss this approach, we start by introducing the very important idea of a null tetrad.

In §10.5, we met the idea of a tetrad $e_i{}^a$ of one timelike and three spacelike vectors. In fact, these tetrads, or frames, possess a formalism of their own called the **frame formalism**, which has proved extremely useful in many applications in general relativity. One of the most important cases is when the tetrad vectors are taken to be **null** vectors. The systematic use of null tetrads is the basis of the 'Newman–Penrose' formalism, or NP formalism for short, which has been used extensively in the study of gravitational radiation, among other topics. In this section, we shall restrict our attention to the definition of a null tetrad.

We start with four linearly independent vector fields $e_i{}^a$, where i serves to label the vectors. Then, working at a point, we define a matrix of scalars g_{ij} called the **frame metric**, by

$$g_{ij} = g_{ab}\, e_i{}^a e_j{}^b. \qquad (20.1)$$

Since $e_i{}^a$ are linearly independent and g_{ab} is non-singular, it follows that the matrix g_{ij} is non-singular and hence invertible. We therefore define its inverse g^{ij}, the contravariant frame metric, by the relation

$$g_{ij}\, g^{jk} = \delta_i^k. \qquad (20.2)$$

We then use the frame metric to raise and lower frame indices in the same way that we use the metric tensor to raise and lower tensor indices. It is then easy to verify that the inverse relationship to (20.1) is

$$g_{ab} = g_{ij}\, e^i{}_a e^j{}_b. \qquad (20.3)$$

Introducing Einstein's Relativity. Second Edition. Ray d'Inverno and James Vickers, Oxford University Press.
© Ray d'Inverno and James Vickers (2022). DOI: 10.1093/oso/9780198862024.003.0020

In §10.5, we took the tetrad to consist of one timelike vector T^a and three spacelike vectors X^a, Y^a and Z^a, say, in which case the orthonormality relations lead to

$$g_{ij} = \eta_{ij} = \mathrm{diag}(1, -1, -1, -1),$$

where the frame metric is the Minkowski metric η_{ij}. We now take

$$e_0{}^a = \ell^a = \frac{1}{\sqrt{2}} \left(T^a + X^a \right),   (20.4)$$

$$e_1{}^a = n^a = \frac{1}{\sqrt{2}} \left(T^a - X^a \right),   (20.5)$$

in which case ℓ^a and n^a are **null** vectors (Fig. 20.1), that is,

$$\ell^a \ell_a = n^a n_a = 0,   (20.6)$$

and satisfy the normalization condition

$$\ell^a n_a = 1.   (20.7)$$

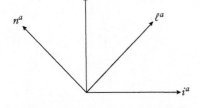

Fig. 20.1 The null vectors n^a and ℓ^a.

Then, if we take $e_2{}^a = Y^a$ and $e_3{}^a = Z^a$, the orthonormality relations (20.1) lead to the frame metric

$$g_{ij} = \begin{bmatrix} 0 & 1 & 0 & 0 \\ 1 & 0 & 0 & 0 \\ 0 & 0 & -1 & 0 \\ 0 & 0 & 0 & -1 \end{bmatrix}.   (20.8)$$

Finally, it is advantageous to introduce a **complex** null vector defined by

$$m^a = \frac{1}{\sqrt{2}} \left(Y^a + iZ^a \right),   (20.9)$$

together with its complex conjugate

$$\bar{m}^a = \frac{1}{\sqrt{2}} \left(Y^a - iZ^a \right).   (20.10)$$

It is then easy to verify (exercise) that the vectors are null,

$$m^a m_a = \bar{m}^a \bar{m}_a = 0,   (20.11)$$

and satisfy the normalizing condition

$$m^a \bar{m}_a = -1.   (20.12)$$

If we choose

$$(e_0{}^a, e_1{}^a, e_2{}^a, e_3{}^a) = (\ell^a, n^a, m^a, \bar{m}^a),\qquad(20.13)$$

then this defines a **null tetrad** with frame metric

$$g_{ij} = \begin{bmatrix} 0 & 1 & 0 & 0 \\ 1 & 0 & 0 & 0 \\ 0 & 0 & 0 & -1 \\ 0 & 0 & -1 & 0 \end{bmatrix}.\qquad(20.14)$$

Thus, writing out (20.3) explicitly, we have decomposed g_{ab} into products of the null tetrad vectors according to

$$g_{ab} = \ell_a n_b + \ell_b n_a - m_a \bar{m}_b - m_b \bar{m}_a.\qquad(20.15)$$

The contravariant form of this equation is

$$g^{ab} = \ell^a n^b + \ell^b n^a - m^a \bar{m}^b - m^b \bar{m}^a.\qquad(20.16)$$

20.2   The Kerr solution from a complex transformation

The Schwarzschild solution in advanced Eddington-Finkelstein coordinates is given by (17.24). The non-zero components of the contravariant metric g^{ab} are found to be (exercise)

$$g^{01} = -1, \quad g^{11} = -\left(1 - \frac{2m}{r}\right), \quad g^{22} = -\frac{1}{r^2}, \quad g^{33} = -\frac{1}{r^2 \sin^2 \theta}.$$
$$(20.17)$$

It is straightforward to check, using (20.16), that the contravariant metric may be written in terms of the following null tetrad:

$$\left.\begin{aligned} \ell^a &= (0,\ 1,\ 0,\ 0) = \delta_1^a, \\ n^a &= \left(-1,\ -\tfrac{1}{2}\left(1 - 2m/r\right),\ 0,\ 0\right) = -\delta_0^a - \tfrac{1}{2}\left(1 - 2m/r\right)\delta_1^a, \\ m^a &= \frac{1}{\sqrt{2}\,r}\left(0,\ 0,\ 1,\ \frac{i}{\sin\theta}\right) = \frac{1}{\sqrt{2}\,r}\left(\delta_2^a + \frac{i}{\sin\theta}\,\delta_3^a\right). \end{aligned}\right\}$$
$$(20.18)$$

The 'trick' starts by allowing the coordinate r to take on complex values and the tetrad is rewritten in the form

$$\left.\begin{aligned} \ell^a &= \delta_1^a, \\ n^a &= -\delta_0^a - \tfrac{1}{2}\left[1 - m\left(r^{-1} + \bar{r}^{-1}\right)\right]\delta_1^a, \\ m^a &= \frac{1}{\sqrt{2}\,\bar{r}}\left(\delta_2^a + \frac{i}{\sin\theta}\,\delta_3^a\right), \end{aligned}\right\}$$
$$(20.19)$$

where throughout this procedure we keep ℓ^a and n^a real and m^a and $\bar{m}^a$ complex conjugate to each other. We next formally perform the complex coordinate transformations

$$v \to v' = v + ia\cos\theta, \quad r \to r' = r + ia\cos\theta, \quad \theta \to \theta', \quad \phi \to \phi',$$
$$(20.20)$$

on the null tetrad. Then, if we require that v' and r' are real, we obtain the following tetrad (exercise):

$$\left.\begin{array}{l}
\ell'^a = \delta_1^a, \\[2ex]
n'^a = -\delta_0^a - \dfrac{1}{2}\left(1 - \dfrac{2mr'}{r'^2 + a^2\cos^2\theta}\right)\delta_1^a, \\[2ex]
m'^a = \dfrac{1}{\sqrt{2}\,(r' + ia\cos\theta)}\left(-ia\sin\theta\,(\delta_0^a + \delta_1^a) + \delta_2^a + \dfrac{i}{\sin\theta}\delta_3^a\right).
\end{array}\right\} \quad (20.21)$$

These are the components of the null tetrad for the promised Kerr solution and the contravariant components of the metric can then be read off using (20.16).

20.3 The three main forms of the Kerr solution

The procedure of the last section gives rise to the following line element (dropping the prime on r)

$$ds^2 = \left(1 - \frac{2mr}{\rho^2}\right)dv^2 - 2\,dv\,dr + \frac{2mr}{\rho^2}\left(2a\sin^2\theta\right)dv\,d\bar{\phi} + 2a\sin^2\theta\,dr\,d\bar{\phi}$$
$$- \rho^2\,d\theta^2 - \left((r^2 + a^2)\sin^2\theta + \frac{2mr}{\rho^2}\left(a^2\sin^4\theta\right)\right)d\bar{\phi}^2, \qquad (20.22)$$

where

$$\rho^2 = r^2 + a^2\cos^2\theta, \qquad (20.23)$$

and, for later convenience, we have replaced ϕ by $\bar{\phi}$. This is obtained by complexifying the advanced Eddington-Finkelstein form of the Schwarzschild solution, and so we shall term (20.22) the **advanced Eddington-Finkelstein** form of the Kerr solution. To obtain the analogue of the Schwarzschild form, we carry out the coordinate transformation (to be explained later) from the old coordinates $(v, r, \theta, \bar{\phi})$ to new coordinates (t, r, θ, ϕ). It turns out to be easier to work with the coordinate differentials rather than the coordinates themselves, in which case the transformation is given by

$$dv = d\bar{t} + dr = dt + \frac{2mr + \Delta}{\Delta}dr, \qquad (20.24)$$

$$d\bar{\phi} = d\phi + \frac{a}{\Delta}dr, \qquad (20.25)$$

where

$$\Delta = r^2 - 2mr + a^2, \tag{20.26}$$

and r and θ remain unchanged. This leads to the form of Kerr's solution called the **Boyer–Lindquist** form, namely,

$$ds^2 = \frac{\Delta}{\rho^2} \left(dt - a\sin^2\theta\, d\phi\right)^2 - \frac{\sin^2\theta}{\rho^2} \left[\left(r^2 + a^2\right) d\phi - a\, dt\right]^2$$

$$- \frac{\rho^2}{\Delta} dr^2 - \rho^2\, d\theta^2. \tag{20.27}$$

In fact, neither (20.22) nor (20.27) was the form in which Kerr originally discovered the solution. He used Cartesian-type coordinates $(\bar{t}, x, y, z)$ to obtain the **Kerr** form

$$ds^2 = d\bar{t}^2 - dx^2 - dy^2 - dz^2$$

$$- \frac{2mr^3}{r^4 + a^2 z^2} \left(d\bar{t} + \frac{r}{a^2 + r^2}(xdx + ydy)\right.$$

$$\left. + \frac{a}{a^2 + r^2}(ydx - xdy) + \frac{z}{r}dz\right)^2, \tag{20.28}$$

where

$$\left.\begin{aligned} \bar{t} &= v - r, \\ x &= r\sin\theta\cos\phi + a\sin\theta\sin\phi, \\ y &= r\sin\theta\sin\phi - a\sin\theta\cos\phi, \\ z &= r\cos\theta. \end{aligned}\right\} \tag{20.29}$$

This line element has the general form

$$ds^2 = \eta_{ab}\, dx^a\, dx^b - \lambda \ell_a \ell_b\, dx^a\, dx^b, \tag{20.30}$$

where the vector ℓ^a is null with respect to the Minkowski metric η_{ab}, that is,

$$\eta_{ab}\, \ell^a \ell^b = 0. \tag{20.31}$$

In the particular case of the Kerr form (20.28), we have

$$\lambda = \frac{2mr^3}{r^4 + a^2 z^2}, \tag{20.32}$$

and

$$\ell_a = \left(1, \frac{rx + ay}{a^2 + y^2}, \frac{ry - ax}{a^2 + y^2}, \frac{z}{r}\right). \tag{20.33}$$

In the special case of the Schwarzschild solution, this reduces to

$$\lambda = 2m/r, \tag{20.34}$$

and

$$\ell_a = (1, x/r, y/r, z/r). \tag{20.35}$$

Indeed, it was precisely by considering metrics of the form (20.30) subject to (20.31) that Kerr originally found the solution; see Adler et al. (1975) for the details. We shall now attempt to gain some physical insight into the Kerr solution, and in so doing we shall make use of all three forms, namely, the Eddington-Finkelstein, Boyer–Lindquist, and Kerr versions of the solution, which is why we have collected them together in this section.

20.4   Basic properties of the Kerr solution

The Boyer–Lindquist form is the most useful one for investigating the elementary properties of the Kerr solution. First of all, it is clear that the solution depends on the two parameters m and a. If we set $a = 0$, we regain the Schwarzschild solution in Schwarzschild coordinates and so m is identified as the **geometric mass**. The metric coefficients in (20.27) are independent of both ϕ and t, and hence the solution is both **axially symmetric** with Killing vector field $\partial/\partial\phi$, and **stationary**, at least in the exterior region $r > r_{S_+}$ where $\partial/\partial t$ is a timelike Killing vector – see equation (20.46). To say that a solution is axially symmetric means that there exists an invariantly defined axis (which in coordinate terms we take to be the z-axis or $\theta = 0$) such that the solution is invariant under rotation about this axis. Equivalently, the orbits of the Killing vector field $\partial/\partial\phi$, namely, the curves $t = $ constant, $r = $ constant, $\theta = $ constant, are circles. These are the only continuous symmetries. As for discrete symmetries, the solution is not symmetric separately under time reflection or ϕ reflection (reflection in the xz-plane), but it is invariant under the simultaneous inversion of t and ϕ, that is, under the transformation

$$t \rightarrow -t, \qquad \phi \rightarrow -\phi. \tag{20.36}$$

This suggests that the Kerr field may arise from a spinning source, since running time backwards with a negative spin direction is equivalent to running time forwards with a positive spin direction. Again, the line element is invariant under

$$t \rightarrow -t, \qquad a \rightarrow -a,$$

which suggests that a specifies a spin direction.

A third property which lends support to the spinning source interpretation is the presence in these canonical (t, ϕ)-coordinates of a cross-term involving $dt d\phi$ (the only cross-term present). Let us consider in New-tonian theory two frames $Oxyz$ and $Ox'y'z'$ whose origins and z-axes coincide, in which the primed frame is rotating relative to the unprimed frame with constant angular velocity $a\mathbf{k}$ (Fig. 20.2). Then a point P has cylindrical coordinates (r, ϕ, z) and (r', ϕ', z') relative to the two frames, where

$$r' = r, \quad \phi' = \phi - at, \quad z' = z. \qquad (20.37)$$

If we take $Oxyz$ to be inertial, then this represents a transformation to a rotating frame. Now write flat space in cylindrical polar coordinates (t, r, ϕ, z), namely,

$$ds^2 = dt^2 - \left(dr^2 + r^2 d\phi^2 + dz^2\right), \qquad (20.38)$$

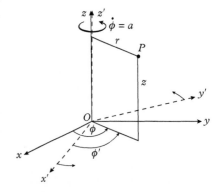

Fig. 20.2 Primed frame rotating about the z-axis of unprimed frame.

and carry out the coordinate transformation (20.37) to a 'rotating frame' (leaving t unchanged). The line element (20.38) becomes (exercise)

$$ds^2 = \left(1 - a^2 r^2\right) dt^2 - 2ar^2 d\phi' dt - \left(dr^2 + r^2 d\phi'^2 + dz^2\right), \qquad (20.39)$$

which, as we see, also possesses a cross-term in $d\phi' dt$. This is somewhat imprecise since we have not discussed rigid rotation in special relativity (nor shall we). The argument presented is merely suggestive of rotation. Nor have we said anything precise yet about the coordinates r and θ in (20.27). Indeed, r is **not** the usual spherical polar radial coordinate except asymptotically (although we shall retain r to agree with standard notation). For, if we take (x, y, z) in (20.28) to be the usual Cartesian coordinates, then the standard spherical polar coordinate R is defined by

$$R^2 = x^2 + y^2 + z^2, \qquad (20.40)$$

and hence, from (20.29),

$$R^2 = r^2 + a^2 \sin^2 \theta. \qquad (20.41)$$

However, for $r \gg a$ (exercise)

$$R = r + \frac{a^2 \sin^2 \theta}{2r} + \cdots, \qquad (20.42)$$

which shows that R and r coincide asymptotically. They also coincide in the Schwarzschild limit $a \to 0$. Further, it follows from the Kerr form (20.28) that

$$g_{ab} \to \eta_{ab} \quad \text{as} \quad R \to \infty,$$

so that the Kerr solution is **asymptotically flat**.

If we return to the idea that the Kerr solution represents a vacuum field exterior to a spinning source, then there are a number of independent arguments to suggest that a is related to the **angular velocity** and ma to the **angular momentum** (as measured at infinity). One argument involves comparing the Kerr solution with a solution due to Lense and Thirring for the gravitational field exterior to a spinning sphere of constant density in the weak-field limit. Another argument is based on the definition of the multipole moments of an isolated source. There are difficulties associated with this latter work because a number of different definitions have been proposed (indeed, an infinitude of them); however, they all lead to the angular momentum being proportional to ma for the Kerr metric. We have already seen that, in the weak-field limit, the $1/R$ term in g_{00} determines the total mass of the field. It is also possible to show that, in certain circumstances, the $1/R$ terms in $g_{0\alpha}$ ($\alpha = 1, 2, 3$) determine the components of the angular momentum. Expanding the Kerr solution (20.28) in powers of $1/R$, we find

$$\mathrm{d}s^2 = \left(1 - \frac{2m}{R} + \cdots\right)\mathrm{d}t^2 - \frac{4ma}{R^3}\left(x\,\mathrm{d}y - y\,\mathrm{d}z\right)\mathrm{d}t + \cdots, \qquad (20.43)$$

which again suggests that the total angular momentum is proportional to ma.

20.5 Singularities and horizons

Calculation of the Riemann invariant $R^{abcd}R_{abcd}$ reveals that this diverges as $\rho \to 0$, indicating that $\rho = 0$ is an intrinsic singularity. Since

$$\rho^2 = r^2 + a^2 \cos^2\theta = 0,$$

this is when $r = 0$ and $\cos\theta = 0$. It then follows from (20.29), (20.40), and (20.41), that this occurs when

$$x^2 + y^2 = a^2, \quad z = 0. \qquad (20.44)$$

This singularity is, rather surprisingly, a **ring** of radius a lying in the equatorial plane $z = 0$. We have only considered how to calculate the gravitational red shift in a static space-time, but it can be shown that the surfaces of infinite red shift in the Kerr solution are again given by the vanishing of the coefficient g_{00}. From (20.27), we find

$$g_{00} = \left(r^2 - 2mr + a^2 \cos^2\theta\right)/\rho^2, \qquad (20.45)$$

and so the surfaces of infinite red shift are

$$r = r_{S_\pm} = m \pm \left(m^2 - a^2 \cos^2\theta\right)^{\frac{1}{2}}. \qquad (20.46)$$

In the Schwarzschild limit, $a \to 0$, the surface S_+ reduces to $r = 2m$, and S_- to $r = 0$. The surfaces are axially symmetric, with S_+ possessing a radius $2m$ at the equator and (assuming $a^2 < m^2$) a radius $m + (m^2 - a^2)^{\frac{1}{2}}$ at the poles, and the surface S_- being completely contained inside S_+. We shall primarily be concerned with the physically more interesting case $a^2 < m^2$, when the spin is small compared with the mass.

The existence of these infinite red-shift surfaces imply the existence of a null event horizon as follows. The Killing vector field

$$X^a = (1, 0, 0, 0),$$

has magnitude

$$X^2 = X_a X^a = g_{ab} X^a X^b = g_{00}.$$

It follows from (20.45) and (20.46) that X^a is timelike outside S_+ and inside S_-, null on S_+ and S_-, and spacelike between S_+ and S_-. In analogy with the Schwarzschild solution, we search for the event horizon by looking for the hypersurfaces where $r = $ constant becomes null, that is, where g^{11} vanishes. From the Boyer–Lindquist form (20.27), we find (exercise)

$$g^{11} = -\frac{\Delta}{\rho^2} = -\frac{r^2 - 2mr + a^2}{r^2 + a^2 \cos^2 \theta},$$

and hence g^{11} vanishes when

$$\Delta = r^2 - 2mr + a^2 = 0,$$

which results in **two null event horizons** (assuming $a^2 < m^2$).

$$r = r_\pm = m \pm \left(m^2 - a^2\right)^{\frac{1}{2}}. \tag{20.47}$$

Then, in a similar way in which the Reissner-Nordström solution is regular in three regions, the Kerr solution is regular in the three regions:
   I. $r_+ < r < \infty$,
   II. $r_- < r < r_+$,
   III. $0 < r < r_-$.

In the Schwarzschild limit, $a \to 0$, and the two event horizons reduce to $r = 2m$ and $r = 0$, from which it follows that, in the Schwarzschild solution, the surfaces of infinite red shift and the event horizons coincide. The event horizon $r = r_+$ lies entirely within S_+, giving rise to a region between the two called the **ergosphere**, the properties of which we shall discuss in §20.11. The various surfaces and the ring singularity are illustrated in Fig. 20.3.

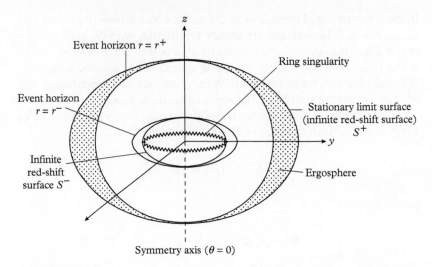

Fig. 20.3 The event horizons, stationary limit surface, and ring singularity of the Kerr solution.

We end this section by summarizing the properties we have met so far. The Kerr vacuum solution:

1. is stationary in the exterior region;
2. is axisymmetric;
3. is invariant under the discrete transformations

$$t \rightarrow -t, \quad \phi \rightarrow -\phi, \quad \text{and} \quad t \rightarrow -t, \ a \rightarrow -a;$$

4. has geometric mass m;
5. represents the field exterior to a spinning source where the spin of the field is related to a and the angular momentum to ma;
6. is asymptotically flat;
7. has a ring singularity at

$$x^2 + y^2 = a^2, \qquad z = 0;$$

8. has two surfaces of infinite red shift S_+ and S_- given by

$$r = m \pm (m^2 - a^2 \cos^2 \theta)^{\frac{1}{2}};$$

9. in the case $a^2 < m^2$, has two event horizons at r_+ and r_- given by

$$r_\pm = m \pm (m^2 - a^2)^{\frac{1}{2}}.$$

20.6 The principal null congruences

The Kerr solution is no longer spherically symmetric and so we no longer expect that there are any curves corresponding to radial null geodesics. This is because, in a loose sense, we expect a rotating source to 'drag' space around with it and consequently drag the geodesics with it. The situation is very different from what happens in Newtonian theory, where, if one was investigating a source rotating about the z-axis, say, one could transfer to a frame rotating with the source and so reduce it to rest. However, one cannot do this in general relativity because it is **not** possible to find a coordinate system which reduces the Kerr solution to the Schwarzschild solution. Put another way, the nonlinear field equations couple the source to the exterior field.

Since the metric is axially symmetric, we might expect to obtain null geodesics which lie in the hypersurface $\theta =$ constant. We therefore search for null geodesics for which

$$\dot{\theta} = ds^2 = 0, \tag{20.48}$$

where the dot denotes differentiation with respect to an affine parameter and where, throughout, θ is kept constant. We use the Boyer–Lindquist form (20.27), and then the fact that the metric coefficients are independent of t and ϕ means that the Euler-Lagrange equations immediately lead to first integrals of the motion. These are

$$\frac{\Delta}{\rho^2}\left(\dot{t} - a\sin^2\theta\,\dot{\phi}\right) + \frac{a\sin^2\theta}{\rho^2}\left[(r^2 + a^2)\dot{\phi} - a\dot{t}\right] = L, \tag{20.49}$$

$$\frac{a\Delta\sin^2\theta}{\rho^2}\left(\dot{t} - a\sin^2\theta\,\dot{\phi}\right) + \frac{(r^2 + a^2)\sin^2\theta}{\rho^2}\left[(r^2 + a^2)\dot{\phi} - a\dot{t}\right] = N, \tag{20.50}$$

where L and N are constants of integration. We have another first integral from the condition $ds^2 = 0$, namely,

$$\frac{\Delta}{\rho^2}\left(\dot{t} - a\sin^2\theta\,\dot{\phi}\right)^2 - \frac{\sin^2\theta}{\rho^2}\left[(r^2 + a^2)\dot{\phi} - a\dot{t}\right]^2 - \frac{\rho^2\dot{r}^2}{\Delta} = 0. \tag{20.51}$$

Finally, we have the Euler-Lagrange equation corresponding to $x^2 = \theta$ and, using the fact that $\ddot{\theta} = 0$ from (20.48), this becomes

$$\frac{a^2\Delta}{\rho^4}\left(\dot{t} - a\sin^2\theta\,\dot{\phi}\right)^2 - \frac{2a\Delta\dot{\phi}}{\rho^2}\left(\dot{t} - a\sin^2\theta\,\dot{\phi}\right)$$
$$- \frac{r^2 + a^2}{\rho^4}\left[(r^2 + a^2)\dot{\phi} - a\dot{t}\right]^2 + \frac{a^2\dot{r}^2}{\Delta} = 0. \tag{20.52}$$

Since (20.49), (20.50), (20.51), and (20.52) represent four equations in the three unknowns $\dot{t}$, $\dot{r}$, and $\dot{\phi}$, it follows that there must exist some constraint between L and N. Some algebra reveals that the constraint is (exercise)

$$\left(N + aL\sin^2\theta\right)\left(N - aL\sin^2\theta\right) = 0, \tag{20.53}$$

where θ is constant. Restricting attention to the condition

$$N - aL\sin^2\theta = 0, \tag{20.54}$$

the system of equations can be solved for $\dot{t}$, $\dot{r}$, and $\dot{\phi}$ to give

$$\dot{t} = \left(r^2 + a^2\right)L/\Delta, \tag{20.55}$$

$$\dot{r} = \pm L, \tag{20.56}$$

$$\dot{\phi} = aL/\Delta. \tag{20.57}$$

We have therefore found two null congruences corresponding to the two signs in (20.56). Moreover, (20.56) shows that r is an affine parameter along each congruence. Choosing $\dot{r} = +L$, we get

$$\frac{\mathrm{d}t}{\mathrm{d}r} = \frac{\dot{t}}{\dot{r}} = \frac{r^2 + a^2}{\Delta}, \tag{20.58}$$

and

$$\frac{\mathrm{d}\phi}{\mathrm{d}r} = \frac{\dot{\phi}}{\dot{r}} = \frac{a}{\Delta}. \tag{20.59}$$

If we restrict our attention to the case $a^2 < m^2$, these equations can be immediately integrated to give (exercise)

$$
\begin{aligned}
t = r + {} & \left(m + \frac{m^2}{(m^2 - a^2)^{\frac{1}{2}}}\right)\ln|r - r_+| \\
& + \left(m - \frac{m^2}{(m^2 - a^2)^{\frac{1}{2}}}\right)\ln|r - r_-| + \text{constant},
\end{aligned} \tag{20.60}
$$

and

$$\phi = \frac{a}{2(m^2 - a^2)^{\frac{1}{2}}}\ln\left|\frac{r - r_+}{r - r_-}\right| + \text{constant}. \tag{20.61}$$

Using the fact that $\Delta > 0$ in regions I and III, and $\Delta < 0$ in region II, then it follows from (20.58) that $\mathrm{d}r/\mathrm{d}t > 0$ in region I, and so this congruence is called the **principal congruence of outgoing null geodesics**. The solution corresponding to $\dot{r} = -L$ is again given by (20.60) and (20.61) if we simply replace t by $-t$ and ϕ by $-\phi$, and so is called the **principal congruence of ingoing null geodesics**. The solutions reduce to the Schwarzschild congruences (17.10) and (17.11), respectively, in the limit $a \to 0$, as we should expect.

These two congruences play the same role as the null radial congruences do in the Schwarzschild solution. They give information about the radial variation of the light cone structure in that the most outgoing and most ingoing null lines – those for which $|dr/dt|$ is a maximum at any point – are members of the principal null congruences. We can draw a space-time diagram of the light cones using these equations and we find in region I a diagram analogous to Fig. 17.7 with the light cones narrowing down as $r \to r_+$. On $r = r_+$, both t and ϕ become infinite, suggesting, as in the Schwarzschild solution, that $r = r_+$ is a coordinate singularity. We therefore proceed as we did in the Schwarzschild solution and look for the analogue of the Eddington-Finkelstein coordinate system.

20.7   Eddington-Finkelstein coordinates

We use the principal null congruences to obtain a coordinate transformation which extends the solution through $r = r_+$. We could work explicitly with the equations of the congruence (20.60) and (20.61), but it turns out to be simpler to work with them in the differential form (20.58) and (20.59), that is,

$$dt = -\frac{r^2 + a^2}{\Delta} dr, \tag{20.62}$$

$$d\phi = -\frac{a}{\Delta} dr, \tag{20.63}$$

for the ingoing congruence. In the Schwarzschild case, we looked for a transformation to new coordinates $(\bar{t}, r, \theta, \phi)$ in which the equations for the ingoing radial null congruence take on the simpler differential form

$$d\bar{t} = -dr, \quad d\theta = d\phi = 0. \tag{20.64}$$

Proceeding similarly in the Kerr case, we search for a transformation to new coordinates $(\bar{t}, r, \theta, \bar{\phi})$ in which the principal ingoing congruence reduces to

$$d\bar{t} = -dr, \quad d\theta = d\bar{\phi} = 0. \tag{20.65}$$

Using (20.58) and (20.59), the requisite transformations are (exercise)

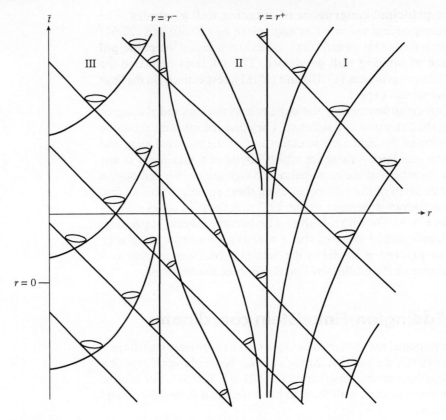

Fig. 20.4 Kerr solution ($a^2 < m^2$) in advanced Eddington–Finkelstein type coordinates.

$$t \rightarrow \bar{t} \quad \text{where} \quad d\bar{t} = dt + \frac{2mr}{\Delta}\, dr, \tag{20.66}$$

$$\phi \rightarrow \bar{\phi} \quad \text{where} \quad d\bar{\phi} = d\phi + \frac{a}{\Delta}\, dr. \tag{20.67}$$

If we define an advanced time coordinate

$$v = \bar{t} + r, \tag{20.68}$$

where $\bar{t}$ is obtained by integrating (20.66), then the Boyer–Lindquist line element is transformed into (20.22), the advanced Eddington-Finkelstein form of the Kerr solution. We will look at how to extend this in §20.9. The two-dimensional space-time diagram for this solution is given in Fig. 20.4 (compare this with the Reissner-Nordström space-time diagram, Fig. 19.4).

20.8   The stationary limit

Consider the set of null curves in the region I given by

$$dr = d\theta = ds^2 = 0. \qquad (20.69)$$

Then the Boyer–Lindquist line element reduces to

$$\frac{\Delta}{\rho^2} \left(dt - a\sin^2\theta d\phi\right)^2 - \frac{\sin^2\theta}{\rho^2} \left[(r^2 + a^2)d\phi - adt\right]^2 = 0,$$

and solving for $d\phi/dt$ produces

$$\frac{d\phi}{dt} = \frac{a\sin\theta \pm \Delta^{\frac{1}{2}}}{(r^2 + a^2)\sin\theta \pm a\Delta^{\frac{1}{2}}\sin^2\theta}. \qquad (20.70)$$

These curves are not geodesics, but are tangent to world-lines of photons initially constrained to orbit the source with fixed r and θ. The positive sign in (20.70) leads to $d\phi/dt > 0$, that is, the photon orbits the source in the same direction as the rotation of the source. We now investigate when it is possible for $d\phi/dt \leqslant 0$, in which case we must restrict attention to the negative sign in (20.70). In region I,

$$r > r_+ \quad \Longleftrightarrow \quad (r^2 + a^2)\sin^2\theta - a\Delta^{\frac{1}{2}}\sin^2\theta > 0,$$

so that the denominator of (20.70) is positive. Hence (exercise),

$$\frac{d\phi}{dt} \leqslant 0 \quad \Longleftrightarrow \quad a\sin\theta - \Delta^{\frac{1}{2}} \leqslant 0 \quad \Longleftrightarrow r \geqslant r_{S_+}. \qquad (20.71)$$

Thus, on S_+, the derivative $d\phi/dt$ is zero, and hence any particle on this hypersurface attempting to orbit the source against its direction of rotation must travel with the local speed of light just to remain stationary (that is, to be precise, stationary relative to a stationary observer at infinity). In the ergosphere, the light cones tip over in the direction of ϕ increasing to such an extent that photons and particles are forced to orbit the source in the direction of its rotation. It is because of this that the infinite red-shift surface S_+ is also termed the **stationary limit surface**. The stationary limit surface is a timelike surface except at the two points on its axis, where it is null and where it coincides with the event horizon $r = r_+$. Where the surface is timelike, the light cone structure reveals that it can be crossed by particles in either the ingoing or the outgoing direction. These properties are most clearly revealed in a spatial diagram of the Kerr solution ($a^2 < m^2$) in the equatorial plane (Fig. 20.5).

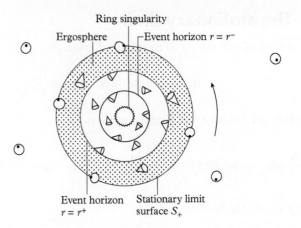

Fig. 20.5 Spatial diagram of the Kerr solution ($a^2 < m^2$) in the equatorial plane.

20.9   Maximal extension for the case $a^2 < m^2$

The Kerr metric can be extended by using advanced and retarded Eddington-Finkelstein coordinates

$$du_\pm = dt \pm \frac{r^2 + a^2}{\Delta} dr, \qquad d\phi_\pm = d\phi \pm \frac{a}{\Delta} dr$$

in a manner analogous to the Reissner-Nordström case, where the maximal extension is built up by a combination of these extensions. The global structure is very similar to that of the Reissner-Nordström solution except that now one can continue through the ring singularity to negative values of r. Figure 20.6 shows the conformal structure of the solution along the symmetry axis ($\theta = 0$) for the case $a^2 < m^2$. Note that, on the symmetry axis, the event horizons and surfaces of infinite redshift coincide. The regions I ($r_+ < r < \infty$) are stationary asymptotically flat regions exterior to the outer event horizon. The regions II ($r_- < r < r_+$) are non-stationary and each point in one is a trapped surface (see §18.5). The regions III ($-\infty < r < r_-$) contain the ring singularity, which is **timelike** and hence avoidable. This region also contains **closed timelike curves**. Such curves violate causality and would seem highly unphysical since, if they represent world-lines of observers, then these observers would travel back and meet themselves in the past! There is no causality violation in the regions I and II. In the limiting case $a^2 = m^2$, the event horizons r_+ and r_- coincide and there are no regions II. The maximal extension is similar to that of the Reissner-Nordström solution when $\varepsilon^2 = m^2$ and its conformal structure along the symmetry axis is shown in Fig. 20.7.

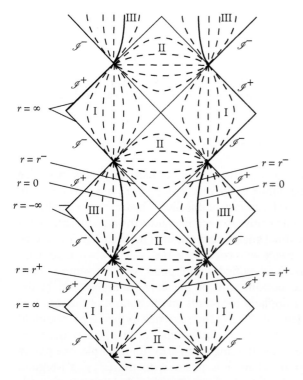

Fig. 20.6 Penrose diagram of the maximal extension of the Kerr solution ($a^2 < m^2$) along the symmetry axis.

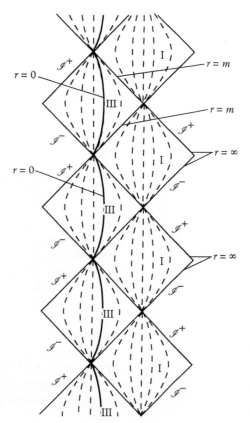

Fig. 20.7 Penrose diagram of the maximal extension of the Kerr solution ($a^2 = m^2$) along the symmetry axis.

20.10 Maximal extension for the case $a^2 > m^2$

In the case $a^2 > m^2$, we find that $\Delta > 0$ and the Boyer–Lindquist form of the Kerr solution (20.27) is regular everywhere except at $r = 0$, where there is a ring singularity. The coordinate r, by (20.29), can be determined in terms of x, y and z from

$$r^4 - (x^2 + y^2 + z^2 - a^2)r^2 - a^2 z^2 = 0.$$

For $r \neq 0$, the surfaces $r = $ constant are confocal ellipsoids in a slice $t = $ constant which degenerate to the disc $x^2 + y^2 \leqslant a^2$, $z = 0$ when $r = 0$. The ring singularity is the boundary of this disc. The function r can be analytically continued from positive to negative values through the interior of the disc to obtain a maximal analytic extension of the solution. To do this, one attaches another surface with coordinates (x', y', z'), where a point on the top side of the disc is identified with a point with the same x- and y- coordinates on the bottom side of the corresponding disc in the (x', y', z')-surface, and similarly for points on the bottom of the disc (see Fig. 20.8). The line element (20.27) then extends to this larger manifold and has the same form on the (x', y', z')-region, but r is now negative. Then on circling twice round the ring singularity, for example, one passes from the (x, y, z)-region, where r is positive, to the (x', y', z')-region, where r is negative, and back to the (x, y, z)-region. At large negative values of r, the space is again asymptotically flat, but this time it has **negative** mass.

For a small value of r near the singularity, the vector $\partial/\partial\phi$ is timelike so the circles $t = $ constant, $r = $ constant, $\theta = $ constant are timelike curves. These closed timelike curves can be deformed to pass through any point of the extended space, so that the solution badly violates causality. The solution is geodesically incomplete at the ring singularity, but the only timelike and null geodesics which reach this singularity are those in the equatorial plane on the positive-r side. This leads to another bizarre property of the solution. The event horizons have now disappeared, but an intrinsic space-time singularity still exists at the ring and now it is possible for information to escape from the singularity to the outside world, provided it spirals around sufficiently (Fig. 20.9). In short, the singularity is visible, in all its nakedness, to the outside world. Such a singularity is called a **naked singularity**. If naked singularities exist, then they open up a whole new realm for wild speculation, so much so that Penrose has suggested the existence of the **cosmic censorship hypothesis**, which would forbid the existence of naked singularities but would only allow singularities to be hidden behind event horizons. There exist various mathematical formulations of this hypothesis and attempts to establish under what conditions, if any, the various cosmic censorship hypotheses hold have been, and remain, an area of active research (Christodoulou 2008).

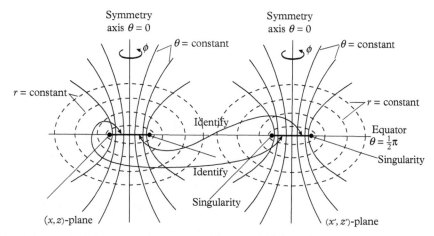

Fig. 20.8 Maximal extension of the Kerr solution ($a^2 > m^2$).

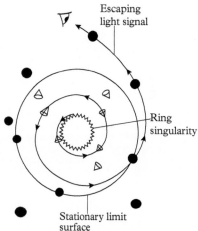

Fig. 20.9 The Kerr solution ($a^2 > m^2$) as a naked singularity.

20.11 Rotating black holes

We consider the ideal case of a rotating star whose exterior field is given by the Kerr solution for $0 < a^2 < m^2$. Intuitively, we may think of the source as a rotating sphere or ellipsoid of matter, but as we have indicated before there is as yet, despite considerable efforts, no known physically realistic interior Kerr solution. (Perhaps the existence of a ring singularity suggests that we might be able to fill in the Kerr solution with a toroidal rather than a spherical source.) Nonetheless, we envisage this source collapsing through the event horizon $r = r_+$ to give rise to a black hole. As before, any observer following the collapse through $r = r_+$ will be unable to return to their original region I. The collapse will necessarily continue through $r = r_-$ and any observer in region II must follow the collapse through to region III. A difference arises in the rotating case, as compared with the non-rotating case, in that the collapse may now halt. The maximal extension then suggests that an observer in region III is able to escape into a new asymptotically flat region I. We shall return to the question of a more physically realistic collapse situation later.

In order to obtain the most general black hole solution, we apply the Newman–Janis trick of §20.2 to the Reissner-Nordström solution in advanced Eddington-Finkelstein coordinates (see Exercise 19.13), namely,

$$ds^2 = \left(1 - \frac{2m}{r} + \frac{\varepsilon^2}{r^2}\right) dr^2 - 2dvdr - r^2(d\theta^2 + \sin^2\theta d\phi^2).$$

We find the result (exercise)

$$ds^2 = \left(1 - \frac{2mr}{\rho^2} + \frac{\varepsilon^2}{\rho^2}\right) d\upsilon^2 - 2d\upsilon dr + \frac{2a}{\rho^2}\left(2mr - \varepsilon^2\right)\sin^2\theta \, d\upsilon d\bar{\phi}$$
$$+ 2a\sin^2\theta \, dr\,d\bar{\phi} - \rho^2\,d\theta^2$$
$$- \left[\left(r^2 + a^2\right)^2 - \left(r^2 - 2mr + a^2 + \varepsilon^2\right)a^2\sin^2\theta\right]\frac{\sin^2\theta}{\rho^2}\,d\bar{\phi}^2, \quad (20.72)$$

which is the **Kerr–Newman** solution in advanced Eddington-Finkelstein coordinates. The solution clearly depends on the three parameters m, a, ε, defining the mass, spin, and charge, respectively. It is stationary and axisymmetric, and possesses a stationary limit surface

$$r = m + \left(m^2 - \varepsilon^2 - a^2\cos^2\theta\right)^{\frac{1}{2}}, \quad (20.73)$$

and, provided that $a^2 + \varepsilon^2 \leqslant m^2$, an outer event horizon

$$r = m + \left(m^2 - \varepsilon^2 - a^2\right)^{\frac{1}{2}}. \quad (20.74)$$

It has properties analogous to the Kerr solution, but we shall not pursue the details further.

If we consider a realistic collapse of a charged rotating black hole, then the Kerr–Newman solution will not represent the true geometry exterior to the star at early times. This is because, if the star has not gone far down the road to collapse, it will not possess the symmetries of stationarity and axisymmetry. Gravitational moments will arise from mountains and other asymmetries. However, if an event horizon develops, then these asymmetries will be radiated away. In fact, a remarkable theorem has been proved which states that, if an event horizon develops in an asymptotically flat space-time, then the solution exterior to this horizon necessarily approaches a Kerr–Newman solution asymptotically in time. Thus, we have remarkably complete information as to the asymptotic state of affairs resulting from a gravitational collapse.

Detailed considerations of gravitational collapse suggest the following picture. A body, or collection of bodies, collapses down to a size comparable to its Schwarzschild radius, after which a trapped surface can be found in the region surrounding the matter. Some way outside the trapped surface there is another surface which will ultimately form the event horizon. But at present this surface is still expanding somewhat. Its exact location is a complicated affair and it depends on how much more matter or radiation falls in. We assume only a finite amount falls in. Then the expansion of the absolute event horizon gradually slows down to stationarity. Thus, when a black hole is created by gravitational collapse, it rapidly settles down to a stationary state that is characterized by the three parameters m, ε, and a. Apart from these three properties, the black hole preserves no other details of the object that collapsed. Wheeler has termed this the theorem that 'a black hole has no hair'. If you've seen one, you've seen them all! Wheeler depicts this rather humorously by a picture in which a vase of flowers and a television set fall into a black hole (Fig. 20.10).

Fig. 20.10 A black hole has no hair.

Once the system has settled down, the only quantities which may have altered are m, ε, and a. All details of the objects swallowed up are obliterated. Considering the time reversal of this situation, we see that, if you happen to be an astronaut travelling in space and you suddenly see a vase of flowers and a television set pop out of nowhere, then you know you are in the vicinity of a **white** hole – a rather 'hairy' prospect!

Penrose has suggested that it might be possible to extract energy from a rotating black hole as follows. A particle is fired into the ergosphere, where it decays into two products, one falling into the black hole and the other escaping outside the stationary limit. Calculations reveal that the escaping component can contain **more** mass-energy than the original particle. This is possible because the angular momentum of the black hole is reduced in the process, which corresponds to a transfer of energy to the matter that escapes. The reduction in energy due to the loss of angular momentum of the rotating black hole is converted into the energy extracted by the particle that escapes. This leads to a fanciful suggestion that an advanced civilization could live near a rotating black hole and develop some mechanism for extracting their energy requirements from the black hole's rotation (Fig. 20.11). The way in which energy can be extracted from a spinning black hole is similar to the quantum mechanical effect of superradiance of electromagnetic radiation from a spinning metal sphere, and in 1971 Zel'dovich suggested that a Kerr solution should produce similar quantum effects and ought to radiate in a similar way. However,

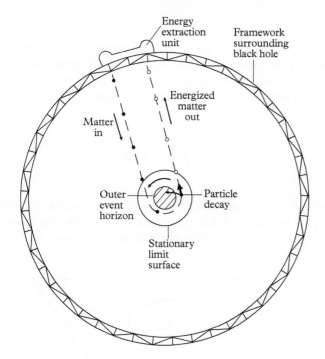

Fig. 20.11 Living off a rotating black hole.

as we will see in §20.14, Hawking showed that rotation is not necessary to produce this effect and that **all** black holes radiate if one takes such quantum effects into account.

20.12   The definition of mass in general relativity

In the previous section, we described the 'no-hair' theorem which says that the most general black hole solution is described in terms of just three quantities, the mass, the angular momentum, and the charge. We looked at these quantities for some special black hole solutions but did not say how to define them for a general space-time. In Newtonian theory, one can compute the mass of an object simply by integrating the mass density ρ. However, as we explain in the next chapter in §21.8, there is no corresponding definition of the mass-energy density in general relativity. However it turns out that it is possible to write down the **total mass** of an asymptotically flat space-time. Before considering the definition of mass, we consider the simpler concept of charge in electromagnetism in order to motivate the construction.

In the presence of charges and currents, we need to modify the first of Maxwell's equations (19.3) to include a source term. This gives

$$\nabla_b F^{ab} = j^a. \tag{20.75}$$

The term j_a encodes the sources of the electromagnetic field and, for an observer with 4-velocity n^a, then $j_a n^a$ is a measure of the charge density. Now suppose we have a spacelike hypersurface Σ extending to infinity and consider a topological two sphere U enclosing a region V on Σ (see Fig 20.12). Then we can compute the total charge Q by integrating the charge density $j_a n^a$ over V

$$Q = \int_V j^a n_a \sqrt{\gamma} \mathrm{d}^3 x. \tag{20.76}$$

Here γ is the determinant of the positive definite 3-metric induced on Σ (see Chapter 14) and n^a is the unit normal to Σ, so that $j^a n_a$ measures the charge density as measured by Eulerian observers. Using (20.75) and the

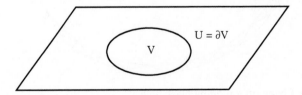

Fig. 20.12 Charge given by an integral of the charge density over the volume V.

divergence theorem, we have

$$Q = \int_V \nabla_b F^{ab} n_a \sqrt{\gamma} \mathrm{d}^3 x \qquad (20.77)$$

$$= \int_U F^{ab} n_a m_b \sqrt{\sigma} \mathrm{d}^2 x, \qquad (20.78)$$

where m^b is a unit normal to U and σ is the determinant of the 2-metric induced on the surface U. If we now consider the limit as U goes out to infinity, we obtain an expression for the total charge that depends on the asymptotic behaviour of F^{ab}

$$Q_{\text{total}} = \int_S F^{ab} n_a m_b \sqrt{\sigma} \mathrm{d}^2 x, \qquad (20.79)$$

where S is now a sphere at infinity. Because the space-time is asymptotically flat, the metric tends to that of Minkowski space, which, in spherical polar coordinates, is just

$$\mathrm{d}s^2 = \mathrm{d}t^2 - \mathrm{d}r^2 - r^2(\mathrm{d}\theta^2 + \sin^2\theta \mathrm{d}\phi^2). \qquad (20.80)$$

This enables us to write (exercise)

$$Q_{\text{total}} = \int_{\theta=0}^{\pi} \int_{\phi=0}^{2\pi} F^{tr} r^2 \sin\theta \mathrm{d}\theta \mathrm{d}\phi. \qquad (20.81)$$

We now show that the above expression gives us the expected result for a point charge q located at the origin in Minkowski space. The solution of Maxwell's equation for a point charge is (exercise) $F^{tr} = q/4\pi r^2$ (in Lorentz–Heaviside units) so we have

$$Q_{\text{total}} = \int_{\theta=0}^{\pi} \int_{\phi=0}^{2\pi} \frac{q}{4\pi r^2} r^2 \sin\theta \mathrm{d}\theta \mathrm{d}\phi = q, \qquad (20.82)$$

as we wanted.

We now describe a similar method for calculating the total mass of a space-time due to Komar. We first suppose that the space-time is **stationary** with timelike Killing vector K^a. If we make an analogy between the electromagnetic field and the gravitational field and consider the object

$$J^a := R^{ab} K_b, \qquad (20.83)$$

one can construct a mass–energy

$$E_{\text{total}} = \frac{1}{4\pi} \int_{\Sigma} J^a n_a \sqrt{\gamma} \mathrm{d}^3 x, \qquad (20.84)$$

where we have included the normalization factor of $1/4\pi$, for reasons that will be clear later. However, as we showed in Exercise 7.14 any Killing vector satisfies the identity

$$\nabla_a(\nabla_b K^a) = R_{ab}K^a. \tag{20.85}$$

Hence $J^a = R^{ab}K_b = \nabla_b(\nabla^a K^b)$. Thus,

$$
\begin{aligned}
E_{\text{total}} &= \frac{1}{4\pi}\int_\Sigma J^a n_a \sqrt{\gamma}\,\mathrm{d}^3 x \\
&= \frac{1}{4\pi}\int_\Sigma \nabla_b(\nabla^a K^b) n_a \sqrt{\gamma}\,\mathrm{d}^3 x \\
&= \frac{1}{4\pi}\int_S (\nabla^a K^b) n_a m_b \sqrt{\sigma}\,\mathrm{d}^2 x,
\end{aligned}
\tag{20.86}
$$

again by the divergence theorem. We now show that, in the case of the Schwarzschild solution, (20.86) gives the mass m. To see this, we note that, for the Schwarzschild solution (in Schwarzschild coordinates), $K^a = (1, 0, 0, 0)$. The normal n^a only has a time component n^0 but satisfies $g_{ab}n^a n^b = 1$ so

$$1 = g_{00}n^0 n^0 = (1 - 2m/r)(n^0)^2.$$

Similarly, m_a only has a radial component and satisfies $g_{ab}m^a m^b = -1$ so that

$$1 = -g_{11}m^1 m^1 = (1 - 2m/r)^{-1}(m^0)^2.$$

Hence,

$$n^a = (1 - 2m/r)^{-1/2}\delta_0^a, \tag{20.87}$$

$$m^a = (1 - 2m/r)^{1/2}\delta_1^a. \tag{20.88}$$

Thus, $(\nabla^a K^b)n_a m_b = n^a m^b \nabla_a K_b = \nabla_0 K_1$. However,

$$\nabla_0 K_1 = \partial_0 K_1 - \Gamma_{01}^a K_a = -\Gamma_{01}^0 K_0 = -g_{00}\Gamma_{01}^0. \tag{20.89}$$

For the Schwarzschild solution, the last term is simply m/r^2 so that substituting into (20.86) we get

$$E_{\text{total}} = \frac{1}{4\pi}\int_{\theta=0}^{\pi}\int_{\phi=0}^{2\pi} \frac{m}{r^2}r^2 \sin\theta\,\mathrm{d}\theta\,\mathrm{d}\phi = m, \tag{20.90}$$

which justifies the factor $1/4\pi$ in the definition (20.84). Thus, the Komar integral gives us the correct expression for the total mass of the Schwarzschild solution. One can also show that it produces the Newtonian expression in the weak-field limit. This gives confidence that it is the correct expression for the mass of a stationary space-time.

The construction of the Komar integral as given above involved the existence of a timelike Killing vector K^a. However, we used the divergence theorem to convert the volume integral into a surface integral at spacelike infinity. We may therefore apply the Komar integral to define the mass of an asymptotically flat space-time where K is now only an asymptotic Killing vector (i.e. a Killing vector of Minkowski space). This enables us to define the Komar mass-energy of an **asymptotically flat** space-time as being given by

$$E_{\text{Komar}} = \frac{1}{4\pi} \int_S (\nabla^a K^b) n_a m_b \sqrt{\sigma} \, \mathrm{d}^2 x, \qquad (20.91)$$

where K^a is a timelike **asymptotic Killing vector** and S is a 2-surface at infinity.

An alternative expression which comes from looking at the ADM Hamiltonian (see §14.11) is the ADM mass-energy. For an asymptotically flat space-time where $g_{ab} \to \eta_{ab}$ as $r \to \infty$, we may write $g_{ab} = \eta_{ab} + h_{ab}$, where $h_{ab} \to 0$ as $r \to \infty$. The ADM mass-energy can also be written in the form of a boundary integral at infinity as (Arnowitt, Deser, and Misner 1959)

$$E_{\text{ADM}} = \frac{1}{16\pi} \int_S \left(\partial_j h^j{}_i - \partial_i h^j{}_j \right) m^i \sqrt{\sigma} \, \mathrm{d}^2 x, \quad \text{where } i,j = 1, \cdots, 3. \qquad (20.92)$$

Provided h_{ij} is time-independent, then this agrees with the Komar result. An important result, first established by Schoen and Yau, is the **positive energy theorem**, which says that, as long as the dominant energy condition holds, any asymptotically flat space-time has non-negative ADM mass-energy and that the only space-time with zero ADM mass is Minkowski space.

20.13 The singularity theorems

Although the Schwarzschild solution was known as early as 1916, it was regarded as describing the exterior region of a star, so neither the coordinate singularity at $r = 2m$ nor the curvature singularity at $r = 0$ were taken seriously. In fact, it was only in 1958 with the work of Finkelstein that the modern interpretation of the Schwarzschild solution as a black hole with a curvature singularity at $r = 0$ was accepted. However, although it was realized that there existed various solutions of the Einstein field equations which had singular behaviour of various kinds, the prevailing view was that these singularities were the result of the high degree of symmetry or were unphysical in some way. This position changed considerably with the work of Penrose, who showed in his 1965 paper that deviations from spherical symmetry could not prevent gravitational collapse. This paper not only introduced the concept of trapped surface, but introduced the notion of geodesic incompleteness to characterize a singular space-time. Thus, a space-time is singular if there exists the world-line of a freely falling particle or photon (i.e. a timelike or null geodesic) which comes to an end after a finite time (or affine parameter for a photon) and cannot be extended. So the space-time comes to an end and cannot be extended for such an observer. Shortly afterwards, Hawking realized that, by considering a trapped surface to the past, one could show that an approximately homogeneous and isotropic cosmological solution must have an initial singularity. There quickly followed a series of papers by Hawking, Penrose, Ellis, Geroch, and others which led to the development of the modern singularity theorems, one of the greatest achievements within modern general relativity. (See the review paper by Senovilla (2012) for details.) The resulting theorems all had the same general framework described by Senovilla as a 'pattern singularity theorem' which showed that a space-time satisfying

- an appropriate initial and/or boundary condition,
- a condition on the curvature,
- a causality condition

contains endless but incomplete causal geodesics and is therefore singular.

To give some flavour of these conditions, we describe them below for the Hawking–Penrose singularity theorem of 1970. The boundary condition is the existence of a trapped surface, that is a closed surface where the gravitational field is so strong that even light rays emitted in an outward direction from the surface are dragged inwards by the gravitational field. The causality condition is that there are no closed timelike curves in the space-time (i.e. curves that would allow travel into ones past). Finally, the curvature condition is the **convergence condition** that

$$R_{ab}V^aV^b \geqslant 0 \quad \text{for every non-spacelike vector } V^a. \tag{20.93}$$

The significance of this condition lies in the effect discovered by Raychaudhuri (see equation (23.44)), which states that, provided this condition is satisfied, whenever a system of timelike geodesics normal to a spacelike hypersurface starts converging, then this convergence inevitably increases along the geodesics until finally the geodesics focus (assuming the geodesics are complete). However, this form of the curvature condition does not have a clear physical interpretation so that one uses the Einstein equations to replace this by the **strong energy condition**

$$T_{ab}V^aV^b \geqslant \tfrac{1}{2} T g_{ab}V^aV^b \quad \text{for every non-spacelike vector } V^a. \tag{20.94}$$

If the energy-momentum tensor has energy density μ and principal stresses p_α ($\alpha = 1, 2, 3$), then, for standard matter, the strong energy condition can be expressed equivalently as

$$\mu + p_\alpha \geqslant 0, \qquad \mu + \sum_{1}^{3} p_\alpha \geqslant 0, \tag{20.95}$$

which, for standard matter, is only violated if there are negative densities or large negative pressures. There is a corresponding focusing effect in the case of **null** geodesics (as used in Penrose's 1965 theorem). This depends on the **null convergence condition** (i.e. equation (20.93) but for null vectors) which is implied by the **weak energy condition** which can be expressed in the form

$$T^{ab}V_aV_b \geqslant 0 \quad \text{for every non-spacelike vector } V^a. \qquad (20.96)$$

For standard matter, this can be expressed equivalently as

$$\mu \geqslant 0, \qquad \mu + p_\alpha \geqslant 0, \qquad (20.97)$$

which is again only violated if there are negative densities or strong negative pressures. There is a sound physical basis for believing both (20.95) and (20.97) and all standard matter satisfies them, but the justification is not quite as compelling as it is for the dominant energy condition (12.64).

In addition, some of the theorems (such as the Hawking-Penrose theorem) require a **genericity** condition, namely,

$$v_{[a}R_{b]cd[e}v_{f]}v^cv^d \neq 0$$

somewhere along every timelike or null geodesic, where v^a is the tangent vector. It is only in very special cases that we might expect this condition to be violated.

The main significance of the theorems is that they show that the presence of space-time singularities in exact models is not just a feature of their high symmetry, but can be expected in generically perturbed models. This is not to say that all solutions are singular; in fact, many exact solutions are known which are complete, that is, maximal and singularity-free. But those which closely resemble the Kerr–Newman collapse models, or the Friedmann cosmological models containing a big bang or big crunch, or colliding plane gravitational waves, must be expected to be singular. The theorems do not, however, say that the singularities need look like those of Kerr–Newman, Friedmann, or colliding plane gravitational waves; in fact, there is some evidence that generic singularities may have a much more complicated structure, but little is known about this. Note, however, that none of the theorems leads directly to the existence of curvature singularities. Instead, one obtains the result that space-time is not geodesically complete in timelike or null directions and, furthermore, cannot be extended to a geodesically complete space-time. The most reasonable explanation would seem to be that space-time is confronted with infinite curvature at its boundary. But the theorems do not quite say this and other types of space-time singularities may be possible.

20.14   Black hole thermodynamics and Hawking radiation

This book is concerned with classical relativity theory, and quantum considerations are beyond its brief. However, we shall make an exception and finish our treatment of black holes by describing in simple terms a quantum effect which suggests that black holes are not the permanent structures that the classical theory suggests. The surface area of the event horizon of a black hole has the remarkable property that it always increases when additional matter or radiation falls into the hole. Moreover, if two black holes collide and merge to form a single hole, the area of the new horizon is greater than the sum of the areas of the colliding holes. These properties suggest there is a resemblance between the area of the event horizon of a black hole and the concept of entropy in thermodynamics. (Entropy can be regarded simply as a measure of the disorder of a system or, equivalently, as a lack of knowledge of its precise state. The second law of thermodynamics states that entropy always increases with time.) Indeed, Hawking and collaborators discovered that the laws of thermodynamics have exact analogues in the properties of black holes. The first law relates the change in mass of a black hole to a change in area of the event horizon. The factor of proportionality involved is a quantity called the surface gravity, which is a measure of the strength of the gravitational field at the event horizon. This suggests that surface gravity is analogous to temperature and, indeed, it is a constant at all points on the event horizon, just as the temperature is the same everywhere in a body at thermal equilibrium.

How, more precisely, can the area of a black hole be related to the concept of entropy? Well, the no-hair theorem implies that a large amount of information is lost in a gravitational collapse. A black hole of given mass, angular momentum, and charge could have been formed by the collapse of any one of a large number of different configurations of matter. If one now takes into account quantum effects, the uncertainty principle requires that the number of configurations, although very large, must be finite. The logarithm of this number is the measure of the entropy of the hole and thus measures the information that was irretrievably lost during the collapse through the event horizon when the black hole was created. It follows that, if this number is finite, then the black hole must have a finite temperature (proportional to its surface gravity), and so it could be in thermal equilibrium with thermal radiation at some temperature other than zero. Yet, according to classical concepts, no such equilibrium is possible, since the black hole would absorb any thermal radiation that fell on it, but by definition would not be able to emit anything in return. This paradox was eventually resolved by Hawking, who discovered that black holes seem to emit particles at a steady rate: this is what is called 'Hawking radiation'. Hawking showed that a black hole would produce black-body radiation with a temperature inversely proportional to the mass of the black hole. In SI units, the radiation from a Schwarzschild black hole of mass M is black-body radiation produced by an object with temperature

$$T = \frac{\hbar c^3}{8\pi G k_B M},$$
(20.98)

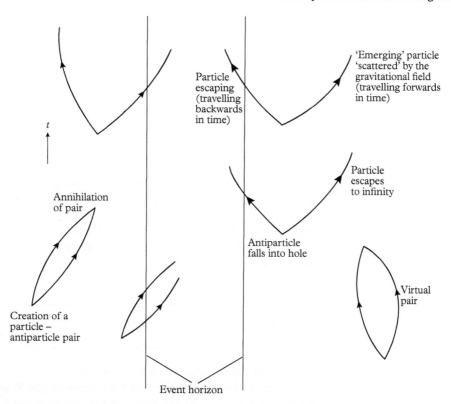

Fig. 20.13 Hawking radiation.

where $\hbar$ is the reduced Planck constant and k_B is the Boltzmann constant (see Chapter 26 for a discussion of black-body radiation). For a black hole of M solar masses, this works out as a temperature of about $(M/M_\odot)$ $\times\, 6 \times 10^{-8}\,\mathrm{K}$.

Quantum mechanics implies that the whole of space is filled with pairs of 'virtual' particles and antiparticles that are constantly materializing in pairs, separating, and then coming together again and annihilating each other. These particles are called virtual because they cannot be observed directly with a particle detector (although they can be measured indirectly by the 'Lamb shift' in the spectrum of hydrogen). Now, in the presence of a black hole, the gravitational attraction will cause one member of a pair to fall into the hole, leaving the other member without a partner with which to undergo annihilation. This particle may also fall into the hole, but it may also escape to infinity, where it appears to be radiation emitted by the black hole. Equivalently, one may regard the member which falls into the hole (the antiparticle, say) as being really a particle travelling backwards in time. Then the motion of the antiparticle can be interpreted as a particle coming out of the hole (travelling backwards in time) and, when it reaches the point at which the particle–antiparticle pair originally materialized, it is scattered by the gravitational field, so that it travels forward in time. Thus, quantum mechanics does allow, in this interpretation, an escape of particles from the hole – a form of quantum-mechanical 'tunnelling' (see Fig. 20.13).

This is a bit like the superradiance due to the Penrose process in which energy is extracted from the black hole and emitted in the form of radiation. However, crucially, it does not require the black hole to be rotating. As a black hole emits particles, its mass and size steadily decrease. This makes it easier for particles to tunnel out, and so the emission will continue at an ever-increasing rate until eventually the black hole radiates itself out of existence. In the long run, every black hole in the universe will evaporate in this way. For large black holes, it will take a very long time indeed (about 10^{67} years for a black hole the mass of the Sun). On the other hand, a primordial black hole formed in the early universe should have almost completely evaporated in the 10 billion years that have elapsed since the big bang. Thus, mini-black holes may be exploding now and may be the source of highly energetic gamma rays. Attempts have been made to quantify this rate of production and to compare the predictions with terrestrial observations of incident gamma radiation, but the results are inconclusive.

Exercises

20.1 (§20.1) Show that the definitions (20.4) and (20.5) lead to (20.6), (20.7), and (20.8). Show also that the definitions (20.9) and (20.10) lead to (20.11), (20.12), (20.14), and (20.15) (see Exercise 8.3).

20.2 (§20.2) Find the covariant metric g_{ab} and contravariant metric g^{ab} for the Schwarzschild line element (16.24) in advanced Eddington-Finkelstein coordinates. Hence confirm (20.17) and (20.18).

20.3 (§20.2) Show that the transformations (20.20) applied to (20.19) lead to (20.21) (keeping v' and r' real). Deduce the line element (20.22) subject to (20.23).

20.4 (§20.3) Apply the transformations (20.24) and (20.25), subject to (20.26) to the line element (20.22) to obtain the form (20.27).

20.5 (§20.3) Apply the transformations (20.29) and (20.24) to the $d\bar{t}^2$ terms in (20.28) (the Cartesian form of Kerr) to confirm they produce the dt^2 term (dropping bars) in (20.27) (the Boyer-Lindquist form of Kerr).

20.6 (§20.3) Show that (20.28) can be written in the form (20.30), subject to (20.31), where λ and ℓ_a are defined by (20.32) and (20.33). Show that, in the Schwarzschild limit, λ and ℓ_a become (20.34) and (20.35). [Hint: $\ell_a \ell_b dx^a dx^b = (\ell_a dx^a)^2$.]

20.7 (§20.4) Show that the transformations (20.37) together with $t' = t$ convert (20.38) into (20.39).

20.8 (§20.4) Show that the definition (20.40) leads to (20.41) and (20.42). Deduce that the Kerr solution is asymptotically flat.

20.9 (§20.5) Confirm Fig. 20.3.

20.10 (§20.5) Show that the stationary limit surface is timelike everywhere except at its poles.

20.11 (§20.5) Find g^{11} for the Boyer–Lindquist form of the Kerr solution (20.27).

20.12 (§20.6) Confirm equations (20.49)–(20.52). Show that they lead to (20.53). Why is it sufficient to consider the condition (20.54)? Check the deductions (20.55), (20.56), and (20.57), and show that r is an affine parameter. Obtain the geodesic equations (20.60) and (20.61).

20.13 (§20.7) Check that the transformations (20.66) and (20.67) map the congruences (20.62) and (20.63) onto (20.65). [Hint: follow the same procedure as in the Schwarzschild case.]

20.14 (§20.7) Confirm Fig. 20.4 and draw the retarded time version of it.

20.15 (§20.8) Show that (20.69) leads to (20.70), and hence deduce (20.71). [Hint: take second term in line element over to the right-hand side and take square roots.]

20.16 (§20.11) Use the Newman–Janis trick to obtain the Kerr–Newman solution (20.72) from the Reissner-Nordström solution. Investigate the surfaces of infinite red shift and the event horizons (where present). [Hint: use (20.16) to obtain g^{ab} and then confirm that $g_{ab}g^{bc} = \delta_a^c$.]

20.17 (§20.12)
(i) Calculate n_a and m_a for Minkowski space and hence show that (20.79) leads to (20.81).
(ii) Show that the solution of Maxwell's equation for a point charge q at the origin is $F^{tr} = q/4\pi r^2$. [Hint: see Exercise 19.2.]
(iii) Deduce that (20.81) leads to (20.82) and hence verify that $Q_{\text{total}} = q$.

20.18 (§20.12) Use (20.92) to calculate the ADM mass for the Schwarzschild solution. Hence show that $E_{\text{Komar}} = E_{\text{ADM}} = m$ for the Schwarzschild solution.

Further reading

Again, the main source for this chapter is the book by Hawking and Ellis (1973). The book by Chandrasekhar (1983) gives a comprehensive discussion of the Kerr solution. The books by Carroll (2004) and Wald (1984) discuss the concept of mass in more detail than other textbooks, and the book by Taylor and Wheeler (2000) gives a nice account of the Penrose process. The book by Christodoulou (2010) gives a detailed mathematical treatment of black holes and cosmic censorship, and the review paper by Senovilla (2011) is an excellent review of the singularity theorems. The article by Wald (2001) gives full details of black hole thermodynamics. See the article by Arnowitt, Deser, and Misner (1959) for details of the ADM mass.

Arnowitt, R., Deser, S., and Misner, C. (1959). Dynamical structure and definition of energy in general relativity. *Physical Review, 116*(5), 1322.

Carroll, S. M. (2004). *Spacetime and Geometry: An Introduction to General Relativity*. Addison Wesley, San Francisco, CA.

Chandrasekhar, S. (1983). *The Mathematical Theory of Black Holes*. Clarendon Press, Oxford.

Christodoulou, D. (2010). 'The formation of black holes in general relativity', in Exner, P., ed., *Proceedings of the XVIth International Congress on Mathematical Physics*. World Scientific Publishing Co. Pte. Ltd., Singapore, 45–55.

Hawking, S. W., and Ellis, G. F. R. (1973). *The Large Scale Structure of Space-Time*. Cambridge University Press, Cambridge.

Senovilla, J. M. M. (2011). 'Singularity theorems in general relativity: Achievements and open questions', in Lehner, C., Renn, J., and Schemmel, M., eds, *Einstein and the Changing Worldviews of Physics*. Birkhäuser, Basel, 305–16.

Taylor, E. F., and Wheeler, J. A. (2000). *Exploring Black Holes: An Introduction to General Relativity*. Addison Wesley Longman, San Francisco, CA.

Wald, R. M. (1984). *General Relativity*. University of Chicago Press, Chicago, IL.

Wald, R. M. (2001). The thermodynamics of black holes. *Living Reviews in Relativity, 4,* 6.

Part E

Gravitational Waves

Linearized gravitational waves and their detection

21

21.1 The linearized field equations

Our consideration of gravitational radiation or gravitational waves (gravity waves for short) starts from the pioneering work of Einstein and is based on the **linearized** form of the field equations. In this approximation, we shall see that plane wave solutions lead to the result that gravitational waves are **transverse** and possess **two** polarization states. Put another way, the gravitational field has two radiation degrees of freedom. In the linearized approximation of the field equations, general relativity is recast as a Lorentz-covariant theory. Considerable caution has to be exercised in doing this because there are associated with it a number of limitations related to the definition of gravitational energy (the full details of which are beyond the brief of this book), but nonetheless it does throw some important light on the general theory.

We begin by assuming that the metric differs only slightly from the Minkowski metric in Minkowski coordinates, that is,

$$g_{ab} = \eta_{ab} + \varepsilon h_{ab}, \tag{21.1}$$

where ε is a small dimensionless parameter and, throughout, **we shall neglect terms of second order or higher in** ε. In addition, we adopt the boundary conditions that space-time is asymptotically flat, that is, if r denotes a radial parameter, then

$$\lim_{r \to \infty} h_{ab} = 0. \tag{21.2}$$

Defining

$$h^{ab} := \eta^{ac} \eta^{bd} h_{cd}, \tag{21.3}$$

then

$$(\eta_{ab} + \varepsilon h_{ab})(\eta^{bc} - \varepsilon h^{bc}) = \delta_a^c, \tag{21.4}$$

from which we get

$$g^{ab} = \eta^{ab} - \varepsilon h^{ab}. \tag{21.5}$$

Introducing Einstein's Relativity. Second Edition. Ray d'Inverno and James Vickers, Oxford University Press.
© Ray d'Inverno and James Vickers (2022). DOI: 10.1093/oso/9780198862024.003.0021

Since η_{ab} is constant, we also have (exercise)

$$\begin{aligned}
\Gamma^a_{bc} &= \tfrac{1}{2}g^{ad}(g_{dc,b} + g_{db,c} - g_{bc,d}) \\
&= \tfrac{1}{2}\varepsilon\eta^{ad}(h_{dc,b} + h_{db,c} - h_{bc,d}) \\
&= \tfrac{1}{2}\varepsilon(h^a{}_{c,b} + h^a{}_{b,c} - h_{bc,}{}^a),
\end{aligned} \tag{21.6}$$

where we make use of the result that, since this term is of order ε, we can, using (21.1) and (21.5), raise and lower indices with the Minkowski metric. The Riemann tensor then becomes

$$R_{abcd} = \tfrac{1}{2}\varepsilon(h_{ad,bc} + h_{bc,ad} - h_{ac,bd} - h_{bd,ac}). \tag{21.7}$$

The Bianchi identities

$$R_{ab[cd;e]} \equiv 0, \tag{21.8}$$

are

$$R_{ab[cd,e]} \equiv 0, \tag{21.9}$$

and are identically satisfied by (21.7).

The Ricci tensor is (exercise)

$$R_{ab} = \eta^{cd}R_{cadb} = \tfrac{1}{2}\varepsilon(h^c{}_{a,bc} + h^c{}_{b,ac} - \Box h_{ab} - h_{,ab}), \tag{21.10}$$

where

$$h := \eta^{cd}h_{cd} = h^c{}_c, \tag{21.11}$$

and $\Box$ is the d'Alembertian operator

$$\begin{aligned}
\Box &:= \eta^{ab}\partial_a\partial_b \\
&= \frac{\partial^2}{\partial t^2} - \nabla^2 \\
&= \frac{\partial^2}{\partial t^2} - \left(\frac{\partial^2}{\partial x^2} + \frac{\partial^2}{\partial y^2} + \frac{\partial^2}{\partial z^2}\right),
\end{aligned}$$

defined previously in (12.40). The Ricci scalar is

$$R = \varepsilon(h^{cd}{}_{,cd} - \Box h), \tag{21.12}$$

and, finally, the Einstein tensor is

$$G_{ab} = \tfrac{1}{2}\varepsilon(h^c{}_{a,bc} + h^c{}_{b,ac} - \Box h_{ab} - h_{,ab} - \eta_{ab}h^{cd}{}_{,cd} + \eta_{ab}\Box h). \tag{21.13}$$

In fact, the Einstein tensor can be found directly from the Lagrangian

$$\mathcal{L}(h^{ab}{}_{,c}) = \tfrac{1}{2}\varepsilon(h^{ab}{}_{,b}h^c{}_{c,a} - h^{ab,c}h_{cb,a} + \tfrac{1}{2}h^{cd,a}h_{cd,a} - \tfrac{1}{2}h^c{}_{c,a}h^d{}_{d}{}^{,a}), \quad (21.14)$$

using (exercise)

$$
\begin{aligned}
G_{ab} &= \frac{\delta\mathcal{L}}{\delta h^{ab}} \\
&= \frac{\partial\mathcal{L}}{\partial h^{ab}} - \left(\frac{\partial\mathcal{L}}{\partial h^{ab}{}_{,c}}\right)_{,c} \\
&= -\left(\frac{\partial\mathcal{L}}{\partial h^{ab}{}_{,c}}\right)_{,c}.
\end{aligned}
\quad (21.15)
$$

21.2 Gauge transformations

Let us consider what happens to the linearized equations under a coordinate transformation of the form

$$x^a \to x'^a = x^a + \varepsilon\xi^a. \quad (21.16)$$

Then

$$\frac{\partial x'^a}{\partial x^b} = \delta^a_b + \varepsilon\xi^a{}_{,b}, \quad (21.17)$$

and, applying this to the transformation formula for g_{ab}, (7.5), we find the consequent transformation of h_{ab} (see exercise 11.1), namely,

$$h_{ab} \to h'_{ab} = h_{ab} - 2\xi_{(a,b)}. \quad (21.18)$$

By analogy with electromagnetic theory (see (12.38)), this is called a **gauge transformation** of h_{ab}. It is easy to establish (exercise) that both the linearized curvature tensor (21.7) and its contractions are **gauge-invariant quantities**, that is, unchanged to first order in ε by transformations of the form (21.18).

   Just as in electrodynamics, we may impose further conditions to fix the gauge. Going back to the field equations, we observe that if **new** variables ϕ_{ab} are defined by

$$\phi_{ab} := h_{ab} - \tfrac{1}{2}\eta_{ab}h, \quad (21.19)$$

then (21.10) becomes

$$R_{ab} = \tfrac{1}{2}\varepsilon(\phi^c{}_{a,bc} + \phi^c{}_{b,ac} - \Box h_{ab}), \quad (21.20)$$

and consequently

$$R = \tfrac{1}{2}\varepsilon(2\phi^{cd}{}_{,cd} - \Box h), \tag{21.21}$$

and

$$G_{ab} = \tfrac{1}{2}\varepsilon(\phi^c{}_{a,bc} + \phi^c{}_{b,ac} - \Box\phi_{ab} - \eta_{ab}\phi^{cd}{}_{,cd}). \tag{21.22}$$

This suggests that our field equations will reduce to **wave equations** if we impose the condition

$$\phi^a{}_{b,a} = 0, \tag{21.23}$$

or, in terms of h_{ab},

$$h^a{}_{b,a} - \tfrac{1}{2}h_{,b} = 0, \tag{21.24}$$

which is called the **Lorentz gauge** (and is also known as the **de Donder** or **harmonic** gauge). A straightforward calculation (exercise) reveals that, under the gauge transformation (21.16),

$$\phi_{ab} \rightarrow \phi'_{ab} = \phi_{ab} - \xi_{a,b} - \xi_{b,a} + \eta_{ab}\xi^c{}_{,c}, \tag{21.25}$$

from which we find

$$\phi'^a{}_{b,a} = \phi^a{}_{b,a} - \Box\xi_b. \tag{21.26}$$

It follows from (21.26) that the gauge transformation (21.16) will transform the equations into the Lorentz gauge, that is,

$$\phi'^a{}_{b,a} = 0,$$

if we choose ξ_a to satisfy

$$\Box\xi_a = \phi^b{}_{a,b}. \tag{21.27}$$

In other words, if we treat the ξ_a as unknowns, then the problem involves solving wave equations with a source term. Then, by (21.22), Einstein's full field equations reduce to (dropping primes)

$$\tfrac{1}{2}\varepsilon\Box\phi_{ab} = -\kappa T_{ab}. \tag{21.28}$$

The gauge is not completely fixed by (21.27) because we can always carry out additional transformations with

$$\Box\xi_a = 0, \tag{21.29}$$

which leaves $\phi^a{}_{b,a}$ unaltered.

The vacuum field equations in the Lorentz gauge reduce to

$$\Box \phi_{ab} = 0, \tag{21.30}$$

and, taking the trace,

$$\eta^{ab}\Box\phi_{ab} = \Box(\eta^{ab}\phi_{ab}) = \Box(h - 2h) = -\Box h = 0, \tag{21.31}$$

by (21.19). Combining this result with (21.30) and (21.19), we find that h_{ab} must also satisfy

$$\Box h_{ab} = 0, \tag{21.32}$$

in the Lorentz gauge (21.24), which, in terms of h_{ab}, is

$$h^a{}_{b,a} - \tfrac{1}{2}h_{,b} = 0. \tag{21.33}$$

21.3 Linearized plane gravitational waves

Before we attempt to solve the linearized field equations, let us consider what theoretical motivation there might be which suggests that gravitational waves exist. We have seen that the linearized vacuum field equations reduce to the wave equations

$$\Box h_{ab} = 0, \tag{21.34}$$

in the Lorentz gauge, from which we might be tempted to conclude that gravitational effects propagate as waves with the velocity of light. However, this is open to the objection that the perturbation h_{ab} is linked to an arbitrary coordinate system and therefore the existence of a non-zero h_{ab} is not an invariant indication of the existence of a gravitational field. A better argument is based on the fact that if (21.34) holds then, by (21.7),

$$\Box R_{abcd} = 0, \tag{21.35}$$

which is a wave equation for the gauge-invariant quantity R_{abcd}. Thus, the Riemann tensor, which gives an absolute criterion for the existence of a gravitational field, itself obeys the wave equation. It follows that, in the linearized theory, gravitational effects propagate with the velocity of light. This does not of itself, however, prove whether or not gravitational radiation exists, since radiation involves energy transfer. We return to this question later in the chapter and also in Chapter 23.

We now look for a special solution of the linearized vacuum field equations which represents an infinite plane wave propagating in the x-direction. We start by introducing the coordinates

$$(x^0, x^1, x^2, x^3) = (t, x, y, z)$$

and adopt the ansatz

$$h_{ab} = h_{ab}(t, x), \tag{21.36}$$

which requires

$$h_{ab,2} = h_{ab,3} = 0. \tag{21.37}$$

This assumption means that the Riemann tensor is highly degenerate and, from (21.7), we find that the twenty independent components fall into the following three groups of terms (exercise):

$$R_{0123} = R_{0223} = R_{0323} = R_{1223} = R_{1323} = R_{2323} = 0, \tag{21.38}$$

$$\left.\begin{aligned}
R_{0101} &= \tfrac{1}{2}\varepsilon(2h_{01,01} - h_{00,11} - h_{11,00}), \\
R_{0102} &= \tfrac{1}{2}\varepsilon(h_{02,01} - h_{12,00}), \\
R_{0103} &= \tfrac{1}{2}\varepsilon(h_{03,01} - h_{13,00}), \\
R_{0112} &= \tfrac{1}{2}\varepsilon(h_{02,11} - h_{12,01}), \\
R_{0113} &= \tfrac{1}{2}\varepsilon(h_{03,11} - h_{13,01}),
\end{aligned}\right\} \tag{21.39}$$

$$\left.\begin{aligned}
R_{0202} &= -\tfrac{1}{2}\varepsilon h_{22,00}, \\
R_{0203} &= -\tfrac{1}{2}\varepsilon h_{23,00}, \\
R_{0212} &= -\tfrac{1}{2}\varepsilon h_{22,01}, \\
R_{0213} &= -\tfrac{1}{2}\varepsilon h_{23,01}, \\
R_{0303} &= -\tfrac{1}{2}\varepsilon h_{33,00}, \\
R_{0313} &= -\tfrac{1}{2}\varepsilon h_{33,01}, \\
R_{1212} &= -\tfrac{1}{2}\varepsilon h_{22,11}, \\
R_{1213} &= -\tfrac{1}{2}\varepsilon h_{23,11}, \\
R_{1313} &= -\tfrac{1}{2}\varepsilon h_{33,11}.
\end{aligned}\right\} \tag{21.40}$$

We now impose the linearized vacuum field equations in the form $R_{ab} = 0$. Then, for example,

$$R_{13} = R^a{}_{1a3} = R_{0103} = 0, \tag{21.41}$$

so that one of the independent components of (21.39) vanishes. In fact, the vacuum field equations result in all the group (21.39) vanishing (exercise). Thus, only the components in the group (21.40) are non-zero and these only involve the components h_{22}, h_{23}, and h_{33}. This means that we can decompose h_{ab} into two parts:

$$h_{ab} = h_{ab}^{(1)} + h_{ab}^{(2)}, \tag{21.42}$$

where

$$h_{ab}^{(1)} = \begin{bmatrix} 0 & 0 & 0 & 0 \\ 0 & 0 & 0 & 0 \\ 0 & 0 & h_{22} & h_{23} \\ 0 & 0 & h_{23} & h_{33} \end{bmatrix}, \qquad (21.43)$$

and

$$h_{ab}^{(2)} = \begin{bmatrix} h_{00} & h_{01} & h_{02} & h_{03} \\ h_{01} & h_{11} & h_{12} & h_{13} \\ h_{02} & h_{12} & 0 & 0 \\ h_{03} & h_{13} & 0 & 0 \end{bmatrix}. \qquad (21.44)$$

The vacuum field equations then lead to the result that the curvature tensor of $h_{ab}^{(2)}$ is identically zero. This suggests that there may exist a coordinate system in which h_{ab} has only h_{22}, h_{23}, and h_{33} components; that is, h_{ab} is a pure $h_{ab}^{(1)}$-type solution. We shall show that we can exploit the gauge freedom to achieve this in the case of a plane wave.

We sharpen our ansatz (21.36) by requiring

$$h_{ab} = h_{ab}(t - x), \qquad (21.45)$$

so that it clearly represents a solution propagating in the x-direction with the speed of light (see Fig. 21.1).

The Lorentz gauge conditions (21.33) then become

$$\left. \begin{aligned} h_{00,0} - h_{01,1} - \tfrac{1}{2}h_{,0} &= 0, \\ h_{01,0} - h_{11,1} - \tfrac{1}{2}h_{,1} &= 0, \\ h_{02,0} - h_{12,1} &= 0, \\ h_{03,0} - h_{13,1} &= 0, \end{aligned} \right\} \qquad (21.46)$$

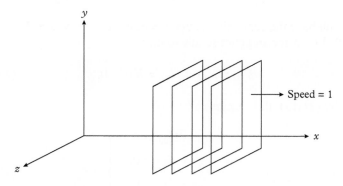

Fig. 21.1 The ansatz (21.45).

or, letting a prime denote differentiation with respect to the argument $t-x$, these can be written

$$h'_{00} + h'_{01} - \tfrac{1}{2}h' = 0,$$
$$h'_{01} + h'_{11} + \tfrac{1}{2}h' = 0,$$
$$h'_{02} + h'_{12} = 0,$$
$$h'_{03} + h'_{13} = 0,$$

These integrate to give

$$\left.\begin{aligned} h_{00} + h_{01} - \tfrac{1}{2}h &= f_1, \\ h_{01} + h_{11} + \tfrac{1}{2}h &= f_2, \\ h_{02} + h_{12} &= f_3, \\ h_{03} + h_{13} &= f_4, \end{aligned}\right\} \tag{21.47}$$

where the fs are all functions of y and z only. However, since the h_{ab} all vanish at spatial infinity by (21.2), it follows that

$$f_1 = f_2 = f_3 = f_4 = 0.$$

Then (21.47) gives

$$h_{12} = -h_{02}, \quad h_{13} = -h_{03}, \quad h_{01} = -\tfrac{1}{2}(h_{00} + h_{11}), \quad h_{33} = -h_{22},$$

that is,

$$h_{ab} = \begin{bmatrix} h_{00} & -\tfrac{1}{2}(h_{00} + h_{11}) & h_{02} & h_{03} \\ -\tfrac{1}{2}(h_{00} + h_{11}) & h_{11} & -h_{02} & -h_{03} \\ h_{02} & -h_{02} & h_{22} & h_{23} \\ h_{03} & -h_{03} & h_{23} & -h_{22} \end{bmatrix}. \tag{21.48}$$

We still have the remaining gauge freedom (21.18), where ξ_a satisfies (21.29). Let us try and choose this so that

$$h'_{00} = h'_{02} = h'_{03} = h'_{11} = 0. \tag{21.49}$$

Then, by (21.18), this requires

$$\left.\begin{aligned} h_{00} - 2\xi_{0,0} &= 0, \\ h_{02} - \xi_{0,2} - \xi_{2,0} &= 0, \\ h_{03} - \xi_{0,3} - \xi_{3,0} &= 0, \\ h_{11} - 2\xi_{1,1} &= 0. \end{aligned}\right\} \tag{21.50}$$

If we assume that

$$\xi_a = \xi_a(t - x), \tag{21.51}$$

then (21.29) is automatically satisfied. We choose

$$(\xi_0, \xi_1, \xi_2, \xi_3) = (F_0(t - x), F_1(t - x), F_2(t - x), F_3(t - x)), \tag{21.52}$$

where, setting $u = t - x$, we see that the functions F_0, F_1, F_2, and F_3 are all functions of u only and are determined by the ordinary differential equations

$$\frac{\mathrm{d}F_0}{\mathrm{d}u} = \tfrac{1}{2}h_{00}(u), \quad \frac{\mathrm{d}F_1}{\mathrm{d}u} = -\tfrac{1}{2}h_{11}(u), \quad \frac{\mathrm{d}F_2}{\mathrm{d}u} = h_{02}(u), \quad \frac{\mathrm{d}F_3}{\mathrm{d}u} = h_{03}(u). \tag{21.53}$$

This choice satisfies (21.51) and (21.50), and, moreover, it leaves h_{11}, h_{22}, and h_{33} unchanged. Hence, dropping primes, we have shown that h_{ab} may be transformed into the **canonical form**

$$h_{ab} = \begin{bmatrix} 0 & 0 & 0 & 0 \\ 0 & 0 & 0 & 0 \\ 0 & 0 & h_{22} & h_{23} \\ 0 & 0 & h_{23} & -h_{22} \end{bmatrix}. \tag{21.54}$$

Clearly, h_{ab} only depends on **two** independent functions, namely,

$$h_{22}(t - x), \quad \text{and} \quad h_{23}(t - x).$$

We consider the physical significance of these two functions in the next section.

21.4 Polarization states of plane waves

We first consider the case $h_{23} = 0$, for which the line element is given by

$$\mathrm{d}s^2 = \mathrm{d}t^2 - \mathrm{d}x^2 - [1 - \varepsilon h_{22}(t - x)]\mathrm{d}y^2 - [1 + \varepsilon h_{22}(t - x)]\mathrm{d}z^2. \tag{21.55}$$

We shall call this an 'h_{22}-wave'. Let us suppose that h_{22} is some oscillatory function of u so that there are values when $h_{22} > 0$ and values when $h_{22} < 0$. Let us investigate what happens when an h_{22}-wave is incident on a distribution of test particles. First of all, consider two neighbouring particles in the yz-plane which initially have coordinates (y_0, z_0) and $(y_0 + \mathrm{d}y, z_0)$ in the plane. Then, using (21.55), the proper distance between them is given by

$$\mathrm{d}\sigma^2 = -\mathrm{d}s^2 = (1 - \varepsilon h_{22})\mathrm{d}y^2. \tag{21.56}$$

The proper distance is a coordinate-independent quantity, and hence, if initially h_{22} changes from zero to $h_{22} > 0$, the particles move closer together and, conversely, if h_{22} changes from zero to $h_{22} < 0$, the particles move further apart. The opposite happens if we consider free particles with coordinates (y_0, z_0) and $(y_0, z_0 + dz)$ and in the plane since now

$$d\sigma^2 = -ds^2 = (1 + \varepsilon h_{22})dz^2. \qquad (21.57)$$

Thus, if an oscillatory plane gravitational wave propagating in the x-direction is incident on a ring of dust particles situated in the yz-plane, then the ring is distorted into a pulsating ellipse whose major axis is, in turn, parallel to the y- and z-axes (see Fig. 21.2). Furthermore, if one considers two neighbouring particles with the same y and z coordinates but separated in the x direction by dx, then the proper distance $d\sigma^2 = dx^2$ remains constant so that the particles remain in the $x = $ constant plane. The **transverse** character of an h_{22}-wave is clear from this. We refer to this state as a wave with **+ polarization**.

Let us turn attention to an 'h_{23}-wave', that is, the case when $h_{22} = 0$, and the line element becomes

$$ds^2 = dt^2 - dx^2 - dy^2 + 2\varepsilon h_{23}(t - x)dydz - dz^2. \qquad (21.58)$$

Let us perform a rotation through $45°$ in the yz-plane given by

$$y \to \bar{y} = \frac{1}{\sqrt{2}}(y + z), \quad z \to \bar{z} = \frac{1}{\sqrt{2}}(-y + z), \qquad (21.59)$$

so that the line element becomes (exercise)

$$ds^2 = dt^2 - dx^2 - [1 - \varepsilon h_{23}(t - x)]d\bar{y}^2 - [1 + \varepsilon h_{23}(t - x)]d\bar{z}^2. \qquad (21.60)$$

Comparing this with (21.55), we see that an h_{23}-wave produces exactly the same effect as an h_{22}-wave but with the axes rotated through $45°$ (see Fig. 21.3). The transverse character of an h_{23}-wave is again clear and we refer to the state as a wave with $\times$ **polarization**.

Clearly, a general wave is a superposition of these two polarization states. The fact that the two polarization states are at $45°$ to each other contrasts with the two polarization states of an electromagnetic wave, which are at $90°$ to each other. (This can be shown to stem from the fact that gravity is represented by the second-rank symmetric tensor h_{ab}, whereas electromagnetism is represented by the vector potential A_a.)

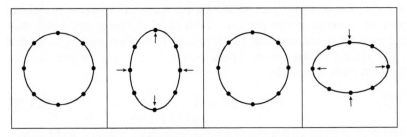

Fig. 21.2 Time sequence showing the transverse effect of an oscillatory linear plane gravitational wave with + polarization.

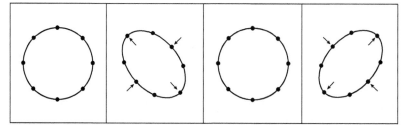

Fig. 21.3 Time sequence showing the transverse effect of an oscillatory linear plane gravitational wave with $\times$ polarization.

21.5   Solving the wave equation

We have shown in equation (21.28) that linearized gravitational waves satisfy the equation

$$\Box\Phi_{ab} = -2\kappa T_{ab}, \tag{21.61}$$

where, for convenience, we have incorporated ϵ into the definition of ϕ_{ab} by defining

$$\Phi_{ab} := \epsilon(h_{ab} - \tfrac{1}{2}\eta_{ab}h).$$

We now show how to solve (21.61).

Since we are working in Minkowski space, we may also work in Minkowski coordinates, in which case each component ϕ (say) of Φ_{ab} satisfies an equation of the form

$$\frac{\partial^2\phi}{\partial t^2} - \nabla^2\phi = f, \tag{21.62}$$

where f is the corresponding component of $-2\kappa T_{ab}$. This is just the standard wave equation in 3-dimensional space for the function $\phi(t, \boldsymbol{x})$ with a source term given by $f(t, \boldsymbol{x})$.

To solve (21.62), we start by considering the case where $f(t, \boldsymbol{x})$ represents a delta function point source at the origin, so that we want to solve

$$\frac{\partial^2\phi}{\partial t^2} - \nabla^2\phi = q(t)\delta^{(3)}(\boldsymbol{x}), \tag{21.63}$$

where $q(t)$ measures how the strength of the source varies with time.

Here the Dirac delta function $\delta^{(3)}(\boldsymbol{x})$ can be loosely thought of as a function on $\mathbb{R}^3$ which is zero everywhere except at the origin, where it is infinite, and has the property that

$$\int_{\mathbb{R}^3} \delta^{(3)}(\boldsymbol{x})\mathrm{d}V = 1,$$

so that, given any function $g(\boldsymbol{x})$ on $\mathbb{R}^3$, we have

$$\int_{\mathbb{R}^3} \delta^{(3)}(\boldsymbol{x})g(\boldsymbol{x})\mathrm{d}V = g(0). \tag{21.64}$$

Since both sides of equation (21.63) are spherically symmetric, we look for a spherically symmetric solution $\phi(t, r)$ where $r = |\boldsymbol{x}|$. Now, for a purely radial function,

$$\nabla^2 \phi = \frac{1}{r}\frac{\partial^2}{\partial r^2}(r\phi),$$

and, away from the origin, the right-hand side of (21.63) is zero so that $\phi(t, r)$ satisfies

$$\frac{1}{r}\frac{\partial^2}{\partial r^2}(r\phi) - \frac{\partial^2\phi}{\partial t^2} = 0, \qquad r \neq 0$$

$$\Rightarrow \quad \frac{\partial^2}{\partial r^2}(r\phi) - \frac{\partial^2}{\partial t^2}(r\phi) = 0, \qquad r \neq 0$$

So, away from the origin, the function $\psi := r\phi$ satisfies the 1-dimensional wave equation

$$\frac{\partial^2\psi}{\partial r^2} - \frac{\partial^2\psi}{\partial t^2} = 0. \tag{21.65}$$

Using d'Alembert's solution to the 1-dimensional wave equation (exercise), we may write

$$\psi(t, r) = u(t - r) + v(t + r), \tag{21.66}$$

and hence

$$\phi(t, r) = \frac{1}{r}u(t - r) + \frac{1}{r}v(t + r). \tag{21.67}$$

The first term represents an **outgoing** spherical wave, while the second represents an **ingoing** wave. In our case, we are interested in the outgoing wave, which represents the so-called **causal solution** in which the gravitational wave is the result of the motion of the source. We therefore take $v \equiv 0$, giving us the solution

$$\phi(t, r) = \frac{1}{r}u(t - r). \tag{21.68}$$

It remains to find the function $u(t - r)$ in terms of the function $q(t)$. To do this, we integrate equation (21.63) over a ball B of radius a and centre the origin. This gives

$$\int_B \left(\frac{\partial^2\phi}{\partial t^2} - \nabla^2\phi\right)\mathrm{d}V = \int_B q(t)\delta^{(3)}(\boldsymbol{x})\mathrm{d}V = q(t), \tag{21.69}$$

using (21.64). Note that, although ϕ diverges like $1/r$, the volume element is $O(r^2)$ so that the integral is well defined. Now, by the divergence theorem,

$$
\begin{aligned}
\int_B \nabla^2 \phi \, dV &= \int_S \nabla \phi . d\mathbf{S} \\
&= \int_S \frac{\partial \phi}{\partial r} dA \\
&= 4\pi a^2 \frac{\partial}{\partial r} \left(\frac{u(t-r)}{r} \right) \Bigg|_{r=a} \\
&= -4\pi u(t-a) - 4\pi a u'(t-a),
\end{aligned}
$$

where S is a sphere of radius a.

On the other hand

$$
\begin{aligned}
\int_B \frac{\partial^2 \phi}{\partial t^2} dV &= \int_{r=0}^a \int_{\theta=0}^\pi \int_{\phi=0}^{2\pi} \frac{u''(t-r)}{r} r^2 \sin\theta dr d\theta d\phi \\
&= 4\pi \int_{r=0}^a u''(t-r) r \, dr \\
&= 4\pi [-u'(t-r)r]_0^a + 4\pi \int_0^a u'(t-r) dr \\
&= 4\pi [-u'(t-r)r - u(t-r)]_0^a \\
&= -4\pi [a u'(t-a) + u(t-a) - u(t)].
\end{aligned}
$$

Substituting from the above two equations into (21.69), we obtain

$$
4\pi u(t) = q(t), \tag{21.70}
$$

so that $\phi = q(t-r)/4\pi r$ is the required solution to (21.63). In particular, if we now take $q(t) = \delta(t)$, we have that

$$
\phi = \frac{1}{4\pi |\mathbf{x}|} \delta(t - |\mathbf{x}|), \tag{21.71}
$$

is a **fundamental solution** of

$$
\frac{\partial^2 \phi}{\partial t^2} - \nabla^2 \phi = \delta(t) \delta^{(3)}(\mathbf{x}). \tag{21.72}
$$

The reason for the term 'fundamental solution' is that we may use it to construct the solution for **any** source $f(t, \mathbf{x})$, as we now show.

We first move the location of the delta function point source from the origin to an arbitrary point $(s, \mathbf{y})$ in Minkowski space-time, giving us

$$\phi(t, \boldsymbol{x}) = \frac{1}{4\pi|\boldsymbol{x} - \boldsymbol{y}|}\delta(t - s - |\boldsymbol{x} - \boldsymbol{y}|), \tag{21.73}$$

as the fundamental solution of

$$\frac{\partial^2 \phi}{\partial t^2} - \nabla^2 \phi = \delta(t - s)\delta^{(3)}(\boldsymbol{x} - \boldsymbol{y}). \tag{21.74}$$

We now use (21.64) to write a general source term $f(t, \boldsymbol{x})$ in the form

$$f(t, \boldsymbol{x}) = \int_{s \in \mathbb{R}} \int_{\boldsymbol{y} \in \mathbb{R}^3} \delta(t - s)\delta^{(3)}(\boldsymbol{x} - \boldsymbol{y})f(s, \boldsymbol{y})\mathrm{d}s\mathrm{d}^3\boldsymbol{y}. \tag{21.75}$$

Since the wave equation is linear, we may use the principle of superposition to add together all the fundamental solutions for the delta functions located at $(s, \boldsymbol{y})$ to obtain the required solution. This shows that the causal solution to (21.62) for a **general source** is given by

$$\phi(t, \boldsymbol{x}) = \frac{1}{4\pi} \int_{s \in \mathbb{R}} \int_{\boldsymbol{y} \in \mathbb{R}^3} \frac{1}{|\boldsymbol{x} - \boldsymbol{y}|}\delta(t - s - |\boldsymbol{x} - v|)f(s, \boldsymbol{y})\mathrm{d}s\mathrm{d}^3\boldsymbol{y}. \tag{21.76}$$

Performing the integration with respect to s, we see that the delta function has the effect of replacing s by the **retarded time** $t - |\boldsymbol{x} - \boldsymbol{y}|$ in $f(s, \boldsymbol{y})$ so that our required solution is given by

$$\phi(t, \boldsymbol{x}) = \frac{1}{4\pi} \int_{\boldsymbol{y} \in \mathbb{R}^3} \frac{f(t - |\boldsymbol{x} - \boldsymbol{y}|, \boldsymbol{y})}{|\boldsymbol{x} - \boldsymbol{y}|}\mathrm{d}^3\boldsymbol{y}. \tag{21.77}$$

This is nothing but the standard (retarded) solution given by the method of **Green's functions** from the theory of partial differential equations.

A case of particular interest is when the source $f(t, \boldsymbol{x})$ oscillates with angular frequency ω so that, without loss of generality, we have

$$f(t, \boldsymbol{x}) = \sin(\omega t)g(\boldsymbol{x}), \tag{21.78}$$

in which case, (21.77) reduces to

$$\phi(t, \boldsymbol{x}) = \frac{1}{4\pi} \int_{\boldsymbol{y} \in \mathbb{R}^3} \frac{\sin(\omega t - \omega|\boldsymbol{x} - \boldsymbol{y}|)}{|\boldsymbol{x} - \boldsymbol{y}|}g(\boldsymbol{y})\mathrm{d}^3\boldsymbol{y}. \tag{21.79}$$

We now assume that the source is compact and is therefore confined within some region of radius R_0, which we may assume is centred on the origin. Then, if we are a long distance from the source compared to the size of the source, that is, $r := |\boldsymbol{x}|$ satisfies

$$r \gg R_0, \tag{21.80}$$

then, since $|\boldsymbol{y}| < R_0$ in the above integral, we may write

$$\frac{1}{|\boldsymbol{x} - \boldsymbol{y}|} \simeq \frac{1}{|\boldsymbol{x}|} = \frac{1}{r}. \tag{21.81}$$

We now further suppose that the wavelength of the oscillation $\lambda = 2\pi/\omega$ is also large compared to R_0, i.e. that

$$\lambda \gg R_0. \tag{21.82}$$

This is called the **long-wavelength approximation**. Then, for this approximation,

$$\omega|\boldsymbol{x} - \boldsymbol{y}| = \frac{2\pi}{\lambda}|\boldsymbol{x} - \boldsymbol{y}| \simeq \frac{2\pi}{\lambda}|\boldsymbol{x}| = \omega r, \tag{21.83}$$

and hence, in (21.79), we may write

$$\omega|\boldsymbol{x} - \boldsymbol{y}| \simeq \omega r. \tag{21.84}$$

Substituting for the above in (21.79) using (21.81) and (21.84) gives

$$\begin{aligned}\phi(t, \boldsymbol{x}) &\simeq \frac{1}{4\pi r} \int_{\boldsymbol{y} \in \mathbb{R}^3} \sin(\omega(t - r))g(\boldsymbol{y})\mathrm{d}^3\boldsymbol{y} \\ &= \frac{1}{4\pi r} \int_{\boldsymbol{y} \in \mathbb{R}^3} f(t - r, \boldsymbol{y})\mathrm{d}^3\boldsymbol{y}.\end{aligned} \tag{21.85}$$

In deriving the above result, we assumed that $f(t, \boldsymbol{x})$ varied sinusoidally in time with angular frequency ω. However, the only mathematical requirement was that the source was spatially compact, we were far from the source, and the wavelength was long compared to the size of the source. So, in this approximation, the solution of the wave equation is given by

$$\phi(t, \boldsymbol{x}) = \frac{1}{4\pi r} \int_{\boldsymbol{y} \in \mathbb{R}^3} f(t - r, \boldsymbol{y})\mathrm{d}^3\boldsymbol{y}. \tag{21.86}$$

We now apply this analysis to the components of (21.61) for a compact source in the 'long wavelength' approximation and far from the source. This gives the following formula for the (low frequency) gravitational waves:

$$\Phi_{ab}(t, \boldsymbol{x}) = -\frac{2\kappa}{4\pi r} \int_{\boldsymbol{y} \in \mathbb{R}^3} T_{ab}(t - r, \boldsymbol{y})\mathrm{d}^3\boldsymbol{y}. \tag{21.87}$$

Working in geometric units (where $G = c = 1$), we have $\kappa = 8\pi$ so that the above becomes

$$\Phi_{ab}(t, \boldsymbol{x}) = -\frac{4}{r} \int_{\boldsymbol{y} \in \mathbb{R}^3} T_{ab}(t - r, \boldsymbol{y})\mathrm{d}^3\boldsymbol{y}. \tag{21.88}$$

21.6 The quadrupole formula

In this section, we show how to more directly relate the linearized gravitational wave Φ_{ab} to the motion of a spatially compact source.

It follows from Einstein's equations that the divergence of the energy-momentum tensor vanishes so that

$$\nabla_a T^{ab} = 0, \tag{21.89}$$

which, in the linear approximation, reduces to

$$\partial_a T^{ab} = 0. \tag{21.90}$$

In particular, taking $b = 0$, we get $\partial_a T^{a0} = 0$ or, more explicitly,

$$\frac{\partial T^{00}}{\partial t} = -\frac{\partial T^{j0}}{\partial x^j}, \tag{21.91}$$

where indices i, j, k etc. are summed over $1, \ldots, 3$. Differentiating the above equation with respect to t gives

$$\frac{\partial^2 T^{00}}{\partial t^2} = -\frac{\partial^2 T^{j0}}{\partial t \partial x^j} = -\frac{\partial}{\partial x^j}\left(\frac{\partial T^{j0}}{\partial t}\right) = \frac{\partial}{\partial x^j}\left(\frac{\partial T^{jk}}{\partial x^k}\right) = \frac{\partial^2 T^{jk}}{\partial x^j \partial x^k}, \tag{21.92}$$

where in the penultimate equality we have used $\partial_a T^{aj} = 0$. Multiplying both sides by $x^\ell x^m$ and integrating by parts over all space, we obtain

$$\int_{\mathbf{x} \in \mathbb{R}^3} x^\ell x^m \frac{\partial^2 T^{00}}{\partial t^2} \mathrm{d}^3\mathbf{x} = \int_{\mathbf{x} \in \mathbb{R}^3} x^\ell x^m \frac{\partial^2 T^{jk}}{\partial x^j \partial x^k} \mathrm{d}^3\mathbf{x}$$

$$= -\int_{\mathbf{x} \in \mathbb{R}^3} \frac{\partial}{\partial x^j}\left(x^\ell x^m\right) \frac{\partial T^{jk}}{\partial x^k} \mathrm{d}^3\mathbf{x}$$

$$= -\int_{\mathbf{x} \in \mathbb{R}^3} \left(\delta_j^\ell x^m + \delta_j^m x^\ell\right) \frac{\partial T^{jk}}{\partial x^k} \mathrm{d}^3\mathbf{x}$$

$$= \int_{\mathbf{x} \in \mathbb{R}^3} \frac{\partial}{\partial x^k}\left(\delta_j^\ell x^m + \delta_j^m x^\ell\right) T^{jk} \mathrm{d}^3\mathbf{x}$$

$$= \int_{\mathbf{x} \in \mathbb{R}^3} \left(\delta_j^\ell \delta_k^m + \delta_j^m \delta_k^\ell\right) T^{jk} \mathrm{d}^3\mathbf{x}$$

$$= \int_{\mathbf{x} \in \mathbb{R}^3} 2 T^{\ell m} \mathrm{d}^3\mathbf{x}.$$

Note that there are no boundary terms when integrating by parts since the source is spatially compact so T^{ij} and its derivatives all vanish at infinity. Changing the integration variable to $\mathbf{y}$ the above calculation shows that

$$\int_{\mathbf{y} \in \mathbb{R}^3} T^{ij} \mathrm{d}^3\mathbf{y} = \frac{1}{2} \int_{\mathbf{y} \in \mathbb{R}^3} y^i y^j \frac{\partial^2 T^{00}}{\partial t^2} \mathrm{d}^3\mathbf{y}. \tag{21.93}$$

Using (21.93) in the solution (21.87) to the wave equation, we thus obtain the following formula for the gravitational perturbation:

$$\ddot{\Phi}^{ij}(t, \boldsymbol{x}) = -\frac{2}{r}\frac{\mathrm{d}^2}{\mathrm{d}t^2}\int_{\boldsymbol{y}\in\mathbb{R}^3} y^i y^j T^{00}(t - r, \boldsymbol{y})\mathrm{d}^3\boldsymbol{y}, \tag{21.94}$$

where $\Phi^{ij} = \delta^{ik}\delta^{j\ell}\Phi_{k\ell}$. If we now write the T^{00} component of the energy-momentum tensor (which represents the mass/energy density) as μ and define the **second mass-moment** I^{ij} by

$$I^{ij}(t) := \int_{\boldsymbol{y}\in\mathbb{R}^3} y^i y^j \mu(t, \boldsymbol{y})\mathrm{d}^3\boldsymbol{y}, \tag{21.95}$$

then we may rewrite (21.94) as

$$\Phi_{ij} = -\frac{2}{r}\ddot{I}_{ij}(t - r), \tag{21.96}$$

(where we use δ_{ij} to lower indices). This is called the **quadrupole formula** and gives an expression for the linearized gravitational perturbation Φ_{ij} in the long-wavelength approximation far from a compact source in terms of the motion of the source.

21.7 The quadrupole generated by a binary star system

Consider two stars of mass m_1 and m_2 in orbit about each other in the xy-plane under their mutual Newtonian gravitational attraction. (Note: as shown in §16.5, the orbits lie in a plane because the total angular momentum is conserved) . If the stars are located at $\boldsymbol{r}_1$ and $\boldsymbol{r}_2$, respectively, the Lagrangian for the system is

$$\mathcal{L} = \tfrac{1}{2}m_1|\dot{\boldsymbol{r}}_1|^2 + \tfrac{1}{2}m_2|\dot{\boldsymbol{r}}_2|^2 + \frac{Gm_1 m_2}{|\boldsymbol{r}_1 - \boldsymbol{r}_2|}. \tag{21.97}$$

Rather than using $\boldsymbol{r}_1$ and $\boldsymbol{r}_2$ as coordinates, it is more convenient to work with the **centre of mass**

$$\tilde{\boldsymbol{r}} := \frac{m_1\boldsymbol{r}_1 + m_2\boldsymbol{r}_2}{m_1 + m_2}, \tag{21.98}$$

and the **relative position**

$$\boldsymbol{r} := \boldsymbol{r}_1 - \boldsymbol{r}_2. \tag{21.99}$$

In terms of these, the Lagrangian becomes (exercise)

$$\mathcal{L} = \frac{1}{2}(m_1 + m_2)|\dot{\tilde{\boldsymbol{r}}}|^2 + \frac{1}{2}\left(\frac{m_1 m_2}{m_1 + m_2}\right)|\dot{\boldsymbol{r}}|^2 + \frac{Gm_1 m_2}{|\boldsymbol{r}|}. \tag{21.100}$$

We see from this that the motion of the centre of mass is just that of a free particle of mass $m_1 + m_2$ moving with constant velocity. Without loss of generality, we may take this to be the origin of our inertial Newtonian coordinate system in which case the Lagrangian reduces to

$$\mathcal{L} = \frac{1}{2} \left(\frac{m_1 m_2}{m_1 + m_2} \right) |\dot{\boldsymbol{r}}|^2 + \frac{G m_1 m_2}{|\boldsymbol{r}|}, \qquad (21.101)$$

or in terms of polar coordinates

$$\mathcal{L} = \frac{1}{2} \left(\frac{m_1 m_2}{m_1 + m_2} \right) (\dot{r}^2 + r^2 \dot{\phi}^2) + \frac{G m_1 m_2}{r}. \qquad (21.102)$$

This is just the Lagrangian of a point particle moving under an inverse square law. Standard Newtonian theory (see also §16.5) then gives the motion as lying on an ellipse and satisfying Kepler's laws. For simplicity, we will take the masses $m_1 = m_2 = m$ to be equal and the orbit to be a circle radius a in the xy-plane. Then conservation of angular momentum gives

$$\dot{\phi} = \Omega = \text{constant.} \qquad (21.103)$$

Substituting in the radial Euler-Lagrange equation for (21.102) then gives

$$\Omega^2 = \frac{2mG}{a^3}. \qquad (21.104)$$

In terms of the binary system, this solution corresponds to the two masses lying on opposite sides of the origin and both orbiting in the same circle of radius R say about their common centre of mass (see Fig. 21.4). Hence the relative displacement is $a = 2R$, so that

$$\Omega = \dot{\phi} = \left(\frac{mG}{4R^3} \right)^{1/2}. \qquad (21.105)$$

Thus, working in units where $G = 1$, we have

$$R = \left(\frac{m}{4\Omega^2} \right)^{1/3}. \qquad (21.106)$$

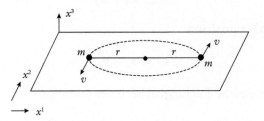

Fig. 21.4 Binary system orbiting about the common centre of mass.

In terms of Cartesian coordinates, we may write the solution for the two stars on opposite sides of the circle as

$$x_1 = R\cos(\Omega t), \qquad y_1 = R\sin(\Omega t),$$
$$x_2 = -R\cos(\Omega t), \qquad y_2 = -R\sin(\Omega t).$$

Given that we model the stars as point masses, this gives a mass distribution for the system as

$$\mu(t, x, y, z) = m\delta(z)\left[\delta(x - R\cos(\Omega t))\delta(y - R\sin(\Omega t))\right.$$
$$\left. + \delta(x + R\cos(\Omega t))\delta(y + R\sin(\Omega t))\right]. \tag{21.107}$$

Inserting this into the formula for the second-mass moment (21.95) gives (exercise)

$$I_{xx} = 2mR^2\cos^2(\Omega t) = mR^2(1 + \cos(2\Omega t)),$$
$$I_{yy} = 2mR^2\sin^2(\Omega t) = mR^2(1 - \cos(2\Omega t)),$$
$$I_{xy} = 2mR^2\cos(\Omega t)\sin(\Omega t) = mR^2\sin(2\Omega t),$$
$$I_{ij} = 0 \quad \text{otherwise,}$$

since the motion is in the plane $z = 0$. Using the quadrupole formula (21.96) this gives us the required formula for the gravitational radiation emitted by a binary system as

$$\Phi_{ij}(t, \boldsymbol{x}) = \frac{8m}{r}\Omega^2 R^2 \begin{pmatrix} \cos(2\Omega(t-r)) & \sin(2\Omega(t-r)) & 0 \\ \sin(2\Omega(t-r)) & -\cos(2\Omega(t-r)) & 0 \\ 0 & 0 & 0 \end{pmatrix}. \tag{21.108}$$

Note that the frequency ω of the gravitational wave is exactly twice the orbital frequency so that

$$\omega = 2\Omega, \tag{21.109}$$

and the amplitude is given by

$$\mathcal{A} = \frac{8\Omega^2 mR^2}{r}. \tag{21.110}$$

In obtaining (21.108), we have applied the various approximations used in deriving (21.96). In particular, we have assumed that $\partial_a T^{ab} = 0$. However, strictly speaking, for a self-gravitating system such as a binary star, $\partial_a T^{ab} = 0$ **does not apply** and it is more appropriate to use the general relativistic formula $\nabla_a T^{ab} = 0$. Nevertheless, although one should use $\nabla_a T^{ab} = 0$ in the **near-zone** to derive the motion of the stars, a careful analysis (see e.g. Misner, Thorne, and Wheeler 1959) shows that, in this situation, it is sufficient to use Newtonian theory to determine the motion of the sources and then apply the quadrupole formula (21.96) far from the binary system to determine the gravitational radiation that is emitted.

21.8 Gravitational energy

Having calculated a formula for the gravitational radiation emitted by a source, it is natural to try and calculate the energy of the gravitational waves. However, it turns out that the notion of the energy associated to gravitational waves is rather problematic. There are two main reasons for this. The first reason is the principle of equivalence which from a mathematical perspective tells us that, at any given point P, one can introduce geodesic coordinates (see §6.6) such that the metric is that of Minkowski space and the partial derivatives of the metric vanishes. Based on the analogy with both Newtonian gravity and electromagnetism, one would expect the energy density to be quadratic in the derivatives of the potentials. However, in general relativity, the role of the gravitational potential is played by the metric and, by the above observation, at any point the derivatives $g_{ab,c}$ vanish in geodesic coordinates. Hence, one cannot construct a tensorial pointwise energy density for the gravitational field using the derivatives of the metric. Thus, any true measure of gravitational energy has to be defined globally rather than from integrating a local energy density. Thus, one can define the total mass of a black hole globally by looking at the asymptotic symmetries of the solution. We will return to this point in Chapter 23.

The other reason why the notion of an energy density is problematic is the relationship between symmetries and conserved quantities. For example, if a mechanical system has a Lagrangian that is preserved under rotations, then the associated angular momentum is conserved and, if the Lagrangian is invariant under time translations, then the energy is conserved. A similar result remains true for field theories such as electromagnetism. The conserved energy and momentum of the Maxwell field as described in §12.5 are a result of the symmetries of Minkowski space. However, in the case of general relativity, there is no fixed space-time geometry on which one considers gravitational perturbations. Any split of a general space-time into a fixed background and a perturbation describing gravitational waves is, to some extent, arbitrary.

However, in this chapter, we are looking at a special situation which makes things simpler. We are considering the weak-field approximation on a background Minkowski space (which can be invariantly defined by the vanishing of the Riemann tensor). In such a situation, we can define an approximate energy density by averaging over a space-time volume of several wavelengths. We start by considering a simple example to show how this works. Consider a cross-polarized wave of frequency ω and amplitude $\mathcal{A}_\times$ propagating in the z-direction. Then, putting $h_{22} = 0$ and $h_{23}(t - z) = \mathcal{A}_\times \sin(\omega(t - z))$ in (21.54), we see that, in the **transverse-traceless** gauge, we have

$$h_{ab} = \mathcal{A}_\times \begin{pmatrix} 0 & 0 & 0 \\ 0 & 0 & \sin(\omega(t - z)) \\ 0 & \sin(\omega(t - z)) & 0 \end{pmatrix}. \qquad (21.111)$$

It therefore follows that the non-zero partial derivatives of h_{ab} are given by terms of the form $\mathcal{A}_\times \omega \cos((t - z))$. From equation (21.14) we see that the Lagrangian, and hence also the Hamiltonian, of the linearized wave is quadratic in the derivatives of h_{ab}. Since the Hamiltonian is closely related to the energy-density, it is not surprising that the energy-density (and hence the energy-flux in the z-direction) is proportional to $\mathcal{A}_\times^2 \omega^2 \cos^2(t - z)$. Averaging the energy-flux over several wavelengths, we find that the average flux is proportional to $\mathcal{A}_\times^2 \omega^2$ (since the average of the $\cos^2(t - z)$ term is just $1/2$). As we show below, a more careful argument shows that the numerical factor is $1/32\pi$, so that the **time-averaged energy flux** $f_\times$ of such a wave is given by

$$f_\times = \frac{\mathcal{A}_\times^2 \omega^2}{32\pi}. \tag{21.112}$$

A similar argument applies to the plus-polarization. So, for a monochromatic gravitational wave of angular frequency ω, the time-average energy-flux in the z-direction is given by

$$f = \frac{(\mathcal{A}_\times^2 + \mathcal{A}_+^2)\omega^2}{32\pi}, \tag{21.113}$$

where $\mathcal{A}_\times$ and $\mathcal{A}_+$ are the amplitudes of the respective polarizations of the wave as measured in the **TT-gauge**.

In calculating the energy of a gravitational wave, the approach we take is to consider a metric of the form

$$g_{ab}(\varepsilon) = \eta_{ab} + \varepsilon h_{ab}^{(1)} + \varepsilon^2 h_{ab}^{(2)},$$

write out the vacuum Einstein equations up to $O(\varepsilon^2)$ and equate coefficients. As $R_{ab}(0) \equiv 0$ since η_{ab} is flat, the first non-zero term is at $O(\varepsilon)$. This is simply (21.10) with h_{ab} replaced by $h_{ab}^{(1)}$, which says that the linearized metric satisfies the linearized Einstein equations. At second order in ε, there are two types of term in $R_{ab}(\varepsilon)$. The first is simply (21.10) with h_{ab} replaced by $h_{ab}^{(2)}$, which we denote $R_{ab}^{(1)}(h^{(2)})$ (to denote the linear part of the Ricci tensor as given by (21.10) but now calculated using $h^{(2)}$). The second involves terms quadratic in the derivatives of $h_{ab}^{(1)}$, which we denote $R_{ab}^{(2)}(h^{(1)})$ (to denote the quadratic part of the Ricci tensor of $\eta_{ab} + h_{ab}^{(1)}$) which is given by (exercise)

$$\begin{aligned}
R_{ab}^{(2)}(h) := \tfrac{1}{2}\eta^{ce}\eta^{df} \big(&h_{ef}h_{cd,ab} - h_{ef}h_{bd,ca} - h_{ef}h_{ad,cb} + h_{ef}h_{ab,cd} + \tfrac{1}{2}h_{cd,a}h_{ef,b} \\
&+ h_{bc,d}h_{ae,f} - h_{bc,d}h_{af,e} + h_{ef,d}h_{ab,c} - h_{cd,e}h_{af,b} - h_{cd,e}h_{bf,a} \big) \\
&+ \tfrac{1}{4}\eta^{cd}h_{,c}\left(h_{ad,b} + h_{bd,a} - h_{ab,d}\right).
\end{aligned} \tag{21.114}$$

Since the $O(\varepsilon^2)$ terms in $R_{ab}(\varepsilon) = 0$ vanish, we must have

$$R_{ab}^{(1)}(h^{(2)}) + R_{ab}^{(2)}(h^{(1)}) = 0. \tag{21.115}$$

If we now take $h_{ab}^{(1)}$ to be a solution of the linearized vacuum Einstein equation (21.10), then the only term in $R_{ab}(\epsilon) = 0$ up to $O(\varepsilon^2)$ is that given by (21.115). We now write this equation in the equivalent but suggestive form

$$R_{ab}^{(1)}(h^{(2)}) - \tfrac{1}{2}\eta^{cd}R_{cd}^{(1)}(h^{(2)})\eta_{ab} = 8\pi t_{ab}, \qquad (21.116)$$

where we have defined t_{ab} by

$$t_{ab} := -\frac{1}{8\pi}\left[R_{ab}^{(2)}(h^{(1)}) - \tfrac{1}{2}\eta^{cd}R_{cd}^{(2)}(h^{(1)})\eta_{ab}\right]. \qquad (21.117)$$

We now regard the term t_{ab} on the right-hand side of (21.116) as the energy-momentum tensor of the gravitational field, as it represents the way in which the linear perturbations provide a source for higher-order perturbations.

The above calculation shows how to associate an energy-momentum tensor with a gravitational wave given by a perturbation h_{ab}. However, there are a number of problems with this. The first is that the above analysis involves separating the space-time metric into two parts; a background and a perturbation h_{ab}. However, in general, it is not obvious how to do this – what is the background space-time and what is the perturbation? Fortunately, in the case of perturbations about flat space, this is not a problem, since Minkowski space is invariantly defined by the vanishing of its curvature. The second problem is that the energy-momentum tensor that one obtains from this procedure is **not gauge invariant**. So that, if we make the coordinate change

$$x^a \mapsto x^a + \epsilon\xi^a, \qquad (21.118)$$

then t_{ab} changes by adding terms involving derivatives of ξ^a. As indicated above, the way round this is to replace the pointwise definition of t_{ab} by a **space-time average** of t_{ab} over a region of space-time of several wavelengths, which is denoted $<t_{ab}>$. In modern terminology, $<t_{ab}>$ is a **quasi-local** quantity. It turns out that $<t_{ab}>$ also has the property of being gauge invariant so that this approach avoids the problems of gauge invariance.

If the background space-time is Minkowski space, we simply define

$$<t_{ab}(p)> = \frac{1}{\mathrm{Vol}(V(p))}\int_{V(p)} t_{ab}(x)\mathrm{d}^4x, \qquad (21.119)$$

where $V(p)$ is a region of several wavelengths centred on the point p. i.e. $<t_{ab}(p)>$ is a moving average of $t_{ab}(x)$ over a region centred on the point p. It follows from the divergence theorem that, under the averaging brackets, divergences vanish, and as a result one may 'integrate by parts' so that, for example,

$$<hh_{ab,c}> = - <h_{,c}h_{ab}>. \qquad (21.120)$$

A straightforward calculation (exercise) shows that, using the above equation, one may write the space-time average of (21.117) as

$$<t_{ab}> = \frac{1}{32\pi} <h_{cd,a}h^{cd}{}_{,b} - \frac{1}{2}h_{,a}h_{,b} - h^{cd}{}_{,c}h_{ad,b} - h^{cd}{}_{,c}h_{bd,a}>, \quad (21.121)$$

where we have used (21.120). We now address the issue of gauge invariance of (21.121). We have seen that under the change of coordinates

$$x^a \mapsto x'^a := x^a + \varepsilon\xi^a,$$
$$h_{ab} \mapsto h'_{ab} := h_{ab} - 2\xi_{(a,b)},$$

and hence

$$h_{ab,c} \mapsto h'_{ab,c} := h_{ab,c} - \xi_{a,bc} - \xi_{b,ac}. \quad (21.122)$$

Inserting the above in (21.121) and then employing a long but straightforward calculation in which we use the rule (21.120) for integration by parts, gives

$$<t'_{ab}> = <t_{ab}>, \quad (21.123)$$

so that (21.121) is indeed gauge invariant. We therefore define the gauge invariant tensor

$$\tilde{T}_{ab} := <t_{ab}> = \frac{1}{32\pi} <h_{cd,a}h^{cd}{}_{,b} - \frac{1}{2}h_{,a}h_{,b} - h^{cd}{}_{,c}h_{ad,b} - h^{cd}{}_{,c}h_{bd,a}> .$$
$$(21.124)$$

This is called the **Isaacson energy-momentum tensor** of the gravitational field. Although it is a gauge-invariant quantity, it is often useful to work in the transverse-traceless gauge (i.e. one in which $h^{ij}{}_{,j} = 0$, $h = 0$, and, in addition, $h_{0\mu} = 0$, $\mu = 1, 2, 3$). If we apply the field equations, which in this gauge are simply $\Box h_{ij} = 0$, then all but one of the terms in (21.121) vanish and we get

$$\tilde{T}_{ab} = \frac{1}{32\pi} <h_{cd,a}h^{cd}{}_{,b}> . \quad (21.125)$$

We now consider a monochromatic plane wave in Minkowski space in the transverse-traceless gauge and where, without loss of generality, we may choose our coordinates so that the wave propagates along the z-axis. For the case of a cross-polarized wave of amplitude $h_\times$ and angular frequency ω,

$$h_{ij} = C_{ij}\sin(\omega(t - z)), \quad (21.126)$$

where

$$C_{ij} = \begin{pmatrix} 0 & 0 & 0 \\ 0 & 0 & h_\times \\ 0 & h_\times & 0 \end{pmatrix}. \quad (21.127)$$

The $\tilde{T}_{ij}$ term in the Isaacson energy-momentum tensor (21.125) is

$$\tilde{T}_{ij} = \frac{1}{32\pi} C_{k\ell} C^{k\ell} < \partial_i(\sin(\omega(t-z)))\partial_j \sin(\omega(t-z)) >$$

$$= \frac{1}{16\pi} h_\times^2 < \partial_i(\sin(\omega(t-z)))\partial_j \sin(\omega(t-z)) > .$$

It then follows that the **energy-density** is given by

$$\tilde{T}_{00} = \frac{1}{16\pi} h_\times^2 \omega^2 < \cos^2(\omega(t-z)) > = \frac{h_\times^2 \omega^2}{32\pi}, \qquad (21.128)$$

since averaging the $\cos^2(\omega(t-z)$ terms over several wavelengths is $1/2$.

We get a similar result for a plus-polarized monochromatic wave of amplitude h_+ and frequency ω

$$\tilde{T}_{00} = \frac{h_+^2 \omega^2}{32\pi}. \qquad (21.129)$$

So, for a wave containing both polarizations, we obtain

$$\tilde{T}_{00} = \frac{(h_\times^2 + h_+^2)\omega^2}{32\pi}. \qquad (21.130)$$

An almost identical calculation (exercise) shows that the radiation flux in the z-direction is given by

$$\tilde{T}_{0z} = \frac{(h_\times^2 + h_+^2)\omega^2}{32\pi}, \qquad (21.131)$$

in agreement with (21.113).

21.9 Gravitational energy-flux from a binary system

We now apply (21.113) to (21.108) to calculate the gravitational flux in the z-direction. We note that (21.108) is already in the TT-gauge with respect to the z-direction so that the gravitational wave luminosity $\mathrm{d}L$ radiated into a solid angle $\mathrm{d}S$ about the z-axis is $\mathrm{d}L = fr^2\mathrm{d}S$ and hence

$$\frac{\mathrm{d}L}{\mathrm{d}S} = \frac{r^2 \omega^2 \mathcal{A}^2}{32\pi}, \qquad \text{for each polarization state.} \qquad (21.132)$$

Noting that (21.108) has frequency $\omega = 2\Omega$ and contains **two** polarization states, each with amplitude $\mathcal{A} = 8\Omega^2 mR^2/r$, we get, for the total luminosity per solid angle in the z-direction

$$\frac{\mathrm{d}L_z}{\mathrm{d}S} = 2\left(\frac{8\Omega^2 mR^2}{r}\right)^2 \frac{(2\Omega)^2}{32\pi} r^2 = \frac{16}{\pi}\left(\Omega^3 mR^2\right)^2. \qquad (21.133)$$

We now wish to calculate the energy-flux per solid angle in an arbitrary direction. We can then integrate this over a sphere to calculate the total gravitational radiation from the binary system. Since the time-averaged energy flux is axisymmetric about the z-axis, we need only consider the angle θ that the direction makes with the z-axis. Our method will be to transform the gravitational perturbation Φ_{ab} to a new basis $e_{\hat{a}}$ so that the direction we want to consider is the new $\hat{z}$-axis. We will then transform to the TT-gauge and use the above formula to calculate the total luminosity per solid angle in the $\hat{z}$-direction.

In the $e_{\hat{i}}$ basis, the gravitational perturbation is given by

$$\Phi_{\hat{i}\hat{j}} = \Phi_{ij} M^i{}_{\hat{i}} M^j{}_{\hat{j}}, \tag{21.134}$$

where

$$M^i{}_{\hat{i}} = \begin{pmatrix} \cos\theta & 0 & \sin\theta \\ 0 & 1 & 0 \\ -\sin\theta & 0 & \cos\theta \end{pmatrix}. \tag{21.135}$$

This gives from (21.108)

$$\Phi_{\hat{i}\hat{j}} = \frac{8m\Omega^2 R^2}{r} S_{\hat{i}\hat{j}},$$

where $S_{\hat{i}\hat{j}}$ is given by

$$\begin{pmatrix} \cos^2\theta\cos(2\Omega(t-r)) & \cos\theta\sin(2\Omega(t-r)) & \cos\theta\sin\theta\cos((2\Omega(t-r)) \\ \cos\theta\sin((2\Omega(t-r)) & -\cos(2\Omega(t-r)) & \sin\theta\sin(2\Omega(t-r)) \\ \cos\theta\sin\theta\cos((2\Omega(t-r)) & \sin\theta\sin(2\Omega(t-r)) & \sin^2\theta\cos(2\Omega(t-r)) \end{pmatrix}.$$

To calculate the flux in the $\hat{z}$-direction, we must write the above expression in the TT-gauge. This simply consists of first projecting in the transverse direction and then subtracting half the trace from the diagonal elements. This gives

$$\Phi_{\hat{i}\hat{j}}^{TT} = \frac{4(\cos^2\theta+1)m\Omega^2 R^2}{r} \begin{pmatrix} \cos(2\Omega(t-r)) & 0 & 0 \\ 0 & -\cos(2\Omega(t-r)) & 0 \\ 0 & 0 & 0 \end{pmatrix}$$

$$+ \frac{8\cos\theta m\Omega^2 R^2}{r} \begin{pmatrix} 0 & \sin(2\Omega(t-r)) & 0 \\ \sin(2\Omega(t-r)) & 0 & 0 \\ 0 & 0 & 0 \end{pmatrix}.$$

So, it consists of two polarized waves, one + and one ×, both of frequency 2Ω, and with amplitudes

$$\mathcal{A}_+ = \frac{4(\cos^2\theta+1)m\Omega^2 R^2}{r},$$

$$\mathcal{A}_\times = \frac{8\cos\theta m\Omega^2 R^2}{r},$$

respectively. Using formula (21.113), this gives the total luminosity per solid angle in the $\hat{z}$-direction to as

$$\frac{dL_{\hat{z}}}{dS} = \frac{(2m\Omega^3 R^2)^2}{2\pi}(\cos^4\theta + 6\cos^2\theta + 1). \qquad (21.136)$$

To calculate the total radiated power, we integrate $dL_{\hat{z}}/dS$ over the whole sphere. This gives

$$L = \frac{(2m\Omega^3 R^2)^2}{2\pi} \int_{\theta=0}^{\pi} \int_{\phi=0}^{2\pi} (\cos^4\theta + 6\cos^2\theta + 1)\sin\theta\, d\theta\, d\phi$$

$$= (2m\Omega^3 R^2)^2 \int_{\theta=0}^{\pi} (\cos^4\theta\sin\theta + 6\cos^2\theta\sin\theta + \sin\theta)\, d\theta$$

$$= 4(m\Omega^3 R^2)^2 \left[-\frac{1}{5}\cos^5\theta - 2\cos^3\theta - \cos\theta \right]_0^{\pi}$$

$$= 8(m\Omega^3 R^2)^2 \left[+\frac{1}{5} + 2 + 1 \right]$$

$$= \frac{128}{5}(m\Omega^3 R^2)^2.$$

Hence the total power radiated is

$$\frac{dE}{dt} = \frac{128}{5}m^2 R^4 \Omega^6. \qquad (21.137)$$

It is convenient to write this in terms of the period T so that, on using (21.106) and $\Omega = 2\pi/T$, we find the rate at which gravitational energy is radiated is given by

$$\frac{dE}{dt} = \frac{128}{5}4^{1/3}\left(\frac{\pi m}{T}\right)^{10/3}. \qquad (21.138)$$

Note that a more general calculation (see exercises) shows that the energy radiated is given by the so-called **quadrupole formula**

$$\frac{dE}{dt} = \frac{1}{5} < \dddot{I}_{ij}\dddot{I}^{ij} >, \qquad (21.139)$$

where

$$I_{ij} = I_{ij} - \frac{1}{3}\delta_{ij}I^k{}_k \qquad (21.140)$$

is the **reduced quadrupole tensor** and the angled brackets in (21.139) indicates the need to take a **space-time average** over several wavelengths.

Using the formula for the second-mass moment of the binary system, we find the reduced quadrupole for the binary system is given by

$$\mathcal{I}_{ij} = mR^2 \begin{pmatrix} \cos(2\Omega t) + \frac{1}{3} & \sin(2\Omega t) & 0 \\ \sin(2\Omega t) & -\cos(2\Omega t) + \frac{1}{3} & 0 \\ 0 & 0 & -\frac{2}{3} \end{pmatrix}. \tag{21.141}$$

Hence

$$\dddot{\mathcal{I}}_{ij} = 8mR^2\Omega^3 \begin{pmatrix} \sin(2\Omega t) & -\cos(2\Omega t) & 0 \\ -\cos(2\Omega t) & -\sin(2\Omega t) & 0 \\ 0 & 0 & 0 \end{pmatrix}, \tag{21.142}$$

which gives

$$\dddot{\mathcal{I}}_{ij}\dddot{\mathcal{I}}^{ij} = 64m^2R^4\Omega^6(2\sin^2(2\Omega t) + 2\cos^2(2\omega t)) = 128m^2R^4\Omega^6. \tag{21.143}$$

Hence

$$\frac{dE}{dt} = \frac{1}{5}\langle\dddot{\mathcal{I}}_{ij}\dddot{\mathcal{I}}^{ij}\rangle = \frac{128}{5}m^2R^4\Omega^6, \tag{21.144}$$

in agreement with (21.137), from which we get, finally,

$$\frac{dE}{dt} = \frac{128}{5}4^{1/3}\left(\frac{\pi m}{T}\right)^{10/3}. \tag{21.145}$$

21.10   Effects of gravitational radiation on the orbit of a binary system

The total energy of the binary system in the Newtonian description is

$$E = 2\left(\tfrac{1}{2}mv^2\right) - \frac{m^2}{2R}$$

$$= mR^2\Omega^2 - \frac{m^2}{2R}$$

$$= -\frac{m}{4}\left(\frac{4m\pi}{T}\right)^{2/3}.$$

Note: The total energy is negative since the gravitational potential energy is negative.

Due to the emission of gravitational energy, the total energy of the system must decrease. This means that the radius R decreases or, equivalently, the period T increases. Differentiating the above expression gives

$$\frac{dE}{dt} = \frac{2m^2\pi}{3T^2} \left(\frac{4\pi m}{T} \right)^{-1/3} \frac{dT}{dt}. \qquad (21.146)$$

Equating the rate of change of the Newtonian energy of the binary system with the rate at which gravitational radiation is emitted, using (21.137) and (21.146), with a minus sign since the energy is decreasing, we obtain

$$\frac{2m^2\pi}{3T^2} \left(\frac{4\pi m}{T} \right)^{-1/3} \frac{dT}{dt} = -\frac{128}{5} 4^{1/3} \left(\frac{\pi m}{T} \right)^{10/3},$$

from which we get

$$\frac{dT}{dt} = 2^{2/3} \left(\frac{96\pi}{5} \right) \left(\frac{2m\pi}{T} \right)^{5/3}. \qquad (21.147)$$

Hence, according to our formula for the gravitational radiation for a binary system, $dT/dt \propto T^{-5/3}$. This may be verified if one is able to accurately record the period of a binary system of a long period of time.

In 1974 Russell Hulse and Joseph Taylor used the Arecibo 305 m radio antenna to observe a pulsar (a rapidly rotating, highly magnetized neutron star). The pulsar rotates on its axis about seventeen times per second so that the pulse period is 59 milliseconds. After timing the radio pulses for some time, they noticed that there was a systematic variation in the arrival time of the pulses and this could be explained if the pulsar was in a **binary system** with another star (subsequently shown to be another neutron star), both orbiting about their common centre of mass. The pulses from the pulsar arrive 3 seconds earlier at some times relative to others, showing that the pulsar's orbit is 3 light-seconds across, approximately two-thirds of the diameter of the Sun. The period of the orbital motion is 7.75 hours and, since this is a binary system, the masses of the two neutron stars can be determined, and they are each around 1.4 times the mass of the Sun.

After observing the binary system for a period of several years, they realized that the size of the orbit was contracting and the period of the orbital motion was slowly decaying. This is in agreement with the predictions of general relativity, where we have seen that the emission of gravitational radiation results in a loss of energy from the binary system and a decay of the orbit. For the Hulse-Taylor binary pulsar PSR B1913+16, one can compare the observational results with those predicted by general relativity as given by equation (21.147). The results are shown in Fig. 21.5.

In 2003 a second neutron star binary system, PSR J0737-3029, this time consisting of two pulsars, was discovered. The graph for this is shown in Fig. 21.6. As can be seen from the graphs, this gives excellent agreement with the predictions of general relativity.

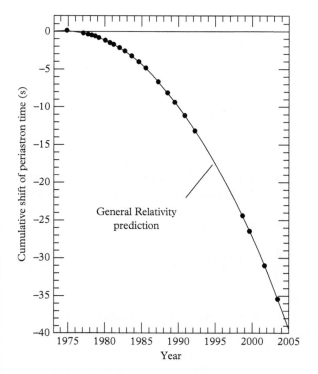

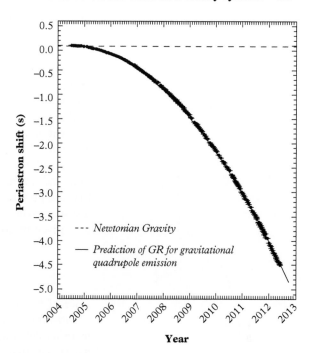

Fig. 21.5 The data for the Hulse-Taylor binary PSR B1913+16 compared to the predictions of general relativity. (Reproduced from *Weisberg et al.* (2010) with permission by the AAS).

Fig. 21.6 The data for the binary pulsar PSR J0737-3029 compared to the predictions of general relativity. (Reproduced from *Tartaglia et al.* (2014) with permission of the authors).

21.11   Measuring gravitational wave displacements

All gravitational wave detectors work by measuring the effect of the wave on the motion of test masses, which we model as particles moving along the geodesics of the curved wave space-time, but whose masses are sufficiently small that we can ignore any influence they have on the space-time curvature. We therefore start by looking at the motion of test masses in more detail.

Consider a situation where a particle is initially at rest in a flat space-time and a gravitational wave is then incident upon it. We will suppose that the gravitational wave is propagating in the x-direction and has the transverse-traceless form given by (21.54). The motion of the particle is then given by solving the geodesic equation

$$\frac{\mathrm{d}V^a}{\mathrm{d}\tau} + \Gamma^a_{bc} V^b V^c = 0, \tag{21.148}$$

where $V^a = \mathrm{d}x^a/\mathrm{d}\tau$ is the 4-velocity.

Now the effect of the gravitational wave will be to change the 4-velocity from its flat space value $V^a_0 = (1,0,0,0)$ (since the particle is at rest in flat-space) to a perturbed value

$$V^a = V^a_0 + \varepsilon U^a.$$

On the other hand, from (21.6), we know that Γ^a_{bc} is of order ε. So, to first order in ε, the geodesic equation becomes

$$\frac{\mathrm{d}V^a}{\mathrm{d}\tau} = -\Gamma^a_{bc} V^b_0 V^c_0 = -\Gamma^a_{00}. \tag{21.149}$$

But, for the metric given by (21.54),

$$\Gamma^a_{00} = \tfrac{1}{2}\varepsilon\eta^{ab}(h_{b0,0} + h_{0b,0} - h_{00,b}) = 0. \tag{21.150}$$

Hence $\mathrm{d}V^a/\mathrm{d}\tau = 0$ and thus

$$\frac{\mathrm{d}^2 x^a}{\mathrm{d}\tau^2} = 0. \tag{21.151}$$

Since the particle is initially at rest $\mathrm{d}x^i/\mathrm{d}\tau = 0$, so that

$$x^i(\tau) = x^i(0), \quad i = 1, 2, 3. \tag{21.152}$$

Therefore, in coordinates in which the gravitational wave takes the transverse-traceless form, the **coordinate position** of the particle remains **constant** as the wave passes through. Thus, the TT-gauge is co-moving with the particles and, since $g_{00} = 1$ and $g_{0a} = 0$, the TT coordinate time is the proper time, as measured by the freely falling test particle. This is one reason why using the transverse-traceless condition makes gravitational wave calculations easier.

Although the coordinate distance between two test particles remains constant, this does not mean that the proper distance remains unchanged as a result of the gravitational wave. We illustrate this by looking at the effect of an h_{22} wave on two test particles lying in a plane orthogonal to the direction of propagation. Without loss of generality, we may take the wave to be travelling in the x-direction and be given by (21.55), and we may take one of the particles to be at $(0,0,0)$ and the other at $(0,L,0)$. Then the proper distance $\ell(t)$ between the two masses is given by

$$\ell(t) = \int_0^L [1 - \varepsilon h_{22}(t)]^{1/2} \, dy \simeq \left(1 - \tfrac{1}{2}\varepsilon h_{22}(t)\right) L. \qquad (21.153)$$

So, as $h_{22}(t)$ changes, so does the proper separation $\ell(t)$.

An alternative method for investigating these results is to consider the equation of geodesic deviation (10.21). If we introduce a local coordinate system adapted to the tetrad so that

$$e_i{}^a \overset{*}{=} \delta_i^a,$$

then, by (10.37), the equation becomes

$$\frac{D^2 \eta^\alpha}{D\tau^2} + R^\alpha{}_{0\beta 0}\eta^\beta = 0. \qquad (21.154)$$

This contains the same information as the metric calculation above but has the advantage that the Riemann tensor is a gauge-invariant quantity in the linearized theory. To show that (21.154) gives the same result as (21.153), we let the connecting vector have tetrad components given by

$$\eta^\alpha = (X, Y, Z),$$

and using (21.38), (21.39), and (21.40), we get

$$\left.\begin{aligned}
\frac{D^2 X}{D\tau^2} &= 0, \\
\frac{D^2 Y}{D\tau^2} + \tfrac{1}{2}\varepsilon(h_{22,00}\, Y + h_{23,00} Z) &= 0, \\
\frac{D^2 Z}{D\tau^2} + \tfrac{1}{2}\varepsilon(h_{23,00}\, Y - h_{22,00} Z) &= 0.
\end{aligned}\right\} \qquad (21.155)$$

Then, for example, an h_{22}-wave leads to

$$\frac{D^2 Y}{D\tau^2} = -\tfrac{1}{2}\varepsilon h_{22}'' Y, \qquad \frac{D^2 Z}{D\tau^2} = \tfrac{1}{2}\varepsilon h_{22}'' Z.$$

For slowly moving particles, we have $\tau = t$ to lowest order, and the connection terms are $O(\varepsilon)$ so that the above becomes

$$\frac{\mathrm{d}^2 Y}{\mathrm{d}t^2} = -\tfrac{1}{2}\varepsilon h''_{22} Y, \quad \frac{\mathrm{d}^2 Z}{\mathrm{d}t^2} = \tfrac{1}{2}\varepsilon h''_{22} Z. \tag{21.156}$$

To linear order in ε, this has solution

$$Y(t) = \left(1 - \tfrac{1}{2}\varepsilon h_{22}\right) Y(0), \quad Z(t) = \left(1 + \tfrac{1}{2}\varepsilon h_{22}\right) Z(0). \tag{21.157}$$

Note that Y and Z are the components of the connecting vector in an orthonormal tetrad so measure the proper distance in agreement with the above and the calculation in §17.10.

We turn now to how one can go about the detection of gravitational waves. As we saw in Chapter 10, one can detect the presence of space-time curvature, and hence gravitational waves, through the equation of geodesic deviation. In practical terms, this involves measuring the relative motion of freely-falling test masses. Although this might, in principle, seem quite straightforward, the extremely small size of any gravitational wave reaching the Earth (see §21.14 for details) makes this an extremely difficult task in practice. The first practical attempt to measure gravitational waves was undertaken by Weber in the 1960s, using a **resonant bar detector**. The idea here was to choose the resonant frequency of the bar to be that of the gravitational waves in order to enhance the signal. Considerable controversy surrounded his claims to be detecting radiation emanating from the centre of the galaxy, since the sensitivity of the bar was considered to be too low to detect radiation at the energy which might be expected. Such signals would probably be swamped by the noise emanating from people, vehicles, aircraft, and so on, passing near the equipment. Moreover, there was also disquiet over the way the results were analysed, and the consensus is that the equipment was probably not detecting gravity waves. However, Weber has played an important part in alerting the instrumentalists to the need to undertake this work, and work on advanced resonant bar detectors still continues.

A simple model of a resonant detector consists of two point particles, each of mass m connected by a spring of natural length L lying on the y-axis. Then, if ξ is the extension of the spring, one can show that, in the absence of a gravitational wave, the motion is that of a damped oscillator with equation (exercise)

$$\ddot{\xi} + 2\gamma\dot{\xi} + \omega_0^2 \xi = 0, \tag{21.158}$$

where ω_0 is the natural frequency and γ is a damping term. The effect of an h_{22}-gravitational wave is to replace the coordinate distance for the extension of the spring by the proper distance for the extension, where the relevant factor in transverse-traceless coordinates is given by (21.153). This adds a source term depending on the gravitational wave to the right-hand side of (21.158) so the equation of motion now reads (exercise)

$$\ddot{\xi} + 2\gamma\dot{\xi} + \omega_0^2\xi = \tfrac{1}{2}L\ddot{h}_{22}. \tag{21.159}$$

If h_{22} is given, for example, by $\mathcal{A}\sin(\omega t)$, the source term becomes $-\tfrac{1}{2}\mathcal{A}\omega^2 L\sin(\omega t)$ and the steady-state solution of (21.159) is given by an oscillation with amplitude

$$a = \frac{\mathcal{A}L\omega^2}{2[(\omega_0^2 - \omega^2)^2 + \gamma^2\omega^2]^{1/2}}. \tag{21.160}$$

If the frequency of the gravitational wave ω is close to the resonant frequency ω_0 of the bar, then this results in a significant amplification of the signal and, for the case where the frequencies are exactly equal, $\omega_0 = \omega$, we obtain

$$a_{\max} = \frac{\mathcal{A}L\omega_0}{2\gamma}. \tag{21.161}$$

So, if the damping γ is very small, there is a significant amplification of the signal.

A more recent (and successful) approach is to attempt to measure small relative displacements of the test masses using the principle of the Michelson interferometer. A number of such **laser interferometers** have been constructed and, by using advanced engineering and optoelectronics, can now be made extremely sensitive. The basic idea is to use three test masses, a beam splitter and two mirrors in an L-shaped configuration (see Fig. 21.7), and to use laser interferometry to measure the phase change as one arm contracts and the other expands in response to the gravitational wave. The light from the laser passes through a beam splitter (S) that is suspended from a wire and is free to move horizontally. The splitter divides the light into two beams running along the perpendicular arms of the interferometer and the light is reflected back from the mirrors attached to the other two test masses, which are also suspended from wires and are free to move horizontally. The reflected light from the two beams is then recombined at (S) and the intensity measured at a photodetector (D). Small differences in the relative lengths of the arms are detected as changes in the amount of interference and hence the intensity of the signal measured by the detector.

For simplicity, assume that the gravitational wave is an h_{22} wave propagating in the z-direction where $\varepsilon h_{22}(t) = \mathcal{A}\sin(\omega t)$. Let both arms have length L (in the absence of the wave) and let the proper length of the arms in the presence of the wave be $L_{(x)}(t)$ and $L_{(y)}(t)$, respectively. Then, according to (21.153), we have

$$L_{(x)}(t) = \left(1 - \tfrac{1}{2}\mathcal{A}\sin(\omega t)\right)L, \quad \text{and} \quad L_{(y)}(t) = \left(1 + \tfrac{1}{2}\mathcal{A}\sin(\omega t)\right)L. \tag{21.162}$$

So that the difference in the proper length of the arms is

$$\Delta L = L_{(y)} - L_{(x)} = \mathcal{A}L\sin(\omega t). \tag{21.163}$$

Let the wavelength of the laser be λ; then, if we measure ΔL in units of λ, so that

$$\Delta L = \alpha\lambda. \qquad (21.164)$$

Then taking into account that the waves travel twice the length of the arms, when $2\alpha = n$, for n an integer, we have constructive interference, whereas, if $2\alpha = (n + \frac{1}{2})$, we have destructive interference. In general, the intensity of the two combined beams is given by $\mathcal{A}(1 + \cos(\pi\alpha))$, where

$$\alpha = \frac{AL}{\lambda}\sin(\omega t), \qquad (21.165)$$

so that the size of the phase shift is proportional to the length of the arms L. As discussed below, the strongest sources are expected to have a value of $\mathcal{A} < 10^{-21}$ so, even if the length of the arms is 1 km, $\Delta L \sim 10^{-18}$ m. A typical infrared laser has a wavelength of 1,000 nm (i.e. 10^{-6} m), which gives a fringe shift of only $\Delta L/\lambda \sim 10^{-12}$. For this reason, it is not possible to detect gravitational waves with a simple Michelson interferometer; instead, a much more sophisticated device using more than one cavity and multiple reflections of the beam to increase the effective length of the beams by many factors is needed. Such a device is shown schematically in Fig. 21.7

In the above calculation, the spatial difference between the test masses was calculated using (21.153), which measures the proper distance between them. However, for an interferometer, it would be more correct

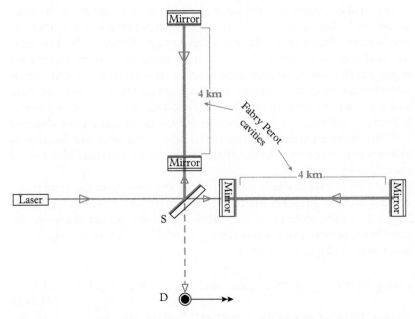

Fig. 21.7 Gravitational wave interferometer: a basic Michelson interferometer with 4 km Fabry Perot cavities.

to calculate the separations on the path of a light ray. In flat space, the light rays are just straight lines so that it makes no difference whether one measures along light rays or uses proper distance. However, in the gravitational wave space-time, this is no longer true and one needs to make a correction to the coordinates of the geodesic from those of a straight line. However, since the straight line is an extremal for the flat space metric, any first-order variation in the path only produces a second-order change in the length (using the flat space metric) so that the only first-order change in length comes from measuring the length of a straight line using the curved metric. This gives the same answer as the proper length as long as the size of the arms of the interferometer are small compared to the gravitational wavelength. This is true for the ground based detectors described above but is not true for the proposed space based detector LISA or for a measurement using a pulsar timing array. We therefore describe below a method of measuring the space-time geometry directly using light rays.

21.12   A direct interferometric measurement

In Chapter 2, we introduced the k-calculus and described how to use light rays and the constancy of the speed of light to measure distance in flat Minkowski space. However, one can use a similar procedure in a curved space-time to measure distance and use this to deduce the curvature of space-time. Let a freely falling observer O emit a pulse of light which reaches a distant point P and is then reflected back to the observer O at a time T later. Since observer O is freely falling, they are in a local inertial frame, and the time measured by their clock is the proper time between the pulse being emitted and received back at O. Because of the assumed constancy of the velocity of light, even in curved space-time, we may take half the propagation time as a measure of the distance to P. If we make sufficiently many such measurements, we can reconstruct the geometry of the space-time. (This process is called 'differential chronometry' by Synge 1960). We illustrate this by using light to calculate the distance, as measured by T, between two freely falling bodies that are initially at rest in a gravitational wave space-time. The value of T is coordinate invariant but it is convenient to calculate it in the transverse-traceless gauge since, as we have seen, freely falling bodies remain at fixed coordinates in this gauge, and the proper time measured by a freely falling clock is just the coordinate time t.

Consider an h_{22}-wave space-time with metric

$$ds^2 = dt^2 - [1 - \varepsilon h_{22}(t - z)]dx^2 - [1 + \varepsilon h_{22}(t - z)]dy^2 + dz^2. \quad (21.166)$$

We want to measure the return time as measured by an observer at the origin O for a photon to emitted at O to reach a point P at a coordinate distance L on the x-axis and then to be reflected back to O (see Fig. 21.8).

We first find the equation of a null geodesic pointing in the x-direction. We have $dy = 0$, $dz = 0$, and $ds^2 = 0$ on the geodesic so that inserting this in (21.166) and using $z = 0$ gives

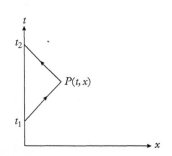

Fig. 21.8 Space-time diagram of a photon emitted from O and reflected back from P.

$$\left(\frac{dt}{dx}\right)^2 = 1 - h_{22}(t). \tag{21.167}$$

To find the equation of the geodesic, we use the coordinate distance x as a parameter along the geodesic (since we know the location of the particles in terms of x) and solve

$$\frac{dt}{dx} = (1 - h_{22})^{1/2}, \tag{21.168}$$

to find $t(x)$. If the photon is emitted from O at time t_0, then it reaches L at coordinate time

$$t_L = t_0 + \int_0^L (1 - \varepsilon h_{22}(t(x)))^{1/2} \, dx. \tag{21.169}$$

Since we are working to first order in ε, this is just

$$t_L = t_0 + \int_0^L \left(1 - \tfrac{1}{2}\varepsilon h_{22}(t(x))\right) dx$$

$$= t_0 + L - \tfrac{1}{2}\varepsilon \int_0^L h_{22}(t(x)) dx. \tag{21.170}$$

Furthermore, we may write $t(x)$ as the flat space result (a straight line) together with an $O(\varepsilon)$ perturbation so that

$$t(x) = t_0 + x + O(\varepsilon), \tag{21.171}$$

and then expanding $h_{22}(x(t))$ in a Taylor series and working to first order in ε we have

$$t_L = t_0 + L - \tfrac{1}{2}\varepsilon \int_0^L h_{22}(t_0 + x) dx. \tag{21.172}$$

Having arrived at P, the photon is reflected back and reaches the origin at time T. A similar argument to the above shows that, working to first order in ε, the time T is given by

$$T = t_L + L - \tfrac{1}{2}\varepsilon \int_L^0 h_{22}(t_L + L - x) dx$$

$$= t_L + L - \tfrac{1}{2}\varepsilon \int_L^0 h_{22}(t_0 + 2L - x) dx$$

$$= t_L + L - \tfrac{1}{2}\varepsilon \int_0^L h_{22}(t_0 + L + \tilde{x}) d\tilde{x},$$

where we have used (21.172) in the second equality (working to first order in ε and, in the last line, $\tilde{x} = L - x$. Dropping the tilde and substituting for t_L from (21.172), we find the return time T as measured by O is given by

$$T = t_0 + 2L - \tfrac{1}{2}\varepsilon \int_0^L h_{22}(t_0 + x)\mathrm{d}x - \tfrac{1}{2}\varepsilon \int_0^L h_{22}(t_0 + L + x)\mathrm{d}x. \quad (21.173)$$

If the size of the detector is small compared to the wavelength of the gravitational wave, then we may approximate both integrands by $h_{22}(t_0)$, which gives a change ΔT in the time of arrival compared to the flat space result of

$$\Delta T = -\varepsilon L h_{22}(t_0). \quad (21.174)$$

Taking into account the fact that the light goes there and back, and to this order of accuracy, we may regard the light ray as moving in Minkowski space, we find that this is in agreement with (21.153) in this approximation.

However, for space based detectors, this approximation is not valid, so that one must use (21.173) rather than (21.163) to calculate the phase shift. Actually, in practice, rather than measure T, one uses the laser as an atomic clock and measures the changes in the time T of the return as a function of the time t when it was emitted. Differentiating (21.173) with respect to t_0 (which we now just call t), we get

$$\begin{aligned}
\frac{\mathrm{d}T}{\mathrm{d}t} &= 1 - \tfrac{1}{2}\varepsilon \int_0^L h'_{22}(t + x)\mathrm{d}x - \tfrac{1}{2}\varepsilon \int_0^L h'_{22}(t + L + x)\mathrm{d}x \\
&= 1 - \tfrac{1}{2}\varepsilon \left[h_{22}(t + x)\right]_0^L - \tfrac{1}{2}\varepsilon \left[h_{22}(t + L + x)\right]_0^L \\
&= 1 - \tfrac{1}{2}\varepsilon \left(h_{22}(t + 2L) - h_{22}(t)\right). \quad (21.175)
\end{aligned}$$

If we again use an approximation in which the size L of the detector is small compared to the wavelength of the gravitational wave, then we may expand $h_{22}(t + 2L)$ in a Taylor series to give

$$h_{22}(t + 2L) \simeq h_{22}(t) + 2L\dot{h}_{22}(t). \quad (21.176)$$

So, in this approximation,

$$\frac{\mathrm{d}T}{\mathrm{d}t} = 1 - \varepsilon L \dot{h}_{22}(t). \quad (21.177)$$

We also see that, differentiating (21.177) again, we get

$$\frac{\mathrm{d}^2 T}{\mathrm{d}t^2} = -\varepsilon L \ddot{h}_{22}(t), \quad (21.178)$$

which, taking into account the fact that the light travels a distance $2L$, is in agreement with the equation of geodesic deviation (21.156).

21.13   The detection of gravitational waves

In this section, we briefly review the various methods suggested for detecting gravitational waves before discussing the successful observation of gravitational waves by the LIGO interferometers in the next section. There have been four sorts of detectors proposed so far: resonant mass detectors, laser interferometers, Doppler tracking, and pulsar timing arrays.

The original **resonant mass** detectors used by Weber consisted of massive metal bars tuned to a particular resonant frequency and, as a result, these are limited to detecting waves in a narrow frequency band. Modern bar detectors consist of solid bars weighing a few tons, suspended in vacuum, and cooled down to temperatures of only a few millikelvin and have achieved sensitivities of around $h \simeq 10^{-20}$. Although a number of such devices have operated for periods of ten or more years, they have not produced reliable evidence of the detection of gravitational waves. A new generation of such detectors but using spheres rather than cylinders has been proposed. These would offer greater sensitivity and improved bandwidth compared to the bar detectors. Although they might potentially be able to measure a supernova in our galaxy, they operate at the wrong frequency to detect many of the most promising sources such as colliding neutron stars or black holes.

As described above, an **interferometer** essentially measures the difference in the return times along two different arms. Ground based interferometers operate at a frequency in the range 20 Hz to 2 kHz and are therefore small enough to use the formula (21.174) that we derived above. As of 2020, there are a number of such detectors in operation, including the two LIGO sites in the USA (Hanford, Washington, and Livingston, Louisiana), the VIRGO detector near Pisa, Italy, and the GEO600 detector near Hannover, Germany. The best such detectors achieve a strain sensitivity of $10^{-23}\,\mathrm{Hz}^{-1/2}$ (see details below). At frequencies below a few Hertz, it is not possible to shield detectors on the Earth from noise. One strategy to avoid ground based noise is therefore to use a detector in space. In 2017, the Laser Interferometer Space Antanae (LISA) was selected by the European Space Agency (ESA) a launch slot and, since then, an international collaboration has been working towards a launch in 2034. It will consist of three freely flying spacecraft arranged in a triangular formation, each containing two test masses and two telescopes and two lasers arranged to point at the other two spacecraft. It will thus form three Michelson-Morely-type interferometers, each centred on one of the spacecraft, with the test masses defining the ends of the arms. The entire arrangement, which is ten times larger than the orbit of the Moon, will be placed in solar orbit at the same distance from the Sun as the Earth, but trailing the Earth by $20°$ (see Fig. 21.9). The sensitivity of LISA will be similar to that of LIGO but at a frequency 10^5 times lower. Since the size of the interferometers used in LISA would be larger than a wavelength of gravitational waves for frequencies above 10 mHz, one cannot use (21.174) to measure the phase shift, but instead one needs a generalization of (21.175) for a gravitational wave making an angle θ to the z-axis which gives (exercise)

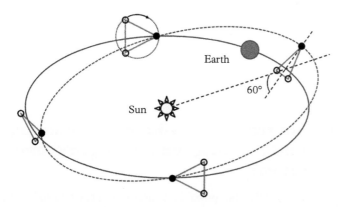

Fig. 21.9 Schematic of LISA's motion about the Sun. The three spacecraft form an equilateral triangle of side length 2.5×106 km. The triangle's center follows a circular path about the Sun with radius 1 AU, trailing Earth by 5×107 km. The plane of the triangle is tilted by $60°$ from the plane of Earth's orbit.

$$\frac{\mathrm{d}T}{\mathrm{d}t} = 1 + \tfrac{1}{2}\left\{(1 - \sin\theta)h_{22}(t + 2L) - (1 + \sin\theta)h_{22}(t)\right.$$

$$\left. + 2\sin\theta\, h_{22}(t + L(1 - \sin\theta))\right\}. \qquad (21.179)$$

Both NASA and ESA use **Doppler tracking** to monitor the position of interplanetery spacecraft in order to look for the effect of gravitational waves. Typically, they monitor the return time of communications to the spacecraft which, for missions to Jupiter and Saturn, for example, are of order $2 - 4 \times 10^3$ s. Any gravitational wave event shorter than this will appear three times in the time delay: once when the wave passes the Earth-based transmitter, once when it passes the spacecraft, and once when it passes the Earth-based receiver. Searches use a form of data analysis using pattern matching. Using two transmission frequencies and very stable atomic clocks, it is possible to achieve sensitivities for h of order 10^{-13}, and even 10^{-15} may soon be reached. This is effectively using a formula like (21.173) to measure the time delay so it is limited by clock accuracies.

Millisecond pulsars are natural examples of extraordinarily regular clocks, with irregularities too small for the best atomic clocks to measure. If one assumes that they emit pulses perfectly regularly, then one can use observations of timing irregularities of single pulsars to set upper limits on the background gravitational wave field by comparing the arrival time with that given by (21.172). The delay is a combination of the effects of waves at the pulsar when the signal was emitted and waves at the Earth when it is received. If one simultaneously observes two or more pulsars, the Earth-based part of the delay is correlated, and this offers a method of actually detecting long-period gravitational waves. However, to take account of the random fluctuations in the pulsar signal, one needs to observe the signal for several years in order to achieve stability of the pulse arrival times. Therefore, such detectors can only be used for looking for strong gravitational waves with periods of several years. Observations are underway at a number of observatories including the Parkes Pulsar Timing Array, the European Pulsar Timing Array, and the American Nanograv collaboration.

21.14   Sources of gravitational radiation and the observation of gravitational waves

Let us discuss briefly the possible sources of gravitational radiation. Thorne distinguishes between three sorts of radiation, namely, **bursts**, **periodic**, and **stochastic**. Known sources of bursts are collisions between binary black holes, and between black holes and neutron stars, or pairs of neutron stars. Examples of all of these bursts have been observed by LIGO. Other possible sources are white dwarf binaries, and extreme-mass-ratio inspirals (EMRI), which is when a compact stellar remnant (white dwarf, neutron star, or stellar mass black hole) is captured and swallowed up by a supermassive black hole. It is hoped that both these will be observed by LISA. It is also believed that occasionally two supermassive black holes will merge. Although such mergers would be rare (one per year is an optimistic estimate), the strength of the signal is such that the signal should be observable by LIGO for any merger in the observable universe. Possible sources of periodic waves include binary star systems, rotating neutron stars, and pulsations of white dwarfs following nova outbursts. Stochastic sources include the relic background signal from the big bang, and a binary background coming from thousands of binary systems emitting gravitational waves continuously in overlapping frequency bands so that the individual signals cannot be resolved.

It is extremely difficult to obtain estimates of the energy output from the various sources, because they often depend on the details of the model which in many cases is uncertain. Furthermore, even if the details are known the equations are very complicated so that they need to be solved using numerical methods. For some of the physically simpler systems, significant advances have been made using **numerical relativity**. For example, numerical codes exist which suggest that a collapsing star may emit up to 1 or 2% of its mass in the form of gravitational waves.

Figure 21.10 illustrates the typical amplitude and wavelengths λ of the various sources as well as the sensitivity of the various detectors. The vertical axis is not the raw 'instantaneous' strain h discussed earlier, but a 'characteristic' strain that one obtains by accumulating the signal over some observational timescale.

In 2005 the two LIGO detectors at Livingston and Hanford reached the target sensitivity for the initial design and, in the sixth science run from July 2009 to October 2010, they managed to take over a years worth of data. Although this did not result in the observation of any gravitational waves, it resulted in some useful upper limits on potential sources of gravitational radiation. The detectors were then shut down for a period in

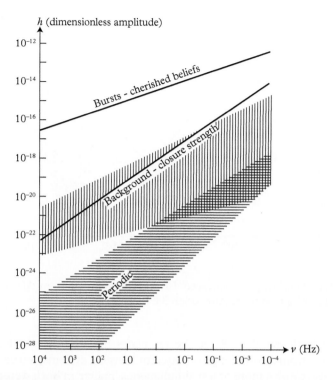

Fig. 21.10 Estimates of the strength of gravity waves reaching the Earth.

order to upgrade the technology to what is called 'Advanced LIGO'. This resulted in considerably enhanced sensitivity, as shown in Fig. 21.11.

In February 2015, the two advanced detectors were brought back on stream in 'engineering mode' in order to fine tune and test the detectors prior to the first scientific observations starting in September. This resulted in the first detection of a gravitational wave on 14th September 2015, which is called GW150914 to indicate the date of detection. The signal was in the frequency range 35 Hz to 250 Hz, lasted about 0.2 seconds and had the distinctive 'chirp' feature expected from numerical simulations of two black holes colliding to form a single rotating black hole. One can see by eye (Fig. 21.12) the similarity of the signals received at Hanford and Livingston (with the 7 millisecond offset of the timing of the signals consistent with the light travel time between the two sites). This figure is also shown on the back cover of the book.

A detailed statistical analysis of the signal confirmed GW150914 as a genuine observation of a gravitational wave, with an estimated significance of at least 5.1σ or a confidence level of 99.99994%. Extracting the astrophysical signal of a gravitational wave from the noise in order to confirm an observation is a difficult task and is undertaken using the technique of **matched filtering**. This involves constructing a a database of **templates** which describe the gravitational waves produced by the merger of two compact objects (either neutron stars or black holes) with a range of different masses and angular momenta. The aim of matched filtering is to see if the observed data contains any signals similar to a

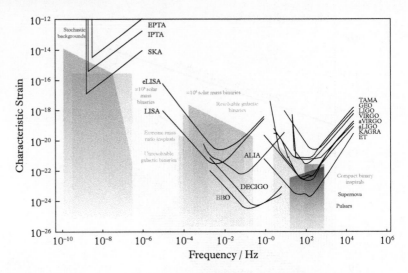

Fig. 21.11 Sensitivity of enhanced LIGO. (Reproduced from *Moore et al.* (2015), Creative Commons Attribution 3.0 License).

template bank member. Since the templates describe the gravitational waveforms from a wide range of different merging systems, a true signal should also cause a more or less simultaneous 'match' in **both detectors**. Since we know gravitational waves travel at the speed of light, the match must occur within 15 milliseconds or less, depending on the direction on the sky from which the gravitational wave signal emanates. The particular template for which there is a match also enables one to determine the astrophysical features of the signal. In the case of GW150914, this analysis showed that the event happened at a distance of 1.4 ± 0.6 billion light years (determined by the amplitude of the signal) and was produced by the merger of two black holes with masses of about 35 times and 30 times the mass of the Sun, resulting in a post-merger Kerr black hole of about 62 solar masses and a spin parameter a of about 0.68. The mass–energy of the missing three solar masses was radiated away in the form of gravitational waves with the power of the radiation peaking at about 3.6×10^{49} watts, or 50 times greater than the combined power of all light radiated by stars in the observable universe. The orbital frequency of 75 Hz (half the gravitational wave frequency) means that the objects were orbiting each other at a distance of only 350 km by the time they merged. The phase changes to the signal's polarization allowed calculation of the objects' orbital frequency and, taken together with the amplitude and pattern of the signal, allowed calculation of their masses and therefore their extreme final velocities and orbital separation when they merged. In constructing the gravitational wave templates, the early phases of the merger can be well described by **post-Newtonian calculations** (Blanchet 2014) but the strong gravitational field merger stage can only be solved in full generality by large-scale numerical relativity simulations. The final ringdown phase of the post-merger object can be calculated using the so-called quasi-normal modes of the Kerr solution (Kokkotas and Schmidt 1999).

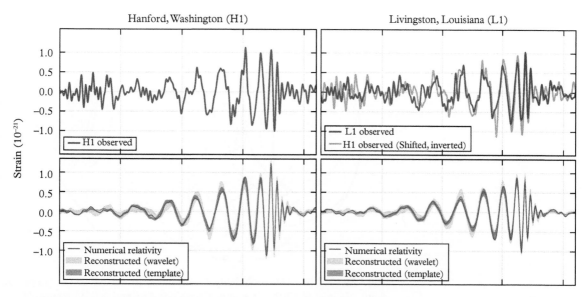

Fig. 21.12 First detection of a gravitational wave signal. (Reproduced from *Abbott et al.* (2016), Creative Commons Attribution 3.0 License).

Unlike optical telescopes, which only have a small field of view, interferometric detectors have good coverage of a large portion of the sky. The downside of this is that a single detector has only a very limited ability to locate the source of the gravitational wave and that one really needs a network of detectors in order to locate the source through triangulation based on the different times of arrival of the signal. For the case of GW150914, the signal was only observed by the two LIGO detectors, and, as a result, the location could only be restricted to an arc of the sky. Since 2015, as well as the two LIGO detectors, the VIRGO detector outside Pisa and the GEO600 detector near Hannover have come on stream. As well as enhancing the ability to detect gravitational waves, this has enabled the network of detectors to have much better ability to locate the source of gravitational waves, as shown by Fig. 21.12 which gives the 90% probability localizations for GW170814. Plans are currently in place to extend the network through building a fourth detector (called KAGRA) in Japan and a fifth LIGO detector in India. There are also plans to improve the strain sensitivity of the detectors by a factor of 10, which would increase the volume by a factor of 1,000.

The LIGO and VIRGO collaborations have in 2020 confidently detected gravitational waves from a total of ten stellar-mass binary black hole mergers and one merger of neutron stars in the first two observing runs. The event GW170729, detected in the second observing run on 29 July 2017, is the most massive and distant gravitational-wave source ever observed. In this coalescence, which happened roughly 5 billion years ago, an equivalent energy of almost five solar masses was converted into gravitational radiation. GW170814 was the first binary black hole merger measured by the all three of the global network formed by the LIGO and

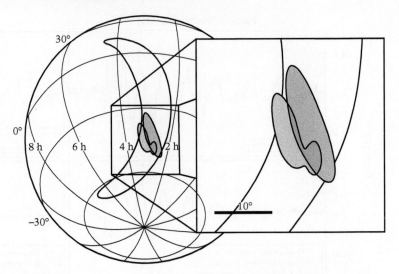

Fig. 21.13 The 90% probability localizations for GW170814. (Reproduced from *Abbott et al.* (2017), Creative Commons Attribution 4.0 License).

Virgo observatories and allowed for the first tests of gravitational-wave polarization (analogous to light polarization), while GW170818, which was again detected by all three detectors and was very precisely pinpointed in the sky. The position of the binary black holes, located 2.5 billion light-years from the Earth, was identified in the sky with a precision of 39 square degrees. That makes it the next best localized gravitational-wave source after the GW170817 neutron star merger, which was the first time that gravitational waves were ever observed from the merger of a binary neutron star system. Furthermore, this collision was also seen optically, marking an exciting new chapter in multi-messenger astronomy, in which cosmic objects are observed simultaneously in different forms of radiation. Details of these and subsequent detections can be found at the website of the LIGO Scientific Collaboration (LSC), www.ligo.org.

Exercises

21.1 (§21.1) Show that, if we work to order ε^2, then (21.1) implies (21.4), (21.5), (21.6), (21.7), and (21.10) (subject to (21.11)), (21.12), and (21.13).

21.2 (§21.1) Show that the Bianchi identities (21.8) can be written in the form (21.9) to order ε^2, and that these equations are satisfied automatically by (21.1).

21.3 (§21.1) Show that the quadratic Lagrangian (21.14) leads to the field equations (21.13). [Hint: the field equations must be symmetric in a and b.]

21.4 (§21.2) Show that h_{ab} transforms according to (21.18) to order ε^2 under the coordinate transformation (21.16). Show also that ϕ_{ab} transforms according to (21.25) under this transformation.

21.5 (§21.2) Show that in the slow-motion approximation for a distribution of dust of proper density ρ_0 that (21.28) reduces to

$$\varepsilon \nabla^2 \phi_{00} = 16\pi\rho_0.$$

Compare this with Poisson's equation in relativistic units to deduce that

$$\varepsilon \phi_{00} = 4\phi,$$

with all other components vanishing. Use (21.19) to deduce that

$$\varepsilon h_{00} = \varepsilon h_{11} = \varepsilon h_{22} = \varepsilon h_{33} = 2\phi,$$

and hence that, in this approximation, the metric is

$$ds^2 = (1 + 2\phi)dt^2 - (1 - 2\phi)(dx^2 + dy^2 + dz^2).$$

Show that this is consistent with the Schwarzschild solution (in isotropic coordinates) in the weak-field limit.

21.6 (§21.2) Confirm equations (21.20), (21.21), and (21.22), and deduce (21.28), (21.30), and (21.32) in the Lorentz gauge. Show that there is an additional gauge freedom (21.18) subject to (21.29).

21.7 (§21.3) Show that the ansatz (21.36) leads to a Riemann tensor satisfying (21.38), (21.39), and (21.40). [Hint: use the identity (6.79) to eliminate R_{0312}.] Show that the linearized vacuum field equations lead to the vanishing of the group of equations (21.39). [Hint: consider R_{02} $= R_{03} = R_{12} = R_{13} = R_{00} - R_{11} + R_{22} + R_{33} = 0$, and remember to raise and lower indices with η_{ab}.]

21.8 (§21.3) Fill in the details of the argument which shows that the ansatz (21.45) leads to the canonical form (21.54). [Hint: be careful about signs.]

21.9 (§21.4) Show that the transformation (21.59) transforms (21.58) to (21.60).

21.10 (§21.5) Complete the details of the argument that an outgoing wave solution of

$$\frac{1}{r}\frac{\partial^2}{\partial r^2}(r\phi) - \frac{\partial^2 \phi}{\partial t^2} = 0,$$

is given by $\phi = u(t - r)/r$, where u is an arbitrary function.

21.11 (§21.7) Show that the Lagrangian $\mathcal{L}$ for the Newtonian two-body problem is given by (21.100).

21.12 (§21.7) Verify the expression for the second-mass moment of the mass distribution of the two stars as given by (21.107).

21.13 (§21.8)

(i) Show that, if we assume

$$g_{ab}(\varepsilon) = \eta_{ab} + \varepsilon h_{ab}^{(1)} + \varepsilon^2 h_{ab}^{(2)},$$

then, to calculate the Einstein equations to $O(\varepsilon^2)$, we need only take g^{ab} to $O(\varepsilon)$, where

$$g^{ab} = \eta^{ab} - \varepsilon h^{(1)ab}.$$

[Hint: check first with this definition that $g_{ab}g^{bc} = \delta_a^c$ to $O(\varepsilon)$ and then use the fact that Γ_{bc}^a involves derivatives of g_{ab} which are of $O(\varepsilon)$, so that any term of $O(\varepsilon^2)$ in g^{ab} would lead to $O(\varepsilon^3)$ terms which we neglect.]

(ii) Show that (ignoring terms of $O(\varepsilon^3)$)

$$\Gamma_{bc}^a = \tfrac{1}{2}\varepsilon\eta^{ad}(h_{dc,b}^{(1)} + h_{bd,c}^{(1)} - h_{bc,d}^{(1)})$$
$$+ \tfrac{1}{2}\varepsilon^2[\eta^{ad}(h_{dc,b}^{(2)} + h_{bd,c}^{(2)} - h_{bc,d}^{(2)}) - h^{(1)ad}(h_{dc,b}^{(1)} + h_{bd,c}^{(1)} - h_{bc,d}^{(1)})].$$

(iii) Compute R_{ab} to $O(\varepsilon^2)$ and show that the $O(\varepsilon^2)$ terms are given by

$$R_{ab}^{(1)}(h^{(2)}) + R_{ab}^{(2)}(h^{(1)}),$$

where the first term $R_{ab}^{(1)}$ is given by (21.10), with h replaced by $h^{(2)}$, and the second term $R_{ab}^{(2)}$ is given by (21.114), with h replaced by $h^{(1)}$. [Hint: this is a long but straightforward calculation and it may be helpful to split up the calculation into three sets of terms: second derivatives of $h_{ab}^{(2)}$, second derivatives of $h_{ab}^{(1)}$, and the remaining first derivative terms.]

21.14 (§21.8)

(i) Show that in the TT-gauge (namely, $\eta^{ab}h_{ac,b} = h_a^a = 0$),

$$R_{ab}^{(2)}(h) = \tfrac{1}{2}\eta^{ce}\eta^{df}(h_{ef}h_{cd,ab} - h_{ef}h_{bd,ca} - h_{ef}h_{ad,cb} + h_{ef}h_{ab,cd}$$
$$+ \tfrac{1}{2}h_{cd,a}h_{ef,b} + h_{bc,d}h_{ae,f} - h_{bc,d}h_{af,e}).$$

(ii) Use the fact that one can integrate by parts under the angled brackets to show that

$$\langle R_{ab}^{(2)}(h) \rangle = \tfrac{1}{4}\langle h_{cd,a}h^{cd}{}_{,b}\rangle.$$

[Hint: you will need the fact that, in the TT-gauge $\Box h_{ab} = 0$.]

(iii) Starting from the definition (21.117) use part (ii) to establish (21.125) in the TT-gauge.

(iv) Deduce from the gauge invariance of (21.121), and the fact that it gives the correct expression in the TT-gauge, that it gives the correct expression for the Isaacson energy–momentum tensor in any gauge.

21.15 (§21.8) Show that $\tilde{T}_{0z}$ is given by (21.131).

21.16 (§21.10) Let n^i be a unit vector in $\mathbb{R}^3$ with the Euclidean metric δ_{ij} and let

$$P_i^j = \delta_i^j - n_i n^j$$

be the projection orthogonal to n^i. If X_{ij} is a symmetric tensor, define X_{ij}^{TT} by

$$X(n)_{ij}^{\mathrm{TT}} = \left(P_i^k P_j^\ell - \tfrac{1}{2}P_{ij}P^{k\ell}\right) X_{k\ell}.$$

Show that the projection $P_i^k P_j^\ell - \tfrac{1}{2}P_{ij}P^{k\ell}$ gives the **transverse-traceless** part of a symmetric tensor X_{ij} in the n^i direction by
(i) showing that $X(n)_{ij}^{\mathrm{TT}} n^i = 0$ and $X(n)_{ij}^{\mathrm{TT}} n^j = 0$, so that $X(n)_{ij}^{\mathrm{TT}}$ is **transverse** to n;
(ii) showing that $\delta^{ij}X(n)_{ij}^{\mathrm{TT}} = 0$, so that $X(n)_{ij}^{\mathrm{TT}}$ is **traceless**.

21.17 (§21.9)

(i) If X_{ij} is a general symmetric tensor then show that

$$X_{ij}^{\mathrm{TT}} X_{\mathrm{TT}}^{ij} = X_{ij}X^{ij} - 2X_i^j X^{ik} n_j n_k + \tfrac{1}{2}X^{ij}X^{k\ell}n_i n_j n_k n_\ell - \tfrac{1}{2}X^2 + XX^{ij}n_i n_j.$$

where $X = \delta^{ij}X_{ij}$ is the trace of X_{ij}.

(ii) The **reduced quadrupole moment** I is defined by

$$\mathit{I} = I_{ij} - \tfrac{1}{3}\delta_{ij}\delta^{k\ell}I_{k\ell}.$$

Show that $\delta^{ij}\mathit{I}_{ij} = 0$.

(iii) Deduce that

$$\mathit{I}_{ij}^{\mathrm{TT}}\mathit{I}_{\mathrm{TT}}^{ij} = \mathit{I}_{ij}\mathit{I}^{ij} - 2\mathit{I}_i^j\mathit{I}^{ik}n_j n_k + \tfrac{1}{2}\mathit{I}^{ij}\mathit{I}^{k\ell}n_i n_j n_k n_\ell.$$

21.18 (§21.9) Using the same argument as in §21.9 it follows from (21.125) that the luminosity in the n^i direction is given by

$$L(n) = \frac{1}{32\pi} < \dot{h}(n)_{ij}^{\mathrm{TT}} \dot{h}(n)_{\mathrm{TT}}^{ij} > .$$

(i) Use the fact that $h_{ij}^{\mathrm{TT}} = \Phi_{ij}^{\mathrm{TT}}$ and (21.96) to show that

$$L(n) = \frac{1}{8\pi r^2} < \ddot{\mathit{I}}(n)_{ij}^{\mathrm{TT}} \ddot{\mathit{I}}(n)_{\mathrm{TT}}^{ij} > .$$

(ii) The total luminosity L is obtained by integrating over a sphere of radius r so that

$$L = \int_{S^2} L(n) r^2 \, d\Omega,$$

where $n^i = x^i/r$ and S^2 is the unit sphere. Use the identities

$$\int_{S^2} d\Omega = 4\pi,$$

$$\int_{S^2} n_i n_j d\Omega = \frac{4\pi}{3}\delta_{ij},$$

$$\int_{S^2} n_i n_j n_k n_\ell d\Omega = \frac{4\pi}{15}\left(\delta_{ij}\delta_{k\ell} + \delta_{ik}\delta_{j\ell} + \delta_{i\ell}\delta_{jk}\right),$$

and the result of Exercise 21.16(iii) to show that

$$L = \frac{1}{5} < \dddot{I}_{ij}\dddot{I}^{ij} > .$$

Hence show the total energy radiated is given by the quadrupole formula

$$\frac{dE}{dt} = \frac{1}{5} < \dddot{I}_{ij}\dddot{I}^{ij} >$$

21.19 (§21.11) Show that the equation of geodesic deviation can be written in the form (21.155). Investigate the equation for an h_{22}-wave and an h_{23}-wave.

21.20 (§21.11)
(i) Let two particles of unit mass that are connected by a spring of natural length L have positions y_1 and y_2 respectively, along the y-axis. The extension of the spring is then given by $\xi = y_2 - y_1 - L$. If the masses experience a Hooke force from the spring of $-k\xi$ and a damping of $-\gamma\dot{\xi}$ then show that Newton's law for the particle at y_1 is

$$\frac{d^2 y_1}{dt^2} = k(y_2 - y_1 - L) + \gamma\frac{d}{dt}(y_2 - y_1 - L).$$

(ii) Write down Newton's law for the second particle and deduce that the extension ξ satisfies the equation for a damped harmonic oscillator

$$\ddot{\xi} + 2\gamma\dot{\xi} + \omega_0^2\xi = 0,$$

where $\omega_0 = \sqrt{2k}$ is the natural frequency.

(iii) Show that, if one replaces the coordinate distance in the expression for F by the proper distance (given by (21.153)), then the equation for the first particle in inertial coordinates (which to this order of accuracy agree with the TT-coordinates) is given by

$$\frac{d^2 y_1}{dt^2} = k(L(t) - L) + \gamma\frac{d}{dt}(L(t) - L),$$

where

$$L(t) = \int_{y_1(t)}^{y_2(t)} [1 - h_{22}(t)]^{1/2}\, dt.$$

Calculate the corresponding equation for the second particle and deduce that (if one ignores terms of $O(h^2)$)

$$\ddot{\xi} + 2\gamma\dot{\xi} + \omega_0^2\xi = \tfrac{1}{2}L\ddot{h}_{22}.$$

21.21 (§21.13) Show that an h_{22} wave with metric (21.166) when rotated through an angle $-\theta$ in the xz-plane i.e.

$$x \rightarrow x\cos\theta - z\sin\theta,$$
$$z \rightarrow x\sin\theta - z\cos\theta,$$

has metric

$$ds^2 = dt^2 - (1 - \varepsilon\sin^2\theta h_{22})dx^2 + 2\varepsilon\cos\theta\sin\theta h_{22}dxdz$$
$$- (1 - \varepsilon h_{22})dy^2 - (1 - \varepsilon\sin^2\theta h_{22})dz^2,$$

where

$$h_{22} = h_{22}(t - z\cos\theta - x\sin\theta).$$

Use the method in §21.12 by considering a null geodesic along the x-axis to show that the analogue of equation (21.173) is

$$T = t_0 + 2L - \tfrac{1}{2}\varepsilon\int_0^L h_{22}(t_0 + x(1+\sin\theta))dx - \tfrac{1}{2}\varepsilon\int_0^L h_{22}(t_0 + x(1+\sin\theta))dx$$

Differentiate with respect to t_0 and set $t_0 = t$ to obtain equation (21.179).

Further reading

Textbooks that deal with gravitational radiation at the level of this book include those by Carroll (2004), Hartle (2003), and Schutz (1985). The classic text by Misner, Thorne and Wheeler (1973) goes into most detail and includes linearized solutions about a general metric (rather than just about Minkowski space). The book by our Southampton colleague Andersson (2020) deals with gravitational wave astronomy. The book by Synge (1960) describes the process of differential chronometry and how to construct a 'five-point curvature detector' to measure the curvature of space-time using light signals.

Abbott, B. P. et al. (2016). Observation of Gravitational Waves from a Binary Black Hole Merger. *Physical Review Letters*, **116**, 061102.

Abbott, B. P. et al. (2017). GW170814: A Three-Detector Observation of Gravitational Waves from a Binary Black Hole Coalescence. *Physical Review Letters*, **119**, 141101.

Andersson N. (2020) Gravitational-wave astronomy: Exploring the dark side of the universe. Oxford University Press, Oxford.

Blanchet, L. (2014) Gravitational Radiation from Post-Newtonian Sources and Inspiralling Compact Binaries, Living Reviews in Relativity 17, 2. https://link.springer.com/article/10.12942/lrr-2014-2

Carroll, S. M. (2004) Spacetime and Geometry: an introduction to general relativity, Addison Wesley, San Francisco.

Hartle, J. B. (2003) Gravity: an introduction to Einstein's general relativity, Addison Wesley, San Francisco.

Kokkotas, K. D. and Schmidt, B. G. (1999) *Quasi-Normal Modes of Stars and Black Holes*, Living Reviews in Relativity 2, 2 https://link.springer.com/article/10.12942/lrr-1999-2

Misner, C. W., Thorne, K. S., and Wheeler, J. A. (1973). Gravitation, Freeman, San Francisco.

Moore, C. J., Cole, R. H. and Berry, C. P. L. (2015). Gravitational-wave sensitivity curves. *Classical and Quantum Gravity*, **32**, 015014.

Schutz, B. F. (1985). A first course in general relativity, Cambridge University Press.

Synge, J. L. (1960) Relativity: The general theory. North-Holland, Amsterdam.

Tartaglia, A., Belfi J., Beverini, N., Di Virgilio, A., Ortolan, A., Porzio, A. and Ruggiero, M. L. (2014). Light and/or atomic beams to detect ultraweak gravitational effects. *EPJ Web of Conferences*, **74**, 3001.

Weisberg, J. M., Nice, D. J. and Taylor J. H. (2010). Timing measurements of the relativistic binary pulsar PSR B1913+16. *Astrophysical Journal*, **722**, 1030.

Exact gravitational waves 22

22.1 Gravitational waves and symmetries

In this chapter, rather than looking at solutions of the linearized Einstein equations, we will look at **exact solutions** of the full vacuum Einstein equations that represent gravitational waves. In looking for exact solutions, it is useful to assume that the metric has a high degree of symmetry. The simplest case is to assume that the space-time is spherically symmetric. However, it follows from Birkhoff's theorem that a spherically symmetric vacuum solution in the exterior region is necessarily static, so it cannot produce gravitational waves. Spherically symmetric metrics have three independent spacelike Killing vectors that generate rotations and have an invariant 2-space. The next simplest assumption is that there exist two independent spacelike Killing vectors V and W which have an invariant 2-space. Since the orbits of the Killing vectors are surface forming, this means that they commute (see equation (10.15)), so that

$$[V, W] = 0. \tag{22.1}$$

This gives rise to two important examples of exact solutions describing gravitational radiation, which depend on the topology of the integral curves. If the integral curves of V and W have topology S^1 and $\mathbb{R}$, respectively, then the space-time has **cylindrically symmetric** gravitational waves, which we discuss in §22.2, whereas, if the integral curves of both Killing vectors have the topology of the real line, then we have **plane symmetric** gravitational waves, which we discuss in §22.3.

22.2 Einstein-Rosen waves

In this section, we consider a cylindrically symmetric vacuum space-time with two commuting spacelike Killing vectors V and W, as described above. Since the integral curves of V have topology S^1, we may associate V with a 2π-periodic coordinate ϕ and write $V \overset{*}{=} \partial/\partial\phi$ so that it generates a rotation. Similarly, the integral curves of W are $\mathbb{R}$ so we may introduce coordinates so that it generates a translation along the z-axis and write $W \overset{*}{=} \partial/\partial z$. Finally, we introduce coordinates t and ρ such that the invariant 2-space is given by $t = $ constant and $\rho = $ constant. In this way, we introduce cylindrically symmetric coordinates $(x^0, x^1, x^2, x^3) = (t, \rho, \phi, z)$, adapted to the Killing vectors such that, in these coordinates, the metric coefficients only depend on t and ρ, so $g_{ab} = g_{ab}(t, \rho)$, and $\partial/\partial\phi$ and $\partial/\partial z$ are the Killing vectors. We now make the further assumption that the

Introducing Einstein's Relativity. Second Edition. Ray d'Inverno and James Vickers, Oxford University Press.
© Ray d'Inverno and James Vickers (2022). DOI: 10.1093/oso/9780198862024.003.0022

metric is invariant under $z \to -z$ and also invariant under $\phi \to -\phi$, which is called by some authors 'whole cylinder symmetry' (e.g. Thorne 1965). This amounts to the condition that, in adapted coordinates, there are no $d\phi dx^a$ cross-terms and no $dz dx^a$ cross-terms. In other words, the metric takes the form

$$ds^2 = g_{AB}(t,\rho)dx^A dx^B - C(t,\rho)d\phi^2 - D(t,\rho)dz^2, \text{ where } A, B = 0, 1,$$

and $C(t,\rho)$ and $D(t,\rho)$ are positive functions. We now introduce new coordinates $\tilde{t}, \tilde{\rho}$ such that

$$\tilde{t} = t,$$
$$\tilde{\rho} = C(t,\rho)^{1/2}D(t,\rho)^{1/2},$$

and define

$$\psi(\tilde{t}, \tilde{\rho}) = \tfrac{1}{2}\ln D(\tilde{t}, \rho(\tilde{\rho}, \tilde{t})).$$

Then, in these coordinates, the metric takes the form

$$ds^2 = \tilde{g}_{AB}d\tilde{x}^A d\tilde{x}^B - \tilde{\rho}^2 e^{-2\psi}d\phi^2 - e^{2\psi}dz^2, \qquad (22.2)$$

where $\tilde{g}_{AB} = \tilde{g}_{AB}(\tilde{t}, \tilde{\rho})$ and $\psi = \psi(\tilde{t}, \tilde{\rho})$. We now look at the metric $d\sigma^2$ of the invariant 2-surfaces given by $\tilde{t} = \text{constant}$ and $\tilde{\rho} = \text{constant}$,

$$d\sigma^2 = \tilde{g}_{00}d\tilde{t}^2 + 2\tilde{g}_{01}d\tilde{t}d\tilde{\rho} + \tilde{g}_{11}d\tilde{\rho}^2. \qquad (22.3)$$

Since this is a 2-dimensional metric, by the second theorem in §6.13, it is conformally flat and we can introduce new coordinates (t', ρ') such that

$$d\sigma^2 = E(t', \rho')(dt'^2 - d\rho'^2).$$

Writing $E(t', \rho')$ in the form

$$E(t', \rho') = e^{2\gamma - 2\psi},$$

and (for convenience) dropping the primes, we see that we may write the general cylindrically symmetric metric in the form

$$ds^2 = e^{2\gamma - 2\psi}(dt^2 - d\rho^2) - \rho^2 e^{-2\psi}d\phi^2 - e^{2\psi}dz^2, \qquad (22.4)$$

where $\psi = \psi(t,\rho)$ and $\gamma = \gamma(t,\rho)$. This is called the **Einstein-Rosen** metric.

If one now calculates the vacuum Einstein equations for this metric, one finds that these are simply given by

$$\frac{\partial^2 \psi}{\partial t^2} - \frac{1}{\rho} \frac{\partial}{\partial \rho} \left(\rho \frac{\partial \psi}{\partial \rho} \right) = 0, \tag{22.5}$$

$$\frac{\partial \gamma}{\partial \rho} = \rho \left[\left(\frac{\partial \psi}{\partial \rho} \right)^2 + \left(\frac{\partial \psi}{\partial t} \right)^2 \right], \tag{22.6}$$

$$\frac{\partial \gamma}{\partial t} = 2\rho \frac{\partial \psi}{\partial \rho} \frac{\partial \psi}{\partial t}. \tag{22.7}$$

Equation (22.5) is nothing but the standard wave equation in cylindrical symmetry. The method of solution is therefore to solve (22.5) for ψ and then use (22.6) and (22.7) to determine γ. Such solutions are called **Einstein-Rosen waves**. These solutions were found by Einstein and Rosen in 1937 and were the first exact solutions describing the propagation of gravitational waves. Note that one requires the integrability condition $\gamma_{,\rho t} = \gamma_{,t\rho}$ in order to solve (22.6) and (22.7) for γ (exercise). However, this is automatically satisfied since differentiating the RHS of (22.6) with respect to t is the same as differentiating the RHS of (22.7) with respect to ρ by virtue of (22.5).

Using the method of separation of variables, solutions of the cylindrical wave equation can be given by a superposition of solutions of the form

$$\psi = A \mathcal{J}_0(\omega\rho) \cos(\omega t) + B N_0(\omega\rho) \sin(\omega t), \tag{22.8}$$

where $\mathcal{J}_0$ and N_0 are Bessel functions of the first and second kind. Since the Bessel function of the second kind diverges at the origin, we take $B = 0$, and (22.6) and (22.7) can then be integrated to give γ. However, despite finding such solutions, Einstein and Rosen did not believe that they were physical, due to the presence of (what turned out to be coordinate) singularities.

A more interesting solution due to Webber and Wheeler (1957) is to superpose such solutions to create a wave with the profile of a pulse:

$$\psi = 2C \int_0^\infty e^{a\omega} \mathcal{J}_0(\omega\rho) \cos(\omega t) \mathrm{d}\omega$$

$$= C \left[(a - it)^2 + \rho^2 \right]^{-1/2} + C \left[(a + it)^2 + \rho^2 \right]^{-1/2}, \tag{22.9}$$

where the final expression is real since the second term is just the complex conjugate of the first. Integrating (22.9) gives

$$\gamma = \tfrac{1}{2} C^2 \left\{ a^{-2} - \rho^2 \left[(a - it)^2 + \rho^2 \right]^{-2} - \rho^2 \left[(a + it)^2 + \rho^2 \right]^{-2} \right.$$

$$\left. -a^{-2}(t^2 + a^2 - \rho^2) \left[t^4 + 2t^2(a^2 - \rho^2) + (a^2 + \rho^2) \right]^{-1/2} \right\}.$$

In analysing this solution, Webber and Wheeler showed that the solution was regular and carried energy, contrary to the initial views of Einstein and Rosen.

22.3 Exact plane wave solutions

In this section, we consider plane symmetric gravitational waves possessing two spacelike Killing vectors whose integral curves are the real line. These can be thought of as translations along the y- and z-axes so that $V \overset{*}{=} \partial/\partial y$ and $W \overset{*}{=} \partial/\partial z$. Our starting point is the linearized solution given by an h_{22} wave. If we introduce double null coordinates defined by

$$u = t - x, \quad v = t + x,$$

then an h_{22}-wave, given by (21.55), has a line element of the form

$$ds^2 = du dv - f^2(u) dy^2 - g^2(u) dz^2, \tag{22.10}$$

where

$$f^2(u) = 1 - \varepsilon h_{22}(u), \quad g^2(u) = 1 + \varepsilon h_{22}(u). \tag{22.11}$$

The functions are squared to ensure the correct signature (which is justified in the linearized approximation by assuming that ε is small in (22.11)).

Let us now choose (22.10) as an ansatz and plug this line element into the full vacuum field equations to see if we can solve them. We find that the non-vanishing components of the connection are (exercise)

$$\Gamma_{22}^1 = 2ff', \quad \Gamma_{33}^1 = 2gg', \quad \Gamma_{02}^2 = f'/f, \quad \Gamma_{03}^3 = g'/g, \tag{22.12}$$

where a prime denotes differentiation with respect to u. The Riemann tensor has two independent components,

$$R_{0202} = ff'', \qquad R_{0303} = gg''$$

and there is only one vacuum field equation, namely,

$$f''/f + g''/g = 0. \tag{22.13}$$

Let us denote the first term by the function $h(u)$, i.e.

$$f''/f = h. \tag{22.14}$$

Then the field equation will be satisfied if g is chosen so that

$$g''/g = -h. \tag{22.15}$$

These last two equations determine f and g in terms of $h(u)$ up to constants of integration. Hence any choice of the arbitrary function $h(u)$ gives rise to a vacuum solution. Such exact solutions are called **linearly polarized plane gravitational waves**. They represent plane-fronted gravitational waves, abstracted away from any sources, propagating in the x-direction.

The form of the line element (22.10) is essentially that due originally to **Rosen**. If we carry out the coordinate transformation

$$U = u, \quad V = v + y^2 ff' + z^2 gg', \quad Y = fy, \quad Z = gz, \qquad (22.16)$$

then the line element is transformed into the **Brinkmann** form (exercise)

$$ds^2 = h(U)(Z^2 - Y^2)dU^2 + dUdV - dY^2 - dZ^2, \qquad (22.17)$$

which shows the explicit dependence on the freely specifiable function h. This function can be shown to represent the amplitude of the polarized wave.

Although such solutions are highly unphysical, being infinite in extent, it may be hoped that they represent some of the properties of real waves from bounded sources in some far-zone limit. In particular, they allow us to investigate the question of the scattering of gravitational waves. For, unlike electromagnetic theory, where the linearity of the theory means that electromagnetic waves pass through each other unaltered, there is, in general, no superposition principle in general relativity. Indeed, we may expect the non-linearity of the theory to reveal itself in the interaction of two gravitational waves. However, (22.17) does reveal a limited superposition principle in that two plane waves moving in the **same** direction can be superposed simply by adding their corresponding h functions. Thus, when moving in the same direction, two such gravitational waves do not scatter one another. To exhibit scattering, we need two waves moving in different directions. If we consider two linearly polarized waves colliding at an angle, we can always find a class of observers who consider the collision to be head on (see e.g. Exercise 4.10). Hence, it is sufficient to work in a coordinate system in which the waves appear to collide head on. We shall consider this question in the limited case of impulsive gravitational waves, which we discuss next.

22.4 Impulsive plane gravitational waves

We start with a mathematical digression. The Heaviside step function $\theta(u)$ is defined by

$$\theta(u) = \begin{cases} 0 & \text{if } u \leqslant 0, \\ 1 & \text{if } u > 0. \end{cases} \qquad (22.18)$$

It is closely related to the Dirac delta function $\delta(u)$. Strictly speaking, δ is not a function but rather a distribution and lives under an integral sign. It will be sufficient for our purposes to define δ by the requirements

$$\delta(u) = 0, \quad \text{if } u \neq 0, \qquad (22.19)$$

$$\int_{-\infty}^{\infty} f(u)\delta(u)du = f(0), \qquad (22.20)$$

for any suitably defined function $f(u)$. Then, with these definitions, we can establish the results (exercise)

$$\theta'(u) = \delta(u), \tag{22.21}$$

$$u\delta(u) = 0, \tag{22.22}$$

$$u\theta'(u) = 0. \tag{22.23}$$

We now consider a line element in the Rosen form defined by

$$f(u) = 1 + u\theta(u), \qquad g(u) = 1 - u\theta(u). \tag{22.24}$$

Then we find, using the above results, that (exercise)

$$f' = -g' = \theta(u), \quad f'' = -g'' = \delta(u), \quad f''/f = -g''/g = \delta(u), \tag{22.25}$$

which means, from (22.13), that (22.24) gives rise to a plane wave. Hence, the Ricci and Einstein tensors vanish, but the Riemann tensor (or, since the solution is vacuum, equivalently the Weyl tensor) does not vanish, having non-vanishing components

$$R_{0202} = -R_{0303} = \delta(u). \tag{22.26}$$

The solution has delta functions in the curvature and hence is non-flat only when $u = 0$. This can be seen more clearly in the Brinkmann form of the solution, which, from (22.14), is obtained by setting

$$h(U) = \delta(U). \tag{22.27}$$

Hence, for $u = U \neq 0$, the line element reduces to

$$ds^2 = dU dV - dY^2 - dZ^2, \tag{22.28}$$

which is Minkowski space-time in double null coordinates. The hypersurface $u = 0$, where the field is concentrated, thus separates two flat regions. It represents a plane wave similar to that of Fig. 22.1, except that now there is just one wave front (Fig. 22.1). Such a solution is called a **shock wave** or **impulsive plane gravitational wave**. Figure 22.2 is a space-time picture (with two dimensions suppressed) of such a solution.

We define a **sandwich wave** to be a non-flat vacuum solution bounded by plane hypersurfaces outside of which the solution is flat (Fig. 22.3). An observer moving on a geodesic will 'feel' the wave passing for a finite period when moving from region I through region II and out into region III. Neighbouring test particles will be accelerated transversely to the direction of propagation of the wave. Then an impulsive gravitational wave can be viewed as a thin sandwich wave in a suitable limit as the thickness goes to zero. Although impulsive waves are yet another idealization, they do prove easier to work with than more general waves at first.

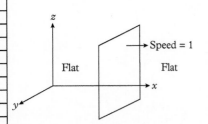

Fig. 22.1 Spatial picture of an impulsive plane gravitational wave.

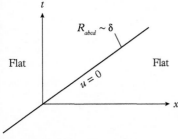

Fig. 22.2 Space-time picture (two dimensions suppressed) of an impulsive plane gravitational wave.

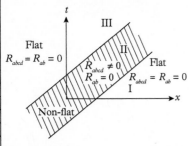

Fig. 22.3 A space-time picture of a sandwich wave.

22.5   Colliding impulsive plane gravitational waves

In the next two sections, we shall outline the pioneering work of Penrose, Khan, and Szekeres on the important problem of colliding plane gravitational waves. We start by generalizing the Rosen form (22.10) to the form

$$ds^2 = \ell\, du\, dv - f^2 dy^2 - g^2 dz^2, \tag{22.29}$$

where $\ell, f,$ and g are now functions of both u and v. This form then allows us to incorporate waves moving in both directions. The explicit vacuum solution of Penrose and Khan is then given by

$$\ell = \frac{m^3}{rw(pq+rw)^2},\ f^2 = m^2\left(\frac{r+q}{r-q}\right)\left(\frac{w+p}{w-p}\right),\ g^2 = m^2\left(\frac{r-q}{r+q}\right)\left(\frac{w-p}{w+p}\right),$$

where

$$p = u\theta(u),\quad q = v\theta(v),\quad r = (1-p^2)^{\frac{1}{2}},\quad w = (1-q^2)^{\frac{1}{2}},\quad m = (1-p^2-q^2)^{\frac{1}{2}}.$$

The space-time diagram is shown in Fig. 22.4.
The solution is only valid in the four regions:

$$\begin{aligned}
&\text{I.}\quad u < 0,\quad v < 0,\\
&\text{II.}\quad 0 < u < 1,\quad v < 0,\\
&\text{III.}\quad u < 0,\quad 0 < v < 1,\\
&\text{IV.}\quad u > 0,\quad v > 0,\quad u^2 + v^2 < 1.
\end{aligned}$$

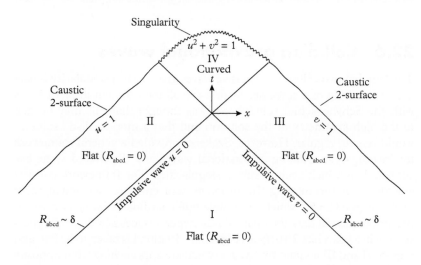

Fig. 22.4 Penrose and Khan space-time picture of two colliding impulsive plane waves.

Regions I, II, and III are flat, and region IV is curved. Region I is separated from region II by an incoming impulsive wave and from region III by another impulsive wave travelling in the opposite direction. They collide at the origin in the figure, and then region IV represents the interaction region between them. If we consider the world-line of the observer $x = 0$, then the two waves collide at $t = 0$, scatter each other, and leave a curved region between them, which, in finite proper time according to the observer, develops a **curvature singularity**. (This is an intrinsic singularity in the usual sense that scalar-invariants in the curvature tensor blow up.) There is also a coordinate singularity in region II at $u = 1$, $v < 0$, and an analogous one in region III at $v = 1$, $u < 0$. These singularities are, in fact, topological singularities, sometimes called fold singularities, and are in this case **caustic 2-surfaces** caused by each wave focusing the other, i.e. they are surfaces where the null geodesics cross. They are not intrinsic curvature singularities. The space-time diagram (Fig. 22.4) is a bit misleading at first sight, since you might think it possible for an observer in region II to cross $u = 1$ and escape. However, the caustic surface is just a 'seam' in the hypersurface $v = 0$, and so the chances of hitting it are remote, and, anyway, any observer getting close will be swept up into region III and end up on the singularity. There is a finite jump in the curvature tensor at $u = 0$, $0 < v < 1$, and at $v = 0$, $0 < u < 1$ (sometimes called a step wave), in addition to the delta function there. Furthermore, inspection of the solution reveals that the waves no longer have planar symmetry after impact.

To summarize, two impulsive plane gravitational waves approaching each other from different directions scatter each other and cease to be plane waves. Eventually, the focusing effect of each wave on the other results in the formation of a spacelike curvature singularity and, whereas timelike singularities are avoidable, spacelike singularities are not.

22.6 Colliding gravitational waves

The fact that two colliding impulsive waves give rise to a singularity is perhaps something of a surprise. At first (recall the situation in black holes with the Schwarzschild solution), it was thought that this may be due to the high symmetry of the solution and that a more realistic solution would remain regular. However, Szekeres provided a general framework for investigating colliding gravitational waves and discovered some exact solutions which again result in singularities. The framework consists essentially of formulating the problem as a characteristic initial value problem (see §23.5), which, in double null coordinates (u, v), consists of prescribing initial data on a pair of null hypersurfaces $u = 0$, $v = 0$, intersecting in a spacelike 2-surface (Fig. 22.5). Region I is taken to be flat, and regions II and III contain two waves which are approaching from opposite directions. Region IV is then the interaction region of the two waves. The problem is well posed in that it can be shown that any given initial data gives rise to a unique solution in region IV. It is convenient to assume that two commuting spacelike Killing vectors $\partial/\partial y$ and $\partial/\partial z$ exist throughout

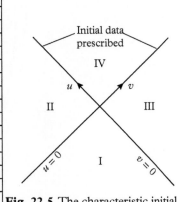

Fig. 22.5 The characteristic initial value problem for colliding waves.

the whole space-time. Szekeres shows that coordinates of the Rosen type exist in which the metric takes on the form

$$ds^2 = e^{-M}dudv - e^{-U}(e^V \cosh Wdy^2 - 2\sinh Wdydz + e^{-V}\cosh Wdz^2),$$

where M, U, V, and W are functions of u and v in general. However, in region II, the functions M, U, V, and W depend on u only; in region III, they depend on v only. If the waves have constant and parallel polarizations, then it can be shown that one can put $W = 0$ globally, and the solution to the initial value problem reduces to a one-dimensional integral for V and two quadratures for M.

Szekeres considered the more realistic case of sandwich waves in regions II and III and again found that they give rise to singularities in region IV. Since the early 1970s, when this work was first reported, there have been a large number of exact solutions found for colliding gravitational waves, including plane gravitational waves and waves coupled with electromagnetic waves, perfect fluids, and null dust (i.e. an energy-momentum tensor of the form (12.1) but where the 4-vector u^a is null). Indeed, there has been considerable controversy over what happens when two planar impulsive gravitational waves, each followed by a distribution of null dust, collide. Do the two distributions pass through each other or do they mix magically to produce a perfect fluid with a 'stiff' equation of state $p = \rho$? These ambiguities make it clear that these sorts of problems, which are a mixture of initial value and boundary value problems, need careful handling and that particular attention needs to be paid to the physical interpretation of the resulting solutions.

Exercises

22.1 (§22.2) Verify that if γ satisfies (22.6) and (22.7) then $\gamma_{,\rho t} = \gamma_{,t\rho}$ by virtue of (22.5).

22.2 (§22.2) Verify that

$$\psi = \left[\rho^2 - (t + ia)^2\right]^{-1/2},$$

is a complex solution of the wave equation (22.5). Deduce that (22.9) is a real solution of the wave equation in cylindrical coordinates.

22.3 (§22.3) Show that the line element (22.10) leads to (22.12) and (22.13).

22.4 (§22.3) Show that (22.16) transforms vacuum solutions in the Rosen form into the Brinkmann form (22.17). What is the inverse form of (22.16)?

22.5 (§22.3) Some authors write the Rosen line element with a 2 in front of the first term, i.e.

$$\mathrm{d}s^2 = \bar{g}_{ab}\mathrm{d}\bar{x}^a\mathrm{d}\bar{x}^b = 2\mathrm{d}\bar{u}\mathrm{d}\bar{v} - \bar{f}^2(\bar{u})\mathrm{d}\bar{y}^2 - \bar{g}^2(\bar{v})\mathrm{d}\bar{z}^2.$$

(i) Show that if

$$\bar{u} = (1/\sqrt{2})u, \quad \bar{v} = (1/\sqrt{2})v, \quad \bar{y} = y, \quad \bar{z} = z,$$

then the line element reduces to the Rosen form (22.10).
(ii) Show that if

$$\bar{u} = u, \quad \bar{v} = v, \quad \bar{y} = \sqrt{2}y, \quad \bar{z} = \sqrt{2}z,$$

then $\bar{g}_{ab} = 2g_{ab}$, where g_{ab} is the Rosen metric (22.10), and deduce that $\bar{g}_{ab}$ gives rise to the same connection, Ricci, and Einstein tensors as g_{ab} does.

22.6 (§22.4) Show that the definitions (22.19) and (22.20) lead to the results (22.21), (22.22), and (22.23). [Hint: use integration by parts to establish (22.21).] Deduce that $u\delta'(u) = -\delta(u)$.

22.7 (§22.4) Show that (22.24) leads to (22.25), (22.26), and (22.27).

Further reading

For a general reference on exact solutions, see the book by Kramer et al. (2009). The paper by Thorne (1965) gives a good discussion of Einstein–Rosen waves, and the book by Griffiths (1991) gives a detailed discussion of both these and plane wave solutions.

Griffiths, J. (1991). *Colliding Waves in General Relativity*. Oxford University Press, Oxford.

Kramer, D., Stephani, H., Herlt, E., and MacCallum, M. A. H. (2009). *Exact Solutions of Einstein's Field Equations* (2nd edn). Cambridge University Press, Cambridge.

Thorne, K. S. (1965). Energy of infinitely long, cylindrically symmetric systems in general relativity. *Physical Review*, *138*, B251.

Radiation from an isolated source

23

23.1 Radiating isolated sources

The extent to which the results of the linearized theory can be trusted is
not clear. The non-linearity of the gravitational field is one of its most
characteristic properties, and it is likely that at least some of the crucial properties of the field should show themselves through the non-linear
terms. Indeed, we have met exact solutions of the Einstein vacuum field
equations corresponding to plane gravitational waves and we have seen
that superposition of them leads to the creation of intrinsic singularities. This result is certainly absent in the linear case, so clearly there are
differences. However, even these solutions are global vacuum solutions
abstracted away from sources and, as such, are physically unrealistic, even
if they may give us important information about how waves behave in
asymptotic regions. What we would really like to do is to be able to investigate gravitational waves from bounded isolated sources, since then
we would be in a position to discuss **energy transfer** and it is this which
determines whether or not gravitational waves behave in the same way
as other forms of radiation. Such a model system consists of an isolated
bounded source which has been quiescent for a semi-infinite period, then
radiates for a finite time, and afterwards becomes quiescent again. If the
resulting waves are real physical waves, in that they carry energy, then
we might expect the source to lose mass (and possibly other multipole
moments may change) in the process.

As discussed in the previous chapter the simplest field due to a
bounded isolated source is spherically symmetric but such solutions cannot emit waves. In the previous chapter, we looked at solutions with two
commuting spacelike Killing vectors which gave rise to cylindrically symmetric and plane gravitational waves. The next simplest assumption is to
consider a system admitting just one spacelike Killing vector field together
possibly with discrete reflection symmetries. This, indeed, was the starting point of Bondi in his pioneering work on gravitational radiation in the
early 1960s in which he considered a source which is axially symmetric
and non-rotating. The symmetry assumptions are that the solution has a
hypersurface orthogonal rotational Killing vector $\partial/\partial\phi$ so that the metric
is invariant under

$$\phi \to \phi' = \phi + \text{constant}, \tag{23.1}$$

$$\phi \to \phi' = -\phi, \tag{23.2}$$

where the reflection symmetry (23.2) prohibits the solution from rotating (why?). Although these assumptions simplify things somewhat, the

Introducing Einstein's Relativity. Second Edition. Ray d'Inverno and James Vickers, Oxford University Press.
© Ray d'Inverno and James Vickers (2022). DOI: 10.1093/oso/9780198862024.003.0023

mathematics is still quite difficult and ultimately recourse has to be made to asymptotic approximation methods to discuss the radiation. It is also possible to employ the additional reflection symmetry in the equatorial plane, namely,

$$\theta \rightarrow \theta' = \pi - \theta, \tag{23.3}$$

but this does not lead to any great simplification and so we shall omit it. We shall return to the definition of the other coordinates in §23.3.

We therefore consider an axially symmetric non-rotating bounded isolated source which is initially static, radiates for a finite period (for example by pulsating axially symmetrically), and subsequently returns to a static configuration (Fig. 23.1). This model assumes that, once a system has radiated, it is possible subsequently for it to become quiescent again. One might expect the non-linearity to cause the waves to interfere, backscatter, and so excite the source, causing it to radiate indefinitely. This is a delicate problem, which is outside the scope of this book, and so, following Bondi, we shall assume a quiescent model is possible and restrict ourselves to outlining the proof of the mass-loss result in this case. We start by considering the surfaces which act as wave fronts in the theory.

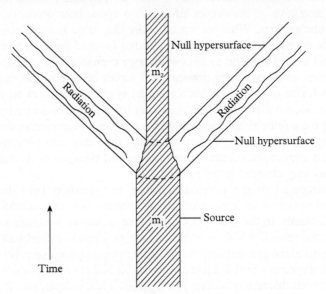

Fig. 23.1 Bondi mass loss: $m_2 < m_1$.

23.2 Characteristic hypersurfaces of Einstein's equations

The field equations of general relativity form a system of hyperbolic partial differential equations. This is most easily seen in the linearized approximation, where, in an appropriate gauge, the equations are simply wave equations. As Bondi has pointed out, hyperbolic equations are very different in character to elliptic or parabolic equations since they allow for 'time-bomb' solutions, that is, solutions which are initially static but then suddenly become dynamic. Such solutions propagate their effects along privileged curves called the **bicharacteristics** of the theory. Moreover, these bicharacteristics lie on privileged surfaces called **characteristic hypersurfaces**, which play the role of **wave fronts** in the propagation of these effects. Along characteristic hypersurfaces, different solutions can meet continuously and, as a consequence, they are defined as those singular hypersurfaces for which the usual Cauchy initial value problem **cannot** be solved.

To find the characteristic hypersurfaces for the vacuum field equations, recall that, in considering the Cauchy problem, we obtained the evolution equations in the form (13.23), namely,

$$g^{00} g_{\alpha\beta,00} = 2M_{\alpha\beta}.$$

Thus, we would be unable to solve for $g_{\alpha\beta,00}$ if and only if $g^{00} = 0$. As we have seen in §17.1, this is the condition for the hypersurface $x^0 = $ constant to be a **null hypersurface**. The normal vector to such a hypersurface is null and, consequently, it is also tangent to the hypersurface. Thus, a null hypersurface is a hypersurface that is locally tangent to the light cone (Fig. 23.2). Not only are null hypersurfaces characteristic surfaces, but they are ruled by **null geodesics** which turn out to be the **bicharacteristics** of the theory (see §23.3). This makes clearer the idea we met in the linearized theory, namely, that gravitational disturbances are propagated along null geodesics with the speed of light. It is clear from these considerations that null hypersurfaces play an important role in the study of gravitational radiation.

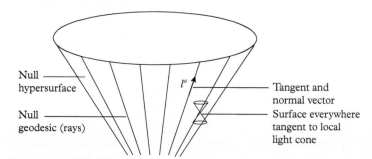

Fig. 23.2 A null hypersurface.

23.3 Radiation coordinates

The discussion of the last section suggests that, in order to investigate radiation, we should introduce the coordinate hypersurfaces

$$x^0 = u = \text{constant}, \tag{23.4}$$

as a family of non-intersecting null hypersurfaces. The normal covariant vector field to these surfaces is therefore

$$\ell_a = \partial_a u = (1, 0, 0, 0) = \delta_a^0, \tag{23.5}$$

and, since it is null,

$$\ell_a \ell^a = g^{ab} \partial_a u \partial_b u = 0, \tag{23.6}$$

and the vector field is both tangent and normal to the null hypersurfaces. The **bicharacteristics** are the integral curves of the contravariant vector field ℓ^a, that is, they have equation

$$x^a = x^a(\rho) \tag{23.7}$$

for some parameter ρ, where

$$\frac{\mathrm{d}x^a}{\mathrm{d}\rho} = \ell^a = g^{ab} \partial_b u. \tag{23.8}$$

Then, taking the absolute derivative of (23.8), we get

$$\begin{aligned}
\frac{\mathrm{D}}{\mathrm{D}\rho}\left(\frac{\mathrm{d}x^a}{\mathrm{d}\rho}\right) &= \frac{\mathrm{D}}{\mathrm{D}\rho}(g^{ab}\partial_b u) \\
&= \frac{\mathrm{d}x^c}{\mathrm{d}\rho}\nabla_c(g^{ab}\partial_b u) \\
&= g^{ab}\frac{\mathrm{d}x^c}{\mathrm{d}\rho}(\nabla_c \partial_b u) \\
&= g^{ab}\frac{\mathrm{d}x^c}{\mathrm{d}\rho}(\nabla_b \partial_c u) \\
&= g^{ab}g^{cd}\partial_d u(\nabla_b \partial_c u) \\
&= \tfrac{1}{2}g^{ab}\nabla_b(g^{cd}\partial_c u \partial_d u) \\
&= 0,
\end{aligned} \tag{23.9}$$

using the symmetry of the connection in the fourth equality, and (23.6) in the last. Hence, the bicharacteristics are **null geodesics** and ρ is an **affine parameter**. These null geodesics are often called **null rays**.

We choose as a second coordinate

$$x^1 = r, \tag{23.10}$$

where r is some radial parameter along the null rays, and we then use the remaining coordinates x^2 and x^3 to label the null rays. Assuming that space-time is asymptotically flat, that is,

$$\lim_{r \to \infty} g_{ab} = \eta_{ab}, \tag{23.11}$$

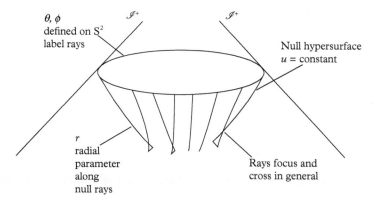

Fig. 23.3 Bondi's radiation coordinates (u, r, θ, ϕ).

we can then take x^2 and x^3 to be the usual spherical polar angles

$$x^2 = \theta, x^3 = \phi, \qquad (23.12)$$

defined on each 2-sphere (u = constant, $r = \infty$) at future null infinity $\mathscr{I}^+$. These coordinates are called **Bondi** or **radiation coordinates**. They are really only defined in a neighbourhood of $\mathscr{I}^+$ because, if we follow the null rays back into the interior, the gravitational field will cause them to focus and cross in general (Fig. 23.3). However, we shall ultimately be working asymptotically and so the coordinate system will be adequate for our needs.

23.4 Bondi's radiating metric

A null ray is one of the coordinate curves

$$u = u_0, \theta = \theta_0, \phi = \phi_0,$$

where u_0, θ_0, and ϕ_0 are constants, and r is varying. The tangent vector to this curve is

$$\frac{\mathrm{d}x^a}{\mathrm{d}r} = (0, 1, 0, 0) = \delta_1^a,$$

and so it must be parallel to ℓ^a, that is, $\ell^a = \lambda \delta_1^a$ for some proportionality factor λ. But, by (23.8) and (23.5),

$$\ell^a = g^{ab} \partial_b u = g^{ab} \delta_b^0 = g^{0a},$$

from which we get

$$g^{00} = g^{02} = g^{03} = 0. \qquad (23.13)$$

These conditions on the contravariant metric are equivalent to the conditions on the covariant metric (exercise)

$$g_{11} = g_{12} = g_{13} = 0. \tag{23.14}$$

Newman and Penrose, in work on gravitational radiation subsequent to Bondi, took x^1 to be an affine parameter, in which case (exercise) $\lambda = 1$ and

$$g_{01} = g^{01} = 1. \tag{23.15}$$

However, Bondi chose $x^1 = r$ to be a **luminosity distance parameter** defined by requiring

$$\begin{vmatrix} g_{22} & g_{23} \\ g_{23} & g_{33} \end{vmatrix} = r^4 \sin^2 \theta. \tag{23.16}$$

The significance of this choice is that the 2-surfaces, $u = $ constant, $r = $ constant, have the usual surface area of a 2-sphere, namely, $4\pi r^2$.

We next impose the symmetry assumptions of **axial symmetry** (23.1), which results in (exercise)

$$\frac{\partial g_{ab}}{\partial \phi} = 0, \tag{23.17}$$

and **azimuth reflection invariance** (23.2), which results in (exercise)

$$g_{03} = g_{13} = g_{23} = 0, \tag{23.18}$$

or, equivalently,

$$g^{03} = g^{13} = g^{23} = 0. \tag{23.19}$$

Putting all these assumptions together, we can write the metric in the particular form of **Bondi's radiating metric** (exercise)

$$ds^2 = \left(\frac{V}{r}e^{2\beta} - U^2 r^2 e^{2\gamma}\right) du^2 + 2e^{2\beta} du dr + 2U r^2 e^{2\gamma} du d\theta$$
$$- r^2 (e^{2\gamma} d\theta^2 + e^{-2\gamma} \sin^2 \theta d\phi^2), \tag{23.20}$$

where V, U, β, and γ are four arbitrary functions of the three coordinates u, r, and θ by (23.17), that is,

$$V = V(u,r,\theta), \;\; U = U(u,r,\theta), \;\; \beta = \beta(u,r,\theta), \;\; \gamma = \gamma(u,r,\theta). \tag{23.21}$$

Bondi's metric is an example of the 2+2 formalism we discussed in §14.13. We have a family of two surface $\mathcal{S}_{u,r}$ on which the metric is that given by $d\sigma^2 = r^2(e^{2\gamma}d\theta^2 + e^{-2\gamma}\sin^2\theta d\phi^2)$. The level surfaces of the u coordinate are the $u = $ constant null hypersurfaces, while the $r = $ constant surfaces are timelike (as depicted in Fig. 14.9). The functions U and β encode the shifts, and V encodes the lapse.

23.5 The characteristic initial value problem

We now consider the initial value problem for Bondi's radiating metric. The situation is different from the Cauchy problem because, this time, the initial data is set on a characteristic or null hypersurface rather than on a spacelike hypersurface. As a consequence, it is called the **characteristic initial value problem**. Bondi showed that the ten vacuum field equations break up into four groups:

(1) three **symmetry conditions**

$$R_{03} = R_{13} = R_{23} \equiv 0; \qquad (23.22)$$

(2) four **main equations**

$$R_{11} = R_{12} = R_{22} = R_{33} = 0; \qquad (23.23)$$

(3) one **trivial equation**

$$R_{01} = 0; \qquad (23.24)$$

(4) two **supplementary conditions**

$$R_{00} = R_{02} = 0. \qquad (23.25)$$

The three components R_{03}, R_{13}, and R_{23} vanish identically as a consequence of the symmetry assumptions. Recall that, in the Cauchy problem, we proved a result which states that, if the dynamical equations hold everywhere and the constraint equations hold on an initial hypersurface, then the contracted Bianchi identities ensure that the constraint equations hold everywhere. There is an analogous result for the characteristic initial value problem, except that, in this case, the 'constraint equations' consist of the trivial equation and the supplementary conditions, and the trivial equation is automatically satisfied as an algebraic consequence.

Lemma: If the main equations hold everywhere, then the contracted Bianchi identities ensure that

(a) the trivial equation holds as an algebraic consequence,

(b) the supplementary conditions hold everywhere if they hold on a hypersurface $r = $ constant.

Hence, the initial value problem reduces to solving the main equations and satisfying the supplementary conditions for one value of r. The main equations break up further into the following:

(2a) one **dynamical equation**

$$R_{33} = 0; \qquad (23.26)$$

(2b) three **hypersurface equations**

$$R_{11} = R_{12} = g^{22}R_{22} + g^{33}R_{33} = 0. \qquad (23.27)$$

The dynamical equation is the only main equation which involves a term differentiated with respect to u and hence propagating into the future (that is, from one null hypersurface to the next). The hypersurface equations only involve differentiation within the hypersurface $u =$ constant.

If we assume that the solution is analytic everywhere, then a detailed analysis of the main equations leads to the following schema for integration. We first prescribe γ on $u = u_0$, that is, on some initial hypersurface N_0, say. The three hypersurface equations then determine β, U, and V on N_0. The dynamical equation serves to determine $\gamma_{,0}$ on N_0, which means that γ is determined on the 'next neighbouring' null hypersurface, N_1, say. We then go through the whole cycle again on N_1 (Fig. 23.4). Proceeding in this way, we can generate a solution of the field equations in some region to the future of N_0. However, we have neglected functions of integration in the schema and it turns out that one of them, called the 'news' function, plays a key role in the analysis.

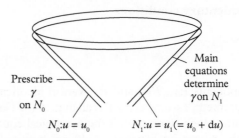

Prescribe γ on N_0

Main equations determine γ on N_1

$N_0 : u = u_0$ $N_1 : u = u_1 (= u_0 + \mathrm{d}u)$

Fig. 23.4 An integration schema for Bondi's solution.

23.6 News and mass loss

In order to proceed further, we need to expand everything in inverse powers of the radial parameter r and carry out an **asymptotic analysis**. We shall outline the procedure. We start by strengthening the condition (23.11), and a detailed analysis reveals that the asymptotic behaviour of the metric is given by

$$g_{00} = 1 + O(r^{-1}),$$
$$g_{01} = 1 + O(r^{-1}),$$
$$g_{02} = O(1), \tag{23.28}$$
$$g_{22} = -r^2 + O(r),$$
$$g_{33} = -r^2\sin^2\theta + O(r).$$

We mention briefly that the coordinate transformations which preserve the form of the metric (23.20) together with the above asymptotic conditions form a group called the **Bondi–Metzner–Sachs**, or **BMS** group. The BMS group is important because it plays the same role asymptotically for an isolated radiative system as the Poincaré group does in special relativity. Bondi adopts a final assumption, namely,

$$\lim_{r\to\infty} \left[\frac{\partial(r^2\gamma)}{\partial r} \right]_{u=\text{const}} = 0, \tag{23.29}$$

in an attempt to prevent radiation coming in from past null infinity and affecting the source. He chose this condition in analogy to the Sommerfield condition in electromagnetic theory, which prevents incoming radiation, but it turns out that (23.29) is not strong enough to prevent the occurrence of sufficiently weak incoming radiation.

We now have sufficient starting assumptions to expand everything in inverse powers of r. It is only necessary to work to a certain limited order in inverse powers to obtain the mass-loss result. For example, to the required order, we get (changing the original notation slightly)

$$\gamma = \frac{n}{r} + \frac{q}{r^3} + O(r^{-4}), \tag{23.30}$$

where $n = n(u, \theta)$ and $q = q(u, \theta)$ are arbitrary functions at this stage. The hypersurface conditions lead to

$$\beta = -n^2/(4r^2) + O(r^{-3}),$$
$$U = -(n_{,2} + 2n\cot\theta)/r^2 + (2d + 3n\,n_{,2} + 4n^2\cot\theta)/r^3 + O(r^{-4}),$$
$$V = r - 2M + O(r^{-1}),$$

$$\tag{23.31}$$

where $d = d(u, \theta)$ and $M = M(u, \theta)$ are also arbitrary. The dynamical equation produces

$$4q_{,0} = 2Mn - d_{,2} + d\cot\theta.$$

The supplementary conditions lead to

$$M_{,0} = -n_{,0}^2 + \tfrac{1}{2}(n_{,22} + 3n_{,2}\cot\theta - 2n)_{,0},$$ (23.32)

$$-3d_{,0} = M_{,2} + 3n\,n_{,02} + 4n\,n_{,0}\cot\theta + n_{,0}n_{,2},$$ (23.33)

where these last two equations are **exact** results by the lemma, since this is the part of the equations which holds on $r = $ constant.

A detailed investigation reveals that, as before, our initial data involves prescribing one function of three variables, namely,

$$\gamma = \gamma(u, r, \theta),$$ (23.34)

on N_0. However, in addition, we must prescribe one function of two variables, namely, the u derivative of n.

$$n_{,0} = n_{,0}(u, \theta),$$ (23.35)

for any value of r. Since we are working asymptotically, we shall prescribe $n_{,0}$ on $\mathscr{I}^+$. Finally, we must prescribe two functions of one variable, namely,

$$M = M(u_0, \theta), d = d(u_0, \theta),$$ (23.36)

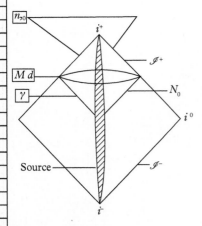

Fig. 23.5 Penrose diagram indicating initial data.

which we prescribe on the intersection of N_0 and $\mathscr{I}^+$ (see Fig. 23.5). With this initial data, **all** other quantities are determined. Clearly, it is the data $n_{,0}$ which determines the evolution of the solution and, as a consequence, it is termed the **news** function. If the solution is static, then the news function vanishes. If we restrict our attention to the periods when the source is static, then it is possible to find a coordinate transformation which relates the quantities M, d, and q to known physical parameters. It turns out that M is intimately connected to the mass and is termed the **mass aspect**. In fact, the quantity

$$m(u) = \tfrac{1}{2}\int_0^\pi M(u, \theta)\sin\theta d\theta,$$ (23.37)

determines the mass of the system at $\mathscr{I}^+$ and is called the **Bondi mass**. It is not obvious that (23.37) determines the mass but it agrees with the ADM and Komar mass for a stationary space-time (see §20.12) and with the Newtonian result in the weak field limit. The term d is termed the **dipole aspect**, and q the **quadrupole aspect**, and these also agree with the Newtonian result in the weak-field limit.

Multiplying (23.32) by $\sin\theta$, integrating with respect to θ from 0 to π, and using (23.37) together with some regularity conditions on the symmetry axis, we find (exercise)

$$m_{,0} = -\tfrac{1}{2}\int_0^{\pi} n_{,0}^2 \sin\theta\,\mathrm{d}\theta. \tag{23.38}$$

The non-positive nature of the right-hand side leads to the promised result.

Theorem: There is mass-loss if and only if there is news.

Thus, if a system remains quiescent, then there is no news and hence the Bondi mass remains constant. If, however, the system radiates, then there is news and the minus sign in (23.38) means there is a consequent mass **loss**. By conservation of energy, the rate of change of mass given by (23.38) corresponds to the rate of change of gravitational radiation emitted by the source $\mathrm{d}E/\mathrm{d}t$ which, unlike the formula derived in §21.9, does not require the gravitational field to be weak. Furthermore, $n_{,0}$ is related to the leading order part of the Weyl tensor as given by the peeling theorem (see §23.8) so that the integrand is essentially $\Psi_{abcd}\Psi^{abcd}$ and (23.38) is a manifestly coordinate independent quantity defined in terms of the curvature tensor. If a radiating system is initially quiescent, then radiates for a period, and then can become quiescent again, this establishes the content of Fig. 23.1. The power of this result is that we have obtained it without having to assume that the gravitational field is weak everywhere and no linearization of the field is needed. In §21.9 we mentioned that the ADM mass is positive. Because of the mass-loss formula, it is not obvious that the same remains true for the Bondi mass. In fact, the positivity of the Bondi mass for an asymptotically flat space-time that satisfies the dominant energy was established by Ludvigsen and Vickers (1981).

We mention that, shortly after Bondi published his results, Sachs dropped the symmetry assumptions and obtained essentially the same result. The calculations are obviously longer, and, in general, it turns out that there are **two** news functions, corresponding to the two gravitational degrees of freedom, but otherwise the argument proceeds along similar lines.

23.7 The Petrov classification

The gravitational field is governed by the Riemann tensor. We can gain considerable insight into the possible types of gravitational field by considering the **algebraic** structure of the Riemann tensor. We restrict ourselves to the **vacuum** case, where the Riemann tensor coincides with the Weyl tensor, because, in four dimensions, by (6.88),

$$C_{abcd} = R_{abcd} - g_{a[c}R_{d]b} + g_{b[c}R_{d]a} - \tfrac{1}{3}g_{a[d}g_{c]b}R, \tag{23.39}$$

and, in the vacuum case, $R_{ab} = R = 0$.

The Weyl tensor has the same symmetries as the Riemann tensor and, in addition, possesses the trace-free property

$$C^a{}_{bac} \equiv 0. \tag{23.40}$$

Since C_{abcd} is skew symmetric on each pair of indices and also symmetric under their interchange, we can start by thinking of it as a 6×6 symmetric matrix (exercise). We can then classify the Weyl tensor algebraically by classifying this 6×6 matrix in terms of its eigenvalues and eigenvectors. So, at first sight, we would expect this to involve classifying the possible roots of a sixth-order or sextic equation. However, the procedure is complicated by the additional symmetries (23.40) and $C_{a[bcd]} = 0$. We shall not pursue the details further, but it turns out that these symmetries reduce the problem to classifying the roots of a **quartic** equation. The resulting classification due to Petrov – and hence called the **Petrov** classification – itemizes the various possibilities of distinct eigenvalues and eigenvectors of the Weyl tensor at a point and gives them a name or **type** as shown in Table 23.1. If we add to this the completely degenerate case of conformally flat space-times in which C_{abcd} vanishes (called type O), then there are six possibilities which can be conveniently arranged in a triangular hierarchy (Fig. 23.6), as suggested by Penrose. In the diagram, the arrows point in the direction of increasing specialization. The Petrov type of a given vacuum space-time is then defined as the type at those points which are highest up the hierarchy. Thus, a solution may be the same type everywhere, or may reduce to lower types at some points or region, but by definition the type cannot move up the hierarchy. A generic solution will be type I, which is called **algebraically general**, whereas all other types are called **algebraically special**.

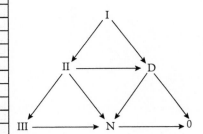

Fig. 23.6 The hierarchy of Petrov types.

Table 23.1

Petrov type:	I	II	D	III	N
Quartic roots:	all distinct	one double	two double	one triple	one four-fold
Distinct eigenvectors:	4	3	2	2	1

A different but equivalent method, due to Debever, consists in classifying certain null vectors, called **principal null directions**, which have a special relationship to the Riemann tensor. The result rests on the following theorem.

Theorem: Every vacuum space-time admits at least one and at most four null directions $\ell^a \neq 0$, $\ell^a \ell_a = 0$, which satisfy

$$\ell_{[a} R_{b]ef[c} \ell_{d]} \ell^e \ell^f = 0.$$

There is a corresponding result for non-vacuum space-times if we replace R_{abcd} by C_{abcd}. The Petrov type then relates to the coincidence of these null directions, as shown in Table 23.2.

Table 23.2

Petrov type:	I	II	D	III	N
Coincidence:	[1 1 1 1]	[2 1 1]	[2 2]	[3 1]	[4]

The coincidence also agrees with the coincidence of roots in the Petrov quartic equation.

The particular vacuum solutions of Schwarzschild, Reissner–Nordström, and Kerr are all algebraically special type D. Plane gravitational waves are type N, and hence the gravitational field from an isolated radiating source is expected to be asymptotically type N. However, any solution which is sufficiently complex to model a realistic solution will be type I. Bondi's radiating vacuum solution, namely (23.20), subject to (23.30) and (23.31), is type I, but asymptotically type N with

$$R_{abcd} \sim n_{,00}/r + O(r^{-2})$$

when $n_{,00} \neq 0$. We add that, in a non-vacuum space-time, the Petrov classification of the Weyl tensor is augmented by an analogous classification of the Ricci tensor called the **Plebanski type**. Moreover, the complete classification of the Weyl tensor and its covariant derivatives (in a canonically defined frame) leads to the **Karlhede classification** mentioned in §13.9.

23.8 The peeling theorem

In the last section, we defined the possible algebraic types of the Riemann tensor in a vacuum space-time. In this section, we consider the physical significance of this classification. Sachs investigated the case of a retarded wave solution emanating from an isolated source in the linearized theory and was able to expand the Riemann tensor in terms of an affine parameter r along each outward null ray (null geodesic) producing the result

$$R = \frac{N_0}{r} + \frac{III_0}{r^2} + \frac{II_0}{r^3} + \frac{I_0}{r^4} + \frac{I'_0}{r^5} + O(r^{-6}), \qquad (23.41)$$

where, for convenience, we have suppressed the indices. Thus, asymptotically, the leading order of the Riemann tensor is type N, then type III, type II, and type I, respectively, at the subsequent orders. In the equation, the 0 denotes a vanishing absolute derivative in the ray direction ℓ^a. Unlike the other coefficients in (23.41), I'_0 does not have a special relationship with ℓ^a since its one principal null direction is not tangent to a null geodesic. Sachs also considered algebraically special fields and found that they do not have an expansion as general as (23.41), but, in generalizing the work of Bondi, he was able to show that the Riemann tensor for an asymptotically flat isolated radiative system has precisely the same form as (23.41). Indeed, starting in the **wave zone**, where the Riemann tensor is type N with a fourfold repeated ray direction ℓ^a, the other principal null directions peel off as we move in towards the source, where terms of a less special nature predominate (Fig. 23.6). This is known as the **peeling-off theorem** or just peeling theorem, for short.

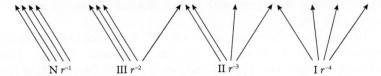

Fig. 23.7 The peeling-off theorem.

Szekeres has investigated the properties of type N, III, and D fields by considering their effect on a cloud of test particles. An observer sets up an orthonormal triad $\{e_1{}^a, e_2{}^a, e_3{}^a\}$ of spacelike vectors adapted to the field in each case. For type N fields, the forces on the ring of particles results in the distortion shown in Fig. 23.8 (compare with Fig. 21.2). This clearly indicates the transverse character of such fields, since $e_1{}^a$ points in the direction of propagation of the field. Szekeres terms this a **pure transverse** gravitational wave. For type III fields, the effect on the particles is still planar, but in this case the plane contains the wave direction $e_1{}^a$, and the axis is tilted through 45° to the wave direction (Fig. 23.9). Szekeres terms this a **longitudinal wave** component. For type D fields, the effect ceases to be planar. In this case, a sphere of particles is distorted into an ellipsoid with the major axis lying in the wave direction (Fig. 23.10). This is precisely the tidal force we discussed before in §17.10 for a radially infalling observer in the Schwarzschild field. Szekeres terms this a **Coulomb-type** field in analogy with electromagnetism. For type I and type II fields, nothing simple emerges.

23.9 The optical scalars

Consider a congruence of null geodesics with tangent vector field ℓ^a. By a change of scale, it is always possible to obtain the geodesic equation in the simple form (exercise)

$$\ell^a{}_{;b}\ell^b = 0.$$

We assume this has been done and define three quantities called **optical**

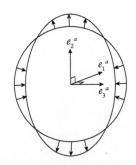

Fig. 23.8 The effects of a type N field on a ring of test particles.

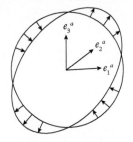

Fig. 23.9 The effects of a type III field on a ring of test particles.

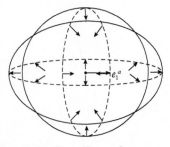

Fig. 23.10 The effects of a type D field on a sphere of test particles.

scalars determined by the congruence ℓ^a as follows:

$$\text{expansion (divergence):} \qquad \theta = \tfrac{1}{2}\ell^a{}_{;a},$$

$$\text{twist (rotation):} \qquad \omega = \left\{\tfrac{1}{2}\ell_{[a;b]}\ell^{a;b}\right\}^{\frac{1}{2}},$$

$$\text{shear (distortion):} \qquad |\sigma| = \left\{\tfrac{1}{2}\ell_{(a;b)}\ell^{a;b} - \theta^2\right\}^{\frac{1}{2}}.$$

Their physical interpretation is embodied in the following result of Sachs. If a small object in a null geodesic congruence casts a shadow on a screen, then all portions hit it simultaneously. The shape, size, and orientation of the shadow depend only on the location of the screen and not on its velocity. If the screen is an infinitesimal distance dr from the object, then the shadow is expanded by θdr, rotated by ωdr, and sheared by $|\sigma|dr$ (Fig. 23.11). The quantity shear turns out to be the most important physically, as is evident from the following theorem.

> **Goldberg–Sachs theorem:** A vacuum solution is algebraically special if and only if it contains a shear-free null geodesic congruence.

In an isolated radiative system, the news function is also intimately connected to the shear.

The Petrov classification, optical scalars, and Killing vectors are three very important tools for classifying vacuum solutions in a coordinate-independent way. In particular, they have been used to find particular exact solutions of the field equations. Indeed, there are known vacuum solutions for each of the four classes of algebraically special Petrov type, determined by the vanishing or otherwise of the expansion and twist. In some of these cases, all possible solutions are known. For example, all vacuum type D solutions have been found. There are also a large number of vacuum type I or algebraically general solutions known. However, few of these solutions are fully understood in the sense that we are able to understand their causal structure, geodesic structure, global structure, and singularity structure. Thus, there are relatively few solutions for which we can draw space-time, spatial, and Penrose diagrams. Moreover, there is evidence to suggest that many of them have a strange singularity structure and, as such, are pathological in nature and unlikely to approximate to any physically realistic solution.

Finally, we mention that, by using the Riemann identity on ℓ^a,

$$\ell^a{}_{;bc} - \ell^a{}_{;cb} = R^a{}_{dcb}\ell^d, \tag{23.42}$$

and the definitions of optical scalars, it is a straightforward matter to derive propagation equations for the optical scalars. For example, setting

$$z = -\theta + i\omega, \tag{23.43}$$

we can deduce

$$\frac{Dz}{Dr} = z^2 + |\sigma|^2 + \tfrac{1}{2}R_{ab}\ell^a\ell^b. \tag{23.44}$$

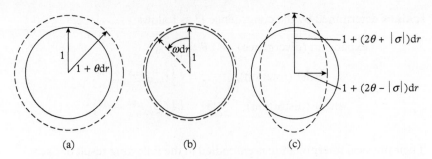

Fig. 23.11 The optical scalars: (a) expansion; (b) twist; (c) shear

This is a null version of the Raychaudhuri equation, and equations such as these play a central role in the proof of the singularity theorems.

Exercises

23.1 (§23.1) Define cylindrical symmetry. What conditions does this impose on the metric coefficients in adapted coordinates $(x^a) = (x^0, x^1, \phi, z)$? Write down the metric of the 2-space ($\phi = $ constant, $z = $ constant). Use the result of Exercise 6.31 to deduce that there exist coordinates in which the line element can be written in the form

$$ds^2 = e^{2\alpha}(dt^2 - d\rho^2),$$

where α is a function of t and ρ only. What are the conditions for the (t, ρ)-plane to be orthogonal to the (ϕ, z)-plane? Assuming these conditions, show that a cylindrically symmetric line element can be written in the canonical form

$$ds^2 = e^{2\gamma - 2\psi}(dt^2 - d\rho^2) - \rho^2 e^{-2\psi} d\phi^2 - e^{2\psi + 2\mu}(dz + \chi d\phi)^2,$$

where γ, ψ, μ, and χ are all functions of t and ρ only.

23.2 (§23.1) What is the condition for a cylindrically symmetric solution to be non-rotating? What effect does this have on the line element of Exercise 23.1?

23.3 (§23.4) Show that if a null ray is given by

$$u = u_0, \theta = \theta_0, \phi = \phi_0,$$

then it leads to the conditions (23.13). Show that (23.13) is equivalent to (23.14). [Hint: consider inverting a general symmetric 4 × 4 matrix with zeros in the positions defined by (23.13).] Show that, if x^1 is an affine parameter, it lead to the conditions (23.15). Show that axial symmetry and azimuth reflection invariance lead to the conditions (23.18) or (23.19).

Let $x^1 = r$ be a luminosity parameter defined by (23.16) and deduce Bondi's radiating metric (23.20) subject to (23.21). [Hint: show first that the conditions lead to a metric in which g_{00}, g_{01}, g_{02}, and g_{22} are four arbitrary functions of u, r, and θ; then the actual forms of these coefficients are chosen to preserve the signature and for later convenience.]

23.4 (§23.4) Show that the surface area of the 2-surface $(u = u_0, r = r_0)$ is $4\pi r_0^2$.

23.5 (§23.4) Find the non-zero components of the metric connection of Bondi's radiating metric. [Hint: use the variational principle approach of §7.6.]

23.6 (§23.5) Use the results of Exercise 23.5 to establish the lemma of §23.5. [Hint: write out the contracted Bianchi identities in terms of Γ^a_{bc} and R_{ab}; do not insert the metric expressions for Γ^a_{bc} and R_{ab} in the identities, but merely consider which quantities are zero and which are not.]

23.7 (§23.5) Evaluate the components of the Ricci tensor which define the four main equations. [Hint: this is a long but straightforward calculation.] Use the results to confirm the integration schema for Bondi's solution.

23.8 (§23.6) The requirement that the Bondi metric remains regular on the symmetry axis $\theta = 0, \pi$ leads to a number of conditions including $n(u, 0) = n(u, \pi) = 0$. Use these conditions together with (23.32) and (23.37) to deduce the mass-loss result.

23.9 (§23.7) Show that the symmetries

$$C_{abcd} = -C_{abdc} = -C_{bacd} = C_{cdab},$$

mean that we can treat C_{abcd} at a point as a symmetric 6×6 matrix.

23.10 (§23.9) Consider a congruence of null geodesics with tangent vector ℓ^a. Write down the geodesic equation that ℓ^a satisfies in general. Show that, if we rescale ℓ^a so that

$$\ell^a \to \bar{\ell}^a = A\ell^a,$$

then we can choose A so that the geodesic equation reduces to

$$\bar{\ell}^b \nabla_b \bar{\ell}^a = 0.$$

23.11 (§23.9) Compute an expression for the expansion for Bondi's radiating line element (23.20), using $\ell^a \stackrel{*}{=} e^{-2\beta} \delta^a_1$.

Further reading

The book by Wald (1984) and the book by Stewart (2008) are fairly advanced, and both deal with the Bondi metric. The 1962 paper by Bondi et al. is also quite readable. See the book by Ludvigsen (1999) for more on the optical scalars and null congruences.

Bondi, H., van der Burg, M. G. J., and Metzner, A. W. K. (1962). Gravitational waves in general relativity, VII. Waves from axi-symmetric isolated system. *Proceedings of the Royal Society of London. Series A. Mathematical and Physical Sciences, 269*(1336), 21–52.

Ludvigsen, M. (1999). *General Relativity: A Geometric Approach*. Cambridge University Press, Cambridge.

Stewart, J. (2008). *Advanced General Relativity*. Cambridge Monographs on Mathematical Physics. Cambridge University Press, Cambridge.

Wald, R. M. (1984). *General Relativity*. University of Chicago Press, Chicago, IL.

Part F

Cosmology

Relativistic cosmology

24

24.1 Preview

Cosmology is the study of the dynamical structure of the universe as a whole. As in most modelling exercises, we shall start by trying to find a very simple model of the universe. This is done by smoothing out all the irregularities in space and in time and concentrating simply on the gross features of the universe. So, to start with, we ignore all details such as the solar system, our own galaxy (the Milky Way), the local cluster of galaxies, and so on; the consideration of these details can then hopefully be introduced at a later stage to yield a more complete or better theory. We shall be concerning ourselves only with the very basics of cosmology, that is, the overall dynamics of the system. We shall see in this chapter that, if one makes some symmetry assumptions, then this is governed by a first-order ordinary differential equation called Friedmann's equation. The resulting solutions are the standard solutions of relativistic cosmology and are called the **Friedmann models**. We shall investigate some of these in the next chapter.

Cosmology as a separate scientific study really only came into existence with the advent of general relativity. It is possible to consider cosmology in a Newtonian framework, but this had not been seriously attempted prior to general relativity, largely because, in as far as there was a generally accepted model of the universe in existence, it was considered devoid of dynamics; that is, the universe was considered **static**. Perhaps surprisingly, it is possible to construct a 'Newtonian cosmology' based on Newtonian theory together with a number of **ad hoc** assumptions which also results in Friedmann's equation. (However, the interpretation of some of the terms in the equation is different.) But it is important to realize that this Newtonian approach only came into existence after general relativity had first tackled the problem. We shall look at a discrete Newtonian model in §24.3. The starting point for both Newtonian and relativistic cosmology is a simplicity principle called the **cosmological principle**, which states, essentially, that the universe is unchanging in space from point to point. This leads to the requirement that space is homogeneous and isotropic (the same in every direction) about each point.

Introducing Einstein's Relativity. Second Edition. Ray d'Inverno and James Vickers, Oxford University Press.
© Ray d'Inverno and James Vickers (2022). DOI: 10.1093/oso/9780198862024.003.0024

In the early decades of cosmology, there were very few reliable observational results and, not surprisingly, different Friedmann models enjoyed periods of fashion, that is, periods when they were considered the best available model for our own universe. However, there was one school of thought that argued vociferously for a simple non-Friedmann model called the **steady-state solution**, based on the perfect cosmological principle that the universe is unchanging in space **and** time. The steady-state solution is no longer regarded as a good description of the universe but the perfect cosmological principle leads to the **de Sitter** solution, which is of considerable theoretical importance and provides an alternative to Minkowski space for the asymptotic structure of space-time (see §25.6 for more details). Although the classical Friedmann solutions have been replaced by more sophisticated models, they are still basic to much of cosmological thinking. We therefore investigate the 'classical' Friedmann models in Chapter 25 before looking at a more modern approach to cosmological models in Chapter 25.

Since the 1960s, one model has emerged as the best available, at least as far as the origins of the universe are concerned, and that is the **hot big bang**. In this model, it is assumed that there occurred a cataclysmic event (some 10^{10} years ago), called the big bang, when the universe sprang into existence and expanded away from a singularity. In the earliest phases, the universe consisted of radiation at incredibly high temperatures and densities. As the universe expanded, the temperature and density fell and protons, electrons, and neutrons emerged from the radiation bath. As the system cooled further, the simple atoms such as hydrogen and helium emerged first, followed later by the heavier elements. This phase can be treated mathematically and one of the great successes of this approach has been the agreement of the theoretical prediction of the abundancies of the heavy elements with the observed abundancies. As the system expands and cools yet further, then conditions become favourable for formation of the stars and galaxies, from the primeval matter. The model then encompasses the dynamics of the interstellar medium including the galaxies and stars up to the present epoch.

The development of the hot big bang model brings out an important point; namely, in modelling the universe in the large, we have made use of our understanding of **local** physical laws. The justification for this is that we are more or less forced to do so – otherwise we would hardly be able to start – and yet it has proved extremely successful, to date, in providing insight into the structure of the universe. However, we cannot rule out the possibility that there exist additional interactions which only reveal themselves on a cosmological scale. One example of this is the cosmological term (Λg_{ab}) which Einstein incorporated into general relativity. Another important point relates to the fact that, in most branches of physics, it is possible to investigate phenomena by repeatedly carrying out experiments in the laboratory in controlled conditions where all but a small number of parameters are held fixed. No such possibility occurs in cosmology. Indeed, cosmology is unlike any other branch of physics in that the system we are studying is unique. Given this constraint, it is perhaps surprising that we are able to construct such apparently successful models. This success is so marked that, in some cosmological circles, the claim is that the universe is well understood after the first 10^{-43} seconds, from its birth, which is when the physics can be described by general relativity and Grand Unified Theories (GUTs) in which the strong, weak, and electromagnetic forces are unified. More modestly after about 10^{-10} seconds, it is belived that the universe is well described by general relativity and the standard model of particle physics.

24.2 Olbers' paradox

The fact that, prior to general relativity, the universe was considered static is perhaps even more surprising when one is confronted by a paradox put forward by Olbers in 1826 which stems from the observation that the sky is dark at night. (In fact, others had considered similar ideas before, but Olbers gave a more precise statement of the paradox.) He assumed that space is Euclidean and infinite and that the average number of stars per unit volume and the average luminosity of each star is constant throughout space and time, provided these averages are taken over sufficiently large regions. He also assumed that the universe has been in existence for an infinite time and that, on the large scale, it is static. Now consider a shell of radius r and thickness dr, and let ℓ denote the product of the average number of stars per unit volume and the average luminosity per star. The intensity at the centre of the shell will be given by the total luminosity produced by the shell divided by its area, that is, approximately

$$\frac{(4\pi r^2 dr)\ell}{4\pi r^2} = \ell dr. \tag{24.1}$$

If we surround any point P by an infinite succession of shells, each of thickness dr, then clearly the intensity at P will be $\int_0^\infty \ell dr$, which is infinite! However, we have omitted to account for the possibility that light from a star may be intercepted on its way by another star (Fig. 24.1). When this is taken into account, it can be shown that the result is no longer infinite but equal to the average luminosity at the surface of a star. Since P is arbitrary, the result must hold everywhere. This leads to a paradox, because the sky is observed to be dark at night. The same conclusion may be reached by thermodynamic arguments. For, if the system is static and of infinite age, then it must have reached thermodynamic equilibrium, which means that each star must be absorbing as much radiation as it emits, and the result follows. Yet another argument is that, if one looks in any direction in an infinite universe in which the average number of stars per unit volume is finite, then the line of sight will eventually end on a star. Since the system is static, the light received from the star is not degraded, and the result again follows.

P
•

Fig. 24.1 Light intercepted by another star.

It is interesting to note that the bulk of this enormous amount of radiation arrives from very distant parts, half, in fact, from regions so distant that the light has only a 50% chance of arriving without being absorbed by other stars. An estimate from observations in our own neighbourhood suggests that half of this radiation should be due to stars more than 10^{20} light years distant.

Olbers tried to resolve this paradox by postulating the existence of a tenuous gas which would absorb the radiation in transit over long distances. This argument will not work, though, because the gas would be heated until it reaches a temperature at which it radiates as much as it receives, and hence it will not reduce the average density of radiation. The same paradox arises even if the assumption that the universe is Euclidean is dropped (exercise). Nor does it make any difference whether the universe is infinite (open) or bounded (closed).

As we look further out into space, we are looking further back in time. One resolution of the paradox rests on assuming that ℓ is a function of time which is sufficiently small in the distant past that the distant regions do not contribute significantly to the radiation density. If it is assumed that the universe is static and that the stars do not start radiating until some finite period in the past, then it is possible to arrange for this period to be short enough to lead to the radiation density we observe today. However, some estimates would then suggest that the universe is younger than the age of the oldest stars. The accepted resolution rests on assuming that the universe is not static but rather undergoing large-scale expansion. Then, because of the Doppler shift, light received from receding stars will be shifted to the red and, if the recessional velocity is large enough, the loss of energy will be sufficient to reduce the radiation density to the observed level.

In summary, assuming that a dark night sky is not just a phenomenon of our current epoch, then Olbers' paradox requires that either the universe is young or it is expanding. In the latter case, the question may be asked as to what happens to the 'lost' energy resulting from the Doppler shift. In fact, it is precisely this energy which is doing the work involved in the expansion of the universe.

24.3   Newtonian cosmology

In this section, we shall introduce Newtonian cosmology by investigating a simple discrete model in which it is assumed that the universe consists of a finite number of galaxies. Let the i-th galaxy have mass m_i and position $\mathbf{r}_i(t)$ as measured from a fixed origin O. We now impose the cosmological principle (see §24.4) in the form that the motion about O must be spherically symmetric, in which case the motion of the galaxies is purely radial, i.e.

$$\mathbf{r}_i(t) = r_i(t)\hat{\mathbf{r}}. \tag{24.2}$$

The kinetic energy T of the system is then

$$T = \tfrac{1}{2} \sum_{i=1}^{n} m_i \dot{r}_i^2.$$

The gravitational potential energy between a pair of galaxies m_i and m_j is given by $-Gm_im_j/|\mathbf{r}_i - \mathbf{r}_j|$, and so the total potential energy V of the system is

$$V = -G \sum_{\substack{i,j=1 \\ (i<j)}}^{n} \frac{m_i m_j}{|\boldsymbol{r}_i - \boldsymbol{r}_j|},\tag{24.3}$$

where the inequality in the double sum means that each pair of particles is only counted once. We also assume that there is a cosmological force acting on the i-th galaxy of the form

$$\boldsymbol{F}_i = \tfrac{1}{3}\Lambda m_i \boldsymbol{r}_i,\tag{24.4}$$

where Λ is a constant called the **cosmological constant**. This yields an additional potential energy, called the cosmological potential energy V_c of the system, given by

$$V_c = -\tfrac{1}{6}\Lambda \sum_{i=1}^{n} m_i r_i^2.\tag{24.5}$$

The total energy E of the system is therefore

$$E = \tfrac{1}{2}\sum_{i=1}^{n} m_i \dot{r}_i^2 - G \sum_{\substack{i,j=1 \\ (i<j)}}^{n} \frac{m_i m_j}{|\boldsymbol{r}_i - \boldsymbol{r}_j|} - \tfrac{1}{6}\Lambda \sum_{i=1}^{n} m_i r_i^2.\tag{24.6}$$

Let us assume that the distribution and motion of the system is known at some fixed epoch t_0. Then the radial motion required by the cosmological principle implies that, at any time t,

$$r_i(t) = S(t) r_i(t_0),\tag{24.7}$$

where $S(t)$ is a universal function of time which is the same for all particles and is called the **scale factor**. This means that the only motions compatible with homogeneity and isotropy are those of uniform expansion or contraction, that is, a simple scaling up or down by a time-dependent scale factor.

The radial velocity of the i-th galaxy is then

$$\dot{r}_i(t) = \dot{S}(t) r_i(t_0) = \frac{\dot{S}(t)}{S(t)} r_i(t),\tag{24.8}$$

by (24.7). We define a quantity called the **Hubble parameter** $H(t)$ by

$$H(t) = \dot{S}(t)/S(t),\tag{24.9}$$

which has dimensions $1/(\text{time})$. Equation (24.8) can then be written as

$$\dot{r}_i(t) = H(t) r_i(t),\tag{24.10}$$

which is called **Hubble's law**. This states that, in an expanding universe, at any one epoch, the radial velocity of recession of a galaxy from a given point is proportional to the distance of the galaxy from the point. The value of the Hubble parameter at our epoch is known as the **Hubble constant**.

If we substitute (24.7) and (24.8) into (24.6), we find (exercise)

$$E = A[\dot{S}(t)]^2 - \frac{B}{S(t)} - D[S(t)]^2, \qquad (24.11)$$

where the coefficients are positive constants defined by

$$A = \tfrac{1}{2} \sum_{i=1}^{n} m_i [r_i(t_0)]^2, \qquad (24.12)$$

$$B = G \sum_{\substack{i,j=1 \\ (i<j)}}^{n} \frac{m_i m_j}{|r_i(t_0) - r_j(t_0)|}, \qquad (24.13)$$

$$D = \tfrac{1}{6} \Lambda \sum_{i=1}^{n} m_i [r_i(t_0)^2] = \tfrac{1}{3} \Lambda A. \qquad (24.14)$$

This is one form of the **cosmological differential equation** for the scale factor $S(t)$. It has a simple interpretation. First of all, consider what happens when Λ vanishes, in which case we can neglect the last term. If the universe is expanding, then the second term on the right-hand side decreases and, since the total energy remains constant, it follows that the first term must decrease as well. Therefore the expansion must slow down. If Λ is positive, then all galaxies experience a cosmic repulsion, pushing them away from the origin out to infinity. In this case, the cosmological term contributes positively to the expansion. If Λ is negative, then the opposite happens and all galaxies experience a cosmic attraction towards the origin. In a later section, we shall go on to consider what solutions of the differential equation are possible for different values of the parameters occurring in them. In particular, we shall investigate whether it is possible for the expansion to slow down, stop, and reverse so that eventually the universe will collapse – the so-called 'big crunch'.

We finish this section by rewriting the differential equation in a form closer to the general relativistic equation. Solving (24.11) for $\dot{S}^2$, we find

$$\dot{S}^2 = \left(\frac{B}{A}\right)\frac{1}{S} + \frac{D}{A}S^2 + \frac{E}{A}$$

$$= \left(\frac{B}{A}\right)\frac{1}{S} + \tfrac{1}{3}\Lambda S^2 + \frac{E}{A}, \qquad (24.15)$$

by (24.14). We now rescale the scale factor $S(t)$ to obtain a new scale factor $R(t)$, where

$$R(t) = \mu S(t). \qquad (24.16)$$

Then, multiplying (24.15) by μ^2, we can write it in the form

$$\dot{R}^2 = \frac{C}{R} + \tfrac{1}{3}\Lambda R^2 - k, \qquad (24.17)$$

where the constants C and k are defined by

$$C = B\mu^3/A, \qquad k = -\mu^2 E/A.$$

If $E = 0$, we choose μ arbitrarily but, if $E \neq 0$, we choose it so that

$$\mu^2 = A/|E|. \qquad (24.18)$$

This choice of rescaling means that k can only have the values $+1$, 0, or -1. In this case, (24.17) has exactly the same form as the Friedmann differential equation of relativistic cosmology. In a similar manner, it is possible to construct a finite **continuum** Newtonian model. Although this model may be taken to be arbitrarily large, it does not apply to an infinite universe.

24.4 The cosmological principle

Cosmology is based on a principle of simplicity, namely, the cosmological principle. It is, in essence, a generalization of the Copernican principle that the Earth is not at the centre of the solar system. In the same spirit, we would not expect the Earth, or the solar system, or our galaxy, or our local group of galaxies to occupy any specially favoured position in the universe. We state the principle in the following form.

The cosmological principle: At each epoch, the universe presents the same aspect from every point, except for local irregularities.

We need to make this statement mathematically precise. We assume that there is a cosmic time t and formulate the principle in each of the space-like slices $t = $ constant. The statement that each slice has no privileged points means that it is **homogeneous**. Technically, a spacelike hyper-surface is homogeneous if it admits a group of isometries which maps any point into any other point (Fig. 24.2). The principle requires that not only should a slice have no privileged points but it should have no privileged directions about any point either. A manifold which has no privileged directions about a point is called **isotropic** and is therefore spherically symmetric about that point. A manifold is globally isotropic if it is isotropic about every point. It can be shown that, if a manifold is globally isotropic, then it is necessarily homogeneous (see §24.7 for more details). Thus, the cosmological principle requires that space-time can be sliced up or 'foliated' into spacelike hypersurfaces which are spherically symmetric about any point in them. The homogeneity of the universe has to be understood in the same sense as the homogeneity of a gas: it does not

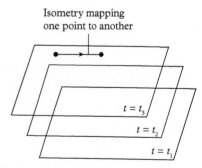

Isometry mapping one point to another

$t = t_3$

$t = t_2$

$t = t_1$

Fig. 24.2 Manifold sliced up into homogeneous 3-spaces.

apply to the universe in detail, but only to a 'smeared-out' universe averaged over cells of diameter over 10^9 light years, which are large enough to include many clusters of galaxies, and are bigger than the largest known structures in the universe, which are superclusters with a length scale of around 3×10^8 light years.

Thus, the cosmological principle is a simplicity principle which leads to the requirement that the universe is both isotropic and homogeneous. What observational evidence is there for each? Observations based on surveys of visible galaxies tend to cover only parts of the sky and, as a result, are somewhat limited. However, observations of radio galaxies, cosmic X-ray sources and quasars are all consistent with isotropy at a level of better than 15%. Associated measurements of the Hubble constant using Type Ia supernovae gives a similar bound. But the greatest support for isotropy came in 1965 with the discovery of the **cosmic microwave background** (CMB) by Penzias and Wilson. They discovered that the universe is currently pervaded by a bath of thermal radiation with a temperature of 2.7 K and, moreover, more recent measurements using the Planck satellite show that this radiation is isotropic to about one part in 100,000. The generally accepted explanation is that this radiation is a thermal remnant of the hot big bang. Spatial homogeneity is also supported by the counts of galaxies and the linearity of the Hubble law.

Despite the high degree of isotropy and homogeneity which we observe now, some cosmologists have considered anisotropic and inhomogeneous models. There are basically three reasons for this. First of all, calculations of statistical fluctuations in Friedmann models suggest that they cannot collapse fast enough to form the observed galaxies. Secondly, although there are strong reasons to support a big bang, there is less reason to suppose that the original singularity has the simple spherically symmetric pointlike structure of a Friedmann singularity. Indeed, calculations by Belinski, Khalatnikov, and Lifschitz – the so-called BKL approach – suggest that a general cosmological singularity would have a quite different structure. Finally, there is the idea that the universe may have been anisotropic and inhomogeneous in the past, but that there is some mechanism such as inflation (see Chapter 25) by which these characteristics would be washed out in the subsequent evolution, regardless of the initial conditions.

Considerable work has been done on the theoretical side in investigating anisotropic and inhomogeneous solutions. One of the biggest group of such solutions is that of the **Bianchi** models, which are spatially homogeneous anisotropic models (technically, they admit a 3-dimensional group of transformations which map any point in a hypersurface of homogeneity into any other point). These are subdivided into classes and labelled I, II, III, IV, V, VI, VII, VIII, and IX. The field equations then reduce to ordinary differential equations with time as the independent variable. These equations can then be studied by either qualitative or numerical methods. These models, in general, have singularities. For example, the vacuum Bianchi I models are described by the **Kasner** solution

$$ds^2 = dt^2 - t^{2p_1} dx^2 - t^{2p_2} dy^2 - t^{2p_3} dz^2,$$

where p_1, p_2, and p_3 are constants satisfying

$$p_1 + p_2 + p_3 = p_1^2 + p_2^2 + p_3^2 = 1,$$

which means that there is only one freely specifiable constant. In general, these solutions have a 'cigar'-like singularity when $t = 0$, that is, a small spatial region which is spherical at some time becomes infinitely long and thin as $t \to 0$. There is also a special case when the initial singularity is apparently of a 'pancake' type where the spherical region becomes an infinitely thin disc. Indeed, if we now include matter, it turns out that most of the Bianchi solutions have physical singularities, in the sense that the density becomes infinite, of these cigar or pancake types. Some special solutions give rise to weaker singularities called 'whimper' singularities, which have the property that the Ricci components in an orthonormal frame parallely propagated along a curve hitting the singularity are unbounded, whereas the components in some other frame are bounded. However, the physical singularities are the generic ones. There is a fair amount known about the qualitative nature of the evolution of these models, but we will not consider them further.

A more radical notion is that there is no 'smeared-out' universe at all, but only clusters of galaxies, and clusters of clusters, and clusters of clusters of clusters, and so on, as in the hierarchical model proposed in 1908 by C. V. I. Charlier. There is, in fact, some observational evidence for superclustering centred on the Virgo cluster, but the hierarchy appears to stop at clusters of clusters of galaxies, and shows little evidence of inhomogeneities on a larger scale.

We shall, from now on, adopt the cosmological principle. The real reason for this is not that it is definitely correct, but rather that it allows us to make use of the limited data provided to cosmology by observational astronomy. Any weaker assumptions, as in the anisotropic models or hierarchical models, would lead to metrics for which there would be insufficient data to determine the unknown functions occurring in them. By making such simplifying assumptions, we have a real chance of confronting theory with observation.

24.5   Weyl's postulate

In 1923, H. Weyl addressed the problem of how a theory like general relativity, based on general covariance, can be applied to a **unique** system like the universe. From one viewpoint, general relativity was specifically designed to deal with the equivalence of the observations of relatively accelerated observers. The universe consists of a single system which looks different to observers in different states of motion. Weyl argued that, in attempting to understand the distant, we must base ourselves, as far as possible, on the theories verified in our neighbourhood. General relativity offers the best available summary of local macroscopic physics and is, accordingly, a suitable theory. Other assumptions are needed such as the cosmological principle. Weyl also added to this the assumption that there is a privileged class of observers in the universe, namely, those associated with the smeared-out motion of the galaxies. The fact that one can work with this smeared-out motion follows from the observation that the relative velocities of matter in each astronomical neighbourhood – each group of galaxies – are small. He then posits the introduction of a 'substratum' or fluid pervading space in which the galaxies move like 'fundamental particles' in the fluid, and assumes a special motion for these particles. This is contained in the following postulate in a formulation due to Robertson.

Weyl's postulate: The particles of the substratum lie in space-time on a congruence of timelike geodesics diverging from a point in the finite or infinite past.

The postulate requires that the geodesics do not intersect except at a singular point in the past and possibly a similar singular point in the future. There is, therefore, one and only one geodesic passing through each point of space-time, and, consequently, the matter at any point possesses a unique velocity. This means that the substratum may be taken to be a **perfect fluid** and this is the essence of Weyl's postulate. Although the galaxies do not follow this motion exactly, the deviations from the general motion appear to be random and less than one-thousandth of the velocity of light. This is to be compared with the relative velocities of the galaxies due to the general motion, which is comparable with the velocity of light. Accordingly, the random motion may be neglected in the first instance. Combined with the observation that the general motion is one of expansion, Weyl's postulate is seen to closely reflect the actual situation in the universe. Note, however, it is possible to construct cosmological models in which the matter is not modeled by a perfect fluid. Thus, the postulate is far from an absolute requirement but is a useful starting point for building cosmological models.

24.6   Standard models of relativistic cosmology

The standard models of **relativistic cosmology** are based on three assumptions, namely:
(1) the cosmological principle,
(2) Weyl's postulate,
(3) general relativity.

Weyl's postulate together with the cosmological principle requires that the geodesics of the substratum are orthogonal to a family of spacelike hypersurfaces since, otherwise, the tangent vectors would give a preferred direction, contradicting the assumption of isotropy. We introduce coordinates (t, x^1, x^2, x^3) such that these spacelike hypersurfaces are given by $t = $ constant and the coordinates (x^1, x^2, x^3) are constant along the geodesics. This means that the spacelike coordinates of each particle are constant along its geodesic and, as a consequence, such coordinates

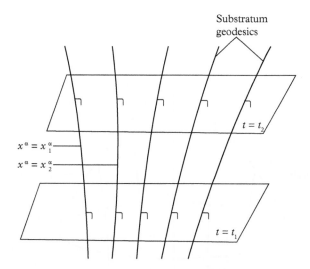

Fig. 24.3 Cosmic time surfaces and substratum geodesics.

are called **co-moving**. This, together with the orthogonality condition, means that t can be chosen so that the line element is of the form

$$ds^2 = dt^2 - g_{\alpha\beta}dx^\alpha dx^\beta,$$

where, as usual, Greek indices run from 1 to 3 and

$$g_{\alpha\beta} = g_{\alpha\beta}(t, x).$$

The coordinate t then plays the role of a **cosmic time** or **world time**.

The world time defines a concept of simultaneity. A **world map** is then the distribution of events on the surfaces of simultaneity (Fig. 24.3). The **world picture** is the aspect of the universe presented to an observer at any instant of world time, that is, it comprises the events seen looking along the observers past light cone (Fig. 24.4). Clearly, events from distant parts of the universe occur at earlier values of the world time than those nearby.

Consider a small triangle formed of three particles at some time t and also the triangle formed by these particles some time later. The second triangle will, in general, differ from the first in many respects. But, when we use the fact that the cosmological principle requires that the 3-spaces are isotropic and homogeneous, so that no point and no direction in the hypersurfaces may be preferential, then it follows that the second triangle must be geometrically similar to the first. Moreover, the magnification factor must be independent of the position of the triangle in the 3-space by similar arguments. It follows then that the time can enter $g_{\alpha\beta}$ only through a common factor in order that the ratios of the distances corresponding to the small displacements may be the same at all times. Hence, the time may only enter $g_{\alpha\beta}$ in the form

$$g_{\alpha\beta} = \left[S^2(t)\right] h_{\alpha\beta}(x^\alpha). \tag{24.19}$$

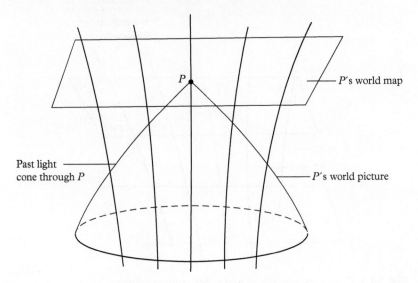

Fig. 24.4 The world map and world picture of an observer.

The ratio of the two values of $S(t)$ at two different times is the magnification factor and because of this it is called the **scale factor**. The scale factor $S(t)$ must be real, for otherwise the lapse of time could change a spacelike into a timelike interval. Next, we have to impose the condition that each slice is homogeneous and isotropic and also independent of time. We show below that this requires that the curvature of $h_{\alpha\beta}$ at any point must be a constant, for otherwise all points would not be geometrically identical. Such a space is called a space of **constant curvature**, which we discuss next.

24.7   Spaces of constant curvature

Mathematically, a space of constant curvature is characterized by the equation

$$R_{abcd} = K(g_{ac}g_{bd} - g_{ad}g_{bc}), \tag{24.20}$$

where K is a constant called the **curvature**. We now show how, for a 3-dimensional metric, the above equation follows from the assumption of isotropy about every point. The starting point is to consider the Einstein tensor of a general n-dimensional metric g_{ab} (of any signature) in the mixed form $G^a{}_b$. At any point, this defines a linear map from vectors to vectors. If the space is isotropic, there must be no special directions for the eigenvectors. This is only possible if $G^a{}_b$ is proportional to the identity matrix δ^a_b for which every vector is an eigenvector. Thus,

$$G^a{}_b = -K\delta^a{}_b,$$

(where we have introduced the minus sign for later convenience) or, lowering the index, we get

$$G_{ab} = -Kg_{ab}, \tag{24.21}$$

where it follows from the contracted Bianchi identities (exercise) that K is a constant. Taking the trace of (24.21) shows that $R = 2nK/(n-2)$, and substituting back in (24.21) gives

$$R_{ab} = \frac{2K}{n-2}g_{ab}. \tag{24.22}$$

In the mathematical literature, spaces that satisfy

$$R_{ab} = \lambda g_{ab}, \quad \text{where } \lambda \text{ is a constant,}$$

are called (somewhat confusingly) **Einstein spaces**. We have therefore shown the following.

> Any space (or space-time) which is isotropic about every point is an Einstein space.

We now show that a 3-dimensional Einstein space is necessarily a space of constant curvature, that is, the 3-metric $h_{\alpha\beta}$ satisfies

$$R_{\alpha\beta\gamma\delta} = K(h_{\alpha\gamma}h_{\beta\delta} - h_{\alpha\delta}h_{\beta\gamma}), \tag{24.23}$$

in accordance with (24.20). We stated in §6.13 that, in three dimensions the Weyl tensor vanishes and then using (6.88) the curvature may be written in terms of the 3-dimensional Ricci curvature $R_{\alpha\beta}$ and scalar curvature R as

$$R_{\alpha\beta\gamma\delta} = h_{\alpha\gamma}R_{\beta\delta} + h_{\beta\delta}R_{\alpha\gamma} - h_{\alpha\delta}R_{\beta\gamma} - h_{\beta\gamma}R_{\alpha\delta}$$
$$+ \tfrac{1}{2}\left(h_{\alpha\delta}h_{\beta\gamma} - h_{\alpha\gamma}h_{\beta\delta}\right)R. \tag{24.24}$$

Substituting in the above, and using the fact that, in three–dimensions (24.22) implies that $R_{\alpha\beta} = 2Kh_{\alpha\beta}$ and $R = 6K$, then gives (24.23) (exercise). We have therefore shown the following

> In three dimensions, a space that is isotropic about every point is a space of constant curvature i.e. $R_{\alpha\beta\gamma\delta} = K(h_{\alpha\gamma}h_{\beta\delta} - h_{\alpha\delta}h_{\beta\gamma})$, where K is a constant.

As we shall see from solving (24.22), the geometries of these spaces are qualitatively different, depending on whether the curvature is positive, negative, or zero. Now, since the 3-space is isotropic about every point, it must be **spherically symmetric** about every point. It follows that the line element will have the form (compare with (15.37))

$$d\sigma^2 = h_{\alpha\beta}dx^\alpha dx^\beta = e^\lambda dr^2 + r^2(d\theta^2 + \sin^2\theta d\phi^2), \tag{24.25}$$

where $\lambda = \lambda(r)$. The non-vanishing components of the Ricci tensor are

$$\overset{(3)}{R}_{11} = \lambda'/r, \quad \overset{(3)}{R}_{22} = \operatorname{cosec}^2 \theta, \quad \overset{(3)}{R}_{33} = 1 + \tfrac{1}{2}re^{-\lambda}\lambda' - e^{-\lambda}, \quad (24.26)$$

and, using (24.22), these reduce to the two equations

$$\lambda'/r = 2Ke^{\lambda}, \qquad 1 + \tfrac{1}{2}re^{-\lambda}\lambda' - e^{-\lambda} = 2Kr^2. \qquad (24.27)$$

The solution of these equations is

$$e^{-\lambda} = 1 - Kr^2. \qquad (24.28)$$

We have shown that the metric for a 3-space of constant curvature is

$$d\sigma^2 = \frac{dr^2}{1 - Kr^2} + r^2(d\theta^2 + \sin^2\theta d\phi^2), \qquad (24.29)$$

where K is positive, negative, or zero. We can introduce a new radial parameter $\bar{r}$ related to r by

$$r = \bar{r}/(1 + \tfrac{1}{4}K\bar{r}^2), \qquad (24.30)$$

in which case the metric takes on the conformally flat form (exercise)

$$d\sigma^2 = (1 + \tfrac{1}{4}K\bar{r}^2)^{-2}[d\bar{r}^2 + \bar{r}^2(d\theta^2 + \sin^2\theta d\phi^2)]. \qquad (24.31)$$

Combining this with the results of the last section, we obtain the line element for relativistic cosmology, namely,

$$ds^2 = dt^2 - [S(t)]^2 \left(\frac{dr^2}{1 - Kr^2} + r^2(d\theta^2 + \sin^2\theta d\phi^2) \right), \qquad (24.32)$$

or, in terms of the barred radial coordinate,

$$ds^2 = dt^2 - [S(t)]^2 \frac{d\bar{r}^2 + \bar{r}^2(d\theta^2 + \sin^2\theta d\phi^2)}{\left(1 + \tfrac{1}{4}K\bar{r}^2\right)^2}. \qquad (24.33)$$

We prefer to write these line elements in an alternative form where the arbitrariness in the magnitude of K is absorbed into the radial coordinate and the scale factor. Assuming $K \neq 0$, we define k by $K = |K|k$, so that k is +1 or −1, depending on whether K is positive or negative, respectively. If we introduce a **rescaled radial coordinate**

$$r^* = |K|^{\frac{1}{2}}r, \qquad (24.34)$$

then (24.32) becomes (exercise)

$$ds^2 = dt^2 - \frac{[S(t)]^2}{|K|} \left(\frac{dr^{*2}}{1 - kr^{*2}} + r^{*2}(d\theta^2 + \sin^2\theta d\phi^2) \right). \qquad (24.35)$$

Finally, we define a rescaled scale function $R(t)$ by (see (24.16))

$$R(t) = S(t)/|K|^{\frac{1}{2}}, \quad \text{if } K \neq 0,$$
$$R(t) = S(t), \quad \text{if } K = 0.$$

Then, dropping the stars on the radial coordinate, we have shown that the line element of relativistic cosmology can be written in the alternative form

$$ds^2 = dt^2 - [R(t)]^2 \left(\frac{dr^2}{1 - kr^2} + r^2(d\theta^2 + \sin^2 \theta d\phi^2) \right), \qquad (24.36)$$

or, in terms of the barred radial coordinate,

$$ds^2 = dt^2 - [R(t)]^2 \frac{[d\bar{r}^2 + \bar{r}^2(d\theta^2 + \sin^2 \theta d\phi^2)]}{[1 + \frac{1}{4}k\bar{r}^2]^2}, \qquad (24.37)$$

where k is now either $+1$, -1, or 0. This second form is called the **Robertson–Walker** line element after the first investigators to obtain it. At any epoch $t = t_0$, the geometry of the slice is given by

$$d\sigma^2 = R_0^2 \left(\frac{dr^2}{1 - kr^2} + r^2(d\theta^2 + \sin^2 \theta d\phi^2) \right), \qquad (24.38)$$

where the constant R_0 is given by $R_0 = R(t_0)$. In the next section, we shall investigate further the geometry of these 3-spaces of constant curvature for the three cases $k = +1$, 0, and -1.

24.8 The geometry of 3-spaces of constant curvature

Case 1: $k = +1$

Notice that, in this case, the coefficient of dr^2 becomes singular as $r \to 1$. We therefore introduce a new coordinate χ where

$$r = \sin \chi, \qquad (24.39)$$

so that

$$dr = \cos \chi d\chi = (1 - r^2)^{\frac{1}{2}} d\chi,$$

and (24.38) becomes

$$d\sigma^2 = R_0^2 [d\chi^2 + \sin^2 \chi (d\theta^2 + \sin^2 \theta d\phi^2)]. \qquad (24.40)$$

We can now embed this 3-surface in a 4-dimensional Euclidean space with coordinates (w, x, y, z), where

$$
\begin{aligned}
w &= R_0 \cos \chi, \\
x &= R_0 \sin \chi \sin \theta \cos \phi, \\
y &= R_0 \sin \chi \sin \theta \sin \phi, \\
z &= R_0 \sin \chi \cos \theta.
\end{aligned}
\tag{24.41}
$$

The embedding is possible because (exercise)

$$
d\sigma^2 = dw^2 + dx^2 + dy^2 + dz^2 = R_0^2 \left[d\chi^2 + \sin^2 \chi (d\theta^2 + \sin^2 \theta d\phi^2) \right],
$$

in agreement with (24.40). Also, from (24.41), we get (exercise)

$$
w^2 + x^2 + y^2 + z^2 = R_0^2,
\tag{24.42}
$$

which shows that the surface can be regarded as a 3-dimensional sphere in 4-dimensional Euclidean space. This is depicted in Fig. 24.5, where one dimension ($y = 0$ or $\phi = 0$) is suppressed. The hypersurface is defined by the coordinate range

$$
0 \leqslant \chi \leqslant \pi, \quad 0 \leqslant \theta \leqslant \pi, \quad 0 \leqslant \phi < 2\pi.
$$

The 2-surfaces $\chi =$ constant, which appear as circles in the pictures, are 2-spheres of surface area (exercise)

$$
A_\chi = \int_{\theta=0}^{\pi} \int_{\phi=0}^{2\pi} (R_0 \sin \chi d\theta)(R_0 \sin \chi \sin \theta d\phi) = 4\pi R_0^2 \sin^2 \chi,
$$

and (θ, ϕ) are the standard spherical polar coordinates of these 2-spheres. Thus, the area of these 2-spheres is zero at the North Pole, increases to a maximum at the equator, and decreases again to zero at the South Pole. The surface has 3-volume given by (exercise)

$$
\begin{aligned}
V &= \int_{\chi=0}^{\pi} \int_{\theta=0}^{\pi} \int_{\phi=0}^{2\pi} (R_0 d\chi)(R_0 \sin \chi d\theta)(R_0 \sin \chi \sin \theta d\phi) \\
&= 2\pi^2 R_0^3 = 2\pi^2 R^3(t_0),
\end{aligned}
\tag{24.43}
$$

which is why $R(t_0)$ is often referred to as the 'radius of the universe'.

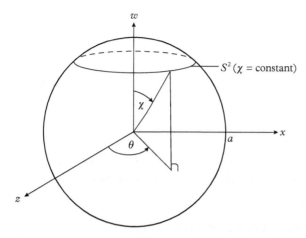

Fig. 24.5 A surface of constant positive curvature embedded in a 4-dimensional Euclidean space ($\phi = 0$).

This 3-space is clearly the generalization of an S^2, or 2-sphere, to a 3-dimensional entity and is called an S^3, or **3-sphere**. The physical space should not really be thought of as embedded in anything else, since it is the totality of everything that exists at any one epoch. Thus, there are no physical points outside it nor does it have a boundary. It may be helpful to think of it as follows. If we introduce yet another radial-type coordinate r', where $r' = R_0\chi$, then (24.40) becomes

$$\mathrm{d}\sigma^2 = \mathrm{d}r'^2 + R_0^2 \sin^2(r'/R_0)(\mathrm{d}\theta^2 + \sin^2\theta\,\mathrm{d}\phi^2),$$

and the surface area of the 2-spheres $\chi = $ constant is given by

$$A_\chi = 4\pi R_0^2 \sin^2(r'/R_0).$$

Notice that, for small r', $\sin r' \sim r'$ and so $A_\chi \sim 4\pi r'^2$. Now choose any point P and consider the surface area of a series of 2-surfaces centred on P of increasing radius r', all at one epoch t_0. For small values of the radius r' (compared with $R(t_0)$), the area is close to the Euclidean value $4\pi r'^2$. As r' increases, the area increases but becomes increasingly less than $4\pi r'^2$. The surface area reaches a maximum value when $r' = \frac{1}{2}\pi$ and decreases from then on until it again becomes zero when $r' = \pi R_0$. In this space, any radial geodesic returns to its starting point. The topology of this space is variously called **closed**, **bounded**, or **compact**. The topology of the whole space-time is called cylindrical, since it is the product $\mathbb{R} \times S^3$, where $\mathbb{R}$ represents the 1-dimensional cosmic time (Fig. 24.6).

Case 2: $k = 0$

If we set

$$x = R_0 r \sin\theta \cos\phi,$$
$$y = R_0 r \sin\theta \sin\phi,$$
$$x = R_0 r \cos\theta,$$

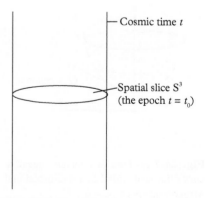

Fig. 24.6 The cylindrical topology $\mathbb{R} \times S^3$ of space-time when $k = +1$.

then (24.38) becomes

$$d\sigma^2 = dx^2 + dy^2 + dz^2,$$

which is clearly 3-dimensional Euclidean space. The 3-space is covered by the usual coordinate range

$$0 \leqslant r < \infty, \quad 0 \leqslant \theta \leqslant \pi, \quad 0 \leqslant \phi < 2\pi.$$

The topology of the space-time is the same as that of 4-dimensional Euclidean space, namely $\mathbb{R}^4$, and is called **open**.

Case 3: $k = -1$

If we introduce a new coordinate χ, where

$$r = \sinh \chi, \tag{24.44}$$

then

$$dr = \cosh \chi d\chi = (1 + r^2)^{\frac{1}{2}} d\chi,$$

and (24.38) becomes

$$d\sigma^2 = R_0^2[d\chi^2 + \sinh^2 \chi(d\theta^2 + \sin^2 \theta d\phi^2)]. \tag{24.45}$$

We can no longer embed this 3-surface in a 4-dimensional Euclidean space, but it can be embedded in a flat Minkowski space with signature +2 (exercise),

$$d\sigma^2 = -dw^2 + dx^2 + dy^2 + dz^2, \tag{24.46}$$

where

$$\left.\begin{aligned} w &= R_0 \cosh \chi, \\ x &= R_0 \sinh \chi \sin \theta \cos \phi, \\ y &= R_0 \sinh \chi \sin \theta \sin \phi, \\ z &= R_0 \sinh \chi \cos \theta. \end{aligned}\right\} \tag{24.47}$$

These equations imply that (exercise)

$$w^2 - x^2 - y^2 - z^2 = R_0^2, \tag{24.48}$$

so that the 3-surface is a 3-dimensional hyperboloid in 4-dimensional Minkowski space. This is depicted in Fig. 24.7, where one dimension ($y = 0$ or $\phi = 0$) is suppressed. The hypersurface is defined by the coordinate range

$$0 \leqslant \chi \leqslant \infty, \quad 0 \leqslant \theta \leqslant \pi, \quad 0 \leqslant \phi < 2\pi.$$

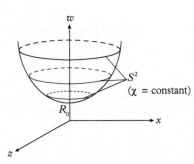

Fig. 24.7 Surface of constant negative curvature embedded in a 4-dimensional Minkowski space ($\phi = 0$).

The 2-surfaces $\chi = $ constant, which appear as circles in the pictures, are 2-spheres of surface area

$$A_\chi = 4\pi R_0^2 \sinh^2 \chi,$$

and (θ, ϕ) are the standard spherical polar coordinates of these 2-spheres. As χ ranges from 0 to ∞, the area of the successive 2-spheres increases from zero to infinity. For large χ, the surface area increases far more rapidly than it would if the hypersurface were flat. The 3-volume of the surface is infinite. The topology is again $\mathbb{R}^4$ and open. In each of the three cases, we have only specified the simplest topology possible; in fact, other topologies are possible by identifying points or regions, but we will not consider the issue further.

24.9   Friedmann's equation

Our three ingredients of relativistic cosmology are as follows.

(1)   The cosmological principle, which as we have shown leads to the Robertson-Walker line element, namely,

$$ds^2 = dt^2 - [R(t)]^2 \frac{[d\bar{r}^2 + \bar{r}^2(d\theta^2 + \sin^2\theta d\phi^2)]}{[1 + \frac{1}{4}k\bar{r}^2]^2}; \qquad (24.49)$$

(2) Weyl's postulate, which requires that the substratum is a perfect fluid, namely,

$$T_{ab} = (\rho + p)u_a u_b - p g_{ab}; \qquad (24.50)$$

(3)   General relativity, with the cosmological term, namely,

$$G_{ab} - \Lambda g_{ab} = 8\pi T_{ab}. \qquad (24.51)$$

Then using the fact that, in our preferred coordinate system,

$$u^a \overset{*}{=} (1, 0, 0, 0),$$

the field equations (24.51) lead to two independent equations (exercise)

$$3\frac{\dot{R}^2 + k}{R^2} - \Lambda = 8\pi\rho, \qquad (24.52)$$

$$\frac{2R\ddot{R} + \dot{R}^2 + k}{R^2} - \Lambda = -8\pi p, \qquad (24.53)$$

where we have used relativistic units and a dot denotes differentiation with respect to time. By homogeneity and isotropy, the density and pressure can only be functions of time t. Together with these equations, we have the requirements that the fluid is physically realistic, as expressed in the dominant energy conditions (12.64). Using our Newtonian analogue, (24.53) involves a second time-derivative of R and so may be thought of as an equation of motion, whereas (24.52) only involves a first time derivative of R and so may be considered an integral of the motion, that is, an **energy** equation. If we differentiate (24.52) with respect to t, multiply through by $1/8\pi$, and add the result to (24.53) multiplied through by $-3\dot{R}/8\pi R$, we get

$$\dot{\rho} + 3p\frac{\dot{R}}{R} = -\frac{3}{8\pi}\frac{\dot{R}}{R}\left(\frac{3\dot{R}^2}{R^2} + \frac{3k}{R^2} - \Lambda\right) = -3\rho\frac{\dot{R}}{R},$$

again using (24.52). Multiplying through by R^3, we can rewrite this in the form

$$\frac{d}{dt}\left(\rho R^3\right) + p\frac{d}{dt}\left(R^3\right) = 0. \qquad (24.54)$$

Consider a set of particles in the substratum enclosing a volume V. Then, clearly, owing to the motion of the substratum, $V \sim R^3(t)$. If we now call the total mass-energy in the volume $E = \rho V$, then equation (24.54) can be written in the form

$$dE + pdV = 0. \qquad (24.55)$$

This is the first law of thermodynamics, or **conservation of energy**, and shows that the pressure does work in the expansion. This is exactly the same equation as results from the conservation equations (exercise)

$$T^{ab}{}_{;b} = 0. \qquad (24.56)$$

Thus, the field equations of the theory contain in them the equation for the conservation of energy. We have met this before in §13.4, and it arises from the fact that the field equations (24.51) satisfy the contracted Bianchi identities

$$\left(G^{ab} - \Lambda g^{ab}\right)_{;b} = 0,$$

which, in turn, leads to the conservation equations (24.56).

It is sometimes useful to eliminate $\dot{R}$ from (24.52) and (24.53). If one multiplies (24.52) by 3 and subtracts (24.53), this gives

$$\frac{\ddot{R}}{R} = -\frac{4\pi}{3}\left(\rho + 3p - \frac{\Lambda}{4\pi}\right).$$

If $\rho + 3p > \Lambda/4\pi$, then the right-hand side is negative so that $\ddot{R} < 0$. This means that the graph of $R(t)$ lies below the tangent at any point. Current observations show that the universe is currently expanding so that the tangent has positive slope and must have crossed the R axis at some finite point in the past. It therefore follows that the graph of $R(t)$ must also have been zero at some finite time in the past (see Fig. 24.8) i.e. there must have been a big bang. For the case where $\Lambda = 0$, then the required condition is just $\rho + 3p > 0$, which is just the strong energy condition. This is sometimes called the **FRW singularity theorem**.

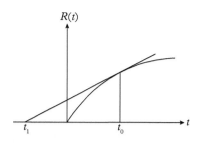

Fig. 24.8 Graph of $R(t)$ when $\rho + 3p > \Lambda/4\pi$.

> If $\Lambda = 0$ and $\rho + 3p > 0$, then $\dot{R}(t_0) > 0$ implies that there was an initial big bang singularity at some finite time in the past.

The pressure p includes all types of pressure, such as that due to the random motion of the stars and galaxies, that due to heat motion of molecules, radiation pressure, and so forth. However, observation reveals that, at the present epoch, the pressure is far smaller than the energy density ρ, due to matter. The ratio of the two quantities is about 10^{-5} or 10^{-6}. Accordingly, as long as only states of the universe differing not too widely from the present one are considered, we may take

$$p = 0, \qquad (24.57)$$

and so we may model the matter in the universe as if it is comprised of dust. Then (24.53) integrates immediately to give

$$R(\dot{R}^2 + k) - \tfrac{1}{3}\Lambda R^3 = C, \qquad (24.58)$$

where C is a constant of integration, and, using (24.52), we find

$$C = \tfrac{8}{3}\pi R^3 \rho. \qquad (24.59)$$

Apart from a numerical factor, this is the energy content E of a volume V of the substratum and is constant immediately by (24.55), which becomes a conservation of mass equation when p vanishes. The value can be remembered as twice the mass of a spherical volume of a Euclidean universe of radius R and density ρ. If we now use (24.59) to eliminate ρ in (24.52), the result can be written in the form

$$\dot{R}^2 = \frac{C}{R} + \tfrac{1}{3}\Lambda R^2 - k. \qquad (24.60)$$

This is **Friedmann's equation** for the time variation of the scale factor in the absence of pressure. Note that it is identical with the Newtonian analogue (24.17). We shall consider the solutions of this equation – the **Friedmann models** – in the next chapter. Some authors refer to these as the **FRW models**, short for Friedmann–Robertson–Walker models. Recall that, in obtaining the Newtonian analogue (24.17), we imposed the assumption (24.7), which is essentially equivalent to Hubble's law (24.10) (see Exercise 24.4). In the rest of this chapter, we shall consider light propagation and distance in relativistic cosmology in order to **deduce** Hubble's law from the premises of the theory.

24.10   Propagation of light

We assume that light propagates in relativistic cosmology in the same way as it does in general relativity. Let us consider how an observer O receives light from a receding galaxy. We use the unbarred form of the Robertson–Walker line element (24.36). Since we assume that the time slices are homogeneous 3-spaces, we can, without loss of generality, take O to be at the origin of coordinates $r = 0$. Inserting the conditions for a radial null geodesic, namely,

$$ds^2 = d\theta = d\phi = 0,$$

into (24.36), we find

$$\frac{dt}{R(t)} = \pm\frac{dr}{(1 - kr^2)^{\frac{1}{2}}}, \tag{24.61}$$

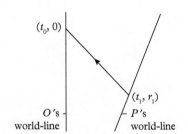

$(t_0, 0)$

(t_1, r_1)

O's
world-line

P's
world-line

Fig. 24.9 Light ray from galaxy P to observer O.

where the + sign corresponds to a receding light ray and the minus sign to an approaching light ray. Consider a light ray emanating from a galaxy P with world-line $r = r_1$ at coordinate time t_1, and received by O at coordinate time t_0 (Fig. 24.9). Using (24.61), we get (exercise)

$$\int_{t_1}^{t_0} \frac{dt}{R(t)} = -\int_{r_1}^{0} \frac{dr}{(1 - kr^2)^{\frac{1}{2}}} = f(r_1), \tag{24.62}$$

where

$$f(r_1) = \begin{cases} \sin^{-1} r_1 & \text{if } k = +1, \\ r_1 & \text{if } k = 0, \\ \sinh^{-1} r_1 & \text{if } k = -1. \end{cases} \tag{24.63}$$

Next, consider two successive light rays emanating from P at times t_1 and $t_1 + dt_1$, respectively, and received by O at times t_0 and $t_0 + dt_0$, respectively (Fig. 24.10). Then, from (24.62),

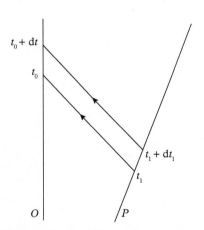

$t_0 + dt$

t_0

$t_1 + dt_1$

t_1

O

P

Fig. 24.10 Successive light rays from galaxy P to observer O.

$$\int_{t_1 + dt_1}^{t_0 + dt_0} \frac{dt}{R(t)} = \int_{t_1}^{t_0} \frac{dt}{R(t)},$$

since each side is equal to the same function $f(r_1)$. Therefore (check),

$$\int_{t_1 + dt_1}^{t_0 + dt_0} \frac{dt}{R(t)} - \int_{t_1}^{t_0} \frac{dt}{R(t)} = \int_{t_0}^{t_0 + dt_0} \frac{dt}{R(t)} - \int_{t_1}^{t_1 + dt_1} \frac{dt}{R(t)} = 0,$$

and, assuming that $R(t)$ does not vary greatly over the intervals dt_1 and dt_0, we can take it outside the integral in the last equation and deduce that

$$\frac{dt_0}{R(t_0)} = \frac{dt_1}{R(t_1)}. \qquad (24.64)$$

All fundamental particles (galaxies) of the substratum have world-lines on which the coordinates r, θ, and ϕ are constant and hence, from (24.36), $ds^2 = dt^2$. It follows that t measures the **proper time** along the substratum world-lines. The intervals dt_1 and dt_2 are the proper time intervals between the rays as measured at the source and observer, respectively. Hence, from (24.64), the interval, as measured by O, is $R(t_0)/R(t_1)$ times the interval measured by P. Thus if a signal is emitted with frequency ν_1 at time t_1 by P and observed with frequency ν_0 at time t_0 by O (see Fig. 24.10), then

$$\nu_1/\nu_0 = R(t_0)/R(t_1). \qquad (24.65)$$

It follows that, in an expanding universe,

$$t_0 > t_1 \implies R(t_0) > R(t_1),$$

the observer O will experience a **redshift** z defined as

$$z := \frac{\nu_1 - \nu_0}{\nu_0},$$

which is given by

$$1 + z = \nu_1/\nu_0 = R(t_0)/R(t_1). \qquad (24.66)$$

This redshift is sometimes called a **Doppler shift**, but is not to be confused with the special relativistic Doppler shift. Clearly, in a contracting universe, O will detect a corresponding blue shift.

If, roughly speaking, P is 'near' to O, then the cosmic times of emission and reception differ only by a small amount, dt, say, that is, $t_0 = t_1 + dt$, and so (24.66) produces

$$1 + z = \frac{R(t_0)}{R(t_0 - dt)} \simeq \frac{R(t_0)}{R(t_0) - \dot{R}(t_0)dt} \simeq 1 + \frac{\dot{R}(t_0)}{R(t_0)}dt, \qquad (24.67)$$

to first order in dt. In addition,

$$\int_{t_1}^{t_0} \frac{dt}{R(t)} = \int_{t_1}^{t_1+dt} \frac{dt}{R(t)} \simeq \frac{dt}{R(t_1)} = \frac{dt}{R(t_0 - dt)} \simeq \frac{dt}{R(t_0)}.$$

But for small r, using (24.63),

$$\int_{t_1}^{t_0} \frac{dt}{R(t)} = f(r_1) \simeq r_1.$$

Combining this with (24.67), we get the result

$$z \simeq \dot{R}(t_0)r_1. \qquad (24.68)$$

Thus, at any one epoch, for small distances, the **redshift z is proportional to the distance r_1**. Interpreting the redshift z as if it were caused by a velocity of recession, we have obtained a velocity–distance relation similar to Hubble's law. To make this more precise, we need to consider how distance is measured, as least theoretically, on a cosmologically interesting scale.

24.11   A cosmological definition of distance

Because we have a world time, it is mathematically easy to define an **absolute** distance between fundamental particles by considering them at the same value of world time and then measuring the geodesic distance between them in the slice (Fig. 24.11). If we set $dt = d\theta = d\phi = 0$ in (24.36), then the absolute distance d_A between O and P at time t is

$$d_A = R(t) \int_0^{r_1} \frac{dr}{(1 - kr^2)^{\frac{1}{2}}}. \qquad (24.69)$$

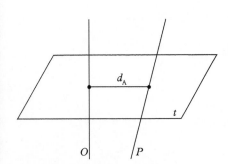

Fig. 24.11 An absolute distance between fundamental particles.

This is of no practical use and so we try another tack. If we know the actual size of a distant object, then we can define an observational distance by

$$d_O = \alpha/\beta, \qquad (24.70)$$

where α is the **actual** diameter of the object and β is the **observed** angular diameter. Such a definition would be satisfactory if some means of determining α were known. Since this is not known, we instead use a definition based on the **apparent** luminosity of what we observe. Let E be the energy radiated per unit time by the distant object and let I be the intensity of the radiation received per unit area per unit time. Then, assuming the energy is distributed uniformly on a sphere in a Euclidean space and neglecting the redshift, the distance can be defined as $(E/4\pi I)^{1/2}$. But, in an expanding universe, the interval of time during which a certain amount of energy is received is longer than the interval of emission by virtue of the Doppler shift, and hence the number of photons received per unit time is reduced by the factor $1 + z$. In addition, the energy of each photon of light is reduced by the same factor (because energy is the time component of a 4-vector and so the transformation from one observer to another introduces the factor $1 + z$). These considerations lead to the definition of a **luminosity distance** d_L, where

$$d_{\mathrm{L}}^2 = \frac{E}{4\pi I(1 + z)^2}. \tag{24.71}$$

The luminosity distance is in essence the distance used by astronomers. However, the detailed way in which astronomical distances are measured is quite complicated and beyond the scope of this book (for further details, see Weinberg 1972), although we mention, without being precise, that one unit of measurement is called the **apparent magnitude**, m. It is related to the energy received, E_R, by the relation

$$m = \text{constant} - 0.4 \log_{10} E_R.$$

Moreover, there is a problem with the definition (24.71), because it involves E, the **absolute** luminosity of the source, which is not observationally measurable. This definition thus appears to suffer from the same defects as (24.70). However, the distances to nearby galaxies may be determined by other means and hence their absolute luminosities may be calculated, and it appears that all galaxies have roughly the same absolute luminosity. So a first approximation is to take all galaxies as having the same absolute luminosity. This assumption is almost certainly wrong, though, because, if we live in an evolutionary universe, then the mean age of the more distant galaxies is much less than the mean age of nearby galaxies and so there is no reason to believe that they will have the same mean luminosity. Certain sources are believed to have luminosities that show little variation in space and time and are referred to as **standard candles**. Examples of these are Cepheid variable stars and Type Ia supernovae. However, even if one accepts that these are standard candles, there still exists the problem of calibration which enables one to convert relative distances to absolute distances. We will not pursue the matter further here, but simply employ (24.71).

24.12   Hubble's law in relativistic cosmology

We start by finding an expression for the luminosity distance in terms of the coordinates of the Robertson–Walker line element in the unbarred form (24.36). Consider light emanating from galaxy P at time t_1 and observed by us 'now' at O at a time $t = t_0$ $(t_1 < t_0)$ (Fig. 24.12). The light will have spread out over the surface of a sphere with centre at the event P_0 $(t = t_0, r = r_1)$ and passing through the event O_0 $(t = t_0, r = 0)$. The surface area of this sphere is the same as that of the sphere centred

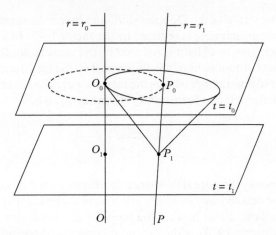

Fig. 24.12 Light from P_1 spreading out on a sphere passing through O_0.

on O_0 passing through P_0 (dotted line in Fig. 24.12), owing to the homogeneity of the 3-sphere. The line element for this sphere ($t = t_0$, $r = r_1$) is, from (24.36),

$$ds^2 = -[R(t_0)r_1]^2 (d\theta^2 + \sin^2 \theta d\phi^2).$$

This is the usual line element for a sphere of radius $R(t_0)r_1$ and so the sphere has surface area $4\pi R^2(t_0)r_1^2$. Hence, the observed intensity is given by

$$I = \frac{E}{4\pi R(t_0)^2 r_1^2 (1 + z)^2},$$

taking into account the double Doppler shift factor. Comparing this with (24.71), we obtain an expression for the luminosity distance in terms of the scale factor, namely,

$$d_L = r_1 R(t_0). \tag{24.72}$$

If we define the **Hubble parameter** by (see (24.9))

$$H(t) = \dot{R}(t)/R(t), \tag{24.73}$$

then (24.68) and (24.72) give

$$z \simeq H(t_0)d_L, \tag{24.74}$$

where $H(t_0)$ is the value of the Hubble parameter at the current epoch and is called **Hubble's constant**. This is the famous **Hubble law** in relativistic cosmology. It states that, for 'nearby' galaxies, the radial velocity of recession as measured by the redshift z is proportional to its distance. The dimension of $H(t)$ is that of inverse time and so, if we define $T = 1/H(t)$, then T has the dimension of time. Current observations give the value

$$T_0 \approx 10^{10} \text{years}, \tag{24.75}$$

which is believed correct to within a factor of 2. Note that, if $\rho + 3p > \Lambda/4\pi$, so the conditions that guarantee a big bang hold, then it is clear from Fig. 24.8 that the age of the universe must be less than the Hubble time T_0 of approximately 10^{10} years.

We stress that Hubble's law is an approximate one in relativistic cosmology and consider a more exact formula in Chapter 25. We now define a dimensionless quantity q called the **deceleration parameter** by

$$q(t) = -R\ddot{R}/\dot{R}^2. \tag{24.76}$$

Then, since $R > 0$ and $\dot{R}^2 > 0$, it follows that

$$\ddot{R} < 0 \quad \Rightarrow \quad q > 0,$$

and so a positive q measures the rate at which the expansion of the universe is slowing down. The value of the deceleration parameter q_0 is uncertain, but most current measurements make it negative (so the rate of expansion is increasing) with a typical range

$$q_0 = -0.6 \pm 0.3. \tag{24.77}$$

We will look at this and the other observational parameters in more detail in Chapter 25. Taking the second-order term into account in the Taylor expansion giving (24.67), we find the relationship (exercise)

$$d_L = zT_0[1 - \tfrac{1}{2}(1 + q_0)z + \cdots]. \tag{24.78}$$

For objects too close to the observer, the random motions which we have excluded from our model do, in fact, obscure the general motion. But there is a good range of nebulae satisfying the velocity–distance law (out to about the 18th magnitude) from which a good determination of T can be made. For more distant observations, the relationship (24.78) must be used, which is crucially dependent on the value of q_0. In Fig. 24.13, we present some data given in a 1970 review by Sandage and give further details on more recent measurements in Chapter 25. It is remarkable to note Hubble first proposed his law in 1929 on the basis of observations of only eighteen nearby galaxies, and this data corresponds to a tiny part of the graph in Fig. 24.13.

Differentiating Friedmann's equation (24.60), we get

$$2\dot{R}\ddot{R} = -\frac{C}{R^2}\dot{R} + \tfrac{2}{3}\Lambda R\dot{R},$$

and multiplying by $-R/2\dot{R}^3$ gives

$$-\frac{R\ddot{R}}{\dot{R}^2} = \frac{C}{2R\dot{R}^2} - \tfrac{1}{3}\Lambda\frac{R^2}{\dot{R}^2}.$$

Then, using (24.76), (24.59), and (24.73), we can write this in the form

$$q = (\tfrac{4}{3}\pi\rho - \tfrac{1}{3}\Lambda)/H^2. \tag{24.79}$$

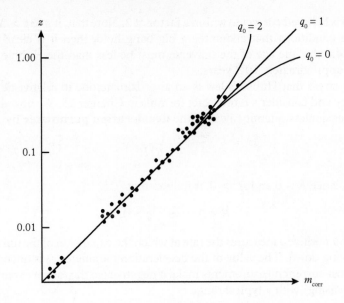

Fig. 24.13 Redshift versus corrected apparent magnitudes (Sandage 1970).

This shows that there is an intimate connection between the deceleration parameter q, the Hubble parameter H, and ρ, the mean density of the universe.

Exercises

24.1 (§24.2) Show that Olbers' paradox remains if we assume space is non-Euclidean but still homogeneous by the cosmological principle.

24.2 (§24.3) Write down an expression for the cosmological potential energy V_{c_i} of the ith particle such that $\mathbf{F}_i = -\text{grad } V_{c_i}$.

24.3 (§24.3) Substitute (24.7) into (24.6) and establish (24.11)–(24.14). Identify A physically. Show that, if $E \neq 0$, the choice (24.18) of μ in (24.16) allows (24.11) to be written in the standard form (24.17) with $k = \pm 1$.

24.4 (§24.3) Integrate Hubble's law (24.10) and deduce the scale factor law (24.7), identifying the function $S(t)$ in terms of $H(t)$.

24.5 (§24.7) Show that, in three dimensions, (24.22) and (24.24) result in the fact that

$$R_{\alpha\beta\gamma\delta} = K(h_{\alpha\gamma}h_{\beta\delta} - h_{\alpha\delta}h_{\beta\gamma}).$$

24.6 (§24.7) Show, by using the contracted Bianchi identities, that if $G_{ab} = -Kg_{ab}$ then K is a constant.

24.7 (§24.7) Work out the non-vanishing components (24.26) of the Ricci tensor for the line element (24.25). [Hint: use the results of Exercise 6.32(iv).] Confirm that, if this line element is a space of constant curvature, then λ is given by (24.28).

24.8 (§24.7) Confirm the results of the following transformations:
(i) (24.30) transforms (24.29) into (24.31);
(ii) (24.34) transforms (24.32) into (24.35).

24.9 (§24.8) Confirm the results of the following transformations:
(i) (24.39) transforms (24.38) with $k = +1$ into (24.40);
(ii) (24.44) transforms (24.38) with $k = -1$ into (24.45).

24.10 (§24.8)
(i) Show that (24.41) is a parametric form of the surface (24.42) in Euclidean 4-space. Confirm that its line element reduces to (24.40).
(ii) Show that (24.47) is a parametric form of the surface (24.48) in Minkowski space with line element (24.46). Confirm that its line element reduces to (24.45).

24.11 (§24.8) Write down the line elements for the 2-spheres χ = constant in the cases $k = 1$ and $k = -1$. By comparing them with the standard line element for a sphere of radius a, namely,

$$ds^2 = a^2(d\theta^2 + \sin^2\theta d\phi^2),$$

confirm the formula for the surface area A_χ in each case. Confirm (24.43) and show that the volume of the 3-surface in the case $k = -1$ is infinite.

24.12 (§24.9) Establish the field equations of relativistic cosmology (24.52) and (24.53). [Hint: this involves working out the Ricci and Einstein tensors for the line element (24.49).]

24.13 (§24.9) Use (24.52) and (24.53) to establish the result (24.55). Confirm that the same equation results from the conservation law (24.56)
(i) by direct computation
(ii) without utilizing expressions for the connection. [Hint: take the covariant derivative of every term in (24.50) and use the results that u^a is a tangent to a geodesic, $u^a \overset{*}{=} \delta_0^a$ and $u^a_{;a} = \frac{1}{2}(d/dt)(\ln g)$ – why?]

24.14 (§24.9) Use (24.52) and (24.53) to obtain (24.60) subject to (24.59) in the case (24.57).

24.15 (§24.10) Confirm the result (24.62) subject to (24.63). Deduce (24.64) and (24.68).

24.16 (§24.12) Confirm Hubble's law in the form (24.74) to first order.

24.17 (§24.12)

(i) Calculate $1 + z = R(t_0)/R(t_0 - \delta t)$ to second order accuracy and hence show that to this order

$$z = \frac{\dot{R}(t_0)}{r(t_0)} \delta t + \left(\frac{\dot{R}^2(t_0)}{R^2(t_0)} - \frac{\ddot{R}(t_0)}{2R(t_0)} \right) (\delta t)^2.$$

(ii) Use the fact that to second order $(dt)^2 = (R^2(t_0)/\dot{R}^2(t_0))z^2$ together with the result of part (i) to show that to second order in z

$$\delta t = T_0 z (1 - (1 + \tfrac{1}{2} q_0) z).$$

(iii) Use the trapezium rule

$$\int_{t_0 - \delta t}^{t_0} \frac{dt}{R(t)} = \frac{1}{2} \left(\frac{1}{R(t_0)} + \frac{1}{R(t_0 - \delta t)} \right) \delta(t),$$

to show that to second order

$$d_L = R(t_0) \left[\frac{\delta t}{R(t_0)} + \frac{\dot{R}(t_0)}{2R^2(t_0)} (\delta t)^2 \right].$$

(iv) Deduce the second order version of Hubble's law (24.78).

Further reading

The main source for this chapter is the classic text by Bondi (1961). Our brief look at Newtonian cosmology can be pursued in the book by Landsberg and Evans (1977). Peter Landsberg was for many years a colleague of ours at Southampton. The book by Weinberg (1972) goes into more detail on many aspects of cosmology and includes a comprehensive discussion of distance in cosmology.

Bondi, H. (1961). *Cosmology*, Cambridge University Press, Cambridge.

Landsberg, P. and Evans, D. A. (1977). *Mathematical cosmology: An Introduction*, Clarendon Press, Oxford.

Weinberg, S. (1972). *Gravitation and cosmology*, Wiley, New York, NY

The classical cosmological models

25

25.1 The flat space models

In this chapter, we will mostly discuss the mathematical features of the classical cosmological models that one obtains by solving Friedmann's equation. We will look at how these relate to the observational evidence as well as considering some more sophisticated models in the next chapter.

Our considerations of the last chapter led to Friedmann's equation

$$\dot{R}^2 = \frac{C}{R} + \tfrac{1}{3}\Lambda\dot{R} - k \qquad (25.1)$$

governing the dynamics of the scale factor in a universe in which the matter is modelled by presureless dust. The task in this chapter is to solve this non-linear first-order ordinary differential equation for different values of the parameters occurring in it. Recall that the values of these parameters are governed by the requirements

$$C > 0, \quad -\infty < \Lambda < \infty, \quad k = -1, 0, +1. \qquad (25.2)$$

There are a number of ways of proceeding. The equation can be solved, in general, by using elliptic functions, or resort can be made to computer plots of numerically generated solutions. However, many of the sub-cases can be integrated directly using elementary functions or, failing that, elementary functions can be used to investigate their qualitative features. We shall not give an exhaustive account of this approach here (but see Landsberg and Evans (1977) for details). Instead, we shall restrict our attention in this section to the important cases of flat space ($k = 0$) and vanishing cosmological constant ($\Lambda = 0$). The techniques employed may be applied to the other cases.

In the flat space case, $k = 0$, (25.1) reduces to

$$\dot{R}^2 = C/R + \tfrac{1}{3}\Lambda R^2. \qquad (25.3)$$

We first assume $\Lambda > 0$ and introduce a new variable

$$u = \frac{2\Lambda}{3C}R^3.$$

Introducing Einstein's Relativity. Second Edition. Ray d'Inverno and James Vickers, Oxford University Press.
© Ray d'Inverno and James Vickers (2022). DOI: 10.1093/oso/9780198862024.003.0025

Differentiating, we get

$$\dot{u} = \frac{2\Lambda}{C} R^2 \dot{R},$$

and, substituting in (25.3), we find

$$
\begin{aligned}
\dot{u}^2 &= \frac{4\Lambda^2}{C^2} R^4 \left(\frac{C}{R} + \tfrac{1}{3}\Lambda R^2 \right) \\
&= \frac{4\Lambda^2}{C} R^3 + \frac{4\Lambda^3}{3C^2} R^6 \\
&= 6\Lambda u + 3\Lambda u^2 \\
&= 3\Lambda(2u + u^2).
\end{aligned}
\tag{25.4}
$$

Taking the positive square root, we have

$$\dot{u} = (3\Lambda)^{\frac{1}{2}} (2u + u^2)^{\frac{1}{2}},$$

which can be integrated by parts. If we assume a big bang model, namely, $R = 0$ when $t = 0$, then $u = 0$ initially, and so integrating gives

$$\int_0^u \frac{du}{(2u + u^2)^{\frac{1}{2}}} = \int_0^t (3\Lambda)^{\frac{1}{2}} dt = (3\Lambda)^{\frac{1}{2}} t.$$

If we complete the square in the u-integral and set $v = u+1$ and $\cosh w = v$ then we get

$$\int_0^u \frac{du}{[(u+1)^2 - 1]^{\frac{1}{2}}} = \int_1^v \frac{dv}{(v^2 - 1)^{\frac{1}{2}}} = \int_0^w \frac{\sinh w dw}{(\cosh^2 w - 1)^{\frac{1}{2}}} = \int_0^w dw = w.$$

In terms of R, the solution becomes

$$R^3 = \frac{3C}{2\Lambda} [\cosh (3\Lambda)^{\frac{1}{2}} t - 1].
\tag{25.5}$$

If $\Lambda < 0$, we introduce a new variable

$$u = -\frac{2\Lambda}{3C} R^3,
\tag{25.6}$$

and then, proceeding as before, we obtain the solution (exercise)

$$R^3 = \frac{3C}{2(-\Lambda)} \{1 - \cos [3(-\Lambda)]^{\frac{1}{2}} t\}.
\tag{25.7}$$

For the case $\Lambda = 0$ equation (25.3) becomes $\dot{R}^2 = C/R$. Taking square roots, the equation is immediately separable, producing

$$R^{\frac{1}{2}} dR = C^{\frac{1}{2}} dt.$$

Integrating, using $R = 0$ when $t = 0$, we get

$$\tfrac{2}{3}R^{\frac{3}{2}} = C^{\frac{1}{2}}t,$$

so that

$$R = (\tfrac{9}{4}Ct^2)^{\frac{1}{3}}. \tag{25.8}$$

This is called the **Einstein–de Sitter model**. The Hubble parameter $H(t)$ and the deceleration parameter $q(t)$ can be easily computed from (25.5), (25.7), or (25.8). For example, in the Einstein–de Sitter case (exercise),

$$H(t) = \dot{R}/R = 2/(3t), \tag{25.9}$$

and

$$q(t) = -R\ddot{R}/\dot{R}^2 = \tfrac{1}{2}. \tag{25.10}$$

Rather than solving (25.1) exactly, we can look at the qualitative behaviour. In the initial stages of a big bang universe, R is small and so the term C/R dominates over $\tfrac{1}{3}\Lambda R^2$ in (25.1). Hence, for small t,

$$\dot{R}^2 \sim C/R, \tag{25.11}$$

and, integrating, we obtain, as in (25.8),

$$R \sim (\tfrac{9}{4}Ct^2)^{\frac{1}{3}}. \tag{25.12}$$

So, in the early stages, all big bang models behave like the Einstein–de Sitter model, namely, they expand at the rate $t^{2/3}$. If we write (25.3) in the form

$$\dot{R}^2 = F(R), \tag{25.13}$$

where

$$F(R) = C/R + \tfrac{1}{3}\Lambda R^2, \tag{25.14}$$

then much of the qualitative behaviour of R can be inferred from the behaviour of $F(R)$. For example,

$$\Lambda < 0 \quad \Rightarrow \quad F(R) = 0 \text{ when } R = R_m = [3C/(-\Lambda)]^{\frac{1}{3}},$$

so that $\dot{R}$ vanishes at $R = R_m$, which is a local minimum (exercise). Conversely, if $\Lambda \geqslant 0$, the solution grows without bound. In the case when

$\Lambda > 0$, then, for large t, the second term on the right dominates in (25.3), and so

$$\dot{R}^2 \sim \tfrac{1}{3}\Lambda R^2, \tag{25.15}$$

and, integrating, we find (exercise)

$$R \sim \exp[(\tfrac{1}{3}\Lambda)^{\frac{1}{2}} t]. \tag{25.16}$$

We now have enough information to sketch the graphs of the three models. We postpone this to §25.3.

25.2 Models with vanishing cosmological constant

In this section, we consider the case when Λ vanishes. Friedmann's equation then becomes

$$\dot{R}^2 = C/R - k. \tag{25.17}$$

To solve this, we need to consider separately the cases $k = +1$ and $k = -1$. In the former case, (25.17) becomes $\dot{R} = C/R - 1$. This time we start with a change of variable given by

$$u^2 = R/C. \tag{25.18}$$

Then $2u\dot{u} = \dot{R}/C$, and, substituting in (25.17), we find

$$\dot{u}^2 = \frac{\dot{R}^2}{4C^2u^2} = \frac{1}{4C^2u^2}\left(\frac{C}{R} - 1\right) = \frac{1}{4C^2u^2}\left(\frac{1}{u^2} - 1\right).$$

Taking positive square roots, the equation is separable, and, integrating with big bang initial conditions, we get

$$2 \int_0^u \frac{u^2}{(1-u^2)^{\frac{1}{2}}} \, du = \frac{1}{C} \int_0^t dt = \frac{t}{C}.$$

To evaluate the u-integral, we set $u = \sin\theta$. Then

$$2 \int_0^u \frac{u^2}{(1-u^2)^{\frac{1}{2}}} \, du = 2 \int_0^\theta \frac{\sin^2\theta \cos\theta \, d\theta}{(1-\sin^2\theta)^{\frac{1}{2}}}$$

$$= 2 \int_0^\theta \sin^2\theta \, d\theta$$

$$= \int_0^\theta (1-\cos 2\theta) \, d\theta$$

$$= \theta - \tfrac{1}{2}\sin 2\theta$$

$$= \theta - \sin\theta \cos\theta$$

$$= \sin^{-1} u - u(1-u^2)^{\frac{1}{2}}.$$

Writing the solution in terms of R, we obtain the result

$$C[\sin^{-1}(R/C)^{\frac{1}{2}} - (R/C)^{\frac{1}{2}}(1-R/C)^{\frac{1}{2}}] = t. \qquad (25.19)$$

Similarly, in the case $\Lambda = 0$, $k = -1$, the solution becomes (exercise)

$$C[(R/C)^{\frac{1}{2}}(1+R/C)^{\frac{1}{2}} - \sinh^{-1}(R/C)^{\frac{1}{2}}] = t. \qquad (25.20)$$

The case $\Lambda = 0$, $k = 0$, is the Einstein–de Sitter model and has already been dealt with in §25.1. Again, the Hubble parameter and deceleration parameter can be computed directly from (25.19) or (25.20). For example, when $k = +1$,

$$H = C^{-1}(R/C)^{-\frac{3}{2}}(1-R/C)^{\frac{1}{2}}, \qquad (25.21)$$

and

$$q = \tfrac{1}{2}(1-R/C)^{-1}, \qquad (25.22)$$

with R determined implicitly in terms of t by (25.19).

As in the last section, if we write (25.17) in the form

$$\dot{R}^2 = G(R), \qquad (25.23)$$

where

$$G(R) = C/R - k, \qquad (25.24)$$

then we find that the model for which $k = +1$ has a local minimum, whereas the other two models grow without bound. When $k = -1$, for large t, we have $\dot{R}^2 \sim 1$, and so $R \sim t$. We again have enough information to sketch the graphs of the models.

25.3 Classification of Friedmann models

In Fig. 25.1, we collect together the graphs of all the various possibilities. They are divided up into three major cases, namely, $k = -1$, 0, or $+1$, and subdivided into three, three, and eight sub-cases, respectively, depending on the sign or value of Λ. We describe the sub-cases briefly.

Case I: $k = -1$

All of these models have open topology.

(i) $\Lambda > 0$. This is an indefinitely expanding model, but it possesses a 'kink' in it where the rate of expansion slows down for a period before picking up again, and asymptotically it approaches $\exp[(\frac{1}{3}\Lambda)^{\frac{1}{2}} t]$. Initially, like **all** big bang models, the rate of expansion goes like that of the Einstein–de Sitter model, namely, like $t^{\frac{2}{3}}$.

(ii) $\Lambda = 0$. An indefinitely expanding model without a kink and which goes like t asymptotically.

(iii) $\Lambda < 0$. In this case, the cosmological force is attractive and eventually halts the expansion and forces the model to collapse, ending in an apocalyptic event called the **big crunch**. It is usually referred to as an **oscillating model**. There is also the possibility that the model is indefinitely oscillating with each cycle followed by another, as in Fig. 25.2. All models for which $\Lambda < 0$ are oscillating models.

Case II: $k = 0$

All of these models have open topology.

(i) $\Lambda > 0$. This is identical in character to the sub-case I(i) above, again possessing a kink and asymptotically approaching $\exp[(\frac{1}{3}\Lambda)^{\frac{1}{2}} t]$.

(ii) $\Lambda = 0$. The Einstein–de Sitter model, where $R \sim t^{\frac{2}{3}}$.

(iii) $\Lambda < 0$. An oscillating model.

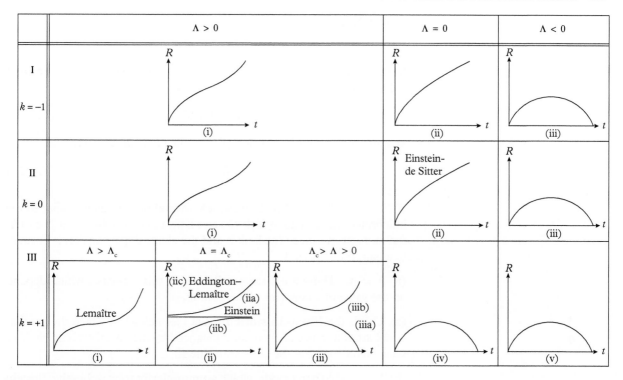

Fig. 25.1 Classification of Friedmann models.

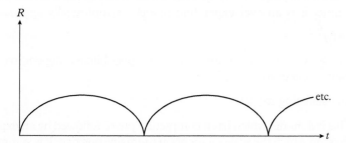

Fig. 25.2 An indefinitely oscillating model.

Case III: $k = +1$

All of these models have closed topology.

In this case, there are more possibilities since there is a positive **critical value** of the cosmological constant Λ_c given by

$$\Lambda_c = 4/(9C^2), \tag{25.25}$$

and an associated critical value of the scale factor R_c given by

$$R_c = \tfrac{3}{2}C. \tag{25.26}$$

(i) $\Lambda > \Lambda_c$ This is called **Lemaître's model** and is again similar to the indefinitely expanding models I(i) and II(i). However, the closer Λ is to Λ_c the more pronounced the kink is and the closer the expansion is brought to a halt in this period.

(ii) $\Lambda = \Lambda_c$. There are three possibilities in this sub-case, which depend on the value of a constant of integration.

(a) This is the **Einstein static model** (see below) in which the gravitational attraction is exactly counterbalanced by the cosmic repulsion. The scale factor then has the constant value R_c.

(b) This is a big bang model which asymptotically approaches the Einstein static model.

(c) This is the **Eddington–Lemaître model**, in which, if time is run backwards, it asymptotically approaches the Einstein static model. In forward time, it is an ever-expanding model, asymptotically approaching $\exp[(\tfrac{1}{3}\Lambda)^{\frac{1}{2}}t]$.

(iii) $\Lambda_c > \Lambda > 0$. There are, again, two possibilities, depending on a constant of integration.

(a) An oscillating model.

(b) This is a model which has a contracting phase followed by an expanding phase in which the scale factor always remains positive. It is symmetric about its point of minimum radius with $R \sim \exp[(\tfrac{1}{3}\Lambda)^{\frac{1}{2}}t]$ as $t \to \infty$ and $R \sim \exp[(\tfrac{1}{3}\Lambda)^{\frac{1}{2}}(-t)]$ as $t \to -\infty$.

(iv) $\Lambda = 0$. An oscillating model.

(v) $\Lambda < 0$. An oscillating model.

25.4   The Einstein static model and the de Sitter model

In this section, we consider two solutions which were influential in the early days of cosmology and, although they are no longer considered to be realistic cosmological models, they have some mathematical features that make them still important in studying cosmology.

As explained earlier when Einstein first formulated his theory of general relativity in 1916 the prevailing assumption was that the universe was both static and homogeneous. If one considers the Einstein equations for a Robertson–Walker metric (24.52) and (24.53) and then sets $\Lambda = 0$ and $R(t) = a = $ constant, one obtains

$$8\pi\rho = \frac{3k}{a^2}, \quad 8\pi p = -\frac{k}{a^2}. \tag{25.27}$$

The density ρ should be non-negative, and the case $k = 0$ just gives Minkowski space. Thus, in order to have a non-trivial static solution without a cosmological constant, $k = 1$ and the pressure has to be **negative** – which is not a feature of standard matter. This was the reason that Einstein introduced the cosmological constant. If one now allows a non-zero cosmological constant, then the above becomes

$$8\pi\rho = \frac{3k}{a^2} - \Lambda, \quad 8\pi p = \Lambda - \frac{k}{a^2}. \tag{25.28}$$

One can obtain a solution with positive energy density and zero pressure by choosing $k = 1$ and $\Lambda = 1/a^2$. This gives a closed, static, homogeneous universe filled with dust of density $\rho = 1/(4\pi a^2)$, which is called the **Einstein static universe**.

Observations of distant galaxies by Slipher, Hubble, and others in the early part of the 20th century showed that the galaxies are moving away from us, indicating that the universe is expanding. As a result of this, Einstein abandoned the cosmological constant and accepted expanding models of the Universe. However, as we show in Chapter 25, the cosmological constant has returned as an important feature of modern cosmology (in the form of 'dark energy') so we will continue to include it in our discussion of cosmological models below. Furthermore, although the Einstein static universe is no longer used as a cosmological model, it remains important to theorists, as it is used to construct conformal Penrose diagrams of realistic models (see §25.10).

The earliest cosmological model that was expanding is the **de Sitter** model, which was discovered as early as 1917. It is not a model of relativistic cosmology because it is devoid of conventional matter. It is obtained by setting $p = \rho = k = 0$ in (24.52) and (24.53). Then (24.52) gives

$$3\dot{R}^2/R^2 - \Lambda = 0,$$

or

$$\dot{R}/R = (\tfrac{1}{3}\Lambda)^{\frac{1}{2}}, \qquad (25.29)$$

which, on integration, becomes

$$R = A \exp\left[(\tfrac{1}{3}\Lambda)^{\frac{1}{2}}t\right],$$

where A is a constant of integration. Since the origin of this curve is arbitrary, let us choose $R = 1$ when $t = 0$, in which case $A = 1$. Alternatively, we can rescale r and absorb the factor A into it. This leads to the **de Sitter model**, for which

$$R = \exp[(\tfrac{1}{3}\Lambda)^{\frac{1}{2}}t]. \qquad (25.30)$$

The graph of the scale factor is shown in Fig. 25.3. This solution is the common limiting case to which all the models I(i), II(i), III (i), III(iic), and III(iiib) tend as $t \to \infty$, so it describes the late time behaviour of a wide range of cosmological models.

From (24.36), (25.30), and the requirement that k vanishes, the line element becomes

$$ds^2 = dt^2 - [\exp 2(\tfrac{1}{3}\Lambda)^{\frac{1}{2}}t][dr^2 + r^2(d\theta^2 + \sin^2\theta d\phi^2)],$$

or, in Cartesian coordinates, the standard form

$$ds^2 = dt^2 - [\exp 2t/\alpha][dx^2 + dy^2 + dz^2], \qquad (25.31)$$

where

$$\alpha = (3/\Lambda)^{\frac{1}{2}}. \qquad (25.32)$$

This line element is invariant under a shift in t and a simultaneous change of scale in the space coordinates (exercise). Note that the metric is completely specified by α. The coordinate range of t is from $-\infty$ to $+\infty$, with the zero of t being conventional. This is because the exponential curve is 'self-similar', that is, one cannot tell where one is along it by intrinsic measurements; it has no natural origin. If we introduce new coordinates $(\bar{t}, \bar{x}, \bar{y}, \bar{z})$, where

$$
\begin{aligned}
\bar{t} &= t - \tfrac{1}{2}\alpha \ln[1 - \alpha^{-2}(x^2 + y^2 + z^2)\exp(2t/\alpha)], \\
\bar{x} &= x\exp(t/\alpha), \\
\bar{y} &= y\exp(t/\alpha), \\
\bar{z} &= z\exp(t/\alpha),
\end{aligned}
\qquad (25.33)
$$

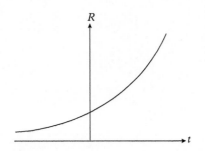

Fig. 25.3 The de Sitter model.

then (25.31) becomes, dropping bars, (exercise)

$$ds^2 = [1 - \alpha^{-2}(x^2 + y^2 + z^2)]dt^2 - dx^2 - dy^2 - dz^2$$
$$- \frac{\alpha^{-2}(xdx + ydy + zdz)^2}{1 - \alpha^{-2}(x^2 + y^2 + z^2)}, \tag{25.34}$$

which is manifestly **stationary** (why?). So that the dynamic behaviour of the metric when written in the form (25.31) is rather misleading and is really a coordinate effect. We shall return to this solution in §25.11.

25.5 Early epochs of the universe

In constructing the simplest possible model of the universe, we have neglected pressure. However, in the early epochs of the universe, one would expect the radiation to dominate completely over matter as a source of gravitation. Let us look briefly at a simple model which includes pressure at the **extreme relativistic condition**

$$3p = \rho, \tag{25.35}$$

which is the equation of state for **radiation**. Then, taking $\Lambda = 0$ in (24.52) and (24.53), the condition (25.35) requires that (exercise)

$$\frac{\ddot{R}}{R} + \frac{\dot{R}^2}{R^2} + \frac{k}{R^2} = 0. \tag{25.36}$$

In the earliest phases, the first two terms will dominate, and so, neglecting the last term in (25.36), we find that, for small t (exercise), $R \sim t^{1/2}$. Comparing this with the small-time behaviour we had previously, namely, $R \sim t^{2/3}$, we see that this corresponds to a more rapid expansion. The effect of the pressure of radiation is that it exerts its own gravitational field, thereby increasing the amount of gravity acting. This increases the rate of expansion, as is clear if we reverse the sense of time and consider the resulting rate of collapse.

Another difference between dust and radiation comes from looking at the thermodynamic equation (24.54). For dust we have $p = 0$ and this immediately gives $\rho R^3 = $ constant so that for dust $\rho \propto 1/R^3$. However, if one puts $p = 3\rho$ in (24.54), then a little algebra shows that

$$\frac{d}{dt}\left(\rho R^4\right) = 0, \implies \rho R^4 = C = \text{constant}. \tag{25.37}$$

Hence, for the case of radiation, we find instead that $\rho \propto 1/R^4$. If, for convenience, we set $C = \frac{8}{3}\pi D$ and substitute for ρ in (24.52), we obtain the **Friedmann equation for radiation**

$$\dot{R}^2 = \frac{D}{R^2} + \frac{1}{3}\Lambda R^2 - k, \tag{25.38}$$

where D is a constant of integration given by

$$D = \tfrac{8}{3}\pi R^4 \rho. \tag{25.39}$$

Again, one sees that, for small $R(t)$, the first term dominates so in the early stages of expansion $R(t) \sim t^{1/2}$. One can systematically look at all the various cases for this, as we did for the case of dust but will not do so here.

25.6   The steady-state theory

In 1948, Bondi, Gold, and, independently in the same year, Hoyle produced a cosmological theory which was, in many ways, radically different from the previous models of relativistic cosmology. It is a theory of charming simplicity, but, unfortunately, one which involves a modification of the law of conservation of energy – a law close to the hearts of many physicists. The theory provides definite answers to cosmological questions and so is more amenable to direct tests. Put another way, since it makes unique predictions, it is easier to disprove. Unfortunately, the theory seems to be at variance with much of present-day observations, and hence many consider it to be of historic interest only. Nonetheless, it leads to a solution that is of interest and of mathematical importance. We shall summarize the original formulation below.

The fundamental assumption of the theory as derived by Bondi and Gold is the following principle.

> **Perfect cosmological principle:** The universe presents an unchanging aspect on the large scale.

As we have seen in the last chapter, the ordinary cosmological principle (which is implied by the perfect cosmological principle), leads to the Robertson–Walker line element. We next use the requirement of **stationarity**, which also follows directly from the perfect cosmological principle to deduce the line element for the steady-state universe. Since the universe is expanding, $R(t)$ must be an increasing function of time. But the curvature of a 3-space of constant curvature in a Robertson–Walker space-time goes like kR^{-2} (exercise), and this is an observable quantity (affecting, for example, the rate of increase of the number of galaxies with distance). The fact that it is observable means that it must be constant and, since R varies with time, we must conclude that $k = 0$. The function $R(t)$ is not directly observable, but the Hubble parameter $\dot{R}(t)/R(t)$ is, and so again must remain constant. Thus, $\dot{R}/R = 1/T_0$, where T_0 is a constant and, proceeding as in (25.29), we get

$$R(t) = \exp(t/T_0), \tag{25.40}$$

and the line element becomes

$$ds^2 = dt^2 - \exp(2t/T_0)[dr^2 + r^2(d\theta^2 + \sin^2\theta)d\phi^2]. \tag{25.41}$$

This is the same as the line element of the **de Sitter model**, which we considered in relativistic cosmology but discarded because it led to an empty universe. Note that (25.41) is completely specified by the scale factor T_0. We leave the question of coordinate range until §25.11.

We can consider light propagation in the same way as we did in §24.10. Then (24.61) becomes, in this case,

$$\frac{\mathrm{d}t}{\exp(t/T_0)} = \pm \mathrm{d}r. \qquad (25.42)$$

For an incoming ray reaching $r = 0$ at $t = t_0$, we get

$$r = T_0(\mathrm{e}^{-t/T_0} - \mathrm{e}^{-t_0/T_0}). \qquad (25.43)$$

The luminosity distance d_L is given by (24.72), which, in this case, becomes

$$d_\mathrm{L} = r_1 \mathrm{e}^{t_0/T_0}, \qquad (25.44)$$

so that the coordinate r is proportional to the luminosity distance. Then, using (24.66), we have

$$1 + z = R(t_0)/R(t_1) = \mathrm{e}^{(t_0 - t_1)/T_0} = 1 + r_1 \mathrm{e}^{t_0/T_0}/T_0 \qquad (25.45)$$

by (25.43). Combining this result with (25.44), we find (exercise)

$$d_\mathrm{L} = z T_0. \qquad (25.46)$$

Thus, in the steady-state theory, Hubble's law is **exact**. It follows from (25.40) that (exercise)

$$q = -1, \qquad (25.47)$$

that is the universe is expanding at an ever increasing rate.

25.7 The event horizon of the de Sitter universe

As we have seen in (25.43), the equation of the past light cone through the point $r = 0$, $t = t_0$, has the equation

$$r = T_0(\mathrm{e}^{-t_0/T_0} - \mathrm{e}^{-t_0/T_0}). \qquad (25.48)$$

At any particular time, information can be received only from events inside an observer's past light cone. If we take the limit in (25.48) as $t_0 \to \infty$, then it follows that an observer whose world-line is $r = 0$ can **never** receive any information from events occurring **outside** the hypersurface

$$r = T_0 \mathrm{e}^{-t/T_0}. \qquad (25.49)$$

In other words, this hypersurface is an **event horizon** for O and has a similar character to the event horizon of black holes. Let us consider what an observer would see while observing a particle of the substratum P with world-line $r = r_1$ (Fig. 25.4). If we set $r = r_1$ in (25.49), then P crosses O's event horizon at the event P_1 at time

$$t_1 = T_0 \ln(T_0/r_1). \qquad (25.50)$$

Observer O can only receive signals from P at events of P's world-line for which $t < t_1$. These signals travel on null geodesics (the dotted line in Fig. 25.4), which reach O at time (exercise)

$$\tau = -T_0 \ln(e^{-t/T_0} - e^{-t_1/T_0}). \qquad (25.51)$$

So, by (24.72), O ascribes to P the luminosity distance

$$d_{\mathrm{L}} = r_1 e^{\tau/T_0} = r_1(e^{-t/T_0} - e^{-t_1/T_0})^{-1}, \qquad (25.52)$$

and a red shift, by (25.46), of

$$z = d_{\mathrm{L}}/T_0. \qquad (25.53)$$

Therefore, as $t \to t_1$ it follows that $\tau \to \infty$ and the light takes longer and longer to reach O from P. In addition, both $d_{\mathrm{L}} \to \infty$ and $z \to \infty$ as P disappears over O's horizon and this happens in a finite proper time as measured by P. At time t, the geodesic distance l from O to P is, by (24.69) with $k = 0$, $R(t) = e^{t/T_0}$

$$l = r_1 e^{t/T_0}, \qquad (25.54)$$

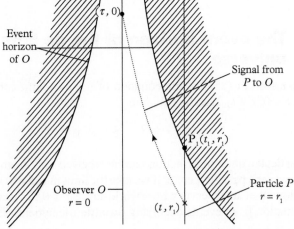

Fig. 25.4 The event horizon of an observer in the de Sitter universe.

which is still finite at the event P_1. The velocity of recession is

$$\frac{\mathrm{d}l}{\mathrm{d}t} = \frac{r_1}{T_0}e^{t/T_0} = e^{(t-t_1)T_0}, \qquad (25.55)$$

by (25.50), and this tends to 1 as $t \to t_1$. Thus, the geodesically measured velocity of recession tends to the velocity of light as the particle approaches the event horizon. So far, we have only considered an observer at $r = 0$, but, by homogeneity of the de Sitter solution, the above conclusions apply to any observer moving with the substratum.

The event horizon is rather like a curtain behind which one can see nothing. However, the curtain can be drawn, but at a price. Consider the world-line of an explorer E who is sent out into space by O and is asked to send back reports on all that E sees (Fig. 25.5). The explorer E will be able to see past O's horizon, but not until passing the event E_1 on this horizon, after which E can never return home to O nor send any information back to O. So we see that O can never receive information about events beyond O's horizon. However, their existence cannot be neglected, since, by travelling around, O's horizon can be changed and some of the forbidden knowledge can be found out – but no return home is then possible. We have met similar event horizons in Minkowski space-time (Chapter 3, Fig. 3.8). In suitably chosen coordinates, the world-line of a uniformly accelerated observer travelling in the x-direction has the equation $x^2 - t^2 = $ constant, $y = z = 0$. It is clear from Fig. 25.6 that light emitted from events in the shaded region will never reach the observer P, who, therefore, has an event horizon. No such horizon exists for inertial observers, of course.

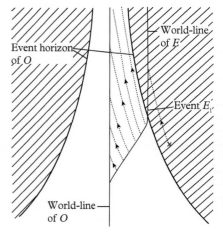

Fig. 25.5 Explorer E draws the curtain.

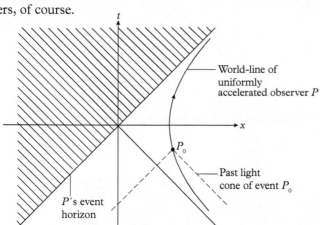

Fig. 25.6 Event horizons in Minkowski space-time.

25.8 Particle and event horizons

We can, in fact, distinguish between different sorts of horizons. Consider the world-line of an observer O moving on a timelike geodesic in a space-time in which $\mathscr{I}^-$ is spacelike (Fig. 25.7). Then, at any point P on O's world-line, the past light cone at P is the set of events in space-time which can be observed by O at that time. The division of particles into those seen by O at P and those not seen by O at P gives rise to the **particle horizon** of O at P. It represents the history of those particles lying at the limits of O's vision. Of course, if $\mathscr{I}^-$ is null (for example, as in Minkowski space-time), then all particles are seen by O at P (Fig. 25.8). Now consider a space-time in which both $\mathscr{I}^-$ and $\mathscr{I}^+$ are spacelike (Fig. 25.9). If we consider the whole history of the observer O, then the past light cone of O at P on $\mathscr{I}^+$ is called the **future event horizon** of O. Events outside this horizon will never be seen by O. Next, consider the case when $\mathscr{I}^+$ is null (Minkowski space-time, for example). If O moves on a timelike geodesic, then O does not possess an event horizon. However, if observer $\bar{O}$ moves with uniform acceleration, then, asymptotically, the speed of the observer approaches the speed of light – which means that the world-line ends up on $\mathscr{I}^+$ – and then $\bar{O}$ possesses a future event horizon (Fig. 25.10). Notice that these event horizons are **observer dependent**. This is to be contrasted with the event horizons of black holes which are more accurately termed **absolute event horizons**, because they are **observer independent**.

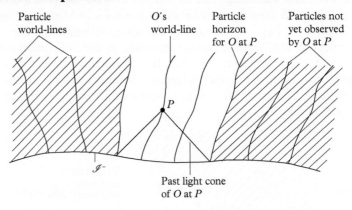

Fig. 25.7 Particle horizons of an observer ($\mathscr{I}^-$ spacelike).

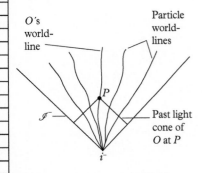

Fig. 25.8 The case when $\mathscr{I}^-$ is null.

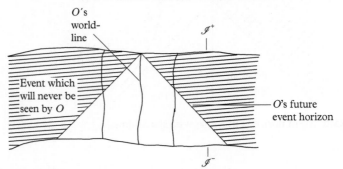

Fig. 25.9 The case when $\mathscr{I}^-$ and $\mathscr{I}^+$ are spacelike.

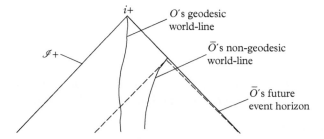

Fig. 25.10 The case when $\mathscr{I}^+$ is null.

25.9 Lorentzian constant curvature space-times

Although, as discussed earlier, the de Sitter model is no longer regarded as a realistic one for the universe as a whole, it is important in describing the final epochs of an ever expanding universe. Since the density of matter scales like R^{-3} and that of radiation like that of R^{-4}, then, if Λ is non-zero, it will tend to be the dominant term in the Friedmann equation for expanding solutions at late times. It is therefore of interest to investigate vacuum solutions with a non-zero cosmological constant such as the de Sitter solution.

In the derivation of the de Sitter solution in §25.6, we used the perfect cosmological principle. Geometrically, this corresponds to it being isotropic in **space-time** about every point. As we have seen, such a space is automatically an Einstein space, a condition which, in turn, is guaranteed by the space being a space of constant curvature. We have already looked at the 3-dimensional Riemannian spaces of constant curvature in §24.8. In this section, we look at 4-dimensional Lorentzian spaces of constant curvature. As discussed in §24.7, a space-time is said to have constant curvature if

$$R_{abcd} = K(g_{ac}g_{bd} - g_{ad}g_{bc}), \tag{25.56}$$

where K is a constant. Note that taking the double trace shows that, in 4-dimensions (exercise),

$$K = R/12, \tag{25.57}$$

where R is the (constant) scalar curvature. Furthermore, it also follows from (25.56) that (exercise)

$$G_{ab} = -3Kg_{ab}, \tag{25.58}$$

So that a constant curvature space-time is a solution of the vacuum Einstein equations with cosmological constant $\Lambda = -3K$, which is isotropic in space-time about every point.

In looking at such solutions, we consider three cases: $K = 0$, $K > 0$, and $K < 0$.

Case I: $K = 0$

Substituting $K = 0$ in (25.56) shows that the curvature tensor R_{abcd} vanishes, so that the case $K = 0$ is nothing but Minkowski space.

Case II: $K > 0$

The constant curvature space with positive curvature is called **de Sitter space**. This space-time is most easily visualized as the hyperboloid

$$-\hat{v}^2 + \hat{w}^2 + \hat{x}^2 + \hat{y}^2 + \hat{z}^2 = T_0^2, \tag{25.59}$$

with topology $\mathbb{R} \times S^3$ embedded in flat **five-dimensional** Euclidean space with a Minkowski-type line element

$$ds^2 = d\hat{v}^2 - d\hat{w}^2 - d\hat{x}^2 - d\hat{y}^2 - d\hat{z}^2. \tag{25.60}$$

One can introduce coordinates $(\hat{t}, \chi, \theta, \phi)$ on the hyperboloid by the relations

$$\left. \begin{aligned}
\hat{v} &= T_0 \sinh(\hat{t}/T_0), \\
\hat{w} &= T_0 \cosh(\hat{t}/T_0) \cos\chi, \\
\hat{x} &= T_0 \cosh(\hat{t}/T_0) \sin\chi \cos\theta, \\
\hat{y} &= T_0 \cosh(\hat{t}/T_0) \sin\chi \sin\theta \cos\phi, \\
\hat{z} &= T_0 \cosh(\hat{t}/T_0) \sin\chi \sin\theta \sin\phi,
\end{aligned} \right\} \tag{25.61}$$

in which case the line element has the form

$$ds^2 = d\hat{t}^2 - T_0^2\cosh^2(\hat{t}/T_0)[d\chi^2 + \sin^2\chi(d\theta^2 + \sin^2\theta d\phi^2)]. \tag{25.62}$$

Apart from coordinate singularities at $\chi = 0$, π, and $\theta = 0$, π, the hyperboloid is covered by the coordinate range

$$-\infty < \hat{t} < +\infty,$$
$$0 \leqslant \chi \leqslant \pi,$$
$$0 \leqslant \theta \leqslant \pi,$$
$$0 \leqslant \phi < 2\pi.$$

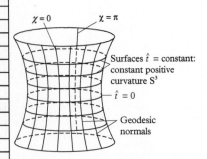

Fig. 25.11 de Sitter space-time embedded in five-dimensional Minkowski space-time.

The surfaces $\hat{t} = $ constant are 3-spheres of constant positive curvature, the particles of the substratum travel on timelike geodesics normal to these surfaces, and the overall topology is cylindrical, being $\mathbb{R} \times S^3$ (Fig. 25.11).

If we then introduce coordinates

$$
\left.
\begin{aligned}
t &= T_0 \ln[(\hat{w} + \hat{v})/T_0], \\
x &= T_0 \,\hat{x}/(\hat{w} + \hat{v}), \\
y &= T_0 \,\hat{y}/(\hat{w} + \hat{v}), \\
z &= T_0 \,\hat{z}/(\hat{w} + \hat{v}),
\end{aligned}
\right\}
\tag{25.63}
$$

then the line element (25.60) reduces to the form (25.31) in Cartesian co-ordinates with $\alpha = T_0$, or (25.41) in the corresponding spherical polar co-ordinates on the hyperboloid. However, the coordinates (t, x, y, z) only cover half the hyperboloid, since t is not defined for $\hat{w} + \hat{v} \leqslant 0$ (Fig. 25.12). In these coordinates, the surfaces $t = $ constant are flat 3-spaces, and the particles of the substratum are geodesics normal to these 3-spaces diverging from a point in the infinite past. Thus, only the portion given by $\hat{w} + \hat{v} > 0$ in the full de Sitter space (25.68) corresponds to the steady-state model.

Case III: $K < 0$

The constant curvature space with negative curvature is called **anti-de Sitter space**. This space-time is again most easily visualized as the hyperboloid

$$
-\hat{u} - \hat{v}^2 + \hat{x}^2 + \hat{y}^2 + \hat{z}^2 = -T_0^2
\tag{25.64}
$$

embedded in a flat **five-dimensional** Euclidean space with line element

$$
\mathrm{d}s^2 = \mathrm{d}\hat{u}^2 + \mathrm{d}\hat{v}^2 - \mathrm{d}\hat{x}^2 - \mathrm{d}\hat{y}^2 - \mathrm{d}\hat{z}^2.
\tag{25.65}
$$

Note the sign changes in (25.64) and (25.65) compared with Case II. In this case, one can introduce coordinates $(\hat{t}, \rho, \theta, \phi)$ on the hyperboloid by the relations

$$
\left.
\begin{aligned}
\hat{u} &= T_0 \sin(\hat{t}/T_0) \cosh \rho, \\
\hat{v} &= T_0 \cos(\hat{t}/T_0) \cosh \rho, \\
\hat{x} &= T_0 \sinh \rho \cos \theta, \\
\hat{y} &= T_0 \sinh \rho \sin \theta \cos \phi, \\
\hat{z} &= T_0 \sinh \rho \sin \theta \sin \phi,
\end{aligned}
\right\}
\tag{25.66}
$$

in which case the line element has the form

$$
n\mathrm{d}s^2 = T_0^2 \left[\cosh^2 \rho \, \mathrm{d}(\hat{t}/T_0)^2 - \mathrm{d}\rho^2 - \sinh^2 \rho (\mathrm{d}\theta^2 + \sin^2 \theta \mathrm{d}\phi^2) \right].
\tag{25.67}
$$

According to (25.66), the coordinate $\hat{t}$ is 2π-periodic. However, one can view this simply as an artefact of how we derived the metric (25.67). Having obtained it, we are free to regard $\hat{t}$ as ranging over the whole of $\mathbb{R}$, and the resulting space has topology $\mathbb{R}^4$.

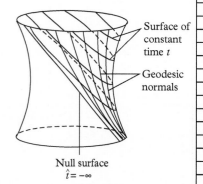

Fig. 25.12 de Sitter space-time in (t, x, y, z) coordinates.

25.10   Conformal structure of Robertson-Walker space-times

We proceed as we did in §24.8 and introduce a new radial coordinate χ so that the Robertson-Walker line element takes the form

$$ds^2 = dt^2 - [R(t)]^2[d\chi^2 + f^2(\chi)(d\theta^2 + \sin^2\theta d\phi^2)], \qquad (25.68)$$

where

$$\text{for } k = 0, \quad r = \chi = f(\chi),$$
$$\text{for } k = +1, \quad r = \sin\chi = f(\chi) \qquad (25.69)$$
$$\text{and for } k = -1, \quad r = \sinh\chi = f(\chi).$$

The coordinate χ runs from 0 to ∞ when $k = 0$ or -1, and from 0 to 2π when $k = +1$. Next, we introduce a new time coordinate τ defined by

$$d\tau = R^{-1}(t)dt,$$

so that (25.68) becomes

$$ds^2 = R^2(\tau)d\bar{s}^2, \qquad (25.70)$$

where

$$d\bar{s}^2 = d\tau^2 - d\chi^2 - f^2(\chi)(d\theta^2 + \sin^2\theta d\phi^2). \qquad (25.71)$$

Let us restrict our attention to the standard models $\Lambda = 0$, in which case $R(\tau)$ has one of the forms (25.8), (25.19), or (25.20). When $k = +1$, the unphysical line element (25.71) is precisely the Einstein static space (18.27). Indeed, all three models can be mapped on to different portions of the Einstein static space depending on the values taken by τ. In the case $k = 0$, the procedure is exactly the same as that employed in obtaining the conformal structure of Minkowski space-time (§18.4) except that now $0 < \tau < \infty$. The solution is therefore conformal to the half $t' > 0$ in Fig. 18.7. When $k = +1$, τ lies in the range $0 < \tau < \pi$. When $k = -1$, it can be shown that the space is conformal to the region

$$-\tfrac{1}{2}\pi \leqslant t' + r' \leqslant \tfrac{1}{2}\pi,$$
$$-\tfrac{1}{2}\pi \leqslant t' - r' \leqslant \tfrac{1}{2}\pi,$$
$$t' \geqslant 0.$$

The various regions of the Einstein static cylinder for each case are depicted in Fig. 25.13.

These conformal diagrams are somewhat different from the others we have met so far, in that part of the boundary is not 'infinity' in the sense it was previously, but represents the initial singularity when $R = 0$. In fact, this makes little difference to the conformal diagrams. The Penrose diagram for the ever-expanding cases $k = 0$ and -1 is given in Fig. 25.14 (two dimensions suppressed). The initial singularity – the big bang – is a spacelike surface. The Penrose diagram for the oscillating universe $k = +1$ is given in Fig. 25.15 (two dimensions suppressed). In this case, both the initial and the final singularity – the big crunch – are spacelike surfaces. It can be shown that matter-filled Robertson-Walker universes are, in fact, inextendible.

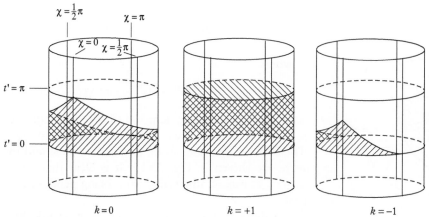

Fig. 25.13 Conformal Robertson-Walker space-times $(\Lambda = 0)$.

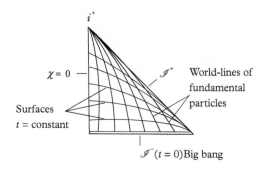

Fig. 25.14 Penrose diagram for $k = 0$ and -1 $(\Lambda = 0)$.

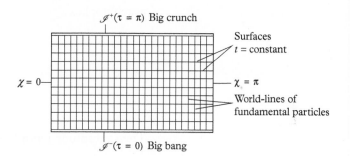

Fig. 25.15 Penrose diagram for $k = +1$ $(\Lambda = 0)$.

25.11 Conformal structure of de Sitter and anti-de Sitter space-time

We can obtain the conformal structure of de Sitter space-time by defining a new time coordinate

$$t' = 2\tan^{-1}[\exp(\hat{t}/T_0)] - \tfrac{1}{2}\pi,$$

which implies that (exercise)

$$\cosh(\hat{t}/T_0) = \frac{1}{\cos t'},$$

where

$$-\tfrac{1}{2}\pi < t' < \tfrac{1}{2}\pi.$$

Then substituting into (25.68) gives

$$ds^2 = T_0^2 \cosh^2(t'/T_0)d\bar{s}^2 = \frac{T_0^2}{\cos^2 t'}d\bar{s}^2, \qquad (25.72)$$

where

$$d\bar{s}^2 = dt'^2 - d\chi^2 - \sin^2 \chi(d\theta^2 + \sin^2 \theta d\phi^2),$$

is the Einstein static line element (18.27) on identifying $r' = \chi$. The region to which de Sitter space is conformal is shown in Fig. 25.16. The Penrose diagrams of de Sitter space-time and the steady-state universe are shown in Fig. 25.17. It is clear that the steady-state theory suffers, at least aesthetically, from being geodesically incomplete in the past.

To derive the conformal structure of anti-de Sitter space-time, we define a new radial coordinate χ by

$$\chi = 2\tan^{-1}(\exp \rho) - \tfrac{1}{2}\pi,$$

which implies

$$\cosh \rho = \frac{1}{\cos \chi}, \qquad (25.73)$$

where

$$0 \leqslant \chi \leqslant \tfrac{1}{2}\pi.$$

Then substituting into (25.68) gives

$$ds^2 = T_0^2 \cosh^2 \rho d\bar{s}^2 = \frac{T_0^2}{\cos^2 \chi}d\bar{s}^2, \qquad (25.74)$$

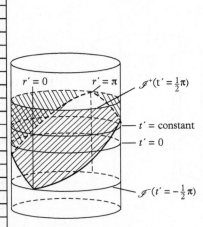

Fig. 25.16 Conformal de Sitter space-time.

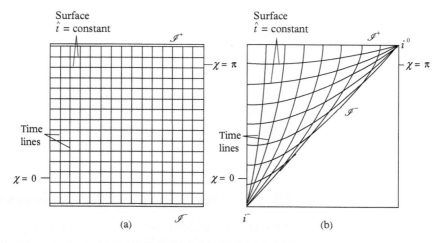

Fig. 25.17 Penrose diagram of (a) de Sitter space-time and (b) the steady-state model.

where $d\bar{s}^2$ is again the Einstein static line element (18.27). Since $0 \leqslant \chi \leqslant \pi/2$, anti-de Sitter space is conformally related to half of the Einstein static cylinder, as shown in Fig. 25.18 using polar coordinates. Since χ only reaches $\pi/2$, rather than π a spacelike slice of anti-de Sitter has topology of the interior of a hemisphere which is $\mathbb{R}^3$ and thus the space-time has the topology $\mathbb{R}^4$, as stated earlier. Infinity in this picture is given by $\chi = \pi/2$, which we see is a timelike hypersurface.

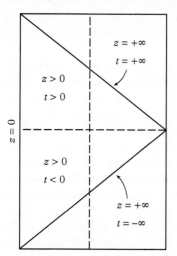
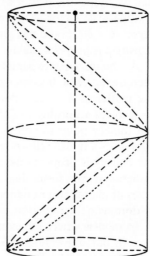

Fig. 25.18 Conformal anti-de Sitter space-time.

Although anti-de Sitter space is not usually considered as a realistic cosmological model, the timelike nature of infinity makes it of great importance in string theory as a result of something called the 'AdS/CFT correspondence'. This is beyond the scope of this book, but it is a remarkable result because it shows that, in an appropriate limit, there is a correspondence between a quantum gravity theory in the interior (the anti-de Sitter space-time) and a conformally invariant field theory (CFT) on the boundary at infinity. We will not go any further into these issues here except to say that it is a very active area of research.

25.12   Our model of the universe

In this chapter, we have examined various spatially isotropic and homogeneous cosmological models. As with all models, they have their own range of validity so we end by summarizing these. As we explained at the start of Chapter 24, there is considerable evidence that, on suitable length scales, the universe is homogeneous and isotropic. However, it is clear that, even on the length scale of galaxies, this is not true so that, although the assumption that the universe is modeled by the Robertson-Walker metric is reasonable, one must remember that it is not exactly true and there may have been earlier epochs where inhomogeneities were important. In the early universe, we expect that the universe is **radiation dominated** so that we model the matter as a perfect fluid with equation of state $p = 3\rho$, and the scale factor $R(t)$ evolves according to (25.38). For the early universe, $\dot{R} \sim D/R^2$ and this gives $R(t) \sim t^{1/2}$, and the corresponding energy density ρ is proportional to $R(t)^{-4}$.

As the universe expands and cools down (see details in Chapter 25), then the universe becomes **matter dominated**. In this phase, we model the matter as a perfect fluid with $p = 0$ (i.e. dust) and the scale factor evolves according to (25.1). These were examined in detail in §25.3 but the key point to note is that, for small R, we have $\dot{R} \sim C/R$ so that $R(t) \sim t^{2/3}$. It also follows from the conservation of energy that, for dust, the density is proportional to R^{-3}. We see from this that, the density of radiation decays faster than that of matter so that whatever the initial densities of matter and radiation, there comes a time where the matter starts to dominate the radiation.

In models where the expansion continues (which we believe to be the case in our universe), we see from (25.1) that, eventually, the cosmological term in the Friedmann equation dominates, however small it is. Although this is beyond the scope of this book, we mention in passing that one potential source of a small cosmological constant that can be considered is something called the quantum mechanical zero-point energy of the vacuum. It turns out that this can be modelled as a perfect fluid with equation of state $\rho = -p$, which corresponds to an energy-momentum tensor of the form $T_{ab} = \Lambda g_{ab}$. This is what one obtains from Einstein's equations if one has no conventional matter but puts the cosmological term on the right-hand side of the equations and interprets it as energy–momentum. For this reason, we describe solutions where the Λ term in the Friedmann equation is dominating as **vacuum dominated**. Since the radiation density falls off like R^{-4} and the matter density like R^{-3}, it is inevitable that, in an expanding universe Λ, however small, will eventually dominate the other two terms. For a vacuum-dominated solution, we get from (24.52) that $\dot{R} \sim \alpha R$, where $\alpha = (\Lambda/3)^{1/2}$, so that $R(t) \sim e^{\alpha t}$ and we have exponential growth.

It is also believed that we have exponential growth in the very early universe due to **inflation**. This is due to a field ϕ called the **inflaton**, which has dynamics given by a potential energy $V(\phi)$ and has an approximate energy–momentum tensor of the form $T_{ab} \simeq V(\phi)g_{ab}$, so behaves like a dynamical cosmological constant and again produces exponential growth. The associated matter is sometimes called **dark energy**. We will give a more detailed description of inflation in Chapter 25. We therefore have the following basic model for the evolution of the universe. There is an initial stage of inflationary exponential growth in the very early universe driven by the inflaton. As a result of the inflation, $V(\phi)$ decreases, the early universe becomes radiation dominated, and the expansion in no longer exponential but becomes a power law. Further expansion results in it moving to the current matter dominated phase where $R(t) \sim t^{2/3}$. The universe continues to expand and cool down until eventually the matter is so dilute that the vacuum-energy dominates and the universe gets closer and closer to the de Sitter solution. Of course, in reality, there is not such a clean-cut division between the various phases and to go from the general picture described above to a more detailed model requires a more careful understanding of both the underlying physics and of the input from observational cosmology, which will be the topic of the next chapter.

Exercises

25.1 (§25.1) Show that taking the negative square root in (25.4) leads to the same result as in (25.5).

25.2 (§25.1) Use the substitution (25.6) to establish the solution (25.7) in the case $k = 0$, $\Lambda < 0$.

25.3 (§25.1)
(i) Confirm (25.9) and (25.10) for the Einstein–de Sitter model.
(ii) Show that in the case $k = 0$, $\Lambda < 0$, $\dot{R}$ vanishes at $R = R_m = [3C/(-\Lambda)]^{\frac{1}{3}}$, which is a local minimum.
(iii) Show that (25.15) leads to (25.16).

25.4 (§25.2) Use the substitution (25.18) to establish the solution (25.20) in the case $\Lambda = 0$, $k = -1$.

25.5 (§25.2) Confirm (25.21) and (25.22) for the model with solution (25.19).

25.6 (§25.3) Show that the general differential equation for $\ddot{R}$ and $\dot{R}$ can be written in the form

$$\ddot{R} = (2/9C)(-x^{-2} + \lambda x),$$

$$\dot{R}^2 = \tfrac{1}{3}(2x^{-1} + \lambda x^2 - 3k),$$

where $R_c = 3C/2$, $\Lambda_c = 4/(9C^2)$, $x = R/R_c$ and, $\lambda = \Lambda/\Lambda_c$. [Hint: let $R = R_c$, $\Lambda = \Lambda_c$ when $\ddot{R} = \dot{R} = 0$]
(i) Deduce that, if $\ddot{R} = \dot{R} = 0$ at some time, then $R = R_c$, $\Lambda = \Lambda_c$, $k = 1$ and $\ddot{R} = \dot{R} = 0$ for all times.
(ii) Show that
(a) if $\Lambda < 0$, then all models are oscillating;
(b) if $\Lambda > 0$, then oscillating models require $k = 1$, and $\Lambda < \Lambda_c$.
[Hint: consider the equations for x and λ in turn when x is small and large, and λ is positive and negative.]

25.7 (§25.3) A straight channel contains a fixed particle of mass M at its origin O, while another particle P of mass m moves under gravitational attraction. Let OP be denoted by x, and take the time to be zero when the particle starts off from O in the positive x-direction. If the particle has velocity v_0 at x_0, then show there exists a value of x, $x = x_1$, say (positive, negative, or infinite), at which the velocity vanishes and find it in terms of x_0 and v_0. Show that the energy equation can be written in the form

$$\dot{x}^2 = 2GM/x - 2GM/x_1.$$

Compare this with Friedmann's equation and hence interpret the types of motion possible for various values of x_1.

25.8 (§25.4) Show that the line element (25.31) is invariant under a shift in t and a simultaneous change of scale in the space coordinates. Apply the transformation (25.33) to the dt^2 term in (25.31) to confirm they produce the dt^2 term (dropping bars) in (25.34).

25.9 (§25.5) Show that, if $\Lambda = 0$, then (24.52) and (24.53), subject to (25.35), lead to (25.36). Neglecting the term involving k, deduce that $R \sim t^{1/2}$ for small t.

25.10 (§25.6) Use the Robertson–Walker line element in the form (24.36) to show that the three-dimensional Ricci scalar curvature of a 3-space $t = t_0$ is $6k[R(t_0)]^{-2}$.

25.11 (§25.6) Confirm the results (25.46) and (25.47) for the steady-state theory.

25.12 (§25.7) Confirm the results (25.51) and (25.54).

25.13 (§25.9) Show that, for a constant curvature space-time in 4-dimensions, $K = R/12$ and hence

$$G_{ab} = -3Kg_{ab}.$$

25.14 (§25.11) Check that (25.61) satisfies (25.59). Show that the line element (25.60) reduces to the forms (25.62) and (25.31) on the hypersurface. [Hint: use (25.61) and (25.63), respectively.]

25.15 (§25.11) Check that (25.66) satisfies (25.64). Show that the line element (25.65) reduces to the form (25.67) on the hypersurface.

25.16 (§25.11) Show that the coordinate transformation

$$t' = 2\tan^{-1}[\exp(\hat{t}/T_0)] - \tfrac{1}{2}\pi,$$

brings the de Sitter solution into the form given by (25.72).

25.17 (§25.11) Show that the coordinate transformation

$$\chi = 2\tan^{-1}(\exp\rho) - \tfrac{1}{2}\pi,$$

brings the anti-de Sitter solution into the form given by (25.74).

Further reading

The main sources for this chapter are again the books by Bondi(1961) and Weinberg (1972). We do not discuss anisotropic cosmologies in this chapter so recommend the book by Hughston and Tod (1990), which has a chapter on anisotropic cosmologies to give you a flavour of this subject.

Bondi, H. (1961). *Cosmology*. Cambridge University Press, Cambridge.

Hughston, L. P., and Tod, K. P. (1990). *An Introduction to General Relativity*. Cambridge University Press, Cambridge.

Weinberg, S. (1972). *Gravitation and Cosmology*. Wiley, New York, NY.

Modern cosmology

26

26.1 Multi-component models

In Chapter 25, we looked at the Friedmann equation in three regimes: radiation dominated, matter dominated and vacuum dominated. We now want to consider situations which allow for all three and relate the theoretical predictions to observational quantities. Before looking at the general situation, we consider (24.52) for the case where the cosmological constant vanishes. Recalling that $H = \dot{R}/R$, this gives (exercise)

$$H^2 = \frac{8\pi}{3}\rho - \frac{k}{R^2}. \tag{26.1}$$

We see from this that, at any epoch, there is a special density required in order that k vanishes, given by $\rho = 3H^2/8\pi$. Note that this varies with time, since $H(t)$ does. In order to give it a definite value, we define the **critical density** to be the current value given by

$$\rho_{\text{crit}} = \frac{3H_0^2}{8\pi}, \tag{26.2}$$

where $H_0 := H(t_0)$ is the current value of the Hubble parameter. If we let $\rho_0 = \rho(t_0)$ be the current density of the universe, then we see that

$$
\begin{aligned}
\rho_0 > \rho_{\text{crit}} &\Rightarrow \quad k = +1 \quad \text{(closed universe)}, \\
\rho_0 = \rho_{\text{crit}} &\Rightarrow \quad k = 0 \quad \text{(open flat universe)}, \\
\rho_0 < \rho_{\text{crit}} &\Rightarrow \quad k = -1 \quad \text{(open universe)}.
\end{aligned}
\tag{26.3}
$$

So that the value of ρ_0 compared to ρ_{crit} discriminates between the three types of universe. Although there is some uncertainty in the value of H_0, it is of the order of 100 km/s per megaparsec (see §26.2) and substituting this into (26.2) gives a value of ρ_{crit} of the order of $10^{-26}\,\text{Kg}\,\text{m}^{-3}$. Although this seems small, there is a lot of empty space between galaxies and, as a very rough estimate, the actual density of the universe is of a similar order to the magnitude of the critical density. We will return to this point in §26.4.

It is standard practice in cosmology to measure the present-day density of the universe, relative to the critical density by setting

$$\Omega := \frac{\rho(t_0)}{\rho_{\text{crit}}} = \frac{8\pi\rho(t_0)}{3H_0^2}. \tag{26.4}$$

Introducing Einstein's Relativity. Second Edition. Ray d'Inverno and James Vickers, Oxford University Press.
© Ray d'Inverno and James Vickers (2022). DOI: 10.1093/oso/9780198862024.003.0026

So that $\Omega > 1$ corresponds to positive spatial curvature ($k = +1$), $\Omega < 1$ corresponds to negative spatial curvature ($k = -1$), and the flat models have $\Omega = 1$ exactly.

We now want to consider the Friedmann equation when we include a cosmological constant. The starting point is to put the cosmological term on the right-hand side of Einstein's equations and regard Λg_{ab} as an energy-momentum tensor. Looking at (24.52) and (24.53), we see that this corresponds to a fluid with constant density given in geometrical units by (exercise)

$$\rho_\Lambda = \frac{\Lambda}{8\pi}, \tag{26.5}$$

and equation of state $p = -\rho$. If we now let ρ_m and ρ_r be the densities of matter and radiation, respectively, then we may write (24.52) in the form (exercise)

$$\frac{\dot{R}^2}{R^2} = \frac{8\pi}{3} \left(\rho_r + \rho_m + \rho_\Lambda \right) - \frac{k}{R^2}. \tag{26.6}$$

We want to look at the relative sizes of the various terms in (26.6), so the first step is to write it in non-dimensional form. For the scale factor R, we measure this compared to the current value and define the non-dimensional quantity $\tilde{R}$ by

$$\tilde{R}(t) := \frac{R(t)}{R(t_0)}. \tag{26.7}$$

For the time, we use the Hubble time $t_H := 1/H_0$ to set the scale, and define a non-dimensional time by

$$\tilde{t} := \frac{t}{t_H} = H_0 t. \tag{26.8}$$

For the density terms, it is useful to measure these relative to the critical density ρ_{crit} given by (26.2). We first look at how the current values compare to the critical density and define

$$\Omega_r := \frac{\rho_r(t_0)}{\rho_{\text{crit}}}, \quad \Omega_m := \frac{\rho_m(t_0)}{\rho_{\text{crit}}}, \quad \Omega_\Lambda := \frac{\rho_\Lambda(t_0)}{\rho_{\text{crit}}}. \tag{26.9}$$

We know from Chapter 25 that conservation of energy for the matter gives $\rho_m \propto 1/R^3$. We therefore have

$$\rho_m(t) = \rho_m(t_0) \left(\frac{R(t_0)}{R(t)} \right)^3 = \frac{\rho_{\text{crit}} \Omega_m}{\tilde{R}^3}. \tag{26.10}$$

Similarly, for radiation, conservation of energy gives $\rho_r \propto 1/R^4$, so that

$$\rho_r(t) = \rho_r(t_0) \left(\frac{R(t_0)}{R(t)} \right)^4 = \frac{\rho_{\text{crit}} \Omega_r}{\tilde{R}^4}. \tag{26.11}$$

Finally, the density of the vacuum energy ρ_Λ is constant, so that

$$\rho_\Lambda(t) = \rho_\Lambda(t_0) = \rho_{\text{crit}}\Omega_\Lambda. \tag{26.12}$$

We now write the general Friedmann equation (26.6) in terms of the non-dimensional quantities and this gives (exercise)

$$\left(\frac{d\tilde{R}}{d\tilde{t}}\right)^2 = \left(\frac{\Omega_m}{\tilde{R}} + \frac{\Omega_r}{\tilde{R}^2} + \Omega_\Lambda \tilde{R}^2\right) - \frac{k}{H_0^2 R_0^2}, \tag{26.13}$$

where $R_0 = R(t_0)$. It is sometimes useful to introduce the quantity

$$\Omega_c = -\frac{k}{H_0^2 R_0^2}, \tag{26.14}$$

in which case it follows from evaluating the Friedmann equation (26.13) at $t = t_0$, where, by definition, $\tilde{R}(t_0) = 1$, that (exercise)

$$\Omega_r + \Omega_m + \Omega_\Lambda + \Omega_c = 1. \tag{26.15}$$

The value of this result is that we can write the final term in (26.13) in terms of potentially measurable quantities as

$$\Omega_c = 1 - (\Omega_r + \Omega_m + \Omega_\Lambda). \tag{26.16}$$

In analysing the dynamics of (26.13), it is helpful to think of this as the formula for conservation of energy for the motion of a particle of unit mass, moving in 1-dimension, in an effective potential $V(\tilde{R})$:

$$\frac{1}{2}\left(\frac{d\tilde{R}}{d\tilde{t}}\right)^2 + V(\tilde{R}) = E, \tag{26.17}$$

where

$$V(\tilde{R}) = -\frac{1}{2}\left(\frac{\Omega_m}{\tilde{R}} + \frac{\Omega_r}{\tilde{R}^2} + \Omega_\Lambda \tilde{R}^2\right), \tag{26.18}$$

and

$$2E = \Omega_c = 1 - (\Omega_r + \Omega_m + \Omega_\Lambda). \tag{26.19}$$

This now gives us a method to construct an FRW cosmological model if we are given Ω_r, Ω_m, Ω_Λ, and H_0. Given Ω_r, Ω_m, and Ω_Λ we first use (26.19) to determine E. We may then (in principle) solve (26.17) by writing it in the form of a separable ODE and integrating to obtain

$$\tilde{t} = \int \frac{d\tilde{R}}{\left[2(E - V(\tilde{R}))\right]^{1/2}}. \tag{26.20}$$

We can then solve this for $\tilde{R}(\tilde{t})$ which can, if desired, be converted back into $R(t)$ using equations (26.7) and (26.8), which give

$$R(t) = R_0 \tilde{R}(\tilde{t}/H_0). \tag{26.21}$$

In fact, one can show from (26.14) (exercise) that, as long as Ω_c is not exactly zero then

$$R(t) = \frac{1}{H_0 |1 - (\Omega_r + \Omega_m + \Omega_\Lambda)|^{1/2}} \tilde{R}(\tilde{t}/H_0). \tag{26.22}$$

We have therefore shown the following result:

> A general FRW cosmology is determined by the four cosmological parameters
> $$\Omega_r, \qquad \Omega_m, \qquad \Omega_\Lambda, \qquad H_0.$$

We discuss in the next section how these parameters may be determined through observation and/or theoretical grounds. However, before doing this, we outline how these parameters may be used to indicate the general behaviour of the solution without having to solve (26.20). The Hubble parameter only provides an overall scaling factor, as shown by (26.22), and since, to a good approximation, the value of Ω_r in our universe is small (current estimates give Ω_r of the order 10^{-4}), we can therefore, for most purposes, assume that $\Omega_r = 0$ and explore the space of possible universes by parameterizing them in terms of Ω_m and Ω_Λ. In this situation, the non-dimensional Friedmann equation is still given by (26.17), where now

$$V(\tilde{R}) = -\frac{1}{2}\left(\frac{\Omega_m}{\tilde{R}} + \Omega_\Lambda \tilde{R}^2\right), \tag{26.23}$$

and

$$2E = \Omega_c = 1 - (\Omega_m + \Omega_\Lambda). \tag{26.24}$$

We now consider the following issues concerning the type of universe:

1. Open or closed

$\Omega_m + \Omega_\Lambda > 1 \quad \Longrightarrow \quad \Omega_c < 0 \quad$ Hence $k = +1$ and the universe is closed,

$\Omega_m + \Omega_\Lambda = 1 \quad \Longrightarrow \quad \Omega_c = 0 \quad$ Hence $k = 0$ and the universe is flat and open,

$\Omega_m + \Omega_\Lambda < 1 \quad \Longrightarrow \quad \Omega_c > 0 \quad$ Hence $k = -1$ and the universe is open.

Thus, the dividing line between an open and closed universe is given by the line $\Omega_\Lambda = 1 - \Omega_m$.

2. Accelerating or non-accelerating

We derived an expression in (24.79) for the deceleration parameter q. We now evaluate this at the present time t_0 and write this in terms of Ω_m and Ω_Λ, using (26.2) and (26.9),

$$q = \frac{4\pi}{3H_0^2}\rho_m(t_0) - \frac{\Lambda}{3H_0^2}$$

$$= \frac{1}{2}\frac{\rho_m(t_0)}{\rho_{\text{crit}}} - \frac{\rho_\Lambda(t_0)}{\rho_{\text{crit}}}$$

$$= \tfrac{1}{2}\Omega_m - \Omega_\Lambda.$$

Thus the dividing line between accelerating and non-accelerating is given by the line $\Omega_\Lambda = \tfrac{1}{2}\Omega_m$.

3. Big bang or no big bang

If $\Omega_m + \Omega_\Lambda \leqslant 1$, then $E \geqslant 0$. Since $V(\tilde{R})$ is negative, then, by (26.17), we see that $\dot{R}$ cannot vanish. So there are no turning points and the solution started from $R = 0$. Thus, all the open models start with a big bang. If $\Omega_m + \Omega_\Lambda > 1$, then $E < 0$. Then whether $\dot{R}$ vanishes depends on the size of V_{max} (the maximum value of $V(\tilde{R})$) relative to E. If $V_{\text{max}} < E$ (see Fig. 26.1), then there are no points where $\dot{R} = 0$ and, again, we have a big bang solution. On the other hand, if $V_{\text{max}} > E$, then there are two points where $\dot{R} = 0$ (see Fig. 26.1). There are then two possibilities. In the first, the universe expands from $R = 0$, reaches a turning point where $\dot{R} = 0$, and then contracts to a 'big crunch' final singularity. In the second, the universe starts out with a large value of R, contracts until it reaches a point where $\dot{R} = 0$, and the re-expands again in a 'bounce'. These two cases are those of $k = 1$, panels (iiia) and (iiib), which are shown in Fig. 25.1 of §25.3.

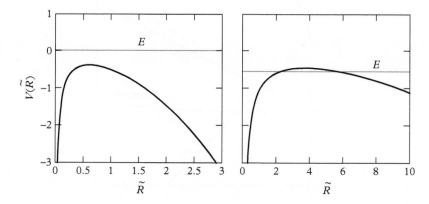

Fig. 26.1 Plots of the effective potential $V(\tilde{R})$ and its relation to E: $V_{\text{max}} < E$ 'on the' left, and $V_{\text{max}} > E$ 'on the' right.

The dividing line between the two cases is when $V_{max} = E$. Differentiating (26.23) gives

$$V'(\tilde{R}) = -\frac{1}{2}\left(-\frac{\Omega_m}{\tilde{R}^2} + 2\tilde{R}\Omega_\Lambda\right),\qquad(26.25)$$

so that this vanishes when $\tilde{R}^3 = \Omega_m/2\Omega_\Lambda$. Substituting for this value of $\tilde{R}$ in (26.23) gives

$$V_{max} = -\frac{3}{4}\left(2\Omega_m^2\Omega_\Lambda\right)^{1/3},\qquad(26.26)$$

so that $E = V_{max}$ when

$$1 - (\Omega_m + \Omega_\Lambda) = -\frac{3}{2}\left(2\Omega_m^2\Omega_\Lambda\right)^{1/3}.\qquad(26.27)$$

If we cube the above equation (and, for simplicity of notation, write $\Omega_m = x$ and $\Omega_\Lambda = y$), we get, after a bit of algebra,

$$4y^3 + (12x - 12)y^2 + (-15x^2 - 24x + 12)y + (4x^3 - 12x^2 + 12x - 4) = 0.\qquad(26.28)$$

Solving this for y as a function of x gives $y = f(x)$, or $\Omega_\Lambda = f(\Omega_m)$. In general, the cubic has three real roots which give rise to three different curves. We see from (26.27) that, when $\Omega_m = 0$, then $\Omega_\Lambda = 1$, so that one of the curves goes through the point $(0, 1)$. This is the curve marked 'no big bang' in the top left corner of Fig. 26.2. We also see from (26.27)

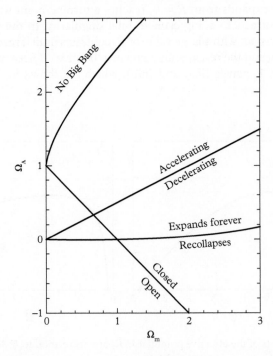

Fig. 26.2 Friedmann–Robertson–Walker models in the $((\Omega_m, \Omega_\Lambda))$-plane. (Reproduced from Liddle (2004), with permission from the publisher.)

that, when $\Omega_\Lambda = 0$, then $\Omega_m = 1$, so that there is another curve that goes through the point $(1, 0)$. This is the curve marked 'expands forever' in Fig. 26.2. The third curve is unphysical as it does not correspond to positive values of Ω_m. It is possible to write down exact formulae giving Ω_Λ as a function of Ω_m for the dividing line between the various cases (Carroll 2001) but they are not very illuminating, so we will content ourselves with drawing a graph.

Using all this information, we plot in the $(\Omega_m, \Omega_\Lambda)$-plane the various regions showing the properties of the various FRW models. This is one of the most important diagrams in modern cosmology as it describes the properties of the universe in terms of physically measurable quantities. In §26.4, we will attempt to use observational information to locate our universe on this diagram.

We end this section by showing how to use (26.17) to calculate the **age of the universe**. The current value of the non-dimensional time $\tilde{t}$ is the value at which $\tilde{R}$ is one, so that it is the solution of $\tilde{R}(\tilde{t}_0) = 1$. Then, by (26.8), we have $t_0 = \tilde{t}_0/H_0$. But, from (26.17),

$$\frac{d\tilde{R}}{d\tilde{t}} = \left[\Omega_r \tilde{R}^{-2} + \Omega_m \tilde{R}^{-1} + \Omega_\Lambda \tilde{R}^2 + (1 - \Omega_r - \Omega_m - \Omega_\Lambda) \right]^{1/2}.$$

This is just a separable equation so that

$$H_0 t_0 = \tilde{t}_0 = \int_0^1 \left[\Omega_r \tilde{R}^{-2} + \Omega_m \tilde{R}^{-1} + \Omega_\Lambda \tilde{R}^2 + (1 - \Omega_r - \Omega_m - \Omega_\Lambda) \right]^{-1/2} d\tilde{R}.$$

$$(26.29)$$

In special cases, this can be integrated. For example, for a matter-dominated flat universe ($\Omega_r = 0$ and $\Omega_c = 0$), then, for the case $\Omega_m < 1$, the above gives (exercise)

$$H_0 t_0 = \frac{2}{3} \frac{1}{\sqrt{1 - \Omega_m}} \sinh^{-1} \left[\frac{\sqrt{1 - \Omega_m}}{\sqrt{\Omega_m}} \right]. \qquad (26.30)$$

In the next section, we will discuss how to determine the four parameters H_0, Ω_r, Ω_m, and Ω_Λ for our universe. It will turn out that determining these is not as straightforward as one would hope but, despite these difficulties, there is consensus about a standard model for the universe.

26.2 Measuring the Hubble constant

The Hubble parameter is defined by $H(t) = \dot{R}/R$, and the Hubble constant $H_0 = H(t_0)$ is the present value of this. The Hubble constant has dimensions of 1/time and the current estimate is usually written as

$$H_0 = 72 \pm 8 \,(\mathrm{Km/s})/\mathrm{Mpc}, \qquad (26.31)$$

Table 26.1 Distance scales in cosmology.

Distance to the Sun	5×10^{-6} pc	1.5×10^{-5} light years
Distance to the nearest star	1 pc	3 light years
Distance to the galactic centre	10^4 pc	3×10^4 light years
Distance to the local group of galaxies	5×10^5 pc	1.5×10^6 light years
Distance to the nearest large cluster	2×10^7 pc	6×10^7 light years
Distance scale of largest structures	10^8 pc	3×10^8 light years
Distance to the edge of the visible universe	1.4×10^{10} pc	5×10^{10} light years

to indicate how the velocity of recession changes with distance. Here the measure of distance is the megaparsec (Mpc) which is the common one used by astronomers to measure intergalactic distances. A parsec (pc) is defined as the distance away for which the width of the orbit of the Earth round the Sun subtends one arcsecond. To give you an idea of scale, one parsec is 3.08×10^{16} m or about 3.26 light years, and is roughly the distance to the nearest star. A megaparsec (Mpc) is of the order of the distance to galaxies in our local group, while the distance to the nearest large cluster is of the order of 20 Mpc. To give an indication of the vast range of distance scales in cosmology, see Table 26.1, where the numbers given are very rough orders of magnitude.

From (24.74) we have the relation $z \simeq H_0 d$, so that, in principle, all that is needed is to measure the redshifts and distances of a large number of galaxies in order to measure H_0. For nearby galaxies, it is just possible to use parallax to measure the distance. However, the problem is that, as well as the overall velocity of galaxies due to the expansion of the universe, galaxies also have individual motions relative to the expansion, which are called 'peculiar' velocities by astronomers. For nearby stars and galaxies, these effects can be comparatively large so that the best way to estimate H_0 is to measure the redshift for distant objects where the velocity due to expansion dominates the peculiar velocities. However, it is precisely these sorts of distant object for which it is hard to estimate d, since they are too far away to use parallax to measure distance. The answer is to use luminosity to measure distance (as explained in §24.11). However, to do this, one needs to know the energy E radiated by the source. The usual method to do this is to use what is known as a **standard candle**, which is a particular type of object that is assumed to have the same properties in all parts of the universe (and at all times, since looking at distant objects involves looking at them as they were in the past). Examples of such standard candles are: main sequence stars, Cepheid variable stars, and Type Ia supernovae. A further problem is that not only do we need to know that all these standard candles have the same brightness but we also need to know its value. This might be hard as there may be no such objects near to us. This is called the 'calibration problem' and is solved by astronomers using a variety of different standard candles and ways of measuring distance to form 'cosmic distance ladders' to enable them to measure extragalactic distances. As a result of improvements in calibration, the estimates for the Hubble constant have come under better control, resulting in the estimate (26.31). However, given the uncertainty in H_0, the value is often written in the astronomical literature as

$$H_0 = 100h \,(\mathrm{km/s})/\mathrm{Mpc}, \qquad (26.32)$$

where h is a measure of the uncertainty in H_0 (not to be confused with Planck's constant), which has a current estimate of 0.72 ± 0.08 with a one-sigma error.

26.3 The cosmic microwave background radiation

The **cosmic microwave background** (CMB) provides important evidence for an initial hot big bang, the isotropy of the universe (which justifies the use of the Robertson-Walker metric) and also gives a way of estimating the cosmological parameter Ω_r. We therefore discuss it in more detail below. However, to go from the current temperature of the CMB as given by (26.33) to an estimation of Ω_r involves some knowledge of results from physics. Indeed, much of the remainder of the chapter involves using physics that goes beyond general relativity, and as such, it is outside the mathematical scope of this book. However, to give a more up-to-date account of cosmology, we really need these ideas. We have indicated this by making most of the remainder of the material in this chapter Level 2. Even if you don't have the background, we suggest you press on and take the results on trust so you should at least get the flavour of the ideas involved. It is the general results, such as the scaling laws (26.36) and (26.38), rather than the precise details that are the important things and at least some familiarity with the ideas should help if you consult other contemporary books or articles on cosmology.

The CMB was first predicted theoretically in 1948 by Alpher and Herman (a prediction later rediscovered by Zel'dovich and also by Dicke in the 1960s). Some of Dicke's colleagues at Princeton started constructing a device to measure it in 1964 but, before it became operational, the CMB was detected by Penzias and Wilson using an antenna constructed for experiments involving microwave communication with satellites. Since that time, the CMB has been measured with increasing precision by experiments made in space such as WMAP in 2001 and the Plank satellite in 2018. These measurements show that the Earth is bathed in a uniform

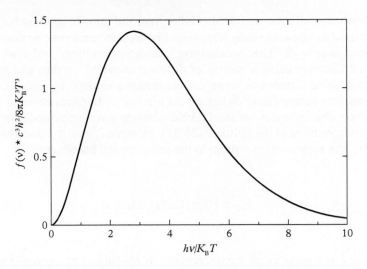

Fig. 26.3 The energy distribution of black-body radiation as given by equation (26.35).

electromagnetic radiation which can be very accurately described as that from a black-body with a temperature of

$$T_0 = 2.725 \pm 0.001 \text{ K.} \qquad (26.33)$$

Furthermore, the CMB is remarkably isotropic with the temperature (after correcting for local sources) varying by less than one part in 10,000. As discussed earlier, this provides extremely strong evidence for the large-scale isotropy of the universe. As we said at the start of this section, in order to go from (26.33) to a calculation of Ω_r, we need some more physics.

The energy of an individual photon of frequency ν, is given by $E = h\nu$, where h is Planck's constant. On the other hand, for the purposes of statistical physics, the kinetic energy of a typical particle at temperature T is given by $k_B T$ where k_B is Boltzmann's constant. For black-body radiation (i.e. radiation emitted from an opaque body in equilibrium with its environment), the energy radiated is given by

$$\epsilon_{\text{rad}} = \int_{\nu=0}^{\infty} f(\nu) d\nu, \qquad (26.34)$$

where $f(\nu)$ measures the energy density in the frequency range ν to $\nu + d\nu$ and is given by

$$f(\nu) = \frac{8\pi h}{c^3} \frac{\nu^3}{(\exp(h\nu/k_B T) - 1)}, \qquad (26.35)$$

with graph shown in Fig. 26.3.

Finding the turning points by setting $f'(\nu) = 0$, we see that the energy distribution of a black-body spectrum peaks when

$$3(\exp(h\nu/k_B T - 1) = h\nu/k_B T,$$

which corresponds to a frequency of $\nu_{max} \simeq 2.7 k_B T/h$ and an energy of $E_{max} \simeq 2.7 k_B T$. As one might expect from the graph, this is also close to the mean energy, which is given by $E_{mean} \simeq 3 k_B T$. This value will be important in modelling the origin of the CMB. It follows from (26.34) and (26.35) that (exercise)

$$\epsilon_{rad} = \alpha T^4. \qquad (26.36)$$

One can actually perform the integral (26.34) and this gives

$$\alpha = \frac{\pi^2 k_B^4}{15 \hbar^3 c^3} = 7.565 \times 10^{-16} \, \text{Jm}^{-3} \, \text{K}^{-4},$$

where J denotes joules. Inserting the value of the temperature of the CMB (26.33) into this gives a value of

$$\epsilon_{rad}(t_0) = 4.17 \times 10^{-14} \, \text{J m}^{-3}.$$

Converting this to a mass density and writing it as a fraction of the critical density we find

$$\Omega_{CMB} \simeq 5 \times 10^{-5}.$$

Actually, there are other forms of radiation, such as massless neutrinos and possibly gravitons, that should really be included in the calculation, but these make no substantial difference to the calculations, which give an upper bound of

$$\Omega_r \lesssim 10^{-4}. \qquad (26.37)$$

On the other hand, as we will see below, a rough estimate of the matter in the universe gives a value of Ω_m, which is several orders of magnitude larger, which justifies ignoring radiation in all but the early universe. There is another important result which follows from (26.36). Since we know that

$$\rho_r \propto \frac{1}{R^4},$$

we must have

$$T \propto \frac{1}{R}. \qquad (26.38)$$

This means that the universe cools as it expands. Although the current temperature is only around 3 K in the early universe, it would have been considerably hotter. For example, when the universe was one-thousandth of its present size, the temperature would be 3,000 K. There is one further point to consider. Since the temperature changes with time, it is not immediately obvious from the fact that it is currently observed as black-body radiation that it was black-body radiation in the past. Although the temperature T scales like $1/R$, the frequency ν also scales in the same way, due to the redshift factor. Hence the ratio ν/T in the exponential term remains unchanged while the ν^3 multiplicative term just scales with 1/volume, as it should. Thus, a black-body spectrum in the past does indeed correspond to one in the present, and vice versa.

We are now in a position to discuss the origin of the CMB. We will consider a simple model of the early universe consisting of matter in the form of hydrogen and radiation in the form of photons. Although this is a somewhat simplified model, it will explain the essential features of the origin of the microwave background. Since $T \propto 1/R$, if we go back to a sufficiently early time, then the universe will be hot enough to fully ionize hydrogen. Then atoms would not exist and the universe would be a sea of photons, free electrons, and nuclei forming an ionized plasma. As the universe expands and cools, the photons lose energy and eventually drop below the ionization energy of around 14 eV. Over a fairly short period of time, the universe moves from a situation where the photons interact with the ionized plasma and is **opaque**, to one where the matter is in the form of atoms so that there is little interaction with the photons and the universe is **transparent**. This process is known as **decoupling**. A crude estimate for the decoupling temperature is given by equating the ionization energy for hydrogen $E = 14$ eV to the mean energy of black-body radiation given by $E = 3k_B T$. Since $k_B = 8.62 \times 10^{-5}\,\mathrm{eV\,K^{-1}}$, this gives

$$T \simeq \frac{14\,\mathrm{eV}}{3k_B} \simeq 50{,}000\,\mathrm{K}. \qquad (26.39)$$

Actually this answer is out by a factor of about 10. There are two reasons for this. First, the mean energy is not really what we need to look at, since we see from Fig. 26.3 that, although most of the energy comes from around $3k_B T$, there are a significant number of higher energy photons coming from the tail of the distribution with higher energy. The second issue is that there are many more photons than hydrogen atoms, so that the relevant calculation involves finding the temperature at which there is roughly one ionizing photon per hydrogen atom. A more sophisticated calculation along these lines gives a temperature for decoupling T_{dec} of

$$T_{\mathrm{dec}} = 3{,}000\,\mathrm{K}. \qquad (26.40)$$

Comparing this with the present temperature of the CMB, we conclude that decoupling happened when the universe was around one-thousandth of its present size. At that point, it was in thermal equilibrium and, as a result, had a black-body spectrum which, as we have seen, is preserved by the evolution of the universe. This explains why the CMB is described so accurately with a black-body spectrum. The hot big bang therefore gives a simple explanation for this fact, in contrast to the steady-state theory, which predicted a different distribution with much more energy in the higher frequencies.

Since decoupling happened when the universe was only one thousandth of its present size, and the CMB photons have been travelling uninterrupted since then, they must have set off at a much earlier time and have therefore travelled a considerable distance. Indeed, they must have originated on a very large sphere centred on us called the **last scattering surface**. It is important to notice that there is nothing special about this surface. Its position just depends on the location in space and time of whoever observes it. It is simply the distance away from us in light years of the time from decoupling to our present epoch. Calculations show that, for us, its radius is of the order of 6,000 Mpc, but photons were emitted from every point at the time of decoupling, and observers at different points in space and different epochs will see different surfaces. The other point to note is that, at the time of decoupling, the radiation had a frequency of about 1.6×10^{14} Hz, but this has been reduced by a factor of about 1,000, due to the expansion of the universe, giving the current value of 160 GHz, which is in the microwave frequency.

26.4 How heavy is the universe?

In attempting to estimate Ω_m, the first thing that one might think of doing is to estimate the total mass of the stars. As long as this is done over a sufficiently large region, this should give a reasonable estimate of the average density of stars in the universe. This has been done by a number of researchers and they come up with a figure of the order

$$\Omega_{\text{stars}} \simeq 0.01.$$

However, this is clearly an underestimate for Ω_m since we know that there is matter that has not been included in this calculation. For example, it is known that there is matter in the form of gas within clusters of galaxies, since, in certain situations, this becomes hot and can be observed in the X-ray regime.

An alternative approach to estimating Ω_m comes from looking at the early stages of the universe when the first atomic nuclei formed in a process called **nucleosynthesis**. Since nuclear binding energies are much larger than those for ionization of electrons, this took place much earlier than decoupling. During the process of nucleosynthesis, the lightest elements in the periodic table, hydrogen, helium, and lithium, were formed. It turns out that the relative abundances of these elements are sensitive to the number density of baryons (i.e. ordinary particles such as protons and neutrons, which make up the vast bulk of atoms) and that, for this to match the observed abundances, the fraction of the critical density in baryons must be around

$$\Omega_{\text{baryon}} \simeq 0.04.$$

This shows that there is a substantial amount of baryonic matter that is not in visible stars and it turns out that this fits well with observations of the motions of galaxy clusters.

Although the above calculation gives a good estimate for the amount of baryonic matter in the universe, it does not preclude a higher value for Ω_m, due to the presence of non-baryonic matter. The first clue to the existence of such matter came from looking at the Newtonian dynamics of objects moving in the gravitational field of a galaxy and using this to estimate the mass. Let $M(r)$ be the mass of the galaxy within a distance r from the centre. Then an object of mass m orbiting the galaxy at a distance r experiences a gravitational force towards the centre of $GmM(r)/r^2$. On the other hand, for a circular orbit, the centripetal acceleration for an object moving with angular velocity ω is $a = r\omega^2$ towards the centre. Using $F = ma$ gives

$$\frac{GM(r)}{r^2} = r\omega^2. \tag{26.41}$$

Using the fact that $v = \omega r$ this shows that the orbital velocity v at a distance r from the centre of the galaxy is given by

$$v = \sqrt{GM(r)/r}. \tag{26.42}$$

By plotting a 'rotation curve' of v against r, one can deduce how the mass varies with r. At large distances, one would expect that $M(r)$ would be more or less constant so that $v \propto 1/\sqrt{r}$. However, the actual rotation curves are fairly flat at large distances, showing that there must be non-luminous matter surrounding the galaxy in a halo. When one computes the total amount of matter in such halos, it is just about possible that this could be baryonic matter, but a common alternative is to suggest that it is some new type of matter that essentially only intersects with baryonic matter through gravitational effects. This is an example of **cold dark matter**, that is to say, non-luminous, non-baryonic matter that only interacts gravitationally with the other matter in the universe.

The best evidence for both cold dark matter and a non-zero cosmological constant come from two other pieces of observational evidence. The first is the 'redshift-magnitude calculation' for Type Ia supernovae. In Chapter 24, we obtained Hubble's law. The key result was (24.72),

$$d_L = r_1 R(t_0),$$

which was used together with (24.68) to obtain $d_L \simeq z/H_0$, which is valid for small values of r_1. However, (24.68) is not exact but involves various approximations. These are not important for comparatively nearby stars and galaxies, but, for more distant objects, such as Type Ia supernovae, this becomes more important. Rather than work with the approximation, we can use (26.13) to give an exact relationship between d_L and z which is valid for all values of r_1. It follows from (24.62) that

$$r_1 = \begin{cases} \sin \chi & \text{if } k = +1, \\ \chi & \text{if } k = 0, \\ \sinh \chi & \text{if } k = -1, \end{cases} \tag{26.43}$$

where

$$\chi := \int_{t_1}^{t_0} \frac{dt}{R(t)}. \tag{26.44}$$

Then, changing variables and using (26.13), we have

$$\chi = \int_{R(t_1)}^{R(t_0)} \frac{dR}{\dot{R}(r)R}$$

$$= \frac{1}{R(t_0)H_0} \int_{R(t_1)/R(t_0)}^{1} \frac{d\tilde{R}}{\tilde{R}(d\tilde{R}/d\tilde{t})}$$

$$= \frac{1}{R(t_0)H_0} \int_{(1+z)^{-1}}^{1} \frac{d\tilde{R}}{\tilde{R}\left[\Omega_c - 2V(\tilde{R})\right]^{1/2}},$$

where $V(\tilde{R})$ is given by (26.23), and we have used (24.66) to write $R(t_1)/R(t_0)$ in terms of z in the last line. If we now define

$$g(z) := \frac{1}{H_0} \int_{(1+z)^{-1}}^{1} \frac{d\tilde{R}}{\tilde{R}\left[\Omega_c - 2V(\tilde{R})\right]^{1/2}}, \tag{26.45}$$

then

$$d_L(z) = \begin{cases} R(t_0) \sin\left(g(z)/R(t_0)\right) & \text{if } k = +1, \\ g(z) & \text{if } k = 0, \\ R(t_0) \sinh\left(g(z)/R(t_0)\right) & \text{if } k = -1. \end{cases} \tag{26.46}$$

Furthermore, for $k \neq 0$, by taking the modulus of (26.14), we have

$$R(t_0) = \frac{1}{H_0|\Omega_c|^{1/2}}, \quad \text{where } \Omega_c = 1 - \Omega_m - \Omega_\Lambda.$$

This gives d_L as a function of z parameterized by Ω_m and Ω_Λ. The resulting redshift-magnitude relation that one obtains can be plotted on the $(\Omega_m, \Omega_\Lambda)$-plane and compared with observation.

The second piece of evidence comes from the CMB anisotropies. The CMB is remarkably isotropic but the very small angular variations in the temperature contain considerable information. The CMB anisotropies

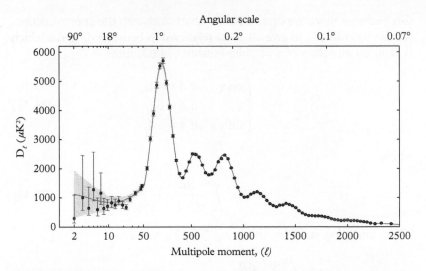

Fig. 26.4 Multipole expansion of the correlation function for the cosmic microwave background, showing peaks corresponding to dark energy, baryonic matter, and dark matter, respectively. (Reproduced from *Ade et al.* (2014), with permission from Astronomy and Astrophysics.)

that we measure today reflect the temperature fluctuations in the last scattering surface. How we see them depends not only on their physical size at the time of last scattering, but also on the geometry of the universe through which the photons have reached us. As a result, we can potentially use information from this to determine the cosmological parameters of the universe. The angular variation of temperature is encoded in the correlation function $C(\theta)$ between the temperatures at a fixed angle θ apart, which can then be broken down into a multipole expansion (which is a three-dimensional generalization of a Fourier series). The results are shown in Fig. 26.4.

The location of the various peaks is determined by the amount of baryonic matter, dark matter, and dark energy and can, again, be used to plot the location of our universe in the $(\Omega_m, \Omega_\lambda)$-plane.

Neither the redshift-magnitude information nor the CMB information individually gives a very precise location of our universe in the $(\Omega_m, \Omega_\lambda)$-plane. Instead, they both give long thin regions for the location of our universe (see Fig. 26.5). Fortunately, these strips intersect more or less at right angles to give a fairly precise value, as shown in Fig. 26.5.

At the time of writing, the above observations give $\Omega_m \simeq 0.3$ and $\Omega_\Lambda \simeq 0.7$, which are also consistent with other observations such as the formation of structure and bulk motions of galaxies (which both seem to require $\Omega_m > 0.2$). Taken together with the negligible value of Ω_r, these give $\Omega_c \simeq 0$, which corresponds to a spatially flat universe. This is called the **ΛCDM model**, where Λ is the cosmological constant, and CDM stands for cold dark matter. At the time of writing, this is the 'standard cosmological model'.

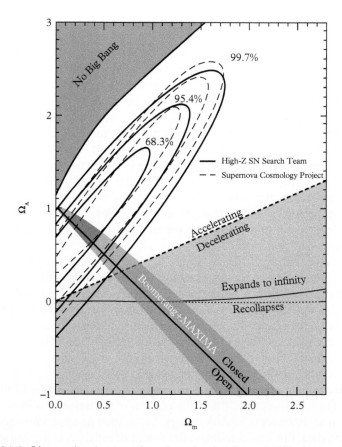

Fig. 26.5 Observational constraints from luminosity red-shift data and data from the cosmic microwave background, plotted in the $(\Omega_m, \Omega_\Lambda)$-plane. (Reproduced from *Liddle* (2004), with permission from the publisher.)

26.5 The ΛCDM model of cosmology

In the previous section, we showed that the cosmological evidence favours the ΛCDM model, which is a spatially flat model ($k = 0$) in which we ignore radiation (except in the early universe) and we take $\Omega_m + \Omega_\Lambda = 1$. The current best estimate is that $\Omega_m \simeq 0.3$ and $\Omega_\Lambda \simeq 0.7$. The virtue of this model is that

- it fits the observed expansion of the universe,
- it fits the existence and structure of the CMB,
- it is consistent with the large-scale structure in the distribution of galaxies,
- it is consistent with the observed abundances of hydrogen (including deuterium), helium, and lithium.

Furthermore, it is one of the models for which we have an exact solution. This is given by (25.5), which we can write in the alternative form (exercise) as

$$R(t) = (\Omega_m/\Omega_\Lambda)^{1/3} \sinh^{2/3}(t/t_\Lambda), \qquad (26.47)$$

where $t_\Lambda := 2/(3H_0\sqrt{\Omega_\Lambda})$. This solution is valid once we reach an era (slightly past decoupling) where the photons no longer interact with the matter. According to the estimates we derive in the next section, this occurs at about $t = 10^{13}$ second (i.e about 10^6 years). The picture of the evolution is shown in Fig. 25.1(i) for the case where $k = 0$, $\Lambda > 0$. In the early stages, this universe initially expands at a rate of $t^{2/3}$; the expansion then slows down, then starts to speed up again, and finally tends to the de Sitter solution at late times, expanding like $e^{\alpha t}$, where $\alpha = H_0\sqrt{\Omega_\Lambda}$. Taking $\Omega_m = 0.7$ in (26.30), we get an age of the universe for about fourteen billion years, which is compatible with some of the lower estimates coming from geological and astrophysical sources. Without the presence of a cosmological constant, the age of the universe would be uncomfortably close to these lower bounds.

However, although this result fits the cosmological data very well, it raises a number of important questions. First, the value of Ω_c is close to zero. It seems unlikely that this has happened by chance. There are a number of theories to explain this, the most commonly accepted (although not the only one) being cosmic inflation, which we discuss in §26.7. Second, as we have seen, the baryonic matter has a value of at most $\Omega_{baryon} \simeq 0.04$, so the remainder of Ω_m must be non-baryonic. Thus, there is nearly ten times as much dark matter in the universe as there is conventional matter. This raises the question, What is the nature of the dark matter? Furthermore, $\Omega_\Lambda \simeq 0.7$, so, that 70% of the content of the universe is in the even more mysterious form of dark energy. So, despite its apparently good fit with the data, this model relies on the theory of inflation to explain spatial flatness, and two of its main ingredients, dark matter and dark energy, are poorly understood. We describe briefly below what these might be.

The obvious place to start comes from considering known astrophysical objects to see if they could be the source of dark matter. There are two possible candidates here. The first is black holes. Although these do not count as being baryonic, they must have been formed from either baryons or dark matter crossing the event horizon. For this reason, the upper bound from nucleosynthesis also applies to black holes formed through, for example, stellar collapse. Thus, if black holes are to provide the source of the cold dark matter, they must have existed prior to nucleosynthesis. A significant number of 'primordial black holes' could potentially give the answer, but it is not clear how many of these should exist and these do not look a promising source for providing all the required cold dark matter. We have already discussed halos in conjunction with rotation curves. This provides a second possible source of cold dark matter, in the form of massive compact halo objects (MACHOs) – large, condensed objects such as neutron stars, white dwarfs, very faint stars, or non-luminous objects like planets which are surrounded by non-baryonic halos of some sort. The search for these objects consists of using gravitational lensing to detect the effects they have on background galaxies. As a result, it is thought that MACHOs have been detected in the Large Magellanic Cloud but most experts believe that these searches also rule out MACHOs as a viable candidate for providing all the required cold dark matter.

The other potential source of dark matter comes from particle physics. One possibility is that neutrinos, rather than being massless, have a small mass. Since this would be in the form of radiation, this is described as 'hot dark matter' but would still be non-baryonic. However, there are experimental bounds on neutrino masses which make this an unlikely source of the required dark matter. Another possibility coming from particle physics are 'axions', which are a conjectured very light particle with a specific type of self-interaction that makes them a suitable candidate for cold dark matter. Axions have the theoretical advantage that their existence solves certain other problems in quantum chromodynamics, but axion particles have only been theorized and never detected. The theory of supersymmetry provides another potential source of dark matter. These are often referred to as WIMPs – weakly interacting massive particles. The search for WIMPs involves attempts at direct detection by highly sensitive detectors, as well as attempts at production of WIMPs by particle accelerators. Despite some claims, so far there have been no conclusive detections of WIMPS. Of course, since there has been no definitive observation of dark matter, it may be that none of the above suggestions are correct and that something else entirely which has not yet been though of is required! For a review of the current state of dark matter, see the article by Amendola et al. (2018).

There are two distinct approaches to 'dark energy'. The first is simply to regard it as the cosmological constant Λ in Einstein's equations, which does not need an explanation any more (or any less) than the value of G. The second is to put it on the right-hand side of Einstein's equations and think of Λg_{ab} as an additional energy momentum term. As we have seen, this means that it corresponds to a perfect fluid with equation of state $p = -\rho$. This is the form of matter that one would get from the quantum mechanical zero-point energy of the vacuum. Unfortunately, the standard model gives a figure that is more than one hundred orders of magnitude too large. However, it may be that some supersymmetric theory (together with some symmetry-breaking mechanism) or string theory will produce a vacuum expectation of the right magnitude. Another possibility is that the dark energy is not constant but is dynamic and related to a scalar field ϕ driven by some potential $V(\phi)$, for example. Such theories are sometimes called 'quintessence'. Other more drastic explanations for the cosmological constant involve modified theories of gravity. We will discuss this in more detail when we look at the theory of inflation in §26.7. It is also important to note that some cosmologists question the observational evidence which results in the need for a cosmological constant. For example, it may be that the assumption that the luminosity of Type Ia supernovae does not vary with age is false. In that case, the luminosity-redshift results used in part to determine Ω_Λ would no longer be valid. A different objection is the use of the Robertson-Walker metric to describe the universe (Ellis 2011). Although we know that, on the large-scale the matter is homogeneous and isotropic, this is not true until one goes to extragalactic scales. Since Einstein's equations are non-linear, it is not at all obvious that the 'average solution' to Einstein's equations is the same as solving Einstein's equations for an averaged source.

26.6 The early Universe

In the previous section, we looked at the ΛCDM model of cosmology, in which we neglected the radiation, since Ω_r was so small. However, as we have previously noted $\rho_r \propto 1/R^4$ while $\rho_m \propto 1/R^3$, so that

$$\frac{\rho_r(t)}{\rho_r(t_0)} = \frac{R^4(t_0)}{R^4(t)}, \quad \text{and} \quad \frac{\rho_m(t)}{\rho_m(t_0)} = \frac{R^3(t_0)}{R^3(t)}. \tag{26.48}$$

Thus,

$$\frac{\rho_r(t)}{\rho_m(t)} = \frac{\rho_r(t_0)}{\rho_m(t_0)} \frac{R(t_0)}{R(t)} = \frac{1}{\tilde{R}} \left(\frac{\Omega_r}{\Omega_m} \right). \tag{26.49}$$

Although the precise number is not important for the general argument, we provide specific numbers in this section and take $\Omega_r = 7 \times 10^{-5}$ and $\Omega_m = 0.3$ so that the current ratio is $\Omega_r/\Omega_m \simeq 2 \times 10^{-4}$. It therefore follows from (26.49) that

$$\rho_m(t) = \rho_r(t) \quad \text{when} \quad \tilde{R}(t) = \tilde{R}_{\text{eq}} \simeq 2 \times 10^{-4}. \tag{26.50}$$

This is known as the time of **matter–radiation equality**. At earlier times, the universe is radiation dominated and afterwards it is matter dominated. Note that this occurs well before the time of decoupling, which, as we saw earlier, takes place at $\tilde{R} = 10^{-3}$. Rather than locate these epochs in terms of size $\tilde{R}$, it is helpful to give them in terms of the time. In order to locate the time of decoupling, we use the fact that the temperature scales as $1/\tilde{R}$ while, in the matter dominated regime, $R \sim t^{2/3}$, which applies from the time of matter–radiation equality onwards. We could, of course, use the more precise formula (26.29) but, since many of the parameters have observational uncertainties, we will content ourselves in this section with just giving order-of-magnitude estimates. Together, these two scaling laws give

$$\frac{T(t)}{T(t_0)} = \frac{R(t_0)}{R(t)} = \left(\frac{t_0}{t} \right)^{2/3}. \tag{26.51}$$

Since, for decoupling, the ratio on the left is approximately 1,000, and the age of the universe is about 4×10^{17} seconds, substituting in the above gives the age at decoupling as

$$t_{\text{dec}} \simeq \times 10^{13} \text{ seconds} \simeq 350,000 \text{ years}. \tag{26.52}$$

We can continue to use (26.51) right down to matter radiation equality. In this case, we have seen that the ratio on the left is approximately 5×10^3, so that substituting in the above gives

$$t_{eq} \simeq 10^{12} \text{ seconds} \simeq 35,000 \text{ years.} \qquad (26.53)$$

We also see from (26.51) that $T_{eq} \simeq 10^4 T_0 \simeq 30,000 \text{ K}$.

Earlier than t_{eq} (and hence at temperatures greater than T_{eq}), the universe is radiation dominated so that $R \propto t^{1/2}$, although, as we saw in §26.1, we still have $T \propto 1/R$. As a result, (26.51) becomes

$$\frac{T(t)}{T_{eq}} = \left(\frac{t_{eq}}{t}\right)^{1/2}, \qquad (26.54)$$

and, substituting in our values of T_{eq} and t_{eq}, we obtain the relation

$$T(t) \simeq \frac{1.5 \times 10^{10}}{\sqrt{t}}, \qquad (26.55)$$

where the temperature T is measured in kelvin and the time t in seconds. Thus, when the universe is about 1 second old, it has a temperature of about $1.5 \times 10^{10} \text{ K}$. If we use $E = k_B T$, this corresponds to a characteristic energy of around 1 MeV, which is comparable to nuclear binding energies. So, earlier than one second, the universe would have been a sea of interacting protons, neutrons and electrons. Of course, at even earlier times, and therefore even higher energy scales, the protons and electrons would have dissociated into their constituent quarks. This is called the quark–hadron phase transition and is supposed to take place at energies of a few hundred MeV corresponding to temperatures of around 2×10^{12}. This takes us back to a time of around 10^{-4} seconds. We summarize the position in Table 26.2. This is also shown graphically in Fig. 26.6.

Table 26.2 Stages in the evolution of the universe (taking $\Omega_m = 0.3$ and $H_0 = 72 \text{ (km/s)/Mpc}$).

Time	Type of matter
$10^{-10}\,\text{s} < t < 10^{-4}\,\text{s}$	Free quarks, neutrinos, photons, and electrons. Everything strongly interacting.
$10^{-4}\,\text{s} < t < 1\,\text{s}$	Free electrons, protons, neutrons, photons, and neutrinos. Quarks no longer free but otherwise strongly interacting.
$1\,\text{s} < t < 10^{12}\,\text{s}$	Free electrons, atomic nuclei, photons, and neutrinos. Radiation dominated.
$10^{12}\,\text{s} < t < 10^{13}\,\text{s}$	Free electrons, atomic nuclei, photons, and neutrinos. Matter dominated.
$10^{13}\,\text{s} < t < t_0$	Atoms have formed and photons no longer interacting. Universe is transparent.
$t_0 = 4 \times 10^{17}\,\text{s}$	Current age of the universe.

Time since
Big Bang

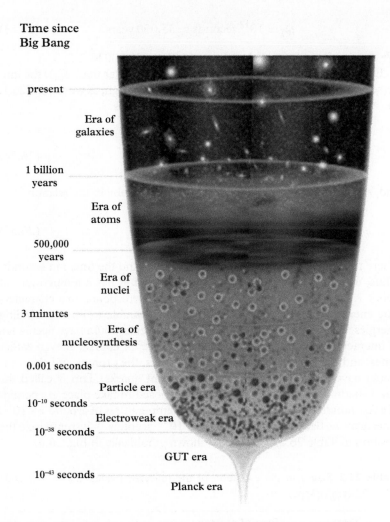

present

Era of
galaxies

1 billion
years

Era of
atoms

500,000
years

Era of
nuclei

3 minutes

Era of
nucleosynthesis

0.001 seconds

Particle era

10^{-10} seconds

Electroweak era

10^{-38} seconds

GUT era

10^{-43} seconds

Planck era

Fig. 26.6 The evolution of the Universe, from the big bang to the present.

26.7 Inflationary cosmology

Despite the success of the ΛCDM model of cosmology, there are various
features it fails to explain. The first issue is what is called the **flatness
problem**. The values of the parameters Ω_m and Ω_Λ are such that Ω_m
$+\Omega_\Lambda \simeq 1$. So, currently, the total amount of matter in the universe is close
to the critical density. If it is exactly at the critical density, then $k = 0$ and
this is preserved by the Friedmann equation. However, what happens if
it is only close to the critical density – which is all that we can conclude
from our measurements? We know from (26.6) that

$$H^2 = \frac{8\pi}{3}\rho_{\text{total}} - \frac{k}{R^2},$$

(26.56)

where

$$\rho_{\text{total}}(t) = \rho_r(t) + \rho_m(t) + \rho_\Lambda(t).$$

If, following (26.2), we define the critical density at time t to be $\rho_{\text{crit}}(t) = 3H^2(t)/8\pi$, then we may write the above as

$$\left| 1 - \frac{\rho_{\text{total}}}{\rho_{\text{crit}}(t)} \right| = \frac{|k|}{H^2(t)R^2}. \tag{26.57}$$

Now, in the matter-dominated period, $R \sim t^{2/3}$ and $H(t) \sim t^{-1}$, so that $H^2(t)R^2 \sim t^{-2/3}$ and hence

$$|1 - \rho_{\text{total}}(t)/\rho_{\text{crit}}(t)| \propto t^{2/3}. \tag{26.58}$$

Current measurements (at $t_0 \simeq 4 \times 10^{17}$ seconds) tell us that, with a very high degree of confidence, we are within 30% of the critical density. Then (26.58) shows that, at the time of matter-radiation equality (at $t_{\text{eq}} = 10^{12}$ seconds), we must have had $|1 - \rho_{\text{total}}(t)/\rho_{\text{crit}}(t)| < 10^{-5}$, which is incredibly close to the critical density. Going back to the quark–hadron phase transition at 10^{-4} second shows that the universe must have been many further orders of magnitude closer to the critical density. The 'flatness problem' is to explain why the universe is flat or so finely tuned that it is extremely close to flat.

The second problem is to explain why the universe is so isotropic (and hence homogeneous). The usual mechanism to explain why something is homogeneous is the fact that the constituent parts have had a period of interaction in which the various components mix, reach some kind of equilibrium, and become fairly uniformly distributed. Think, for example, of dropping some ink into a glass of water which, over time, becomes uniformly distributed in the water to produce something homogeneous. The problem with this explanation for the homogeneity of the universe is the existence of particle horizons similar to those discussed in §25.7. We calculate the size of the particle horizon for a spatially flat Robertson-Walker metric. Consider a photon moving on a radial trajectory; then

$$ds^2 = d\theta = d\phi = 0,$$

and so, from (24.38),

$$ds^2 = dt^2 - R^2(t)dr^2 = 0.$$

If it is emitted at time t_1, then, at time t_2, it will have travelled a coordinate distance

$$\int_{r_1}^{r_2} dr = \int_{t_1}^{t_2} \frac{dt}{R(t)}. \tag{26.59}$$

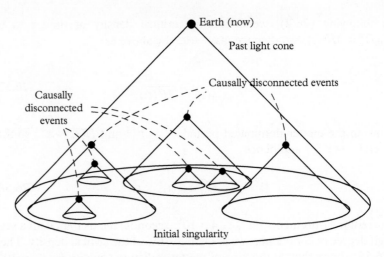

Fig. 26.7 Picture of the horizon problem.

Using (24.38), we have $d\sigma = R(t_2)dr$ at time t_2, so this corresponds to a proper distance

$$\Delta r = R(t_2) \int_{t_1}^{t_2} \frac{dt}{R(t)}. \tag{26.60}$$

If we take $t_1 = 0$ and $t_2 = t$, this gives the maximum proper distance a photon could have travelled since the big bang. We call this the radius of the **particle horizon**, since this is the largest proper distance any particle could have travelled in that time. If one now assumes a radiation dominated universe and takes $R(t) \propto t^{1/2}$, i.e. $R(t) = At^{1/2}$ for some constant A, then this gives

$$\Delta r(t) = At^{1/2} \int_{u=0}^{t} \frac{du}{Au^{1/2}} = t^{1/2} \left[2u^{1/2} \right]_0^t = 2t. \tag{26.61}$$

Now consider a pair of photons from the CMB that have reached us from opposite directions. They originated from points P and Q on opposite sides of the last scattering surface S. However, if one looks at the causal past of P and the causal past of Q all the way back to the big bang, then these do not intersect. This means that there are no events that can influence **both** P and Q. This is lack of causal contact is called the **horizon problem** (see Fig. 26.7). Indeed, one can show that CMB photons need to have an angular separation of less then around 2° for them to have originated from points on the last scattering surface that had causal contact (i.e their past light cones intersected).

Cosmic **inflation** provides a mechanism to get round both the flatness and horizon problems. Suppose that, rather than the $R \sim t^{1/2}$ expansion in the early universe of the FRW radiation dominated solution, one had instead a period of **exponential expansion** so that $R(t) \propto e^{\tilde{H}t}$ where $\tilde{H}$

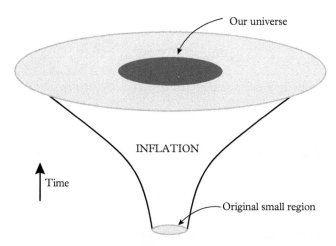

Our universe

INFLATION

Time

Original small region

Fig. 26.8 Schematic picture of inflation, showing how a small thermalized region is inflated to be larger than our observable universe.

is a constant. Then (26.60) now gives

$$\Delta r(t) = e^{\tilde{H}t} \int_{u=0}^{t} e^{-\tilde{H}u} du = e^{\tilde{H}t} \left[-e^{\tilde{H}u}/\tilde{H} \right]_{0}^{t} = \frac{e^{\tilde{H}t} - 1}{\tilde{H}}, \qquad (26.62)$$

so that the horizon now grows exponentially with t rather than linearly as it does in a radiation dominated universe. This enables a small patch of the universe, small enough to become homogeneous through thermalization, to grow exponentially in an inflationary phase and expand to a size greater than our observable universe (see Fig. 26.8). Inflation also stretches out the spatial size of any irregularities and dilutes them. Together, these mechanisms can explain the high degree of isotropy that we observe in the universe today.

We now return to the flatness problem. Since we now have $R(t) = Ae^{\tilde{H}t}$, then $\dot{R} = \tilde{H}R$ so that $H(t) = \tilde{H}$ (i.e. $\tilde{H}$ is just the Hubble constant for an inflationary universe). Substituting into (26.57) gives

$$|1 - \rho_{\text{total}}(t)/\rho_{\text{crit}}(t)| \propto |k| e^{-2\tilde{H}t}. \qquad (26.63)$$

The right-hand side rapidly tends to zero as t gets larger, which means that ρ_{total} quickly becomes extremely close to the critical density. This solves the flatness problem. Thus, inflation provides a good explanation as to why the universe is so homogeneous and isotropic and so close to the borderline between being open and closed. We now turn to a possible source for inflation coming from a scalar field sometimes called the 'inflaton'.

A possible mechanism for producing inflation is provided by the energy-momentum tensor of a scalar field ϕ moving in a suitable potential $V(\phi)$. The action for such a field is

$$I = -\int_{\Omega} \left[\tfrac{1}{2} g^{ab} \nabla_a \phi \nabla_b \phi - V(\phi) \right] \sqrt{-g} d^4 x. \qquad (26.64)$$

So the Lagrangian density is

$$\mathcal{L} = (-g)^{1/2} \left[-\tfrac{1}{2} g^{ab} \nabla_a \phi \nabla_b \phi + V(\phi) \right].$$ (26.65)

The field equations are obtained from $\delta \mathcal{L}/\delta \phi = 0$, which gives (exercise)

$$\Box \phi + V'(\phi) = 0.$$

For a spatially homogeneous field $\phi(t)$ in a Robertson-Walker metric, this is (exercise)

$$\ddot{\phi} + 3H(t)\dot{\phi} + V'(\phi) = 0.$$ (26.66)

The energy-momentum tensor is obtained using

$$T_{ab} = -\frac{2}{\sqrt{-g}} \frac{\delta \mathcal{L}}{\delta g^{ab}},$$ (26.67)

which gives (exercise)

$$T_{ab} = \nabla_a \phi \nabla_b \phi + g_{ab} \left[-\tfrac{1}{2} g^{cd} \nabla_c \phi \nabla_d \phi + V(\phi) \right].$$ (26.68)

For a Robertson-Walker metric, this becomes

$$T_{ab} = \delta_a^0 \delta_b^0 \dot{\phi}^2 + g_{ab} \left[-\tfrac{1}{2} \dot{\phi}^2 + V(\phi) \right].$$ (26.69)

If we think of (26.66) as describing the motion of a particle moving in a potential V, then the $H\dot{\phi}$ term is just a damping term. So, if there is a period where H is sufficiently large compared to $V'(\phi)$ (as for example in the 'slow-roll' potential shown in Fig. 26.9), then the motion of the particle will be damped, $\dot{\phi}$ will be very small, and ϕ will be approximately constant. As a result we will get

$$T_{ab} \simeq V(\phi_\star) g_{ab},$$ (26.70)

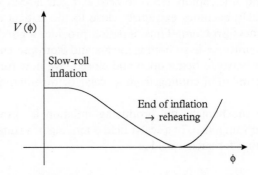

Fig. 26.9 Potential energy $V(\phi)$ for a slowly rolling scalar field.

where $\phi_\star$ is the approximately constant value of ϕ. Provided that $V(\phi_\star)$ is sufficiently large that (26.70) dominates the other contributions to the energy-momentum at that epoch, then, as we saw when looking at the de Sitter solution, we get inflation with

$$R(t) \propto e^{\tilde{H}t}, \quad \text{where } \tilde{H} = (V(\phi_\star)/3)^{1/2}. \qquad (26.71)$$

However, the potential does not actually remain constant but slowly rolls on to the steeper part of the potential, where $V(\phi)$ decays to something small and the universe returns to a radiation-dominated phase. The slow-roll picture of inflation we described here is not the only model available, but is simply illustrative of a potential mechanism. Inflation remains an active area of research and we will not go into further details of other inflationary models here, except to say that the mechanism to provide an inflationary phase takes place at very high energies (and hence temperatures) corresponding to times as early as 10^{-34} seconds. See the article by Deffayet et al. (2015) for a review of the current situation.

Although inflation was designed to explain the flatness and horizon problems, it turns out that it also provides a potential explanation of the anisotropies in the CMB and the large-scale structure of the galaxies we see today. Without going into details (which are beyond the scope of this book), the idea is that quantum fluctuations $\delta\phi$ in the 'inflaton' field ϕ lead to fluctuations in the density

$$\delta\rho = V(\phi_\star + \delta\phi) - V(\phi_\star) \simeq V'(\phi_\star)\delta\phi, \qquad (26.72)$$

which are inflated to provide the anisotropies in the CMB. This inflation mechanism also applies to quantum fluctuations of other fields and leads to predictions of the form of the power spectrum as shown in Fig. 26.4. This provides a fairly direct confirmation of the inflationary mechanism as well as putting constraints on the potential V.

26.8   The anthropic principle

The idea of cosmic inflation was developed in part to explain why the universe was so homogeneous and isotropic (i.e. why the cosmological principle is true). An alternative answer that is sometimes invoked is that, if **we** are to exist, then it could hardly be otherwise. Put another way, a non-smooth universe would not have allowed us humans to have developed. This is an example of the **anthropic principle**, which, in simple terms, states the following.

The anthropic principle: we see the universe the way it is because we exist.

The principle comes in two versions, the weak and the strong, which we consider in turn.

> **The weak anthropic principle:** the conditions for the development of life are only met in certain regions of the universe.

This form of the principle can be used to 'explain' why the big bang occurred some ten thousand million years ago; namely, because it takes that long for sentient beings to emerge. More precisely, this is the time needed for all the intervening processes, such as the condensation of the galaxies from the primeval matter, the subsequent formation of the heavy elements (in supernovae), the eventual birth of our own galaxy, the formation of the solar system, the cooling of Earth, and the slow process of evolution up to the present day.

The earliest epochs of the universe really involve quantum ideas, and this leads to the area of **quantum cosmology**. As is well known, quantum theory involves deep problems of interpretation – see the book by Penrose (1989) for an intriguing viewpoint on this issue and the book by Hawking and Penrose (2015) for a discussion on the nature of a quantum theory of gravity. One interpretation leads to the 'many worlds' of Everett and Wheeler, in which the universe is bifurcating from one instant to the next into many (indeed infinite) disjoint new universes (Fig. 26.10). Or again, the universe may consist of many different regions, each with its own initial configuration and perhaps with its own set of laws of science. If we consider these disjoint regions as different universes, then the strong version of the anthropic principle can be stated as follows.

> **The strong anthropic principle:** the conditions for the development of life are only met in a few universes.

The laws of science involve a number of fundamental constants (such as the charge on the electron) which, at present, cannot be predicted from theoretical considerations, but can only be found by observation. Moreover, their actual values seem to be very finely adjusted. The slightest alteration of these values would lead to very different universes, most of which could not support life. One can interpret them in two ways: as evidence of a divine purpose or Creator (the argument from design in theology) and with it the choice of a particular set of laws of science, or as support for the strong anthropic principle. Although it is not clear the extent to which Einstein believed in a personal God, it is worth remarking that Einstein believed profoundly in the argument of a divine purpose. He considered that God could not have created the universe in any other way.

There are a number of objections to the strong formulation of the principle. If all these universes are really separate from us, in what sense can they be said to exist? If what happens in another universe has no observable consequences on ours then, on simplicity grounds alone, we can ignore them. If, on the other hand, they are different but accessible regions of our universe, then they are just the result of different initial configurations and so the strong anthropic principle would reduce to the weak one. Another objection is that the principle runs counter to the way that

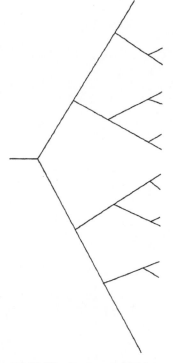

Fig. 26.10 The 'many-worlds' interpretation of Everett and Wheeler.

ideas have developed throughout the history of science, which has continuously demoted the special importance of humankind in the scheme of
things. For example, the cosmological principle leads us to believe that we
live in a typical part of the universe, attached to a typical star, in a typical
galaxy, belonging to a typical cluster, and so on. Yet the strong form of
the anthropic principle turns this on its head and says that the whole giant
structure exists simply for our sake.

The attempt to find a model of the universe in which many different
initial configurations could have evolved into something like the present
universe led to the idea of inflation. So inflation, together with the weak
form of the anthropic principle, may be used to explain why the universe
looks the way it does now.

The anthropic principle can also be used to throw light on whether the
three **arrows of time** agree or not. These are the thermodynamic arrow
(as expressed in the idea that disorder or entropy is always increasing),
cosmological time (world time), and psychological time (as perceived by
humans). For further development of these ideas, see the book by Hawking (1988) (on which this account is based) and, for a more technical
account, see the book by Barrow and Tippler (1986). It seems appropriate to end with a reference to Hawking, given that one of the goals of
the book is to make contact with Hawking and Ellis (1973). It also seems
appropriate to finish with an amusing representation of the development
of life subsequent to the big bang in the universe (the big U) by Wheeler
(Fig. 26.11), because this is reminiscent of the surrealistic pictures by
Hugh Lieber in Lillian Lieber's book – which is where we came in.

26.9   Final questions

We have seen in this chapter that inflation, together with the ΛCDM
model of cosmology, provides a remarkably good model of the universe
that fits the current high-precision measurements of the cosmological
parameters. We have not discussed the formation of structure, but the
basic mechanism is that the inhomogeneities in the density lead to clustering through the mechanism of **gravitational instabilities**. This occurs
where a region has a slightly higher density, so that matter is attracted
to it, thus increasing the density, and where another region has a slightly
lower density, in which case matter moves away from it, thus lowering
the density. We will not go into further detail except to say that computer
simulations are now able to reproduce the observed clustering of galaxies
in our universe, but that dark matter is an essential ingredient to achieve
this. However, despite all this, success in matching the observations the
model brings with it a number of unanswered questions concerning the
underlying physics. Starting with inflation, there are the questions What is
the scalar field ϕ? and, more importantly, Why should the potential $V(\phi)$
have the correct properties to produce inflation? In one sense, we have just
replaced the questions about homogeneity and flatness by ones about the
physics of inflation. For the ΛCDM model, there are also the fundamental unanswered questions, What is the origin of dark matter? and Why
do we have a cosmological constant? or, alternatively, What is the origin

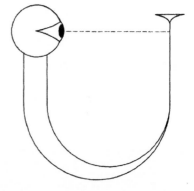

Fig. 26.11 Wheeler's 'big U' from the
big bang (upper right) to the development of the human eye.

of dark energy? Here again, we seem to have answered the cosmological question but posed difficult questions about fundamental physics in the process. Perhaps it will take new ideas from areas such as string theory, alternatives to inflation such as Penrose's ideas about conformal cyclic cosmology, or maybe new observations from ultra-high energy particle accelerators, or advanced gravitational detectors in space which observe gravitational waves from both the early universe, black holes and neutron stars to answer these questions.

The ΛCDM model starts with a big bang singularity, has a period of inflation, and then expands as a flat Friedmann-Robertson-Walker model with a positive cosmological constant – first in a radiation dominated phase, then in a matter dominated phase, and finally in a Λ dominated phase, where it slowly expands exponentially to a de Sitter solution. In studying these cosmological solutions, we have seen the importance of horizons and singularities: both phenomena we have met before in our considerations of classical black holes in Part D. The big bang singularity is a very drastic one, in which both the density and temperature increase without bound as $R \to 0$ and, indeed, space-time itself becomes singular at $R = 0$, where it is squeezed out of existence. However, the results have been deduced from the assumptions of exact spatial homogeneity and spherical symmetry. Although these assumptions may be reasonable on the large scale, they certainly do not hold locally. One might expect that, if one traced the evolution of the universe back in time, the local irregularities would grow and could prevent the occurrence of a singularity – causing the universe to 'bounce' instead. Yet, once again, the singularity theorems of Hawking and Penrose reveal that the occurrence of singularities is generic and, as a consequence, there is good evidence to believe that the physical universe was singular in the past.

There is another difference about the initial singularity of cosmology, compared with the black hole singularities, in that the big bang singularity is, in principle, **observable**. And it is observation that is the linchpin of cosmology. The past decades have brought about huge advances in determining the cosmological parameters. The ability to make measurements from space has brought significant advances in X-ray astronomy, for example. With the detection of gravitational waves, we are now in the era of **multi-messenger** astronomy, in which signals from gravitational waves are combined with those in the radio, optical, and X-ray regimes to survey the universe. The planned launch of LISA to detect gravitational waves from space will provide a new window on the universe, enabling us to detect gravitational signals going back to the very early universe, and allowing us to put our cosmological theories more rigorously to the test.

It is a natural consequence of our inquisitive nature that we should wish to understand our own origins and that of the universe we inhabit. The hot big bang theory would appear to be a great stride forward in our search for this understanding. Whether or not the universe had this singular origin is perhaps the central question of cosmology. The mathematical basis of this question and the attempt to answer it is the principal problem dealt with in the book of Hawking and Ellis (1973). In turn, it has been one of the main objectives in writing this book to make their book, or at least parts of it, more accessible and the hope is that some readers may make this their next port of call.

And so, we end our considerations of cosmology and, with it, we end the book. There are many topics in general relativity which have not been mentioned, and even those that we have met have been covered in a largely introductory manner. None the less, we have acquainted ourselves with the essential components of the precursor to the general theory, namely, special relativity, we have looked carefully at the principles behind general relativity, and we have investigated both the formulation of the theory and its principal consequences. In particular, we have reached the three end points we had promised ourselves, namely, classical black holes, gravitational waves, and cosmology. In the process, it is hoped that some of the richness and beauty of the theory and some of its absorbing and bizarre consequences have been revealed. At the start of this book, we set out on a long journey of discovery. It would seem that we have come a long way, but the journey is really only just begun.

Exercises

26.1 (§26.1) Show that if $\Lambda = 0$ then

$$H^2 = \frac{8\pi}{3}\rho - \frac{k}{R^2}.$$

[Hint: Use (24.52)]

26.2 (§26.1) Use the estimate $1/H_0 \simeq 10^{10}$ years to obtain an estimate for ρ_{crit} in Kg/m^3. [Hint: you need to convert years to seconds, put back the factor of G in (26.2), and use the value $G \simeq 6.7 \times 10^{-11}\,\text{m}^3\,\text{kg}^{-1}\,\text{s}^{-2}$.]

26.3 (§26.1) Show that, if one takes $T_{ab} = \Lambda g_{ab}$, then this corresponds to a fluid with constant density $\rho_\Lambda = \Lambda/8\pi$, and equation of state $p = -\rho$. [Hint use (24.52) and (24.53).]

26.4 (§26.1) Show that (24.52) may be written as

$$\frac{\dot{R}^2}{R^2} = \frac{8\pi}{3}(\rho_r + \rho_m + \rho_\Lambda) - \frac{k}{R^2}.$$

which has non-dimensional form

$$\left(\frac{d\tilde{R}}{d\tilde{t}}\right)^2 = \left(\frac{\Omega_m}{\tilde{R}} + \frac{\Omega_r}{\tilde{R}^2} + \Omega_\Lambda \tilde{R}^2\right) - \frac{k}{H_0^2 R_0^2}.$$

26.5 (§26.1) Show, by evaluating the non-dimensional Friedmann equation at $t = t_0$, that

$$\Omega_r + \Omega_m + \Omega_\Lambda + \Omega_c = 1.$$

26.6 (§26.1) Use the non-dimensional scalings of R and t and (26.14) to show that, as long as Ω_c is not exactly zero then

$$R(t) = \frac{1}{H_0|1 - (\Omega_r + \Omega_m + \Omega_\Lambda)|^{1/2}} \tilde{R}(\tilde{t}/H_0).$$

26.7 (§26.3) Show that, for a spatially flat, matter dominated, open universe, the age is given by (26.30).

26.8 (§26.3) By making the substitution $y = h\nu/k_B T$ in (26.34), show that

$$\epsilon_{\rm rad} = \alpha T^4,$$

where

$$\alpha = \frac{8\pi k_B^4}{h^3 c^3} \int_0^\infty \frac{y^3 \, dy}{e^y - 1}.$$

26.9 (§26.3) In non-geometrical units, the above equation is

$$\rho_r c^2 = \alpha T^4,$$

and the Friedmann equation for a flat, radiation dominated universe is

$$H^2(t) = \frac{8\pi G}{3}\rho_r.$$

Calculate $H(t)$ for a radiation-dominated universe, using the approximate solution $R(t) \propto t^{1/2}$, and hence use the above equations to compute the temperature of the universe when it was one second old. [Hint: you will need to use the numerical values of α, G, and c.]

26.10 (§26.5) Show that the $k = 0$, $\Lambda > 0$ solution given by (25.5) can be written in the alternative form

$$R(t) = (\Omega_m/\Omega_\Lambda)^{1/3} \sinh^{2/3}(t/t_\Lambda),$$

where $t_\Lambda := 2/(3H_0\sqrt{\Omega_\Lambda})$.
Hint: Verify that the above equation satisfies the Friedmann equation

$$\frac{\dot{R}^2}{R^2} = H_0^2 \left(\frac{\Omega_m}{R^3} + \Omega_\Lambda\right).$$

26.11 (§26.6) Estimate ρ_r/ρ_m at the time of decoupling.

26.12 (§26.7)
(i) Show that, for a scalar field with Lagrangian density $\mathcal{L}$ given by (26.65) the field equations are $\Box\phi + V'(\phi) = 0$.

(ii) Show, for a spatially homogeneous field $\phi(t)$ in a Robertson-Walker metric, this gives

$$\ddot{\phi} + 3H(t)\dot{\phi} + V'(\phi) = 0.$$

(iii) Show that, for a spatially homogeneous scalar field in a Robertson-Walker metric, the energy-momentum tensor is given by

$$T_{ab} = \delta^0_a \delta^0_b \dot{\phi}^2 + g_{ab} \left[-\tfrac{1}{2}\dot{\phi}^2 + V(\phi) \right].$$

Further reading

The main source for this chapter is the book by Liddle (2004), which contains more details of the underlying physics such as black–body radiation discussed in §26.3. The books by Weinburg (1972) and Hartle (2003) are also useful. The articles by Lahav and Suto (2004) and Jones and Lasenby (1998) respectively describe the two main sources of measuring the universe, redshift surveys and an analysis of the CMB, while the review by Amendola et al. (2018) discusses both dark matter and dark energy. The classic 1977 video 'Powers of 10' by Charles and Ray Eames gives a nice insight into cosmological length scales and how these relate to those on Earth. The books by Penrose (1989), Hawking (1988), Barrow and Tipler (1986), and Hawking and Penrose (2015) relate to the discussion in the last two sections of the book.

Ade, P. A. R. *et. al.* (2014). Planck 2013 results. XV. CMB power spectra and likelihood. Astronomy & Astrophysics 571, A15.

Amendola, L., Appleby, S., Avgoustidis, A., Bacon, D., Baker, T., Baldi, M., Bartolo, N., Blanchard, A., Bonvin, C., Borgani, S., and Branchini, E. (2018). Cosmology and fundamental physics with the Euclid satellite. *Living Reviews in Relativity, 21, 2.*

Barrow, J. D., and Tipler, F. J. (1986). *The Anthropic Principle.* Clarendon Press, Oxford.

Deffayet, C., Peter, P., Wandelt, B., Zaldarriaga, M., and Cugliandolo, L. F. (2015). *Post-Planck Cosmology.* Lecture Notes of the Les Houches Summer School, vol. 100. Oxford University Press, Oxford.

Eames, C., and Eames, R. (1977). 'Powers of 10', YouTube, https://www.youtube.com/watch?v=0fKBhvDjuy0.

Ellis, G. F. R. (2011). Inhomogeneity effects in cosmology. *Classical and Quantum Gravity, 28,* 164001.

Hartle, J. B. (2003). *Gravity: An Introduction to Einstein's General Relativity.* Addison Wesley, San Francisco, CA.

Hawking, S. W. (1988). *A Brief History of Time*. Bantam Press, London.

Hawking, S. W., and Ellis, G. F. R. (1973). *The Large Scale Structure of Space-Time*. Cambridge University Press, Cambridge.

Hawking, S. W., and Penrose, R. (2015). *The Nature of Space and Time*. Princeton University Press, Princeton, NJ.

Jones, A. W., and Lasenby, A. N. (1998). The cosmic microwave background. *Living Reviews in Relativity*, *1*, 11.

Lahav, O., and Suto, Y. (2004). Measuring our universe from galaxy redshift surveys. *Living Reviews in Relativity*, *7*, 8.

Liddle, A. (2004). *An Introduction to Modern Cosmology*. Wiley, New York, NY.

Penrose, R. (1989). *The Emperor's New Mind*. Oxford University Press, Oxford.

Weinberg, S. (1972). *Gravitation and Cosmology*. Wiley, New York, NY.

Answers to exercises

2.1

$$x = x' + v_1 t'$$
$$y = y'$$
$$z = z'$$
$$t = t'$$

$$x' = x - v_1 t$$
$$y' = y$$
$$z' = z$$
$$t' = t$$

Interchange primes and unprimes and replace v_1 by $-v_1$.

$$x'' = x' + v_2 t'$$
$$y'' = y'$$
$$z'' = z'$$
$$t'' = t'$$

$$x'' = x + (v_1 + v_2)t$$
$$y'' = y$$
$$z'' = z$$
$$t'' = t$$

2.2

(i)

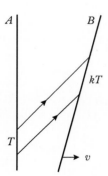

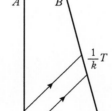

(ii)

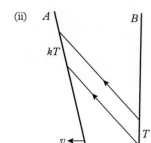

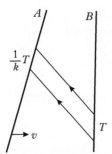

2.3 Blue shift.

2.6

Draw circle centre O, radius OG and two light rays entering and leaving G which cut the circle at points P and Q, as shown.

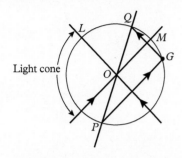

Then POQ is the world-line of an inertial observer who considers O and G to be simultaneous (since $PO = OQ$). Observers whose world-lines through O intersect LQ consider that G occurs later than O, and observers whose world-lines intersect QM consider that G occurs before O.

2.7

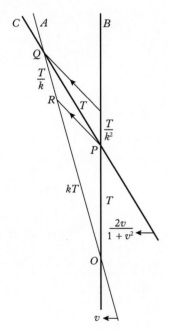

According to B, the coordinates (t, x) of the three events are

O $(0, 0)$

R $(\frac{1}{2}(k^2 + 1)T, \frac{1}{2}(k^2 - 1)T)$

Q $(\frac{1}{2}(k^2 + 1)(1 + 1/k^2)T, \frac{1}{2}(k^2 - 1)(1 + 1/k^2))$

Thus, whereas A's clock has elapsed by $(k + 1/k)T$ between events O

and Q, the time lapse of B's clock is $\frac{1}{2}(1 + k^2)(1 + 1/k^2)T$ (which for $k > 1$ is greater than A's time lapse).

2.9 $v = \pm T(x^2 + T^2)^{-\frac{1}{2}}$.

2.10 $s^2 = -(x_1 - x_2)^2 - (y_1 - y_2)^2 - (z_1 - z_2)^2 = -\sigma^2$

3.2 $(2/3)^{\frac{1}{2}}$.

3.6

Take the room to be in the frame S' moving along the x-axis of the rest frame of the pole with speed $-v$, as shown (not to scale):

(i)

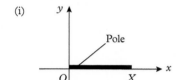

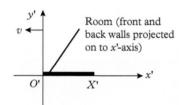

(ii)

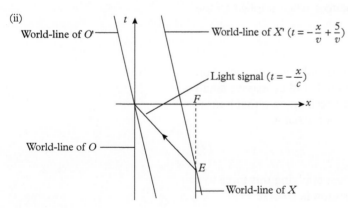

Then in S's frame
When O and O' coincide, S will 'see' X at F, as a result of a light signal from event E.

3.9
   (a) 7.5×10^{-5}s.
   (b) 17 min.

3.10

   3.4×10^9 light years.

   940 years.

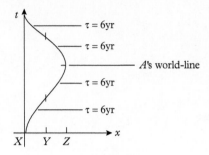

3.11

   $\nu = [(1 - u/c)/(1 + u/c)]^{1/2}\nu_0,$

   $\nu = [(1 + u/c)/(1 - u/c)]^{1/2}\nu_0.$

3.12 0.32c.

4.1

   One possibility is to define a **unit of force** F^1 as that which results in a standard mass m_s undergoing an acceleration g_L that is

$$F^1 = m_s g_L, \tag{1}$$

   where g_L is the acceleration due to gravity at a given latitude. We can then use Newton's second law to compare any other force F by measuring the acceleration a_S^F this produces when applied to the standard mass, that is

$$\frac{F}{F^1} = \frac{m_s a_S^F}{m_s g_L} = \frac{a_S^F}{g_L}, \tag{2}$$

   We could then define unit mass m^1 as that mass which, when acted on by a unit force F^1, suffers a unit acceleration 1. Other masses could then be defined by either (i) measuring the acceleration a that a mass experiences under the influence of the unit force, that is

$$\frac{F^1}{F^1} = \frac{ma}{m^1 1}, \tag{3}$$

   or (ii) using (2) to measure a force F and then applying this force to a mass m and measuring the resulting acceleration a, so that

$$m = F/a. \tag{4}$$

4.2 The kinetic energy of the initial particle in motion.

4.3 $(\bar{m}_0^2 + 2m_0\bar{m}_0\gamma + m_0^2)^{\frac{1}{2}}$ where $\gamma = (1 - u^2/c^2)^{-\frac{1}{2}}$.

4.8 $P = 2Mp_0/(m_0 + M), p = (m_0 - M)p_0/(m_0 + M).$

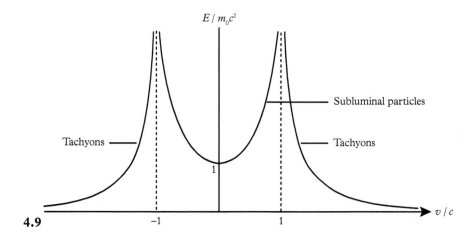

4.9

4.10 $-c \cos \theta$.

4.11

$ch\nu / (h\nu + m_0 c^2)$,
$(m_0^2 + 2h\nu m_0 / c^2)^{1/2}$.

5.1 (i) (a) $x = a \cos \phi, y = a \sin \phi, z = 0, (0 \leqslant \phi \leqslant 2\pi)$.
  (b) $x^2 + y^2 - a^2 = 0, z = 0$.
  (ii) (a) $x = a \sin \theta \cos \phi, y = a \sin \theta \sin \phi, z = a \cos \theta$,
    $0 \leqslant \theta \leqslant \pi, 0 \leqslant \phi \leqslant 2\pi$.
  (b) $x^2 + y^2 + z^2 - a^2 = 0$.

5.2

$(x^a) = (x^1, x^2, x^3) = (x, y, z)$
$(x'^a) = (x'^1, x'^2, x'^3) = (r, \theta, \phi)$
$(x^a) \longrightarrow (x'^a)$:
$r = (x^2 + y^2 + z^2)^{1/2}$,
$\theta = \tan^{-1}[(x^2 + y^2)^{1/2}/z]$.
$\phi = \tan^{-1}(y/x)$.
$(x'^a) \longrightarrow (x^a)$:
$x = r \sin \theta \cos \phi$,
$y = r \sin \theta \sin \phi$,
$z = r \cos \theta$.

$$\frac{\partial x^a}{\partial x'^b} = \begin{pmatrix} \sin \theta \cos \phi & r \cos \theta \cos \phi & -r \sin \theta \sin \phi \\ \sin \theta \sin \phi & r \cos \theta \sin \phi & r \sin \theta \cos \phi \\ \cos \theta & -r \sin \theta & 0 \end{pmatrix}$$

$$\frac{\partial x'^a}{\partial x^b} = \begin{pmatrix} \sin \theta \cos \phi & \sin \theta \sin \phi & \cos \theta \\ \cos \theta \cos \phi / r & \cos \theta \sin \phi / r & -\sin \theta / r \\ -\sin \phi / (r \cos \theta) & \cos \phi / (r \sin \theta) & 0 \end{pmatrix}$$

$J' \to 0$ when $r \to \infty$,
$J' \to \infty$ when $r = 0$ and $\theta = 0, \pi$.

5.6

$(x^a) \longrightarrow (x'^a)$:
$R = (x^2 + y^2)^{1/2}$,
$\phi = \tan^{-1}(y/x)$.

$$\frac{\partial x'^a}{\partial x^b} = \begin{pmatrix} \cos\phi & \sin\phi \\ -\sin\phi/R & \cos\phi/R \end{pmatrix}.$$

$X^a = \dfrac{\mathrm{d}x^a}{\mathrm{d}\phi} = (-a\sin\phi, a\cos\phi)$,

$X'^a = (0, 1)$.

5.7

$$X'^{ab}_c = \frac{\partial x'^a}{\partial x^d}\frac{\partial x'^b}{\partial x^e}\frac{\partial x^f}{\partial x'^c}X^{de}_f.$$

5.15 $\delta^a_a = \delta^a_b\delta^b_a = n.$

5.17 (i) $X'^a = (\cos\phi, -\sin\phi/R)$.

(ii) $\dfrac{\partial}{\partial x} = \cos\phi\dfrac{\partial}{\partial R} - \dfrac{\sin\phi}{R}\dfrac{\partial}{\partial\phi}$,

$\dfrac{\partial}{\partial y} = \sin\phi\dfrac{\partial}{\partial R} + \dfrac{\cos\phi}{R}\dfrac{\partial}{\partial\phi}$.

$\dfrac{\partial}{\partial R} = \dfrac{x}{(x^2+y^2)^{1/2}}\dfrac{\partial}{\partial x} + \dfrac{y}{(x^2+y^2)^{1/2}}\dfrac{\partial}{\partial y}$.

$\dfrac{\partial}{\partial\phi} = -y\dfrac{\partial}{\partial x} + x\dfrac{\partial}{\partial y}$.

(iii) $X^a\partial_a = \dfrac{\partial}{\partial x}$.

$X'^a\partial'_a = \cos\phi\dfrac{\partial}{\partial R} - \dfrac{\sin\phi}{R}\dfrac{\partial}{\partial\phi}$.

(iv) $Y'^a = (\sin\phi, \cos\phi/R)$,
$Z'^a = (0, 1)$,

$Y = \dfrac{\partial}{\partial y} = \sin\phi\dfrac{\partial}{\partial R} + \dfrac{\cos\phi}{R}\dfrac{\partial}{\partial\phi}$,

$Z = -y\dfrac{\partial}{\partial x} + x\dfrac{\partial}{\partial y} = \dfrac{\partial}{\partial\phi}$.

(v) The Lie brackets are given in the table below (the vector in the column being the first entry):

	X	Y	Z
X	0	0	Y
Y	0	0	-X
Z	-Y	X	0

6.2

$L_X Z_{bc} = Z_{bc,d}X^d + Z_{dc}X^d_{,b} + Z_{bd}X^d_{,c}$

$L_X(Y^a Z_{bc}) = X^d(Y^a Z_{bc})_{,d} - Y^d Z_{bc}X^a_{,d} + Y^a Z_{dc}X^d_{,b} + Y^a Z_{bd}X^d_{,c}$

6.15

$g_{ab} = \mathrm{diag}(1, 1, 1)$, $g^{ab} = \mathrm{diag}(1, 1, 1)$, $g = 1$.

$g_{ab} = \mathrm{diag}(1, R^2, 1)$, $g^{ab} = \mathrm{diag}(1, R^{-2}, 1)$, $g = R^2$.

$g_{ab} = \mathrm{diag}(1, r^2, r^2 \sin^2\theta)$, $g^{ab} = \mathrm{diag}(1, r^{-2}, r^{-2}(\sin\theta)^{-2})$,

$g = r^4 \sin^2\theta$.

6.16

$T_{ab} = g_{ac}g_{bd}T^{cd}$.

6.17

$g'_{ab} = \dfrac{\partial x^c}{\partial x'^a} \dfrac{\partial x^d}{\partial x'^b} g_{cd}$.

6.18

$$\dfrac{\mathrm{d}^2 R}{\mathrm{d}u^2} - R \left(\dfrac{\mathrm{d}\phi}{\mathrm{d}u} \right)^2 = 0.$$

$$\dfrac{\mathrm{d}^2\phi}{\mathrm{d}u^2} + \dfrac{2}{R}\dfrac{\mathrm{d}R}{\mathrm{d}u}\dfrac{\mathrm{d}\phi}{\mathrm{d}u} = 0.$$

$$\dfrac{\mathrm{d}^2 z}{\mathrm{d}u^2} = 0.$$

6.22

(i) -2.

(ii) Yes.

(iii) Yes.

6.23

(i) $(x^1, x^2, x^3) = (r, \theta, \phi)$.

(ii) Yes.

6.32 (i) $g_{ab} = \mathrm{diag}(e^\nu, -e^\lambda, -r^2, -r^2 \sin^2\theta)$,

$g = -e^{\nu+\lambda}r^4 \sin^2\theta$,

$g^{ab} = \mathrm{diag}(e^{-\nu}, -e^{-\lambda}, -r^{-2}, -r^{-2}\sin^{-2}\theta)$.

(ii) Non-zero independent components:

$\Gamma^0_{00} = \frac{1}{2}\dot\nu$, $\Gamma^0_{01} = \frac{1}{2}\nu'$, $\Gamma^0_{11} = \frac{1}{2}e^{\lambda-\nu}\dot\lambda$,

$\Gamma^1_{00} = \frac{1}{2}e^{\lambda-\nu}\nu'$, $\Gamma^1_{01} = \frac{1}{2}\dot\lambda$, $\Gamma^1_{11} = \frac{1}{2}\lambda'$,

$\Gamma^1_{22} = -re^{-\lambda}$, $\Gamma^1_{33} = -re^{-\lambda}\sin^2\theta$,

$\Gamma^2_{12} = r^{-1}, \Gamma^2_{33} = -\sin\theta\cos\theta,$

$\Gamma^3_{13} = r^{-1}, \Gamma^3_{33} = \cot\theta.$

(iii) Non-zero independent components:

$R_{0101} = -\frac{1}{2}e^{\nu}\nu'' + \frac{1}{4}e^{\lambda}\dot{\lambda}^2 - \frac{1}{4}e^{\lambda}\dot{\nu}\dot{\lambda} + \frac{1}{2}e^{\lambda}\ddot{\lambda} - \frac{1}{4}e^{\nu}\nu'^2 + \frac{1}{4}e^{\nu}\nu'\lambda',$

$R_{0202} = -\frac{1}{2}re^{\nu-\lambda}\nu',$

$R_{0212} = -\frac{1}{2}r\dot{\lambda},$

$R_{0303} = -\frac{1}{2}re^{\nu-\lambda}\nu'\sin^2\theta,$

$R_{0313} = -\frac{1}{2}r\dot{\lambda}\sin^2\theta,$

$R_{1212} = -\frac{1}{2}r\lambda',$

$R_{1313} = -\frac{1}{2}r\lambda'\sin^2\theta,$

$R_{2323} = r^2 e^{-\lambda}\sin^2\theta - r^2\sin^2\theta.$

(iv) Non-zero independent components:

$R_{00} = \frac{1}{2}e^{\nu-\lambda}\nu'' - \frac{1}{4}\dot{\lambda}^2 + \frac{1}{4}\dot{\nu}\dot{\lambda} - \frac{1}{2}\ddot{\lambda} + \frac{1}{4}e^{\nu-\lambda}\nu'^2 - \frac{1}{4}e^{\nu-\lambda}\nu'\lambda' + r^{-1}e^{\nu-\lambda}\nu',$

$R_{01} = r^{-1}\dot{\lambda},$

$R_{11} = -\frac{1}{2}\nu'' + \frac{1}{4}e^{\lambda-\nu}\dot{\lambda}^2 - \frac{1}{4}e^{\lambda-\nu}\dot{\nu}\dot{\lambda} + \frac{1}{2}e^{\lambda-\nu}\ddot{\lambda} - \frac{1}{4}\nu'^2 + \frac{1}{4}\nu'\lambda' + r^{-1}\lambda',$

$R_{22} = -\frac{1}{2}re^{-\lambda}\nu' + \frac{1}{2}re^{-\lambda}\lambda' - e^{-\lambda} + 1,$

$R_{33} = \sin^2\theta R_{22}.$

$R = e^{-\lambda}\nu'' - \frac{1}{2}e^{-\nu}\dot{\lambda}^2 + \frac{1}{2}e^{-\nu}\dot{\nu}\dot{\lambda} - e^{-\nu}\ddot{\lambda} + \frac{1}{2}e^{-\lambda}\nu'^2 - \frac{1}{2}e^{-\lambda}\nu'\lambda' + 2r^{-1}e^{-\lambda}\nu'$
$\quad - 2r^{-1}e^{-\lambda}\lambda' + 2r^{-2}e^{-\lambda} - 2r^{-2}.$

$G_{00} = r^{-1}e^{\nu-\lambda}\lambda' - r^{-2}e^{\nu-\lambda} + r^{-2}e^{\nu},$

$G_{01} = r^{-1}\dot{\lambda},$

$G_{11} = r^{-1}\nu' - r^{-2}e^{\lambda} + r^{-2},$

$G_{22} = \frac{1}{2}re^{-\lambda}\nu' - \frac{1}{2}re^{-\lambda}\lambda' + \frac{1}{2}r^2 e^{-\lambda}\nu'' - \frac{1}{4}r^2 e^{-\nu}\dot{\lambda}^2 + \frac{1}{4}r^2 e^{-\nu}\dot{\nu}\dot{\lambda} - \frac{1}{2}r^2 e^{-\nu}\ddot{\lambda}$
$\quad + \frac{1}{4}r^2 e^{-\lambda}\nu'^2 - \frac{1}{4}r^2 e^{-\lambda}\nu'\lambda',$

$G_{33} = \sin^2\theta G_{22}.$

(v) Non-zero components:

$$G^0{}_0 = r^{-1}e^{-\lambda}\lambda' - r^{-2}e^{-\lambda} + r^{-2},$$

$$G^0{}_1 = r^{-1}e^{-\nu}\dot{\lambda},$$

$$G^1{}_0 = -r^{-1}e^{-\lambda}\dot{\lambda},$$

$$G^1{}_1 = -r^{-1}e^{-\lambda}\nu' - r^{-2}e^{-\lambda} + r^{-2},$$

$$G^2{}_2 = \tfrac{1}{2}r^{-1}e^{-\lambda}\lambda' - \tfrac{1}{2}r^{-1}e^{-\lambda}\nu' - \tfrac{1}{2}e^{-\lambda}\nu'' + \tfrac{1}{4}e^{-\nu}\dot{\lambda}^2 - \tfrac{1}{4}e^{-\nu}\dot{\nu}\dot{\lambda} + \tfrac{1}{2}e^{-\nu}\ddot{\lambda}$$
$$\qquad - \tfrac{1}{4}e^{-\lambda}\nu'^2 + \tfrac{1}{4}e^{-\lambda}\nu'\lambda',$$

$$G^3{}_3 = G^2{}_2.$$

7.1 $\Phi_{;a} = \Phi_{,a} - \Phi\,\Gamma^b_{ba}.$

7.6
 (i) $y'' - y = 0.$

 (ii) $2y_1 y_1'' + y_1'^2 - y_2'^2 - 3xy_1^2 - y_2 = 0,$
 $2y_1 y_2'' + 2y_1'y_2' - y_1 = 0.$

7.9 $\dfrac{\partial}{\partial y}.$

7.11

$$X^1 = \frac{\partial}{\partial x}, \quad X^2 = \frac{\partial}{\partial y}, \quad X^3 = \frac{\partial}{\partial z},$$

$$X^4 = y\frac{\partial}{\partial z} - z\frac{\partial}{\partial y}, \quad X^5 = z\frac{\partial}{\partial x} - x\frac{\partial}{\partial z}, \quad X^6 = x\frac{\partial}{\partial y} - y\frac{\partial}{\partial x}.$$

	X^1	X^2	X^3	X^4	X^5	X^6
X^1	0	0	0	0	$-X^3$	X^2
X^2	0	0	0	X^3	0	$-X^1$
X^3	0	0	0	$-X^2$	X^1	0
X^4	0	$-X^3$	X^2	0	$-X^6$	X^5
X^5	X^3	0	$-X^1$	X^6	0	$-X^4$
X^6	$-X^2$	X^1	0	$-X^5$	X^4	0

7.14 $(\nabla_c\nabla_b - \nabla_b\nabla_c)X_a = R_{adcb}X^d.$

8.5 (a) $\tfrac{1}{2}n(n-1)\,\omega_{ab}, \; n\,t_a.$

 (b) $6\,\omega_{ab}$: 3 spatial rotations, 3 boosts.
  $4\,t_a$: 3 translations, 1 time translation.

8.8

$$\ddot{x} = \frac{(m_1 - m_2)}{(m_1 + m_2)}g,$$

$$p = (m_1 + m_2)\dot{x},$$

$$H(p, x) = \frac{p^2}{2(m_1 + m_2)} - m_1 g x - m_2 g(\ell - x).$$

8.10 Zeroth component gives the rate of work done by force $\boldsymbol{F}$, viz.

$$\frac{dE}{dt} = \boldsymbol{F}.\boldsymbol{u},$$

8.11

(i) $u'_x = \dfrac{u_x - v}{1 - u_x v},\quad u'_y = \dfrac{u_y}{\beta(1 - u_x v)},\quad u'_z = \dfrac{u_z}{\beta(1 - u_x v)}.$

(ii) $E' = \beta(E - vp_x),\quad p'_x = \beta(p_x - vE),\quad p'_y = p_y,\quad p'_z = p_z.$

(iii) $F'_x = \dfrac{F_x - v\boldsymbol{F}.\boldsymbol{u}}{1 - u_x v},\quad F'_y = \dfrac{F_y}{\beta(1 - u_x v)},\quad F'_z = \dfrac{F_z}{\beta(1 - u_x v)}.\qquad$ Yes

(iv) $\boldsymbol{F}' = (F'_x, F'_y, F'_z) = (F, 0, 0) = \boldsymbol{F}.$

9.1 (i) $\tan^{-1}(a/g)$.

The inertial observer will see the mass accelerate in the direction of motion due to a tension in the rod whose horizontal component produces the acceleration and whose vertical component counterbalances the weight. A non-inertial observer in the car will consider that the pendulum mass experiences two forces, the weight mg down and an inertial force ma in the opposite direction to motion.

(ii) $\tan^{-1}(a/g)$.

(iii) 0.

9.2 (ii) Four inertial forces, namely:-

(a) a linear accelerative force as discussed in Ex. 9.1

(b) a velocity-dependent Coriolis force

(c) a centrifugal force

(d) a non-uniform rotational force (analogous tangentially to (a)).

9.5

(i)   The released body undergoes uniform motion (due to Newton's first law) and the rocket accelerates with acceleration g, so the inertial observer sees the floor of the rocket ship come up and hit the body with relative acceleration g.

(ii)   The rocket and the body have no forces acting on them and therefore, by Newton's first law, both undergo uniform motion i.e. travel with the same constant velocity.

(iv)   The body and the lift both fall under gravity with the same acceleration.

9.6 Ellipsoid (see **§17.10**).

9.7

$$\frac{\mathrm{d}^2 R}{\mathrm{d}u^2} - R\left(\frac{\mathrm{d}\phi}{\mathrm{d}u}\right)^2 = 0,$$
$$\frac{\mathrm{d}^2 \phi}{\mathrm{d}u^2} + \frac{2}{R}\frac{\mathrm{d}R}{\mathrm{d}u}\frac{\mathrm{d}\phi}{\mathrm{d}u} = 0,$$
$$\frac{\mathrm{d}^2 z}{\mathrm{d}u^2} = 0.$$

Inertial force (centrifugal and Coriolis components).

9.8 (i) Straight line.
    (ii) Parabola (projectile motion).

9.9 One example:

$$\nabla_b T^{ab} + R^a{}_{mnc} R^{emn}{}_d \nabla_e T^{cd} = 0.$$

9.10 $\nabla_{[a} F_{bc]} = 0.$

9.11

(i)
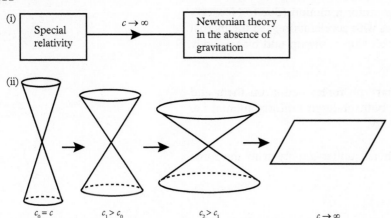

In the limit, the null cones degenerate into planes of simultaneity. That is, all observers, irrespective of their motion, agree that events occurring in one of these planes do so simultaneously.

10.1

$$f(\pmb{x} + \pmb{h}) = f(\pmb{x}) + \left(h_1 \frac{\partial}{\partial x} + h_2 \frac{\partial}{\partial y} + h_3 \frac{\partial}{\partial z} \right) f(\pmb{x})$$

$$+ \frac{1}{2} \left(h_1^2 \frac{\partial^2}{\partial x^2} + 2 h_1 h_2 \frac{\partial^2}{\partial x \partial y} + 2 h_1 h_3 \frac{\partial^2}{\partial x \partial z} + \right.$$

$$\left. + h_2^2 \frac{\partial^2}{\partial y^2} + 2 h_2 h_3 \frac{\partial^2}{\partial y \partial z} + h_3^2 \frac{\partial^2}{\partial z^2} \right) f(\pmb{x}) + \cdots.$$

10.11
(i) Principle of equivalence,
(ii) Principle of equivalence,
(iii) Correspondence principle,
(iv) Correspondence and covariance principles,
(v) Principle of equivalence,
(vi) Correspondence principle.

10.12
Principle of equivalence, principle of minimal gravitational coupling, Mach's principle (?) and correspondence principle.

11.7 (i) $\nabla^a (\sqrt{-g} G_{ab}) = 0,$
(ii) $-\nabla_a (\sqrt{-g} G^{ab}) = 0,$
(iii) $\nabla^a (\sqrt{-g} G_{ab}) = 0.$

11.12
$$\mathcal{L}_{uv} = \sqrt{-g} \left[g_{ud} R^{abcd} R_{abcv} + g_{vd} R^{abcd} R_{abcu} - \tfrac{1}{2} g_{uv} R^{abcd} R_{abcd} \right].$$

11.13
$$T_{ab} = \nabla_a\phi\nabla_b\phi - \tfrac{1}{2}g_{ab}g^{cd}\nabla_c\phi\nabla_d\phi + g_{ab}V(\phi).$$

12.5

$$E'_x = E_x,$$
$$E'_y = \beta(E_y - vB_z),$$
$$E'_z = \beta(E_z + vB_y),$$
$$B'_x = B_x,$$
$$B'_y = \beta(B_y + vE_z),$$
$$B'_z = \beta(B_z - vE_y),$$
$$\rho' = \beta(\rho - vj_x), \quad j'_x = \beta(j_x - v\rho), \quad j'_y = j_y, \quad j'_z = j_z.$$

12.7 $\phi \to \bar{\phi}_a = \phi_a + \partial_a\psi,$
where ψ must be a solution of $\Box\psi = 0$.

13.3 $\mu = -\tfrac{1}{4}.$
$$R_{ab} - \tfrac{1}{4}g_{ab}R = 2(-F_{ac}F^c{}_b + \tfrac{1}{4}g_{ab}F_{cd}F^{cd}).$$

15.1
One example: a particle falls in from infinity at time $t = -\infty$ reaching
the origin at time $t = 0$ whereupon the motion is reversed.

15.7 (a) $ds = ad\theta.$
  $ds = a\sin\theta d\phi.$

15.12 $[G] = M^{-1}L^3T^{-2}.$

15.13

$$R_{0101} = 2mr^{-3}$$

$$R_{0202} = -\left(1 - \frac{2m}{r}\right)\frac{m}{r}$$

$$R_{0303} = -\left(1 - \frac{2m}{r}\right)\frac{m}{r}\sin^2\theta$$

$$R_{1212} = \left(1 - \frac{2m}{r}\right)^{-1}\frac{m}{r}$$

$$R_{1313} = \left(1 - \frac{2m}{r}\right)^{-1}\frac{m}{r}\sin^2\theta$$

$$R_{2323} = -2mr\sin^2\theta.$$

15.16 (1), (2), (3), (4), (5) and (6).

16.2 Motion in a straight line.

16.7 The laws become modified to:

K1': Each planet moves in an ellipse about the centre-of-mass as one of the foci.

K2': The radius vector from the centre-of-mass to the planet sweeps out equal area in equal times.

K3': $\tau = \dfrac{2\pi}{(G(m_{\text{planet}} + m_{\text{sun}}))^{1/2}} a^{3/2}$.

16.8 $\left(1 - \dfrac{2m}{r}\right)^{-1} \ddot{r} - \left(1 - \dfrac{2m}{r}\right)^{-2} \dfrac{m}{r^2}\dot{r}^2 + \dfrac{m}{r^2}\dot{t}^2 - r\dot{\theta}^2 - r\sin^2\theta\dot{\phi}^2 = 0.$

16.9

$$\left(\frac{du}{d\phi}\right)^2 + u^2 = \frac{k^2 - 1}{h^2},$$

$$\frac{du^2}{d\phi^2} + u = 0.$$

16.16

$$\pm\tau \simeq (r^2 - D^2)^{1/2} + 2m\cosh^{-1}\left(\frac{r}{D}\right) - m\frac{(r^2 - D^2)^{1/2}}{r} + \text{constant}$$

$$= (r^2 - D^2)^{1/2} + 2m\ln\left(\frac{r + (r^2 - D^2)^{1/2}}{D}\right) - m\frac{(r^2 - D^2)^{1/2}}{r} + \text{constant}.$$

17.1 This is the Schwarzschild solution under the renaming of the coordinates:

$$(\theta, \phi, t, r) \longrightarrow (t, r, \theta, \phi).$$

17.2 $\dfrac{4\bar{t}^2\bar{r}^3\cos\bar{\theta}}{\bar{r}\cos\bar{\theta} + 2m} - (\bar{r}\cos\bar{\theta} + 2m)^2\bar{r}^2\bar{\theta}^2\bar{\phi}^2\sin^2(\bar{\phi}\bar{t}).$

17.3 (i) t timelike; ρ, z, ϕ spacelike.
    (ii) u null; x, y, z spacelike.

17.4 $ds^2 = A(t)dt^2 - B(t)dx^2 - C(t)dy^2 - D(t)dz^2.$

17.5

$$ds^2 = \left(1 - \frac{2m}{r}\right)dt^2 - \left[1 + \frac{2mx^2}{r^2(r - 2m)}\right]dx^2 - \frac{4mxy}{r^2(r - 2m)}dxdy$$

$$- \frac{4mxz}{r^2(r - 2m)}dxdz - \left[1 + \frac{2my^2}{r^2(r - 2m)}\right]dy^2$$

$$-\frac{4myz}{r^2(r-2m)}\mathrm{d}y\mathrm{d}z - \left[1 + \frac{2mz^2}{r^2(r-2m)}\right]\mathrm{d}z^2,$$

where $r = (x^2 + y^2 + z^2)^{\frac{1}{2}}$.

17.6

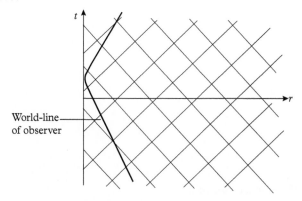

World-line of observer

17.7 $r = \pm ku + c$, c constant.

17.8 -1.

17.15 (Roughly)

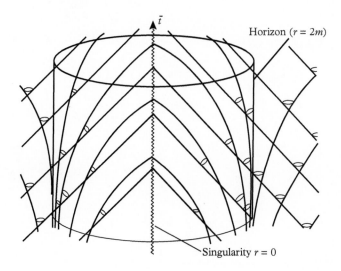

Horizon $(r = 2m)$

Singularity $r = 0$

17.17

A non-rotating white hole consists of a visible singularity situated at the origin of coordinates, which suddenly erupts into a star whose radius increases inexorably through its Schwarzschild radius.

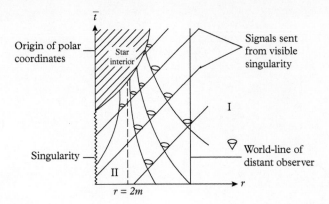

18.4 Region II.

They cannot escape from region II, but are ultimately crushed out of existence by the singularity.

18.7

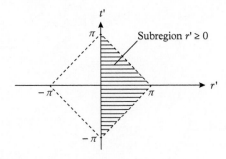

18.8

$$t' = \tan^{-1}(t+r) + \tan^{-1}(t-r),$$

$$r' = \tan^{-1}(t+r) - \tan^{-1}(t-r),$$

$$t = t_0 \leftrightarrow \tan\tfrac{1}{2}(t'+r') + \tan\tfrac{1}{2}(t'-r') = 2t_0,$$

$$r = r_0 \leftrightarrow \tan\tfrac{1}{2}(t'+r') - \tan\tfrac{1}{2}(t'-r') = 2r_0.$$

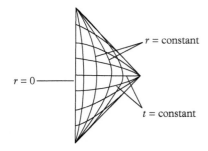

$r = $ constant

$r = 0$

$t = $ constant

18.9

(a)

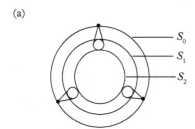

S_0
S_1
S_2

(b)

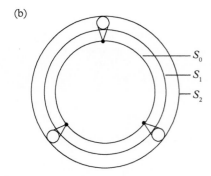

S_0
S_1
S_2

18.10
Figure 18.10 is the Penrose diagram of the Schwarzschild solution in the absence of a source. The introduction of a source suppresses regions I′ and II′.

19.4

$$T_{ab} = \frac{\varepsilon^2}{8\pi} \text{diag} \left[2e^{-\lambda} - e^{\nu}, e^{\lambda} - 2e^{-\nu}, r^2, r^2 \sin^2 \theta \right].$$

19.6

I: t timelike, r spacelike.
II: t spacelike, r timelike.
III: t timelike, r spacelike.
$r = r_{\pm} = m \pm (m^2 - \varepsilon^2)^{1/2}$.

19.7

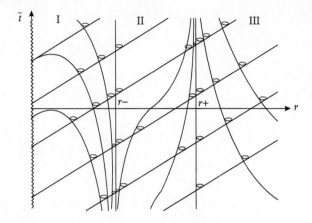

19.8

$$t + r + \frac{r_+^2}{r_+ + r_-} \ln\left(r - r_+\right) - \frac{r_-^2}{r_+ - r_-} \ln\left(r - r_-\right) = \text{constant.}$$

19.9

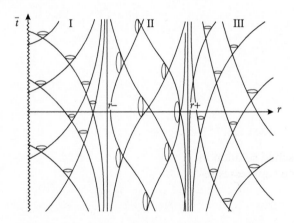

$\varepsilon^2 = m^2$ there is no region II.

19.13
$$\mathrm{d}s^2 = \left(1 - \frac{2m}{r} + \frac{\varepsilon^2}{r^2}\right) \mathrm{d}v^2 - 2\mathrm{d}v\,\mathrm{d}r - r^2\left(\mathrm{d}\theta^2 + \sin^2\theta\,\phi^2\right).$$

19.14
$r = \varepsilon^2/m.$
$r = -m + \left(m^2 + \varepsilon^2\right)^{\frac{1}{2}},$
$- 1.$

20.2

$$
g_{ab} = \begin{pmatrix} 1 - 2m/r & -1 & 0 & 0 \\ -1 & 0 & 0 & 0 \\ 0 & 0 & -r^2 & 0 \\ 0 & 0 & 0 & -r^2\sin^2\theta \end{pmatrix},
$$

$$
g^{ab} = \begin{pmatrix} 0 & -1 & 0 & 0 \\ -1 & -(1 - 2m/r) & 0 & 0 \\ 0 & 0 & -r^{-2} & 0 \\ 0 & 0 & 0 & -r^{-2}\sin^{-2}\theta \end{pmatrix}.
$$

20.11

$$
g^{11} = -\frac{(r^2 - 2mr + a^2)}{(r^2 + a^2\cos^2\theta)}.
$$

20.12 The other condition is identical, expect that the sign of L is reversed and both signs of L are considered in (20.56) and the sequel.

20.14

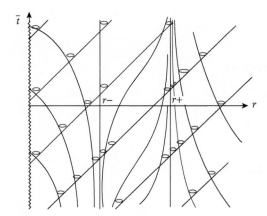

20.17

Working in spherical polar coordinates $n^a = (1, 0, 0, 0)$ and $m^a = (0, 1, 0, 0)$ and $\sqrt{\sigma} = r^2 \sin\theta$. Substituting these into (20.79) gives (20.81).

22.4

$$u = U,$$

$$v = V - Y^2 f'/f - Z^2 g'/g,$$

$$y = Y/f,$$

$z = Z/g,$

where $f = f(U)$ and $g = g(U)$.

23.1

A solution is cylindrically symmetric if it admits a symmetry axis and is invariant under both rotations about the axis and translations parallel to it.

23.2 Invariant under $\phi \to -\phi$.

No cross term in $d\phi dz$, i.e. $\chi = 0$.

23.5 Non-zero independent components are:

$$\Gamma^0_{00} = r^2 U^2 e^{2\gamma - 2\beta} \gamma_{,1} + r^2 U e^{2\gamma - 2\beta} U_{,1} + r U^2 e^{2\gamma - 2\beta} + 2\beta_{,0}$$
$$- r^{-1} V\beta_{,1} - \tfrac{1}{2} r^{-1} V_{,1} + \tfrac{1}{2} r^{-2} V,$$

$$\Gamma^0_{02} = -r^2 U e^{2\gamma - 2\beta} \gamma_{,1} - \tfrac{1}{2} r^2 e^{2\gamma - 2\beta} U_{,1} - r U e^{2\gamma - 2\beta} + \beta_{,2},$$

$$\Gamma^0_{22} = r^2 e^{2\gamma - 2\beta} \gamma_{,1} + r e^{2\gamma - 2\beta},$$

$$\Gamma^0_{33} = -r^2 e^{-(2\gamma + 2\beta)} \sin^2 \theta \gamma_{,1} + r e^{-(2\gamma + 2\beta)} \sin^2 \theta,$$

$$\Gamma^1_{00} = r^2 U^3 e^{2\gamma - 2\beta} \gamma_{,2} + r^2 U^2 e^{2\gamma - 2\beta} \gamma_{,0} + r^2 U^2 e^{2\gamma - 2\beta} U_{,2} - r U^2 V e^{2\gamma - 2\beta} \gamma_{,1}$$
$$- r U V e^{2\gamma - 2\beta} U_{,1} - U^2 V e^{2\gamma - 2\beta} - r^{-1} U V\beta_{,2} - \tfrac{1}{2} r^{-1} U V_{,2} - r^{-1} V\beta_{,0}$$
$$+ \tfrac{1}{2} r^{-1} V_{,0} + r^{-2} V^2 \beta_{,1} + \tfrac{1}{2} r^{-2} V V_{,1} - \tfrac{1}{2} r^{-3} V^2,$$

$$\Gamma^1_{01} = -\tfrac{1}{2} r^2 U e^{2\gamma - 2\beta} U_{,1} - U\beta_{,2} + r^{-1} V\beta_{,1} + \tfrac{1}{2} r^{-1} V_{,1} - \tfrac{1}{2} r^{-2} V,$$

$$\Gamma^1_{02} = -r^2 U^2 e^{2\gamma - 2\beta} \gamma_{,2} - r^2 U e^{2\gamma - 2\beta} \gamma_{,0} - r^2 U e^{2\gamma - 2\beta} U_{,2} + r U V e^{2\gamma - 2\beta} \gamma_{,1}$$
$$+ \tfrac{1}{2} r V e^{2\gamma - 2\beta} U_{,1} + U V e^{2\gamma - 2\beta} + \tfrac{1}{2} r^{-1} V_{,2},$$

$$\Gamma^1_{11} = 2\beta_{,1},$$

$$\Gamma^1_{12} = \tfrac{1}{2} r^2 e^{2\gamma - 2\beta} U_{,1} + \beta_{,2},$$

$$\Gamma^1_{22} = r^2 U e^{2\gamma - 2\beta} \gamma_{,2} + r^2 e^{2\gamma - 2\beta} \gamma_{,0} + r^2 e^{2\gamma - 2\beta} U_{,2} - r V e^{2\gamma - 2\beta} \gamma_{,1} - V e^{2\gamma - 2\beta},$$

$$\Gamma^1_{33} = r^2 U e^{-(2\gamma + 2\beta)} \cos\theta \sin\theta - r^2 U e^{-(2\gamma + 2\beta)} \sin^2 \theta \gamma_{,2} - r^2 e^{-(2\gamma + 2\beta)} \sin^2 \theta \gamma_{,0}$$
$$+ r V e^{-(2\gamma + 2\beta)} \sin^2 \theta \gamma_{,1} - V e^{-(2\gamma + 2\beta)} \sin^2 \theta,$$

$$\Gamma^2_{00} = r^2 U^3 e^{2\gamma - 2\beta} \gamma_{,1} + r^2 U^2 e^{2\gamma - 2\beta} U_{,1} + r U^3 e^{2\gamma - 2\beta} - U^2 \gamma_{,2} + 2 U\beta_{,0}$$
$$- 2 U\gamma_{,0} - U U_{,2} - U_{,0} - r^{-1} U V\beta_{,1} - \tfrac{1}{2} r^{-1} U V_{,1} + \tfrac{1}{2} r^{-2} U V$$
$$+ r^{-3} V e^{2\beta - 2\gamma} \beta_{,2} + \tfrac{1}{2} r^{-3} e^{2\beta - 2\gamma} V_{,2},$$

$$\Gamma^2_{01} = -U\gamma_{,1} - \tfrac{1}{2} U_{,1} - r^{-1} U + r^{-2} e^{2\beta - 2\gamma} \beta_{,2},$$

$$\Gamma_{02}^2 = -r^2 U^2 e^{2\gamma-2\beta}\gamma_{,1} - \tfrac{1}{2}r^2 U e^{2\gamma-2\beta} U_{,1} - r U^2 e^{2\gamma-2\beta} + U\beta_{,2} + \gamma_{,0},$$

$$\Gamma_{12}^2 = \gamma_{,1} + r^{-1},$$

$$\Gamma_{22}^2 = r^2 U e^{2\gamma-2\beta}\gamma_{,1} + r U e^{2\gamma-2\beta} + \gamma_{,2},$$

$$\Gamma_{33}^2 = -r^2 U e^{-(2\gamma+2\beta)}\sin^2\theta\,\gamma_{,1} + r U e^{-(2\gamma+2\beta)}\sin^2\theta - e^{-4\gamma}\cos\theta\sin\theta + e^{-4\gamma}\sin^2 2\theta\,\gamma_{,2},$$

$$\Gamma_{03}^3 = -\gamma_{,0},$$

$$\Gamma_{13}^3 = -\gamma_{,1} + r^{-1},$$

$$\Gamma_{23}^2 = \cot\theta - \gamma_{,2}.$$

23.7

$$R_{11} = -2(\gamma_{,1})^2 + 4r^{-1}\beta_{,1},$$

$$\begin{aligned}
R_{12} = {}& -r^2 e^{2\gamma-2\beta}\beta_{,1}U_{,1} + r^2 e^{2\gamma-2\beta}\gamma_{,1}U_{,1} \\
& + \tfrac{1}{2}r^2 e^{2\gamma-2\beta}U_{,11} + 2r e^{2\gamma-2\beta}U_{,1} + 2\cot\theta\,\gamma_{,1} \\
& - \beta_{,12} - 2\gamma_{,1}\gamma_{,2} + \gamma_{,12} + 2r^{-1}\beta_{,2},
\end{aligned}$$

$$\begin{aligned}
R_{22} = {}& -\tfrac{1}{2}r^4 e^{4\gamma-4\beta}U^2_{,1} + r^2 U e^{2\gamma-2\beta}\cot\theta\,\gamma_{,1} \\
& + 2r^2 U e^{2\gamma-2\beta}\gamma_{,12} + r^2 e^{2\gamma-2\beta}\gamma_{,1}U_{,2} \\
& + r^2 e^{2\gamma-2\beta}\gamma_{,2}U_{,1} + 2r^2 e^{2\gamma-2\beta}\gamma_{,01} \\
& + r^2 e^{2\gamma-2\beta}U_{,12} + r U e^{2\gamma-2\beta}\cot\theta + 2r U e^{2\gamma-2\beta}\gamma_{,2} \\
& - r V e^{2\gamma-2\beta}\gamma_{,11} + 2r e^{2\gamma-2\beta}\gamma_{,1} \\
& - r e^{2\gamma-2\beta}\gamma_{,1}V_{,1} + 3r e^{2\gamma-2\beta}U_{,2} - V e^{2\gamma-2\beta}\gamma_{,1} \\
& + 3\cot\theta\,\gamma_{,2} - 2(\beta_{,2})^2 + 2\beta_{,2}\gamma_{,2} - 2\beta_{,22} \\
& - 2(\gamma_{,2})^2 + \gamma_{,22} + 1 - e^{2\gamma-2\beta}V_{,1},
\end{aligned}$$

$$\begin{aligned}
R_{33} = {}& -r^2 U e^{-(2\gamma+2\beta)}\cos\theta\sin\theta\,\gamma_{,1} - 2r^2 U e^{-(2\gamma+2\beta)}\sin^2\theta\,\gamma_{,12} \\
& + r^2 e^{-(2\gamma+2\beta)}\cos\theta\sin\theta\,U_{,1} \\
& - r^2 e^{-(2\gamma+2\beta)}\sin^2\theta\,\gamma_{,1}U_{,2} - r^2 e^{-(2\gamma+2\beta)}\sin^2\theta\,\gamma_{,2}U_{,1} \\
& - 2r^2 e^{-(2\gamma+2\beta)}\sin^2\theta\,\gamma_{,01} \\
& + 3r U e^{-(2\gamma+2\beta)}\cos\theta\sin\theta - 2r U e^{-(2\gamma+2\beta)}\sin^2\theta\,\gamma_{,2} \\
& + r V e^{-(2\gamma+2\beta)}\sin^2\theta\,\gamma_{,11} \\
& - 2r e^{-(2\gamma+2\beta)}\sin^2\theta\,\gamma_{,0} + r e^{-(2\gamma+2\beta)}\sin^2\theta\,\gamma_{,1}V_{,1} \\
& + r e^{-(2\gamma+2\beta)}\sin^2\theta\,U_{,2} \\
& + V e^{-(2\gamma+2\beta)}\sin^2\theta\,\gamma_{,1} - 2e^{-4\gamma}\cos\theta\sin\theta\,\beta_{,2} \\
& + 3e^{-4\gamma}\cos\theta\sin\theta\,\gamma_{,2} \\
& + 2e^{-4\gamma}\sin^2\theta\,\beta_{,2}\,\gamma_{,2} - 2e^{-4\gamma}\sin^2\theta\,(\gamma_{,2})^2 \\
& + e^{-4\gamma}\sin^2\theta\,\gamma_{,22} + e^{-4\gamma}\sin^2\theta \\
& - e^{-(2\gamma+2\beta)}\sin^2\theta\,V_{,1}.
\end{aligned}$$

23.10 $\ell^a_{;b}\ell^b = \lambda\ell^a.$

23.11 $\theta = e^{-2\beta}/r$.

24.2 $V_{C_i} = -\dfrac{\Lambda}{6} m_i\, r_i^2$.

24.3

A is a half of the sum of the moments of inertia of the system about a set of orthogonal axes situated at the origin at epoch t_0 (i.e. a half of the trace of the inertia tensor).

24.10

$(k = +1)$ $d\sigma^2 = R_0^2 \sin^2 \chi (d\theta^2 + \sin^2\theta d\phi^2)$.

$(k = -1)$ $d\sigma^2 = R_0^2 \sinh^2 \chi (d\theta^2 + \sinh^2 2\theta d\phi^2)$.

25.6 (ii) Oscillating model.

26.2 $H_0^2 \simeq 10^{-35}\mathrm{s}^{-2}$ and hence $\rho_{\mathrm{crit}} \simeq 1.8 \times 10^{-26} \mathrm{Kgm}^{-3}$.

26.9

$H(t) = 1/2t$. So the two equations give $T^4 = 3c^2/(32\pi G\alpha t^2)$. Substituting $t = 1$ and the numerical values for the constants gives $T \simeq 2 \times 10^{10}$ kelvin.

26.11

$\rho_r(t_{\mathrm{dec}})/\rho_{\mathrm{crit}} \simeq 0.04$ where t_{dec} is the time of decoupling which we have taken to be $\sim 10^{13}$s.

Selected bibliography

We have divided the bibliography into three sections – Biographies of Einstein, general textbooks on general relativity and differential geometry, and more specialised research articles.

Biographies of Einstein (1879–1955)

There are a large number of books and articles of a biographical nature on Einstein. Most of the post-1955 biographies are listed below, together with a few of the more important earlier biographies. The Schilpp volume is particularly significant because it includes Einstein's autobiography.

Bernstein, J. (1976). *Einstein, Modern Masters Series*. Penguin, London.

Born, M. (1971). *Born-Einstein Letters*, 1916-1955. Macmillan, London.

Burke, T. F. (ed.) (1984). *Einstein: a portrait*, Pomegranate.

Cahn, W. (1955). *Einstein: A Pictorial Biography*, Citadel, New York.

Clark, R. W. (1973). *Einstein: The Life and Times*, Hodder and Stoughton, London.

Cuny, H. (1963). *Albert Einstein: The Man and His Theories*, Souvenir, London.

Dank, M. (1983). *Albert Einstein, Impact Biography Series*, Watts.

de Broglie, L., Armand, L., and Simon, P. (1979). *Einstein*, Peebles, New York.

Flückiger, M. (1974). *Albert Einstein in Bern*, Paul Hampt Verlag, Bern.

Frank, P. (1947). *Albert Einstein: His Life and Times*, Knopf, New York.

Hoffman, B. (1972). *Albert Einstein: Creator and Rebel*, Viking, New York.

Hunter, N. (1987). *Einstein, Great Lives Series*, Bookwright, Watts.

Infeld, L. (1950). *Albert Einstein: His Work and His Influence on the World*, Charles Scribner's Sons, New York.

Kirsten, C. and Treder, H. J. (eds.) (1979). *Albert Einstein in Berlin 1913–1933*, Akademie Verlag, Berlin.

Levinger, E. E. (1962). *Albert Einstein*, Dodson, London.

Moszkowski, A. (1921). *Einstein the Searcher*, Methuen, London

Pais, A. (1982). '*Subtle is the Lord …*' the Science and Life of Albert Einstein, Oxford University Press.

Reisner, A. (1930). *Albert Einstein, a Biographical Portrait*, A and C Boni, New York.

Schilpp, P. A. (ed.) (1949). *Albert Einstein: Philosopher-scientist*, Vol 8, The Library of Living Philosophers, Evanston, Illinois.

Seelig, C. (1960). *Albert Einstein*, Europa Verlag, Zürich.

Whitrow, G. J. (ed.) (1967). *Einstein: The Man and His Achievement*, Dover, New York.

Texts on Differential Geometry, Relativity, and Cosmology

This section contains the main textbooks on relativity. It is supplemented by some important references to differential geometry and some specialist texts on cosmology.

Adler R., Bazin M., and Schiffer M. (1975). *Introduction to General Relativity (2nd edn)*, McGraw-Hill, New York.

Andersson N. (2020). *Gravitational-wave astronomy: Exploring the dark side of the universe*, Oxford University Press, Oxford.

Bondi H. (1967). *Assumption and Myth in Physical Theory*, Cambridge University Press, Cambridge.

Bondi H. (1961). *Cosmology*, Cambridge University Press, Cambridge.

Carmeli, M. and Malin, S. (1976). *Representation of the Rotation and Lorentz Groups: An Introduction*, Dekker, New York.

Carroll S. M. (2004). *Spacetime and Geometry: An Introduction to General Relativity*, Addison Wesley, San Francisco.

Chandrasekhar S. (1983). *The Mathematical Theory of Black Holes*, Clarendon Press. Oxford.

Choquet-Bruhat, Y., De Witt-Morette, C., and Dillard-Bleick, M. (1977). *Analysis, Manifolds and Physics*, North-Holland, Amsterdam.

Dixon W. G. (1978). *Special Relativity, the Foundation of Modern Physics*, Cambridge University Press, Cambridge.

Einstein A (2015). *Relativity: The Special and General Theory (100th anniversary edition)*, Princeton University Press, Princeton.

Hartle J. B. (2003). *Gravity: An Introduction to Einstein's General Relativity*, Addison Wesley, San Francisco.

Hawking S. W. and Ellis G. F. R. (1973). *The Large Scale Structure of Space-time*, Cambridge University Press, Cambridge.

Hawking S. W. (1988). *A Brief History of Time*, Bantam Press, London.

Hawking, S. W. and Penrose, R. (2015). *The Nature of Space and Time*, Princeton University Press, Princeton.

Hawking, S. W. and Israel, W. eds. (1987). *300 Years of Gravitation*, Cambridge University Press, Cambridge.

Hestenes, D. (2015). *Space-time Algebra (2nd edn)* Birkhäuser, Basel.

Hughston, L. P. and Tod, K. P. (1990). *An Introduction to General Relativity*, Cambridge University Press.

Kramer, D., Stephani, H., Herlt, E., and MacCallum, M. A. H. (2009). *Exact Solutions of Einstein's Field Equations (2nd edn)* Cambridge University Press, Cambridge.

Landau, L. D. and Lifshitz, E. M. (1971). *The Classical Theory of Fields*, Pergamon, Oxford.

Landsberg P. and Evans D. A. (1977). *Mathematical Cosmology: An Introduction*, Clarendon Press, Oxford.

Liddle A. (2004). *An Introduction to Modern Cosmology*, Wiley, New York.

Lieber L. R. (2008). *The Einstein Theory of Relativity (reprinting of 1945 edition)* Paul Dry Book Inc., Philadelphia.

Lightman, A. P., Press, W. H., Prices, R.H., and Teukolsky, S.A. (1975). *Problem Book in Relativity and Gravitation*, Princeton University Press, Princeton.

Ludvigsen M. (1999). *General Relativity: A Geometric Approach*, CUP, Cambridge.

Marder L. (1968). *An Introduction to Relativity*, Longman, London.

Misner, C. W., Thorne, K. S., and Wheeler, J. A. (1973). *Gravitation*, Freeman, San Francisco.

O'Neil, B. (1983). *Semi-Riemannian Geometry: With Application to Relativity*, Pure and Applied Mathematics Series, Academic Press, New York.

Penrose, R. (2004). *The Road to Reality*, Vintage, London.

Penrose, R. and Rindler, W. (1986). *Spinors and Space-time. Vols 1 and 2*, Cambridge University Press.

Rindler, W. (1982). *Introduction to Special Relativity*, Oxford University Press, Oxford.

Schrödinger E. (1950). *Space-time Structure*, Cambridge University Press, Cambridge.

Schutz, B. F. (1985). *A First Course in General Relativity*, Cambridge University Press.

Schutz, B. F. (1980). *Geometrical Methods in Mathematical Physics*, Cambridge University Press.

Sciama, D. W. (1971). *Modern Cosmology*, Cambridge University Press, Cambridge.

Stewart J (2008). *'Advanced General Relativity'*, *Cambridge Monographs on Mathematical Physics*, CUP, Cambridge.

Synge J. L. (1960). *Relativity: The General Theory*. North-Holland, Amsterdam.

Synge, J. L. and Schild, A. (1949). *Tensor Calculus*, University of Toronto Press.

Taylor, E. F. and Wheeler, J. A. (1966). *Spacetime Physics*, Freeman, San Francisco.

Taylor, E. F. and Wheeler, J. A. (2000). *Exploring Black Holes: An Introduction to General Relativity*, Pearson.

Trautmann, A., Pirani, F. A. E. and Bondi, H. (1964). *'Lectures on General Relativity'*, in Brandeis 1964 Summer Institute on Theoretical Physics, vol 1. Prentice-Hall, New Jersey.

Wald R. M. (1984). *General Relativity*, University of Chicago Press, Chicago.

Weinberg S. (1972). *Gravitation and Cosmology*, Wiley, New York.

Will C. M. (1993). *Theory and Experiment in Gravitational Physics (revised edition)*, Cambridge University Press, Cambridge.

More specialist and advanced references

This section contains the more detailed references. These are mainly to published papers and chapters from research monographs as well as to a number of 'Living reviews in relativity' articles. The living reviews are all open access on the web and a listing by topic is available at https://www.springer.com/gp/livingreviews/relativity/lrr-articles

Akiyama et al (2019). 'First M87 Event Horizon Telescope Results. I. The Shadow of the Supermassive Black Hole', *Astrophysical J. Letters* **875**, 1.

Amendola, L. et al. (2018). 'Cosmology and Fundamental Physics with the Euclid Satellite', *Living Reviews in Relativity* **21**, 2. https://doi.org/10.1007/s41114-017-0010-3

Arnowitt, R.; Deser, S.; Misner, C. (1959). 'Dynamical Structure and Definition of Energy in General Relativity', *Physical Review.* **116**, 1322–1330. https://journals.aps.org/pr/abstract/10.1103/PhysRev.116.1322

Barrow, J. D. and Tipler, F. J. (1986). The Anthropic Principle, Clarendon Press, Oxford.

Blanchet, L. (2014). 'Gravitational Radiation from Post-Newtonian Sources and Inspiralling Compact Binaries', *Living Reviews in Relativity*, **17**, 2. https://link.springer.com/article/10.12942/lrr-2014-2

Bondi H.,van der Burg, M G J and Metzner A W K (1962). 'Gravitational Waves in General Relativity. VII. Waves from Axi-Symmetric Isolated Systems', *Proc. R. Soc. A*, **269**.

Brady P. R., Droz S., Israel W and Morsink S M (1996) 'Covariant Double Null Dynamics: (2+2) Splitting of the Einstein Equations', *Class. Quant. Grav.* **13**, 2211. https://doi.org10.1088/0264-9381/13/8/015

Cardoso, V. and Pani, P. (2019). 'Testing the Nature of Dark Compact Objects: A Status Report', *Living Reviews in Relativity* **22**, 4. https://link.springer.com/article/10.1007%2Fs41114-019-0020-4

Christodoulou, D. (2010). 'The Formation of Black Holes in General Relativity' in *Proceedings of XVIth International Congress on mathematical Physics*, ed. Exner, World Scientific.

Clarke, C. J. S. (1993). *The Analysis of Space-time Singularities*, Cambridge University Press, Cambridge.

Deffayet et al (2015). *Post-Planck Cosmology: Lecture Notes of the Les Houches Summer School: Volume 100* Oxford Scholarship Online. DOI:10.1093/acprof:oso/9780198728856.001.0001.

d'Inverno, R. (1980). 'Algebraic Computing in General Relativity' in *General Relativity and Gravitation (One Hundred Years After the Birth of Albert Einstein), Vol. 1* ed. Held Plenum, New York.

d'Inverno, R.A. (1992). Editor *Approaches to Numerical Relativity*, Cambridge University Press, Cambridge.

d'Inverno, R.A. (1997). '2+2 Formalism and Applications', in *Relativistic Gravitation and Gravitational Radiation*, eds. J-A. Marck and J-P. Lasota, CUP, Cambridge.

d'Inverno, R A and Stachel, J (1978). 'Conformal Two-structure as the Gravitational Degrees of Freedom in General Relativity', *J Math. Phys.* **19**, 2447. https://aip.scitation.org/doi/10.1063/1.523650

d'Inverno, R. A. and Smallwood, J. (1980). 'Covariant 2+2 Formulation of the Initial-value Problem in General Relativity', *Phys. Rev. D*, **22**, 1233. https://doi.org/10.1103/PhysRevD.22.1233

d'Inverno, R. A. and Vickers, J. A. (1995). '2+2 Decomposition of Ashtekar Variables Class'. *Quantum Grav.* **12**, 753.

Ellis, G. F. R. (2011). 'Inhomogeneity Effects in Cosmology', *Classical and Quantum Gravity*, **28**, 164001.

Gourgoulhon, E. (2012). 'The 3+1 Formalism in General Relativity', in Lecture Notes in Physics, Springer, Berlin.

Hall G. S. and Pulham J. R. Eds. (1996). *General Relativity*, IOP publishing, London.

Hawking, S. W. (1979). 'Euclidean Quantum Gravity', in *Recent Developments in Gravitation, Cargèse 1978 NSSB volume 4*, Plenum press, New York. 145-173.

Jones, A. W. and Lasenby, A. N. (1998). 'The Cosmic Microwave Background', *Living Reviews in Relativity*, **1**, 11. https://link.springer.com/article/10.12942/lrr-1998-11

Kiefer, C. (2005). *Quantum Gravity: General Introduction and Recent Developments*. https://arxiv.org/abs/gr-qc/0508120v2

Kokkotas, K. D. and Schmidt, B. G. (1999). 'Quasi-Normal Modes of Stars and Black Holes', *Living Reviews in Relativity*, **2**, 2 https://link.springer.com/article/10.12942/lrr-1999-2

Kunzinger, M., Steinbauer, R. and Vickers, J. A. (2015). 'The Penrose Singularity Theorem in Regularity $C^{1,1}$', *Class. Quantum Grav.* **32**. 155010. https://iopscience.iop.org/article/10.1088/0264-9381/32/15/155010

Lahav, O. and Suto, Y. (2004). 'Measuring our Universe from Galaxy Redshift Surveys', *Living Reviews in Relativity*, 7, 8. https://link.springer.com/article/10.12942/lrr-2004-8

Ludvigsen, M. and Vickers, J. A. (1981). 'The Positivity of the Bondi Mass', *J. Phys. A: Math. Gen.* **14**, L389. https://iopscience.iop.org/article/10.1088/0305-4470/14/10/002

MacCallum, M. A. H., Skea, J.E.F., McCrea, J.D. and McLenaghan (1994). *Algebraic Computing in General Relativity*, Clarendon Press, Oxford

Penrose, R. (1968). 'Structure of Spacetime' in *Battelle Rencontres 1967 Lectures in Mathematics and Physics*, Eds. DeWitt C. M. and Wheeler J.A, W. A. Benjamin, New York.

Penrose, R. (1968). 'Twistor Quantisation and Curved Space-time' *IJTP* **1**, 61-99.

Penrose, R. (1989). *The Emperor's New Mind*, OUP, Oxford.

Penrose R. (2016). *Fashion, Faith and Fantasy in the New Physics of the Universe*, Princeton University Press, Princeton.

Perlick, V. (2004). 'Gravitational Lensing from a Spacetime Perspective', *Living Reviews in Relativity* 7, 9. https://link.springer.com/article/10.12942/lrr-2004-9

Ringström, H. (2009). *The Cauchy Problem in General Relativity*, European mathematical Society, Zurich.

Schiff, L. I. (1960). 'Motion of a Gyroscope According to Einstein's Theory of Gravitation', *Proc. Nat. Acad. Sci.* **46**, 871.

Senovilla, J. M. M. (2011). 'Singularity Theorems in General Relativity: Achievements and Open Questions' in *Einstein and the Changing Worldviews of Physics* eds Lehner, Renn and Schemmel, Birkhäuser.

Smarr, L. (1979). *Sources of Gravitational Radiation*, CUP, Cambridge.

Vickers, J. A. (2011). 'Double null hamiltonian dynamics and the gravitational degrees of freedom', *Gen Rel. Grav.* **43**, 3411. https://link.springer.com/article/10.1007/s10714-011-1242-2

Will, C. M. (2014). 'The Confrontation between General Relativity and Experiment', *Living Reviews in Relativity* **17**, 4. https://link.springer.com/article/10.12942/lrr-2014-4

Index